ECOLOGY AND MANAGEMENT OF FOREST SOILS

FOURTH EDITION

Dan Binkley
Colorado State University, USA

Richard F. Fisher

WILEY-BLACKWELL
A John Wiley & Sons, Ltd., Publication

This edition first published 2013 © 2013 by John Wiley & Sons, Ltd

Wiley-Blackwell is an imprint of John Wiley &Sons, formed by the merger of Wiley's global Scientific, Technical and Medical business with Blackwell Publishing.

Registered office: John Wiley & Sons, Ltd, The Atrium, Southern Gate, Chichester, West Sussex, PO19 8SQ, UK

Editorial offices: 9600 Garsington Road, Oxford, OX4 2DQ, UK
The Atrium, Southern Gate, Chichester, West Sussex, PO19 8SQ, UK
111 River Street, Hoboken, NJ 07030-5774, USA

For details of our global editorial offices, for customer services and for information about how to apply for permission to reuse the copyright material in this book please see our website at www.wiley.com/wiley-blackwell.

Library of Congress Cataloging-in-Publication Data

Binkley, Dan.
  Ecology and management of forest soils / Dan Binkley, Richard F. Fisher. –
4th ed.
      p. cm.
  Authors' names in reverse order on previous edition.
  Includes bibliographical references and index.
  ISBN 978-0-470-97947-1 (cloth) – ISBN 978-0-470-97946-4 (pbk.) 1.   Forest
soils. 2.   Soil ecology. 3.   Soil management. 4.   Forest management. I.
Fisher, Richard F.   II. Title.
  SD390.F56 2012
  577.5'7–dc23
                                                      2012020078

A catalogue record for this book is available from the British Library.

Wiley also publishes its books in a variety of electronic formats. Some content that appears in print may not be available in electronic books.

Cover image: Photo by Cristian Montes.

Cover design by Sandra Heath

Set in 9/12pt, Meridien-Roman by Thomson Digital, Noida, India.
Printed and bound in Singapore by Markono Print Media Pte Ltd.

First Impression 2013

Dedication

We dedicate this book to Earl Stone, a great integrative forest soil scientist and a wonderful educator about soils and about life.

We also dedicate the book to its readers, past and future, who will continue to advance our understanding of forest ecosystems and the vital, fascinating roles of soils.

Contents

Preface

The sheer volume of information on forest soils has probably increased by ten-fold since the previous edition of this book was drafted. Much of the information added new numbers to our understanding of the typical rates of soil processes. Other changes have been more fundamental, such as the shift in the conceptualization of humified soil organic matter: long-lasting soil organic matter may not depend so much on chemical recalcitrance as on intricate soil environments and ecological interactions (including aggregate formation).

A textbook cannot provide a systematic review for all the important topics addressed; authors need to pick and choose examples that have strong educational value. Compiling examples into representative tables and histograms is more powerful than any single example, but these compilations are intended to illustrate commonly observed situations rather than provide reliable conclusions about the population of all forest soils.

This edition retains many of the examples that illustrated important ideas in the earlier editions, for two reasons: many studies that are older than a decade still have as much educational value as more recent cases, and many studies that are older than two decades tend to be hidden from online Internet sites (particularly for studies not published in journals). Many historical studies have a great deal to offer, not just on a particular topic, but for the background value of gaining insights on how our understanding of forest soils developed from our community of inquisitive colleagues over long spans of time. We're sorry that we couldn't include more of the great examples of new forest soil studies from all our colleagues.

Forests and soils are whole systems, so it's very arbitrary to pull out any single feature, such as soil chemistry, to examine in a single chapter when it's actually relevant to every other chapter. For this reason, we consider chapter topics to be general themes, with diverse soil subjects entering each chapter, and with each topic arising in more than one chapter. For example, a chapter on soil organic matter relies, in part, on the subjects of soil chemistry and soil biota, but it also overlaps somewhat with a chapter on carbon sequestration. We hope readers will find the weaving of topics within and among the chapters to be helpful in stitching together a coherent, holistic understanding of forest soils.

Dan Binkley and Richard F Fisher

Acknowledgments

This book was made possible by the career-long curiosity and work of scientists and land managers who have been drawn to the study of soils, developing techniques that shed light below ground. Throughout our careers, we've been delighted, too, each time we opened up a new journal article or a time-worn book and learned new things about soils and forests; the only thing better than learning by reading has been learning directly from our students, colleagues, and friends – in meetings, in the field, and in soil pits. For help with various parts of these chapters, we especially thank Peter Attiwill, Charles Davey, Laurence Morris, Bob Powers, Cindy Prescott, Chuck Rhoades, Dan Richter, Jose Stape, and Victor Timmer.

In Memoriam

Richard F. Fisher, Jr. 1941–2012

Dick Fisher grew up in Urbana, Illinois, where his ancestors were clever enough to realize that treeless prairie soils could make great farm soils. He combined his undergraduate forestry major with a minor in the history and philosophy of science (B.Sc. 1964); this broad and deep curiosity about how things connect, and how things got to be the way they are, characterized Dick's work with forest soils. He worked with Earl Stone at Cornell University for his PhD in soil science (1968), and again he picked up additional insights by minoring in chemistry and geomorphology. Dr. Fisher then began a series of academic adventures, first an assistant professor at his alma matter, the University of Illinois, from 1969–1972. He moved to the University of Toronto as an associate professor (1972–1977), and then joined the University of Florida as a professor and the Director of the Cooperative Research in Forest Fertilization program (1977–1982). Dick next headed west, to become a professor and head of the Department of Forest Science at Utah State University (1982–1990). His final academic positions were at Texas A&M University (1990–1999), where he was a professor, the head of the Forest Science Department, and the director of the Institute of Renewable Natural Resources. Retirement from A&M freed Dick's time to dive into applied aspects of forest soils, which he supported by working as the director of research and development for Temple-Inland Forest Products Corporation in Dibol, Texas (1999–2006).

Dick's career included a wide and deep set of contributions to his profession, students, and colleagues. He was an instructor in the Organization for Tropical Studies field courses in Central America from 1970 through 1999; some experiments he established in Costa Rica are continuing to advance our understanding of the connections between tree species and soils (see Chapter 11). He served both the Soil Science Society of America (including chairing the Forest, Range and Wildland Soils Division) and the Society of American Foresters (he was elected a Fellow of the SAF). Dick authored and coauthored over 100 refereed publications, co-authored three editions of this book, and served as co-editor-in-chief of the journal *Forest Ecology and Management* for 18 years (overseeing and engineering a 5-fold increase in the Journal's size).

The field of forest soils has been greatly enriched by Dick Fisher; he has a large and impressive legacy; and he is missed.

PART I
Introduction to Forest Soils

CHAPTER 1

History of Forest Soil Science and Management

OVERVIEW

In this text, forest soils are considered to be soils that presently support forest cover. These soils differ in many ways from agronomic soils: they have O horizons, organic layers that cover the mineral soil; they have diverse fauna and flora that play major roles in their structure and function; they are often wet or steep, shallow to bedrock, or have a high stone content. Soil layers that occur at great depth are important to forests. The influence of soils on forests and other vegetation was known to the ancients, although our current understanding of soils did not begin to develop until the nineteenth century. The study of forest soils is as old as soil science itself. Researchers working on forest soils made many of the early discoveries that form the foundation of modern soil science.

In the broadest sense, a forest soil is any soil that has developed primarily under the influence of a forest cover. This view recognizes the unique effects of the deep rooting of trees, the role of organisms associated with forest vegetation, and the role the litter layer (forest floor, or O horizon) and the eluviation promoted by the products of its decomposition have on soil genesis. By this definition, forest soils can be considered to cover approximately one-half of the Earth's land surface area. Essentially, all soils except those of tundra, marshes, grasslands, and deserts were developed under forest cover and have acquired some distinctive properties as a result. Of course, not all of these soils support forests today.

Perhaps as much as one-third of former forest soils are now devoted to agricultural, urban, or industrial use. A better definition of forest soils might be those soils that are presently influenced by a forest cover. Currently, forests of various types cover about one-third of the world's land surface.

The need for a separate study of forest soils is sometimes questioned on the assumption that a forest soil is no different from a soil supporting other tree crops, such as citrus, pecans, olives, or even a soil devoted to agronomic crops. Persons who are not well acquainted with natural ecosystems and who have failed to note even the most obvious properties of soils associated with forests generally make this assumption. Upon close observation of forests, one notices many unique properties of forest soils. The forest cover and its resultant O horizon provide a microclimate and a spectrum of organisms very different from those associated with cultivated soils or horticultural plantations. Such dynamic processes as nutrient cycling among components of the forest community and the formation of soluble organic compounds from decaying debris, with the subsequent eluviation of mineral ions and organic matter, give a distinctive character to soils developed beneath forest cover.

When European settlers arrived in what became North America, forests covered half of the land area, another two-fifths was grassland, and the remainder was desert or tundra. The eastern seaboard was almost entirely forested, and, largely as a consequence of the difficulties of clearing new land, agricultural settlement was mostly confined to the Atlantic slope until the end of the eighteenth century. Slowly, settlement began expanding

---

Ecology and Management of Forest Soils, Fourth Edition. Dan Binkley and Richard F. Fisher.

westward, and extensive forest clearing for agriculture began in the central portion of the continent. By the middle of the nineteenth century, more than a million hectares of virgin forests had been cleared and the land converted to permanent agricultural uses.

This area of cleared forest land gradually increased until well into the twentieth century. As new and better farmlands were opened in the Midwest and West, millions of acres of former croplands were abandoned. Especially in the eastern regions, large areas of degraded farmland reverted to forests, and by 1950 forested areas had increased until they again covered nearly one-third of the total land area. The forest land base has continued to grow, and is projected to continue to grow slightly despite the relentless pressures of increasing urban and industrial development (Alig and Butler, 2004).

Few truly virgin forests exist today in populated regions of the globe. The conversion of forests to croplands and back to forests has gone through many cycles in sections of central Europe and Asia, as well as eastern North America and portions of South America. Large areas of non-forest soil now support forests, especially in Australia, New Zealand, and southern Africa. Deforestation continues at a rapid pace, particularly in the tropics. However, even there, degraded lands that were cleared of forest for agriculture are being reforested. Many European forestlands have been managed rather intensively for centuries. At the other extreme, relatively short-term shifts in land use occur in the tropics, where "swidden" agriculture or shifting cultivation, a form of crop rotation involving 1 to 3 years of cultivated crops alternating with 10 to 20 years of forest fallow, is practiced. Such practices alter many properties of the original forest soil.

In recent years, intensively managed plantation forests have been created in several countries of the world. Among these forests are several million hectares of exotic pine and eucalypt forests in the Southern Hemisphere and an even larger area of plantations employing native species in the Northern Hemisphere. The latter includes some 8 million ha of pine plantations in the coastal plains of the southeastern United States and large areas of Douglas fir plantations in the Pacific Northwest.

Because of the alteration in certain properties of forest soils as a result of intensive management, the distinction between forest soils and agronomic soils has become progressively less evident in some areas. Although some properties acquired by soil during its development persist long after the forest cover has been removed and the soil cultivated, other characteristics are drastically modified by practices associated with intensive land use. In this text, we will generally treat forest soils in the narrow sense as soils that presently support a forest cover, but we will also address the soils of intensively managed forest plantations, many of which were not developed under forests or have seen long periods of use in agriculture. Only cursory attention will be given to genesis and classification of forest soils. Emphasis will be placed on understanding various physical, chemical, and biological properties and processes and how they influence forest dynamics and the management of forests.

FOREST SOILS DIFFER IN MANY WAYS FROM CULTIVATED SOILS

The soil is more than just a medium for the growth of land plants and a provider of physical support, moisture, and nutrients. The soil is a dynamic system that serves as a home for myriad organisms, a receptor for Nature's wastes, a filter for toxic substances, and a storehouse for scarce nutrient ions. The soil is a product as well as an important component of its environment. Although it is only one of several environmental factors controlling the distribution of vegetation types, soil can be the most important one under some conditions. For example, the farther removed a tree is from the region of its climatic optimum, the more discriminating it becomes with respect to its soil site. This means that the range of soil conditions favorable to the growth of a species narrows under unfavorable climatic conditions for that species.

Many properties and processes characteristic of forest soils will be discussed in detail in later chapters. At this time, it will suffice to point out a few properties of forest soils that differ from those of cultivated soils. These differences derive, in part, from the fact that often the most "desirable" soils have been selected for agronomic use and the remainder left for native vegetation such as forests and grasslands. Fortunately, soil requirements for forest crops generally differ from those for agronomic crops. It is not unusual to find that productive forest sites are poor for agronomic use. Nevertheless, they may be used for agriculture because of their location with respect to markets or centers of population. Poor drainage, steep slopes, or the presence of large stones are examples of soil conditions that favor forestry over agriculture. However, the choice of land use often results from differences in crop requirements. Good examples are the wet flatlands of many coastal areas around the world. These important forest soils cannot be effectively used for agricultural purposes without considerable investments in water control, lime, and fertilizers.

Not all forest soils are nutrient poor. Some soils with excellent productive capacity for both trees and agronomic crops remain in forests today. This is generally because of location, size of holdings, ownership patterns, or landowner objectives. The fact that many forest soils contain a high percentage of stones by volume has a profound effect on both water and nutrient relations. Water moves quite differently through stony soil than it does through stone-free soil, and the volume of stones reduces proportionally the volume of water retained per meter of soil depth. Likewise, although stones contain weatherable minerals that release nutrient ions to the soil, the volume of stones reduces proportionally the soil's ability to provide nutrients for plant growth.

Forest trees customarily occupy a site for many years. Their roots frequently penetrate deeply into the subsoil and even into fractured bedrock (Fisher and Stone, 1968). During this long period of site occupancy, considerable amounts of organic material are returned to the soil in the form of fallen litter and decaying roots. As a result, a litter layer forms and exerts a profound influence on the physical, chemical, and biological properties of the soil.

The tree canopy of a forest shades the soil, keeping the soil cooler during the day and warmer during the night than cultivated soils. The presence of forest vegetation and the litter layer also results in more uniform moisture conditions, producing a soil climate nearly maritime in nature.

The physics of both overland and subsurface flow of water in steep forest soils are quite different from those in cultivated soils. Steep slopes under forests have their surfaces protected by the litter layer, their shear strength increased by the presence of roots, and their infiltration capacity enhanced by old root channels.

The more favorable climate of forest soils also promotes more diverse and active soil fauna and flora than are to be found in agronomic soils. The role of these organisms as mixers of the soil and intermediaries in nutrient cycling is of much greater importance in forest soils than in agronomic soils.

The deep-rooted character of trees leads to another unique feature of forest soils. Although the great majority of roots occur at or near the soil surface, deep roots also take up both moisture and nutrients. Thus, deep soil horizons, of little importance to agronomic crops, are of considerable importance in determining forest site productivity.

Agronomic soils may be described as products of human activity, in contrast to forest soils, which are natural bodies and exhibit a well-defined succession of natural horizons. This was certainly a valid contrast a few decades ago, and it continues to be valid in most areas today. But the contrast has diminished greatly in the exotic forests of the Southern Hemisphere and in the short-rotation forests of the southeastern United States, the Pacific Northwest, and much of western Europe. Clear-cut harvesting of trees disturbs the surface litter, resulting in short-term changes in the temperature and moisture regimes of the surface soil. Seedbed preparation by root raking or shearing, disking or plowing, and sometimes bedding incorporates the litter layer with the mineral soil, often enhancing microbial activity. Fertilization increases the nutrient level of the surface soil but may also affect the rate of breakdown of the organic layer. These practices influence the characteristics of forest soil and render it more like agronomic soil. However, most of these changes are relatively temporary, existing only until a forest cover becomes well established. With the development of the forest canopy and a humus layer, the soil again acquires the properties that distinguish it from cultivated soils.

FOREST SOIL SCIENCE IS AS OLD AS SOIL SCIENCE ITSELF

That trees and other plants are intimately related to the soil on which they grow seems rather obvious, but it took a long time for science to perceive this and even longer for us to understand the relationship. It is difficult to say just when perception occurred, but our scientific understanding of soils began with Aristotle (384–322 B.C.) and later Theophrastus (372–297 B.C.), both of whom considered soil in relation to plant nutrition. Cato (234–149 B.C.), Varo (116–27 B.C.), and Virgil (70–19 B.C.) also considered the relationship between plant growth, including trees, and soil properties. However, the knowledge of soils accumulated by Aristotle and the natural philosophers who followed him vanished with the fall of Rome and was unknown to Western scholars for over 1500 years.

Pliny the Elder (A.D. 23–79) gave the most complete account of the ancients' understanding of soil as a medium for plant growth. Pliny spoke not only of crops but also of forests and grasslands. He, like Cato, was a chronicler and recounted folk wisdom and other empirically derived information, much of which is incorrect in light of modern discoveries. Oddly, while the wisdom of the natural philosophers was lost to Western science, much of the traditional knowledge that Pliny chronicled persisted among the agrarian population throughout the Dark and Middle Ages. For example, many farmers recognized that certain crops were more productive on certain sites or soils, and the value of legumes in improving soil was widely known. It was not until the fifteenth century that science (natural philosophy) again turned its attention to the mystery of plant growth. Even then, much of the work in this regard was carried out by the learned nobility, and there was little sharing of knowledge among the various thinkers. In 1563 Bernard de Palissy published his landmark treatise *On Various Salts in Agriculture*, in which he stated that soil is the source of mineral nutrients for plants. However, his work was little known at the time.

The period 1630 to 1750 saw the great search for the principle of vegetation. During this time, any one of five "elements" – fire, water, air, earth, and niter – was considered, from time to time, to be the active ingredient in vegetable matter. It was during this period that Van Helmont (1577–1644) conducted his classic experiment with a willow (*Salix*) tree. He grew 164 pounds of willow tree in 200 pounds of soil, and only 2 ounces of soil were consumed. Since only water had been added during the experiment, he concluded that the 164 pounds of willow had come from water alone. Van Helmont may not have been the first to conduct this experiment and draw the wrong conclusion – some believe that a similar experiment was conducted and similar conclusions drawn by Nicholas of Cusa (1401–1446) – nor was he the last. Robert Boyle repeated the experiment with *Cucurbita*, obtained similar results, and reached a similar erroneous conclusion. It was not until 1804 that de Saussure successfully explained the experiment when he found that most of the mass of the plant was carbon derived from the carbon dioxide of the air.

Despite the experiments of Van Helmont and Boyle, the quest for the principle of vegetation continued. John Woodward (1699) grew *Mentha* in rainwater, River Thames water, Hyde Park Conduit effluent, and effluent plus garden mold. He found that the plants grew better as the amount of "sediment" increased and concluded that "certain peculiar terrestrial matter" was the principle of growth. Frances Home experimented and reached similar conclusions in the 1750s; however, he noted that "exhausted soils recovered from exposure to the air alone" and therefore concluded that the air must be the ultimate source of the essential materials.

The work of de Saussure and Boussingault in France in the early nineteenth century began a period of rapid scientific advancement. In 1840, Justus von Liebig in Germany published *Chemistry Applied to Agriculture and Physiology*, and modern soil science began. Liebig helped dispel the theory that plants obtained their carbon from the soil and developed the concept that mineral elements from the soil and added manure are essential for plant growth. However, he continued to believe, erroneously, that plants received their nitrogen from the air. It remained for de Saussure to show that plants' source of nitrogen was the soil. Lawes and Gilbert put these European theories to test at the now-famous Rothamsted Experiment Station in England and found them to be generally sound.

Following the work of these chemists, scientists in many different fields including geology (Dokuchaev, Hilgard, Glinka), microbiology (Beijerinck, Winogradsky), and forestry (Grebe, Ebermayer, Muller, Gedroiz) contributed to the development of what today is termed "soil science." Most of the early research on soils was directed to its use for agricultural purposes because Europe had a critical food shortage; however, the importance of soils to the natural forest ecosystem was recognized by several early scientists.

In 1840, Grebe, a German forester, recognized the importance of soil to forest growth, stating, "In short, almost all of the forest characteristics depend on the soil, and hence, intelligent silviculture can only be based upon a careful study of the site conditions." Pfeil echoed this thought in 1860, and it became a central theme of European forestry (Fernow, 1907). Hilgard (1906) recognized a relationship between vegetation and soils in North America similar to the relationship that had been noted in Europe. The work of Grebe, Pfeil, Hilgard, and others, perhaps unintentionally, laid the foundations of forest ecology. In North America, early ecologists such as Merriam (1898), Cowels (1899), and Clements (1916) knew that soil was important in vegetation dynamics, but none of them understood soils well. Toumey (1916) noted the importance of soils in American silviculture for the first time.

Early research in forest soils was dominated by basic scientific studies. In fact, some very important early soil science research was done on forest soils. Ebermayer's work on forest litter and soil organic matter (1876) had a strong influence on soil science and agriculture. Muller's work on humus forms (1879) marked the beginning of the study of soil biology and biochemistry. Gedroiz (1912) did pioneering work on soil colloids, suggested that soils performed important exchange reactions, and laid the groundwork for modern soil chemistry. These scientists were interested not only in soil properties, but also in the processes that led to the existence of

the properties. Much of the early research on forest soils in North America was similarly basic in nature.

As wood production became a pressing problem in Europe late in the nineteenth century, and the restoration of degraded forestlands abandoned after agriculture became a necessity in North America in the twentieth century, forest soils research became quite applied. Studies of species selection for reforestation, methods of site preparation for reforestation, methods for estimating the site quality of forestlands that no longer supported forest vegetation, tree nutrition and response to fertilization, and soil changes under intensive forest management dominated forest soils research for most of the twentieth century. However, this situation began to change as concern over the impact of acidic deposition and global change arose, and there is currently a broad spectrum of forest soils research. Today, applied research is still necessary, but some of our most pressing environmental problems require more thorough knowledge of the basic processes that take place in soils and their relationship to other ecosystem processes.

H. Cotta in Germany had made the importance of soils to forest production clear as early as 1809. Thus, soils education became an important part of forestry education in Europe. Portions of forestry textbooks were devoted to soil science by Grebe (1852) and by B. Cotta (1852), and in 1893 Raman published a text called *Forest Soil Science* in German. In 1908, Henry published the first forest soils text in French.

The science of forest soils was slow to develop in North America because of the lack of any compelling need for soils information during the early period of forest exploitation. Only after World War I did the ideas of managing selected forests for sustained yields, reforestation of abandoned farmlands, and the establishment of shelterbelts in the Midwest begin to take hold. Perhaps the greatest impetus to the scientific study of forest soils was the publication of textbooks on the subject by Wilde (1946; 1958) and Lutz and Chandler (1946). These books were widely used by students throughout the United States for many years, and North American forest soil scientists today are largely the "academic offspring" of these intrepid scholars.

CHAPTER 2
Global Patterns in Forest Soils

OVERVIEW

Forest soils are variable at all spatial scales – from beneath one tree to another, across slopes and landforms, and across major gradients in climate. Some generalizations are possible, but not many; on average, tropical soils are no poorer or richer than temperate soils. The major types of forests around the world have some correspondence to soil types, but, again, the variations in soil types within a forest type are striking. Major challenges and opportunities in understanding and managing forest soils depend on local details rather than on regional generalizations.

───────────────

Forests cover about one-third of the Earth's land surface, and trees play a major role in most ecosystems. Forests are assemblages of biotic and abiotic features, with trees dominating the interactions between the environment and the biota. Trees profoundly affect the soils on which they grow, including the microclimate of the soil. In turn, the survival and growth of trees are affected by the soil.

Differences in temperature, precipitation, and physiography produce great diversity in ecosystem productivity and species composition around the Earth. Adequate nutrients, along with a proper balance between temperature and precipitation, are of prime importance in determining the suitability of an area for living organisms. The composition of a plant community is strongly influenced by both climate and soil. The boreal forests of North America are quite similar in climate, soils, and vegetation to the boreal (taiga) forests of northern Europe and Asia. Such broad, common characteristics have led

us to group vegetation and soils into major ecosystem types or *biomes*.

Forest soils are shaped by the influence of plants and animals over time on soil parent material, with strong influence from environmental features such as temperature regime and water supply. Broad patterns of soil development and climate can lead to some useful generalizations, but local variations in soil-forming factors can result in an astonishing variety of soils within relatively small areas. For example, 9 of the 12 orders in the U.S. soil classification system occur on the island of Puerto Rico (it's too moist for Aridisols, has no recent volcanic activity, so no Andisols, and has no permafrost for Gelisols). Ten of the 12 orders occur on the island of Hawaii (no Spodosols or Gelisols). Soil formation is mostly a local story that scales up to a global story only in vague generalities, For example, Oxisols occur only in warm and wet environments, but most soils in such environments are *not* Oxisols.

SOILS COMMONLY DIFFER AS MUCH WITHIN REGIONS AS AROUND THE GLOBE

Many early ideas about patterns in soil properties were developed before much information was at hand (Richter and Babbar, 1991), and, unfortunately, some disproven ideas live on. For example, it is still reported that tropical ecosystems have most of their nutrients tied up in biomass, with relatively little stored in mineral soils because rapid decomposition prevents accumulation of soil organic matter. This description is true for some tropical sites, just as it is for some temperate sites, but many

Ecology and Management of Forest Soils, Fourth Edition. Dan Binkley and Richard F. Fisher.
© 2013 John Wiley & Sons, Ltd. Published 2013 by John Wiley & Sons, Ltd.

Table 2.1　Soil chemical pools for representative soils in the tropics.

Feature	Young Alluvial Soil Venezuela 1.7 m/yr Precipitation	Old Weathered Terrace in Colombia 3.0 m/yr Precipitation	Weathered Ash Hawaiian Soil 4.0 m/yr Precipitation	Weathered Metamorphic Rock Nigeria 1.4 m/yr Precipitation	Unusually Poor Pleistocene Sediments Venezuela 2.4 m/yr Precipitation
O horizon mass (Mgha)	2	15–50	9	5–15	50–120
Soil C (Mgha)	58	81	80	63	47
Soil N (Mgha)	7	5	7	6	2
Soil P (Mgha)	1.4	0.7	–	6	2
Exch. K (kgha)	460	120	–	650	70
Exch. Ca (kgha)	7100	31	–	2600	7
Exch. Mg (kgha)	960	40	–	290	16
% acid saturation	3	90	–	25	93

tropical soils are quite rich in nutrients and organic matter (Table 2.1). No general relationship has been found between overall soil nutrient stores and latitude. Low-nutrient soils are as common at high latitudes as in the tropics.

The variations in soil properties across relatively small regions are illustrated by comparing the total phosphorus and extractable calcium in a set of representative subtropical soils from New South Wales in Australia and from the temperate Sierra Occidental region of Chile. These soils were chosen by the authors to illustrate the common suite of soils in each region (Turner and Lambert, 1996; Fölster and Khanna, 1997). On average, the Chilean soils had about two-thirds more exchangeable calcium

than the Australian soils (Figure 2.1), and about two and a half times more phosphorus (because of the presence of high-phosphorus volcanic ash soils in Chile). However, the calcium content of the Australian soils spanned more than an eight-fold range and phosphorus more than a three-fold range. The Chilean soils were similarly variable, spanning a seven-fold range for calcium and an eighteen-fold range for phosphorus. One may accurately state that the Chilean soils, on average, had higher nutrient levels than the Australian soils, but this would not be a very useful statement given the large overlapping ranges of values for individual sites. On average, the old tropical Australian soils had lower quantities of these nutrients than the

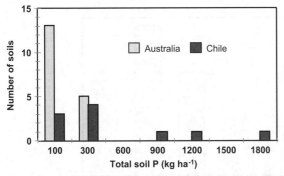 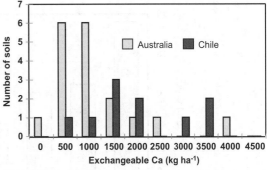

Figure 2.1　Total soil phosphorus and extractable calcium (estimated to a depth of 20 cm) for 17 representative soils of the Sierra Occidental in Chile (from Fölster and Khanna, 1997) and from New South Wales in Australia (from Turner and Lambert, 1996). On average, the old tropical Australian soils had lower quantities of these nutrients than the younger temperate soils from Chile, but the overlap among regions is too large to provide much insight into individual sites between the regions.

younger temperate soils from Chile, but the overlap among regions is too large to provide much insight into individual sites between the regions.

Soil variability is typically high at all scales: from region to region, within regions, and even within single stands or compartments (Fölster and Khanna, 1997). At the scale of a 5000 ha forestry operation in East Kalimantan, Indonesia, exchangeable soil calcium (to a depth of 0.5 m) ranged from 60 to $3240 \, kg \, ha^{-1}$ of calcium across compartments. At the level of a single compartment in the Jari plantations in the Amazon basin of Brazil, exchangeable soil calcium (to 0.6 m in depth) varied from 40 to $290 \, kg \, ha^{-1}$.

The long-term fertility of soils may depend most strongly on the accumulation and turnover of soil organic carbon and nitrogen. Tropical soils tend to cycle carbon and nitrogen faster than temperate soils, but the faster turnover does not result in lower total pool sizes. The storage of soil carbon and nitrogen among 1500 soils around the world is related somewhat to soil moisture regime (wetter soils have more carbon and nitrogen), but not to latitude (Figure 2.2). Despite the efforts of scientists to dispel the myth of "poor soils in the tropics" (Sanchez *et al.*, 1982; Duxbury *et al.*, 1989; Richter and Babbar, 1991), it has taken on a life of its own.

Although soil properties do not relate well to latitude, these properties do vary consistently among soil taxonomic groups (such as soil orders in the U.S. classification system). Motavalli *et al.*

(1995) found that allophanic (amorphous clay) soils had more total nitrogen than soils with smectitic clays, which exceeded the nitrogen in kaolinitic or oxic (amorphous sesquioxide) clays. Interestingly, the mineralizability of nitrogen from each of these different soil types was essentially constant; soils with more nitrogen mineralized proportionally more nitrogen.

The U.S. system of soil classification places all the soils of the world into 12 classes called *orders*. The soil order with the greatest coverage of the globe is Aridisols (Table 2.2), but these soils are unimportant for forests because they are too dry to support trees.

Together, Entisols and Inceptisols provide much of the forestland of the temperate and boreal regions. Spodosols commonly form under conifers (and sometimes under hardwoods too), typically on coarser materials under high-rainfall regimes. Andisols are volcanic soils restricted primarily to the margins of continents. These soils tend to be high in amorphous clays and soil organic matter and often support very productive forests. Alfisols are common under hardwood forests in areas with well-developed soil profiles. Ultisols take tens of thousands of years to form, and they dominate many tropical areas, as well as some old, unglaciated areas with high precipitation such as the U.S. Pacific Northwest and western Canada. Oxisols are highly weathered soils with high concentrations of clays that have low exchange capacity; they occur primarily on ancient landforms in the tropics. Histosols

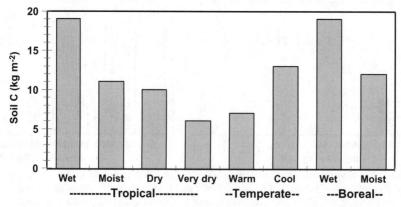

Figure 2.2 The carbon content of tropical soils is higher on wetter sites, and the carbon content of soils shows no particular pattern with latitude (from over 1500 soil profiles, from data summarized by Post *et al.*, 1982; see also Figure 4.9).

Table 2.2 Global occurrence of soil orders.

Soil Orders	Similar Categories in Canadian and FAO Classification Systems	Land Area (%)
Aridisols, very dry areas	Solonetzic, Xerosols, Solonetz, Solochaks, Yermosols	23.4
Inceptisols, B horizon developing	Brunisolic, Cambisols	16.0
Alfisols, B horizon with clay accumulation	Luvisolic, Luvisol	13.5
Entisols, under-developed horizons	Regosolic, Regosol, Fluvisol	11.0
Oxisols, extremely weathered soils	Latisols, Ferralsols	8.7
Ultisols, acid soils with clay accumulation	Acrisols	8.3
Gelisols, soils underlain by permafrost	Cryosolic, Cryosols, Cambisols	6.0
Mollisols, high carbon in A horizon	Chernozemic, Chernozem, Kastanozem	4.1
Spodosols, B horizon with Fe, Al, organics	Podzols	3.6
Vertisols, with Vertisolic, Shrink-swell clays	Vertisols	2.4
Andisols, developed in volcanic ash	Andosols	1.8
Histosols, organic soils	Organic, Histosols	1.2

comprise major forest soils in areas of poor drainage in temperate and boreal areas; forest growth is commonly improved by draining Histosols.

MAJOR FOREST TYPES OCCUR ON A VARIETY OF SOILS

The forests of the world span a vast range of climatic conditions and soil properties, and the variations across relatively small landscapes may be almost as large as the variations across hemispheres. This wide range of situations is reflected in the major soil types that are found within major types of forests, and we conclude our discussion of soil and forest associations with a table that summarizes the major orders that are found most commonly among forest types (Table 2.3).

Table 2.3 Major soil types in forest regions.

Forest Types	Most Common Soil Orders
Boreal forests	Gelisols, Spodosols, Histosols, and Inceptisols
Temperate coniferous, mixed, hardwood, and montane forests	Alfisols, Inceptisols, Ultisols, and Entisols
Tropical rain, monsoon, and dry forests	Ultisols, Inceptisols, Oxisols, and Andisols

TROPICAL FORESTS ARE DIVERSE AND OCCUR ON A DIVERSE RANGE OF SOILS

Tropical forests are arbitrarily defined as all forests between the Tropic of Cancer and the Tropic of Capricorn, an area about one-third of the Earth's land surface (4.6 billion ha) that contains about 50% of the world's population. It is the region of most rapid population growth and, in recent years, an area of rapid changes in land-use patterns. Trade winds, monsoons, and mountain ranges provide enormous variations in the amount and distribution of rainfall, producing a wide range of vegetation, from permanently humid evergreen rainforests through various types of semi-deciduous and deciduous forests to savanna, semi-desert, and desert.

Some 42% (1.9 billion ha) of tropical land areas contain significant forest cover. Of this amount, 58% (1.1 billion ha) is closed forest in which the forest canopy is more or less continuous. The remaining 42% (0.8 billion ha) is open forest or woodland in which the canopy is not continuous and the trees may be widely scattered.

Tropical forests are among the Earth's most diverse ecosystems. These forests are supplying an increasing proportion of the world's demand for wood products. Native tropical forests are being harvested and landscapes converted to second-growth forests, agricultural fields, and pastures. Intense pressure from people seeking food, fuel wood, forage, and timber is changing tropical forests on a global scale. In some areas, tropical forests are recovering on former cropland and pastureland, with trees reestablishing on soils altered by previous land management.

TROPICAL CLIMATES ARE WARM YEAR ROUND, AND INCLUDE BOTH DRY AND WET REGIONS

The equatorial belt between the Tropics of Cancer and Capricorn possesses great diversity in climate and soil conditions, ranging from cold highlands to hot lowlands and from marshes to deserts. The most common features in this portion of the Earth are relatively uniform air temperature between the warmest and coolest months of the year and little variation in soil temperatures. These constants apply to all tropical areas, regardless of location. Although mean annual air temperatures decrease, on average, by about 0.6°C for every 100-m increase in elevation, they remain fairly constant from month to month at any specific location. The mean monthly air temperature variation is 5°C or less between the three coolest and the three warmest months, with the least variation closest to the equator.

Soil temperatures are isothermic, with less than a 5°C difference between mean summer and mean winter temperatures at 50-cm depth (Soil Survey Staff, 1993). Surface soil temperatures beneath forest canopies approximate those of the air, but with the removal of the canopy, surface soil temperatures rise dramatically. For example, a daily maximum temperature of 25°C at the soil surface beneath a canopy may increase to 50°C or more after canopy removal, especially during dry periods (Chijicke, 1980).

Rates of tree growth are high in many areas of the tropics because of extended growing seasons that are limited (if at all) by periods of drought rather than by cold seasons. Wood growth rates of $10\,Mg\,ha^{-1}$ annually are common in tree plantations of the tropics; rates that are matched only by those of the most productive temperate plantations (Binkley et al., 1997).

Annual rainfall can vary from near zero to more than 10 m in tropical areas, generally decreasing with increasing latitude but greatly influenced by local relief and seasonal wind patterns. The seasons in the tropics are nominally classified as wet and dry rather than winter and summer. The length of the dry season increases with the increase in latitude, being essentially zero at the equator. The period of heaviest rainfall usually comes during periods of greatest day length (or when the sun is directly overhead). On average, there is essentially no moisture limitation to tree growth in approximately one-third of the tropics, but there are slight to severe moisture limitations during two to six months or more in the remaining areas.

Precipitation interacts strongly with soil properties in determining forest productivity. Along the central Atlantic coast of Brazil, differences in geology have led to development of a variety of soils in the local (< 100 km) landscape, from sandy Entisols to sandy or clayey Ultisols to sandy Oxisols (Figure 2.3).

This diversity of soil types is overlain by a gradient in precipitation, with areas near the coast receiving about twice the annual precipitation of inland sites. At the dry end of the precipitation gradient, the productivity of *Eucalyptus* plantations is similar among soil types. Productivity increases with increasing precipitation, but the gains from extra rainfall are more pronounced for the Ultisols than for the Oxisols.

RAINFORESTS HAVE NO PROLONGED DRY SEASON

Tropical rainforests have developed where year-round rainfall and temperatures are favorable for a high level of biological activity. Three distinct formations of tropical rainforests are (1) American, (2) African, and (3) Indo-Malaysian. These three formations are remarkably similar in structure and general appearance. Trees in these formations are largely evergreen and have a wide variety of leaf

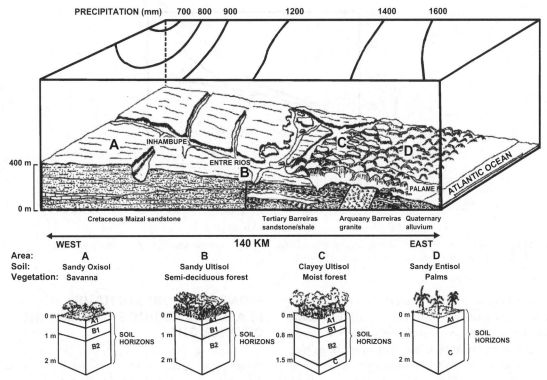

Figure 2.3 Soils near the Atlantic coast of Bahia, Brazil, range from marine-derived Entisols to Oxisols formed in Cretaceous sandstones over a distance of about 100 km. Precipitation drops by half over the same distance. "Tropical soils" cannot be represented by simple generalizations any better than temperate forests can (modified from Pessotti *et al.*, 1983, used by permission of Copener Florestal, S.A.).

forms, and the dominant species often develop large flank buttresses on the lower trunk. Trees support a variety of plants that can survive without soil contact (epiphytes), such as orchids, bromeliads, ferns, mosses, and lichens. A dense, mature rainforest has a compact, multistoried canopy that allows little light to penetrate to the ground. The understory may be rather open in these dense forests. Rainforests that have been opened up, either by nature or by humans, quickly grow into dense, second-growth forests. Similar dense, impenetrable jungles occur naturally along edges of streams and other natural clearings.

The high temperatures and humidity of rainforests ensure that litter reaching the O horizon decomposes rapidly, sustaining rapid nutrient cycles. The soils of the rainforests and other tropical formations are extremely variable (Richter and Babbar, 1991) and have been less well studied than temperate soils. Furthermore, the dissemination of information concerning their properties and management has been impeded by the large variety of systems used in their classification. Sanchez (1976) contrasted the 1938 USDA, French (ORSTROM), Belgian (INEAC), Brazilian, and United Nations FAO systems of classifying tropical soils with the U.S. Soil Taxonomy (Soil Survey Staff, 1999). He summarized the distribution of tropical soils at the suborder level from a generalized map (Aubert and Tavernier, 1972) based on the U.S. taxonomy.

Parent material weathering in tropical Oxisols is deeper than in any other group of soils (Figure 2.4). In many areas, weathering and leaching gradually remove a large part of the silica from the silicate minerals, forming a thick subsurface oxic horizon

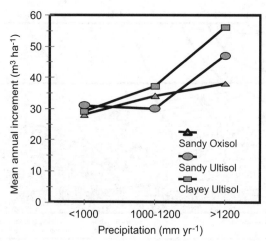

Figure 2.4 Interaction between soil type and precipitation in determining productivity of *Eucalyptus* (clones of *E. grandis* x *urophylla*) in Bahia, Brazil (data from Stape *et al.*, 1997).

that is high in hydrous oxides of iron and aluminum. Less intensive weathering produces Ultisols. Oxisols and Ultisols are the most commonly occurring soils in rainforests, and both types have high acidity and frequently low supplies of phosphorus, potassium, calcium, nitrogen, magnesium, and various micronutrients. These soils are generally high in exchangeable aluminum but low in effective cation exchange capacity, resulting in a moderate to high leaching potential. The capacity to immobilize phosphorus fertilizer is often high, but phosphorus fertilization is still very effective.

Floodplain areas are often occupied by Inceptisols that may be rich or poor in nutrients, depending on the source of the sediment from which they are formed. Andisols also occur in rainforests, as well as in isotemperate montane forests in the tropics. On less well drained sandy soils, Spodosols (Aquods) may develop. Areas of impeded drainage are not unusual in tropical rainforests, but deep peat accumulations appear to be rather rare. On the other hand, the organic matter content of mineral soils is generally higher than might be expected in an area of such high biological activity. This underscores the importance of distinguishing at least two general pools of soil organic matter: relatively young, labile pools with high throughput, and large, stabilized pools of organic matter with very slow rates of turnover.

MONSOON FORESTS HAVE SEASONAL PERIODS OF RAIN AND OF DROUGHT

In tropical areas where significant seasonal droughts are experienced, the rainforest gives way to "rain green" or monsoon deciduous forests. The formation is not as distinct as that of the evergreen rainforest and, in fact, consists of a transitional zone of semi-deciduous seasonal forest as well as deciduous seasonal forest. The transitional forest is similar in structure and appearance to the rainforest, except that the trees are not as tall and the forest may have only two canopy tiers. Monsoon forests are found in areas with relatively mild drought periods, usually no more than five months, with less than 9 cm of rain. In areas with more distinctive drought periods, the forest becomes predominantly seasonal deciduous. Deciduous trees such as teak (*Tectona grandis*), with seasonal opening of the canopy and a subsequent show of flowers in the lower tier, are common to both the transitional semi-deciduous and seasonal deciduous communities with this formation. Lianas and epiphytes also become less abundant in areas with distinctive dry periods. Formations of monsoon forests have been described in Central and South America, Indo-Malaysia, and Africa (Eyre, 1963). The American formation is typically found in a belt skirting both the northern and southern fringes of the Amazon Basin, along the east coast of Central America, and some West Indies islands. The Indo-Malaysian formation stretches from northeast India and Indo-china through Indonesia to northern Australia. In the wetter portions of the areas, teak and a deciduous legume, *Xylia xylocarpa*, are dominant; in drier sections, *Dipterocarpus tuberculatus* (eng tree), bamboo, *Acaela, Butea,* and other leguminous species are abundant. The African formation is rather distinct. Except for a discontinuous belt along the northern edge of the West African tropical rainforest, the other African communities with this formation merge rapidly into savanna woodlands, perhaps kept in check by frequent burning. Soils with these formations may be less highly weathered than those of the tropical rainforests. They generally include various groups of Ultisols, Oxisols, and Vertisols.

MANY TROPICAL FORESTS ARE DRY

The deciduous seasonal forests of Asia and America often grade into drier regions with no sharp change in vegetation. In these drier areas, many dominants disappear and those that persist have a lower, spreading canopy. Many of these low-growing, bushy trees and shrubs possess thorns, and where the woody plants form a closed canopy, the formation is called dry tropical forest. Dry tropical forests are found in northern Venezuela and Colombia, much of northeast Brazil, the West Indies, and Mexico. They also show up in India, Myanmar (formerly Burma), and Thailand in Asia and in vast areas of Africa and Australia. Both deciduous and evergreen species are found in this forest type. The leaves of both types are usually small and cutinized and possess other xeromorphic adaptations. Although great attention has been focused on rates of loss of tropical rainforests through conversion to other land uses, a far larger proportion of the dry tropical forests has been lost.

Soils developed in these dry tropical forests are often light in color, low in organic matter, and not severely leached. They may even have an accumulation of calcium salts in the subsurface horizon. These soils are mostly classified as Inceptisols and Alfisols.

Adjacent to the tropical thorn forests but in drier areas, semi-desert shrub and savanna woodlands grade into true desert areas. In these formations, trees occupy a small percentage of the area but their effects may be profound (Rhoades, 1997), especially for nitrogen-fixing trees (Rhoades *et al.,* 1998). Soils beneath these scattered trees are frequently Argids and generally have a darker-colored surface layer than adjoining soils. This darkening is created by the organic matter produced by the trees and by the entrapment of wind-blown organic debris beneath the trees.

TROPICAL MONTANE FORESTS HAVE MODERATE, UNIFORM CLIMATES

At higher altitudes in the tropics, montane forests occur in isotemperate climates. Tropical montane forests are species rich and contain many unique species, including those from common temperate-forest families such as Pinaceae, Cupressaceae, Betulaceae, Cornaceae, Fagaceae, Magnoliaceae, and Ulmaceae. Tropical montane forests in the Southern Hemisphere are often dominated by species of Araucariaceae and Podocarpaceae. Common soils are Andisols, Inceptisols, Alfisols, and Ultisols.

PLANTATION FORESTRY IS EXTENSIVE THROUGHOUT THE TROPICS

Intensively managed plantations of one or more tree species are attractive development projects for the tropics, and the area of tropical plantations increased from less than 1 million ha in 1950 to 45 million ha in the 1990s (Brown *et al.,* 1997). Plantations offer opportunities to relieve pressures on the mixed natural tropical forests while providing a source of wood products including pulp, charcoal, fuel, and others. Much of this expansion of tropical forest plantations has involved reforestation of agricultural fields, and some has followed clearing of native forests.

The intensive management of introduced tree species may provide more rapid growth and a greater economic return than would native species.

Successful plantations in tropical areas produce four to ten times the amount of usable wood produced by a natural forest. However, introduced species must be carefully matched to local soil, climate, and other environmental conditions. *Pinus caribaea, P. oocarpa, P. elliotti, P. taeda,* and *Gmelina arborea* are used operationally in forestry in many tropical countries. Short rotations of fast-growing and coppicing hardwoods, such as *Tectona grandis* and species of *Eucalyptus, Acacia,* and *Leucaena* have been established for the production of pulpwood, charcoal, fuel wood, and other forest products (Nambiar and Brown, 1997). Plantation forestry also can be labor intensive; an important consideration in low-employment areas of the tropics (Evans, 1992).

A number of concerns have arisen about the establishment of plantation monocultures in the humid tropics (West, 2006). For example, intensive plantation management often involves methods of harvesting, site preparation, and planting requiring heavy machinery. These operations may be suited to large blocks of relatively level land with sandy to sandy loam soils, but may damage more fragile or sloping soils. The use of wheeled vehicles can compact fine-textured soils, reduce infiltration rates, increase runoff, and promote erosion on sloping lands. Bulldozers with cutting and clearing blades are sometimes used to eliminate harvest debris and residual vegetation from areas to be planted.

If the harvest debris is burned prior to blading, a portion of the nutrients contained in the vegetation may be left on the site as ashes instead of being concentrated into windrows. However, a significant part of the ash, along with some of the nutrient-rich topsoil, is often pushed into the slash piles even when conscientious efforts are made to keep the blade above the soil surface. Losses of nutrients during harvesting and site preparation operations may be sufficient to create the need for mineral fertilizer additions to sustain rapid growth of plantations.

SOME TEMPERATE BROADLEAVED FORESTS ARE EVERGREEN

Evergreen broadleaved forests extend outside the tropics in some areas of Japan, southern China,

parts of Australia, and the southeastern United States. Evergreen oaks make up most of the species of Asian broadleaved formations. Vast forests of broadleaved evergreen trees once covered southwestern and southeastern Australia before European colonization. These forests were composed almost entirely of *Eucalyptus* species. The cut-over forests are still widespread, with substantially altered species composition. The eucalypts are quite different from the broadleaved evergreens of the northern mid-latitudes in both appearance and habitat. Many soils of Australia are very old, with low precipitation. Widely varying properties of these soils result from some distinctive geomorphic characteristics of the country. Australia has suffered only limited uplift, dissection has been modest, basement rocks have been subjected to little juvenile weathering, and only minor glaciation occurred on the continent. Despite a wide range in parent materials, climatic environment, and age, most of the forested soils are very old Ultisols (Udults), with some Alfisols (Ustalfs) and Vertisols. Interestingly, Oxisols are absent from Australia.

TEMPERATE RAINFORESTS ARE AMONG THE LARGEST FORESTS IN THE WORLD

Northwestern coastal North America is an area of high rainfall, well distributed throughout the year, and mild temperatures. This marine climate favors the development of dense stands of unusually tall and massive conifers, with aboveground biomass of more than $1500 \, \text{Mg ha}^{-1}$. The dominant species include Sitka spruce (*Picea sitchensis*), which ranges from Alaska through British Columbia and southward along the coasts of Washington and Oregon. These northern coastal rainforests also include western red cedar (*Thuja plicata*), western hemlock (*Tsuga heterophylla*), and Douglas fir (*Pseudotsuga menziesii*). In northern coastal California and extreme southern coastal Oregon, the giant coastal redwoods (*Sequoia sempervirens*) are a common forest dominant. The landscapes that contain redwood forests are less affected by summer droughts as a result of frequent fog. Western species of hemlock, pine, spruce, firs, and false cypress are also found in this coastal region.

Other temperate rainforest formations are found along the western side of the South Island of New Zealand, in the Valdevian forests of southern Chile, and in southeastern Australia. These Southern Hemisphere formations are largely broadleaved evergreens. Apart from this fact, the beech (*Nothofagus*) forests of New Zealand and the *Eucalyptus regnans* forests of Australia resemble in size and general appearance the temperate rainforests of other regions and contain some of the tallest trees in the world.

Soils developed under temperate rainforests commonly accumulate large amounts of organic matter and have a moderate abundance of layer lattice clay and argillic horizons. Common soils include some Sposodols (especially on coarse-textured parent materials), extensive Alfisols (Udalfs and Ustalfs) and some Inceptisols (particularly Umbrepts) and Ultisols (including Humults).

LOWLAND CONIFEROUS FORESTS ARE MAJOR SOURCES OF TIMBER

These forests are well represented in North America, where they make up several distinct formations, all of which are largely the result of human activity. Extending across the lake states into southern Ontario and northern New England, a belt of white spruce, white pine, red pine, and hemlock flourished until the middle of the nineteenth century. After this mid-continent forest was cut over for its valuable timber, much of the land was diverted to agricultural use. Pines occur in pure stands on dry sands, but they are usually mixed with oaks, aspen, and other hardwoods.

The pine forest of the coastal plain and piedmont of the southeastern United States is composed of several species collectively called southern pines. The dominant species are loblolly (*Pinus taeda*), shortleaf (*P. echinata*), longleaf (*P. palustris*), and slash (*P. elliottii*) pines. Pine forests in this region often have frequent fires that limit the development of hardwood trees. In recent years, southern pines have been widely used in intensively managed

plantations for pulp, veneer, and sawn-log products. These forests occur on highly weathered low-base status soils, mostly Ultisols and the Aquod group of Spodosols. Aquods have developed on sandy soils in humid areas under a wide range of temperature conditions, from temperate to tropical zones.

TEMPERATE MIXED FORESTS INCLUDE CONIFERS AND HARDWOODS

Except on the west coast of North America, high-altitude coniferous forests of the Northern Hemisphere generally grade into broadleaved deciduous forests at lower elevations. This change is usually achieved through a transition zone where the two forest formations exist side by side. These zones, with diverse mixtures of species, vary from a few miles to several hundred miles in width and are found in Europe and Asia as well as in North America. Species of birch, beech, maple, and oak are commonly found growing with the conifers. Mollisols, Inceptisols, and Alfisols are commonly found under these mixed forests. Since these productive soils are well suited to agriculture, most have been cleared, plowed, and cropped, and in some cases abandoned and reforested. In several lowland areas of the Southern Hemisphere there are mixed forests of broadleaved evergreens and conifers. They occur in several climatic regions in Chile, southern Brazil, Tasmania, northern New Zealand, and South Africa. The species in this formation vary rather widely among the continents, as might be expected, but they are quite similar in life forms and general appearance. Such genera as *Araucaria*, *Podocarpus*, and *Agathis* are found in many of these mixed evergreen forests. The predominant soils in these areas are Alfisols and Ultisols.

The broad-leaved, winter-deciduous forests of eastern and north-central North America, western and central Europe, and northern China generally lie south of the mixed coniferous-deciduous forests. These areas are heavily populated and intensively farmed; few relatively undisturbed forests remain.

In France and the British lowlands, *Quercus robur*, *Fagus sylvatica*, and *Fraxinus excelsior* are the dominant species, with other species of oak, birch, and elm becoming important in local areas. The tree associations gradually change when moving to southern Europe, and *Quercus lusitanica*, Q. *pubescens*, *Acer platenoides*, *Castanea sativa*, and *Fraxinus ornus* become more plentiful. Some of the best soils from the northern Ukraine across to the southern Urals once carried deciduous forests dominated by various oak species. Except for shelterbelts and windbreaks, these forests have been converted to agricultural uses.

A similar but more diverse formation of hardwoods is found in the United States, extending from the Appalachian Mountains west beyond the Mississippi River, from the mixed forests of New England and the lake states southward to the southern coniferous and mixed forests of the coastal plain. It reaches westward into the prairies along the valleys of several rivers. Although noted for its diversity, the American formation is dominated by species of oak and hickory (*Carya*), with basswood (*Tilia americana*), maple, beech (*Fagus sylvatica*), and ash (*Fraxinus*) species found in the eastern regions. Along the lower slopes of the southern Appalachians, several other species of oak, sweet gum (*Liquidambar styracfiua*), and yellow poplar (*Linodendron tulipifera*) appear, often with an understory of wild cherry (*Prunus serotina*), and dog-wood (*Cornus fiorida*).

The East Asian formation of deciduous hardwoods is very similar to the formations of Europe and North America. However, northern and central Chile have a formation that is more closely related to the nearby evergreen forest than to the deciduous genera so common in the Northern Hemisphere formations. In southern Argentina and in the Valdevian forests of southern Chile, hardwood forests are common on Andisols.

Precipitation is relatively heavy and well distributed throughout the year in many areas supporting temperate forests of deciduous hardwoods. Summers are generally warm and humid, and winters are cool to cold, with heavy snowfall in the coldest regions. Differences in nutrient demands of the various species, plus the diversity of parent material on which they grow, have resulted in a wide range of soil properties in these forests. Spodosols have developed on parent materials of granites and sands in many northern hardwood forests in North America, as well as in some oak–birch forests in France and other parts of Europe. Mollisols are common on the calcium-rich clays of central and southern Europe, and Alfisols are sometimes found in a narrow zone along the margins of the wooded steppes in the central European basins and in the Ukraine.

Most Alfisols of the northern part of the United States, as well as the Ultisols in the south, were initially fertile and easily cultivated. They were exploited early in the settlement of the country, and then often abandoned when market conditions no longer allowed row crops to be profitable. After abandonment, these areas were often planted to pine or reverted to pine–hardwood mixtures.

TEMPERATE MONTANE CONIFER FORESTS ENDURE COLD, SNOWY WINTERS

Alpine ecosystems occupy high-elevation territories where trees cannot grow, so by this definition, the highest-elevation forests are called subalpine forests. Subalpine forests around the Northern Hemisphere are commonly composed of spruce species, sometimes alone or mixed with fir, birch, and occasionally pines. Below these forests are found a great diversity of montane forests, with conifers in many cooler or drier regions, and hardwoods in warmer and moister regions. Although these patterns are typical, they often do not apply to specific situations.

Pine, spruce, larch, and fir comprise the subalpine forests of western and central Europe. Spruce and fir are found at 1000 to 2000 m in the Alps and at 1500 to 2500 m in the Pyrenees. Similar associations are found at subalpine heights on the mountains of central and eastern Asia, extending southward to the Himalayas. Subalpine forests of Europe extend southwest into the Mediterranean peninsulas and even as far as the Atlas Mountains in North Africa and the mountains of Turkey and Spain. Pines (*Pinus sylvestris* and *P. nigra*) and firs are common dominant species, with cedars (*Cedrus* species) being of local importance.

In North America there is an almost continuous belt of subalpine and montane forest extending the whole length of the Cascade and Sierra Nevada Mountains from Canada to California in the west and along the length of the Rocky Mountains as far south as New Mexico. The altitude of the lower limit of the subalpine forest varies from 600 m in southern Alaska to 1000 m in British Columbia to 2700 m in New Mexico. Engelmann spruce (*Picea engelmannii*), subalpine fir (*Abies lasiocarpa*), and lodgepole (*Pinus contorta*), bristlecone (*P. aristata*), and limber (*P. ftexilis*) pines are widely distributed in these western areas. The montane forests of the Pacific Northwest occur on the middle slopes of mountains from the Rockies westward, usually interposed between the subalpine forest and the temperate rainforest along the coast. At lower elevations in the Southwest they give way to scrub communities. The dominant species are ponderosa (*Pinus ponderosa*), western white (*P. monticola*), and lodgepole pines, Douglas fir (*Pseudotsuga menziesii*), and grand fir (*Abies grandis*).

The Appalachian province consists of a series of mountain ranges and high plateaus extending from Newfoundland in the north to Georgia and Alabama in the south. The Appalachian subalpine forests are dominated by black and white spruce, balsam fir, and tamarack (*Larix laricina*) in the north and by red spruce (*Picea rubens*), fraser fir (*Abies fraseri*), and eastern white pine (*Pinus strobus*) farther south. The altitudes of these forests rise southward, from sea level in Newfoundland to 200 m in the Adirondacks, 900 m in West Virginia, and nearly 1800 m in Georgia.

The soils of subalpine forests are often of glacial or colluvial origin and generally are weakly developed. Spodosols and Inceptisols predominate in the higher latitudes. In the lower latitudes Spodosols are infrequent, and Alfisols and Ultisols become more common. In the Rocky Mountains, Inceptisol, Alfisol, Spodosol, and Mollisol soils occur beneath subalpine and montane forests.

BOREAL FORESTS SPAN VAST AREAS

This vast northern forest, composed largely of conifers, extends southward from the tundra treeline across Canada, Alaska, Siberia, and Europe and is referred to as boreal forest or simply taiga. Although boreal forests are characterized by conifers, hardy species of *Populus, Betula*, and *Salix* extend throughout the forested boreal landscape.

Within the boreal forest, species are distributed according to landforms and landscape position, with soil drainage being the key factor. In North America, a gradual transition in species spans the continent. Red pine (*Pinus resinosa*), eastern hemlock (*Tsuga canadensis*), yellow birch (*Betula alleghaniensis*), and maple (*Acer*) of southern and eastern Canada are gradually replaced by a predominantly forested belt across the middle of Canada composed chiefly of white spruce (*Picea glauca*), black spruce (*P. mariana*), balsam fir (*Abies balsamea*), jack pine (*Pinus banksiana*), white birch (*Betula papyrifera*), and aspen (*Populus tremuloides*).

These species give way to white spruce, black spruce, and tamarack (*Larix laricino*) near the tundra. In a somewhat analogous fashion, pine (*Pinus sylvestris*), spruce (*Picea abies*), and European larch (*Larix decidua*) predominate in the European boreal forests, and to the east in Siberia the dominant species become *Abies siberica, Larix siberica*, and *Picea siberica*.

The mean annual rainfall of most boreal forests is less than 900 mm yr^{-1}, with many forests receiving less than half of this maximum. Evapotranspiration rates are low owing to the short growing season, and boreal soils tend to be moist (and often saturated) except on well-drained landforms or in very dry summers. Some boreal forest soils are underlain by permanently frozen parent material called permafrost; the zone that freezes in winter and thaws in summer is called the active layer. The depth of the active layer may increase following fires that remove canopies and insulating O horizons, although this is not a universal pattern (Swanson, 1996). The O horizons are often thick, with well-developed layers of humus overlain by carpets of moss or lichen. The organisms that decay the litter produce organic acids that aid in leaching minerals important for plant growth from the upper soil (E) horizon. The sesquioxides of iron and aluminum are removed from this eluviated horizon during the podzolization process, and the resultant sandy horizon is light gray in color. The iron and aluminum compounds that are removed from the upper horizons are mostly redeposited in the illuviated horizons lower in the profile. Soils of the boreal forest are predominantly classified as Spodosols and Inceptisols to the south and Gelisols to the north, with Histosol soils found in bogs or fens.

A positive water balance results in a high water table over large expanses of nearly level terrain in the boreal forest. Where the water table remains within about 0.5 m of the surface for a large part of the year, trees are stunted or absent. Extensive areas of the boreal zone are covered by organic soils (Histosols), with various mosses (*Sphagnum* and *Polytrichum* species), flowering plants (*Eriophorum* and *Trichophorum* species), and dwarf shrubs (*Vaccinium* and *Calluna* species). Various names are used for these mires in different countries: bogs, muskegs, and fens.

Wind-throw mounds are numerous on some Spodosols of the boreal zone, occupying up to 30% of the O horizon area in parts of the Great Lakes region of North America (Buol *et al.*, 2003). They are formed by the uprooting of trees in storms, and they typically have relief of as much as 0.5 to 1 m and a diameter of 1 to 3 m. This disturbance creates a characteristic pit and mound surface morphology and complicates the characterization of boreal soils (Fisher and Miller, 1980).

The major effect of humans on most boreal forests has been through the ignition of fire, but increasingly large portions of the boreal forests are being harvested for wood products. Very little of the boreal forest realm has been cleared and plowed for agriculture.

PART II

Composition of Forest Soils

CHAPTER 3
Soil Formation and Minerals

OVERVIEW

The development of soil and associated forest vegetation is a complex and continuing process. Soils play vital roles in the development of forests, and forests likewise play vital roles in the development of soils. All pedogenic processes appear to operate to some degree in all soils; however, they operate at different rates at different times and the dominant processes in any one soil body cause it to develop distinctive properties. These processes involve some phase of (1) addition of organic and mineral materials to the soil as solids, liquids, and gases; (2) loss of these materials from a given horizon or from the soil; (3) bioturbation and translocation of materials from one point to another in the soil; or (4) transformations of mineral and organic matter within the soil.

The mineralogical makeup of the soil's parent material plays a dominant role in the character of soil, and an understanding of the soil's parent material is nearly always necessary to understand the soil. For example, soils that develop from acidic and basic parent materials are nearly always quite different. As a variety of differing plant communities occupies a soil, they have different effects on the soil's character. In forests, bioturbation, or the lack of it, plays an important role in soil development. In a systems context, climate, parent material, and the effects of topography are viewed as driving variables, which affect the state of a system and may change over time, but are not affected by the system. Animals and vegetation are viewed as state variables, features of the system of interest affected by driving variables and by each other, and time

becomes the dimension over which soil development occurs.

———————————

The development of soil and associated forest vegetation is a complex and continuing process. It takes place over thousands of years through a complex sequence of interrelated events. Although a number of factors are involved in the development of both soils and forests, none is more important than climate. Climate paints the globe with biomes with a broad brush. Other factors, such as physiography, soil parent material, and biota, then play a role in determining the distribution of ecosystems and their characteristic soils.

Soils, like climate, play vital roles in the development of forests. They provide water, nutrients, and support for trees and other forest vegetation. Soils are derived from parent materials of different mineral composition, and these differences result in soil properties that influence both the composition of the forest and its rate of growth.

As parent rocks weather, a soil develops, influenced not only by the mineralogy of the rocks and the physical factors of the environment, but also by the biota of the area that contributes to mineral weathering and the buildup of organic material in the soil. Eventually, a series of horizons form in a typical well-developed forest soil: an organic layer (O), an organically enriched mineral layer (A), an eluviated layer (E), an illuviated layer or zone of accumulation (B), and a mineral layer little altered by soil-forming processes (C). The formation of these distinctive soil horizons results from a series

Ecology and Management of Forest Soils, Fourth Edition. Dan Binkley and Richard F. Fisher.
© 2013 John Wiley & Sons, Ltd. Published 2013 by John Wiley & Sons, Ltd.

of complex reactions and simple rearrangements of matter termed pedogenic processes (Buol *et al.*, 2003).

PEDOGENIC PROCESSES OPERATE SIMULTANEOUSLY AT VARYING RATES

All pedogenic processes appear to operate to some degree in all soils; however, they operate at different rates at different times, and the dominant processes in any one soil body cause it to develop distinctive properties. All pedogenic processes involve some phase of (1) addition of organic and mineral materials to the soil as solids, liquids, and gases; (2) losses of these materials from a given horizon or from the soil; (3) translocation of materials from one point to another in the soil; or (4) transformations of mineral and organic matter within the soil.

Additions to the soil occur in both organic and inorganic forms. Wind and rain act as agents of transport for dust particles, aerosols, and organic compounds washed from the forest canopy. These additions are generally not large, but they can be substantial as a result of catastrophic storms or volcanic action. The most significant addition to forest soil is organic matter. The death and decay of live roots contribute a significant amount of organic matter directly to the soil in each growing season (Gill and Jackson, 2000; Silver and Miya, 2001). The surface O horizon contributes to the accumulation of soil organic matter and exerts considerable influence on the underlying mineral soil, as well as on the associated populations of microorganisms and soil animals. The process by which organic matter is added to the A horizon, thus darkening it, is called melinization.

Processes involving losses include leaching of mobile ions such as calcium, magnesium, potassium, and sulfate, as well as surface erosion by water or wind, solifluction, creep, and other forms of mass wasting. The latter are generally not problems in forests except on steep slopes and in mountainous areas. The dominant processes in forest soils are those related to translocations or transformations within the soil body. Eluviation is the movement of material out of a portion of a soil profile. It is a major process in the development of the A horizon and is the

dominant process in the development of the E horizon. Illuviation is the movement of material into a portion of a soil profile. It is the dominant process in the development of most forest soil B horizons. More specifically, decalcification is a series of reactions that removes calcium carbonate from a soil horizon, and lessivage is the downward migration and accumulation of clay-sized particles, producing an argillic or clay-enriched horizon (Chesworth, 2008).

Probably the two most important soil-forming processes in forest areas are podzolization and desilication. Podzolization is a complex series of processes that brings about the chemical migration of iron, aluminum, and/or organic matter, resulting in a concentration of silica in the eluviated albic (E) horizon and an accumulation of organic matter and/or sesquioxides in the illuviated spodic (Bh or Bs) horizon (Stobbe and Wright, 1959; Ponomareva, 1964; Brown, 1995; Lundström *et al.*, 2000).

Podzolization is the dominant soil-forming process in the boreal climatic zone, but it is not restricted to a specific climatic regime. Extensive areas of Aquods (Groundwater Podzols) occur on the coastal plain of the southeastern United States, and "giant podzols" with extremely thick E and Bs horizons occur in the tropics (Eyk, 1957; Klinge, 1976). The processes leading to podzolization intensify certain types of vegetation, especially hemlocks (*Tsuga*), spruces (*Picea*), pines (*Pinus*), and heath (*Calluna vulgaris*) in northern latitudes and kauri (*Agathis australis*) in New Zealand. Organic compounds leached from the foliage and litter (Bloomfield, 1954) or produced during litter and soil organic matter decomposition (Fisher, 1972) lead to rapid mineral weathering (Herbauts, 1982) and translocation of organic complexes downward in the soil profile (Ugolini *et al.*, 1977; van Breemen *et al.*, 2000).

Desilication (ferrilization) is the dominant soil-forming process in forested areas of the intertropical zone, where high temperatures and high rainfall favor rapid silica loss and the concentration of iron immobilized as ferric oxides under oxidizing conditions. In a well-drained environment, this process leads to the formation of an oxic horizon, a zone of low-activity clay that lacks weatherable primary minerals and is resistant to further alteration. In a soil with a fluctuating water table, this process leads to the formation of plinthite, which may occur as scattered cells of material or may form a continuous zone in the soil. When exposed to repeated wetting and drying, plinthite becomes indurated to iron-stone, which has also been termed laterite (Alexander and Cady, 1962; Soil Survey Staff, 1999; Eswaran *et al.*, 2003).

Pedoturbation is a third process that often plays a significant role in the development of forest soils. Pedoturbation is the biological or physical (freeze–thaw or wet–dry cycles) churning and cycling of soil materials, thereby mixing or even inverting soil layers (Bockheim *et al.*, 1997; Schaetzl and Anderson, 2005). In many forests, wind-throw of mature trees causes significant mixing of soil horizons. Discontinuous horizons are often developed in this process, and soil horizons may even be inverted in some cases. A characteristic "pit and mound" surface condition is also often developed (Lyford and MacLean, 1966).

EXTERNAL FACTORS GUIDE SOIL FORMATION

The pedogenic processes involved in the genesis of soils are conditioned by a number of external factors. Simply stated, these factors are the forces of weather and of living organisms, modified by topography, acting over time on the parent rock (Jenny, 1980). These factors were first outlined by Dokuchaev, the Russian scientist who laid the foundation for our understanding of soil genesis in 1883. He remarked:

'The still young discipline of these relations is of an exceptional inspiring scientific interest and meaning. Each year it makes greater and greater strides and conquests; gains daily more and more of active and energetic followers, eager to devote themselves to its study with the passionate love and enthusiasm of adepts.' (Quoted by Jenny, 1961a.)

Dokuchaev conceived of relating soil-forming factors in the form of an equation:

$$S = f(cl, o, p)to$$

where S = soil, cl = climate, o = organisms, p = geologic substrate, and to is a measure of relative age (this version of the equation was related by Jenny, 1961b). A somewhat similar approach was proposed independently by an American soils professor, Charles Shaw (1930):

$$S = M(C + V)T + D$$

where S = soil, M = parent material, C = climate, V = vegetation, T = time and D = deposition or erosion.

In 1941, Jenny popularized these ideas in the form of an equation that he hoped could someday be solved quantitatively. Although no one ever solved this equation, it focused thinking on how to deal quantitatively with soil formation. Systems thinking and simulation modeling developed in the 1950s and 1960s, offering better ways to conceptualize and quantify the formation of soils. This view allowed reciprocal interactions to be included, whereas Jenny's approach assumed that soil-forming factors worked independently. In a systems context, climate, parent material, and the effects of topography are viewed as driving variables, variables that affect the state of a system and may change over time but are not affected by the system. Animals and vegetation are viewed as state variables, features of the system of interest affected by driving variables and by each other, and time becomes the dimension over which soil development occurs (Figure 3.1). This approach is not as simple as Jenny's equation appeared to be, but it provides a more realistic

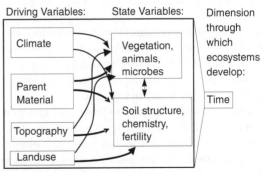

Figure 3.1 Diagrammatic representation of a systems model of soil development.

framework for the way soils really develop (Bryant and Arnold, 1994).

PARENT MATERIAL CAN BE MINERAL OR ORGANIC

Parent materials consist of mineral material, organic matter, or a mixture of both. The organic portion is generally dead and decaying plant remains and is the major source of nitrogen in soil development. Mineral matter is the predominant type of parent material, and it generally contains a large number of different rock-forming minerals in either a consolidated state (granite, conglomerate, mica-schist, etc.) or an unconsolidated state (glacial till, marine sand, loess, etc.). Knowledge of the parent material of youthful soils is essential for understanding their properties and management. As soils age, the influence of parent material on their properties and management decreases but never disappears (Schaetzl and Anderson, 2005).

Although a thorough discussion of rock-forming minerals (Klein and Dutrow, 2008) is impossible in this book, we will outline the structure and chemical composition of the main types in order to better explain the behavior of parent material as it is subjected to soil-forming processes (Table 3.1).

NONSILICATES ARE IMPORTANT SOIL-FORMING MINERALS

Oxides, hydrous oxides, and hydroxides are among the most abundant and important soil-forming minerals. Hematite hydrates rather easily to form limonite and turgite, which are important in many forest soils as coatings on particles and as cementing agents in hardpans. Magnitite is a heavy mineral resistant to weathering and often occurs in soils as a black sand with strong magnetism.

Goethite, gibbsite, and boehmite are common minerals in soils. They occur as coatings on other mineral or rock particles, and gibbsite and boehmite often accumulate in the spodic horizons. In tropical forest soils, limonite, hematite, gibbsite, boehmite, and other iron and aluminum oxides and

hydroxides accumulate in the B horizon and may form plinthite.

Sulfates, sulfides, and chlorides are uncommon in forest soils. Pyrite is the most common and important of these minerals. It occurs in igneous and metamorphic rocks and is often abundant in mine spoil materials. Pyrite alters to a variety of iron oxides or hydroxides and releases sulfur as sulfuric acid. Thus, soil or spoil material high in pyrite is highly acidic in reaction and may become more acidic as weathering progresses.

Carbonates are the major source of calcium and magnesium in soils. Calcite and dolomite occur in igneous, sedimentary, and metamorphic rock and are commonly added to agricultural soils to alter the soil's acidity; however, they are uncommon constituents and are generally ineffective as amendments in acid forest soils. Siderite is common in some forest soils and is an important cementing agent in some hardpans. Carbonates are generally present only in the lower horizons of humid zone forest soils. In more arid forests, secondary and even primary carbonate minerals may occur at or near the soil surface.

Phosphorus is one of the mineral nutrients that are commonly deficient in forest soils. Phosphorus occurs as apatite in all classes of rocks, but it is much less abundant in highly siliceous rocks. Apatite does not persist long, particularly in acid soils. Weathering leads to increasing amounts of phosphorus bonding to iron and aluminum ions through complex ion-exchange reactions. The chemical properties of phosphate in these forms resemble those of variscite, strengite, fluoroapatite, hydroxyapatite, and octacalcium phosphate. The high stability of these minerals and their very low secondary minerals, such as solubility at low pH lead to the low phosphorus status of many forest soils.

SILICATES ARE THE DOMINANT SOIL-FORMING MINERALS

Silicates account for 70–90% of the soil-forming minerals. These minerals have complex structures in which the fundamental unit is the silicon-oxygen tetrahedron. This is a pyramidal structure with a

Table 3.1 Soil-forming minerals.

Crystal Mineral Structure	Group	Mineral Species	Chemical Formula
Nonsilicates	Oxides	Hematite	Fe_2O_3
		Magnetite	Fe_3O_4
	Hydrous oxides	Limonite	$2Fe_2O_3 \cdot H_2O$
		Turgite	$2Fe_2O_3 \cdot H_2O$
	Hydroxides	Gibbsile	$Al(OH)_3$
		Boehmitc	$AlO(OH)$
		Goethite	$FeO(OH)$
	Sulfates	Gypsum	$CaSO_4 \cdot H_2O$
	Sulfides	Pyrite	FeS_2
	Chlorides	Halite	$NaCl$
	Carbonates	Calcite	$CaCO_3$
		Dolomite	$(Ca,Mg)CO_3$
		Siderite	$FeCO_3$
	Phosphates	Apatite	$Ca(CaF \text{ or } CaCl)(PO_4)_3$
		Variscite	$AlPO_4 \cdot 2H_2O$
		Sytrengite	$FePO_4 \cdot 2H_2O$
		Barrandite	$(Al,Fe)PO_4 \cdot 2H_2O$
Ortho- and ring silicates	Epidote	Epidote	$4CaO \cdot 3(Al,Fe)_2O_3 \cdot 6SiO_2 \cdot H_2O$
	Garnet	Almandite	$Fe_3Al_2Si_3O_{12}$
	Zircon	Zircon	$ZrO_2 \cdot SiO_2$
	Olivine	Olivine	$2(Mg,Fe)O \cdot SiO_2$
Chain silicates	Amphibole	Hornblende	$Ca_3Na_2(Mg,Fe)_8(Al,Fe)_4Si_{14}O_{44}(OH)_4$
		Tremolite	$2CaO\ 5MgO\ 8SiO_2 \cdot H_2O$
		Actinolite	$2CaO_5(Mg,Fe)O$
	Pyroxene	Enstatite	$MgO \cdot SiO_2$
		Hypersthene	$(Mg,Fe)O \cdot SiO_2$
		Diopside	$CaO \cdot MgO \cdot 2SiO_2$
		Augite	$CaO \cdot 2(Mg,Fe)O \cdot (Al,Fe)_2O_3 \cdot 3SiO_2$
Sheet silicates	Mica	Muscovite	$K_2Al_6Si_6O_{20}(OH)_4$
		Biotite	$K_2Al_2Si_6(Fe,Mg)_6O_{20}(OH)_4$
	Serpentine	Serpentine	$Mg_3Si_2O_5(OH)_4$
	Clay minerals	Hydrous mica	$KAL_4(Si_7Al)O_{20}(OH)_4 n H_2O$
		Kaolinile	$Si_4Al_4O_{10}(OH)_8$
		Vermiculite	$Mg_{0.55}(AL_{0.5}Fe_{0.7}Mg_{0.48})O_{22}(OH)_n \cdot n H_2O$
		Montniorillonite	$Ca_{0.4}(AL_{0.3}Si_{7.7})Al_{2.6}(Fe_{0.9}Mg_{0.5})O_{20}(OH)_4 \cdot n H_2O$
Block silicates	Feldspar	Orthoclase	$(Na,K)_2OAl_2O_36SiO_2$
		Plagioclase	$[Na_2 \text{ or } Ca]\ OAl_2O_36SiO_2$
	Quartz	Quartz	SiO_2

base of three oxygen ions (O^{2-}), an internal silicon ion (Si^{4+}), and an apex oxygen ion (Figure 3.2).

The four positive charges of the silicon ion are balanced by four negative charges from the oxygen ions, and the tetrahedral unit (SiO_4^{4-}) has four excess negative charges. In nature these tetrahedra are linked in a number of different patterns by silicon ions sharing oxygen ions in adjoining tetrahedra. This leads to the average composition of silica being SiO_2.

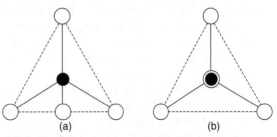

Figure 3.2 Diagram of the silicon–oxygen tetrahedron: (a) side view; (b) top view. Open circles are oxygen and solid circles are silicon.

Varying patterns of tetrahedral linkage lead to a variety of distinctive and characteristic mineral forms and are the basis for classification. In addition, the type of linkage is an important factor in determining the mineral's resistance to weathering. Isomorphous (ion-for-ion) substitution of aluminum for silicon in the tetrahedral structure leads to structural and electrostatic imbalances, the inclusion of cations such as sodium, potassium, calcium, and magnesium in the mineral, and a change in the resistance to weathering. The silicate minerals are commonly divided into four broad classes based on crystalline structure: ortho- and ring silicates, chain silicates, sheet silicates, and block silicates.

Ortho- and ring silicates are the most variable silicate minerals. Olivines have the simplest structure, with separate silicon-oxygen tetrahedra arranged in sheets and linked by magnesium and/or iron ions. Olivines are common and are a frequent constituent of many basic and ultrabasic igneous rocks; however, olivines are the most easily weathered silicates and do not persist in soils. Zircon is another important mineral in which a cation provides linkage for silica tetrahedra. Each zirconium ion is surrounded by eight oxygens, resulting in a very strong structure and great resistance to weathering. Because of its persistent nature, zircon is often used as a marker in soil genesis studies. Garnets are complex silicates with alumina octahedra and iron linkages. Many species of this group are found in soils, where they are fairly persistent.

Chain silicates are divided into the amphiboles and the pyroxenes. The latter are composed of tetrahedra linked together by sharing two of the three basal oxygens to form continuous chains (Figure 3.3). These chains are then variously linked by cations such as calcium, magnesium, iron, sodium, and aluminum to form a variety of mineral species. The pyroxenes are fairly easily weathered by cleavage at the cation linkage.

Amphiboles are composed of double chains of silicon-oxygen tetrahedra (Figure 3.4) linked together by calcium, magnesium, iron, sodium, or aluminum. These minerals are only slightly resistant to weathering. The chain silicates are, in reality, ferromagnesian metasilicates and occur in both

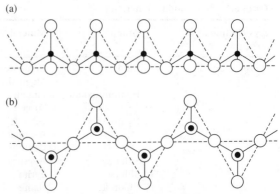

Figure 3.3 Diagram of the pyroxene chain: (a) side view; (b) top view. Open circles are oxygen and solid circles are silicon.

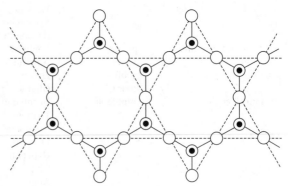

Figure 3.4 Diagram of the amphibole chain. Open circles are oxygen and solid circles are silicon.

igneous and metamorphic rocks. They are primary sources of calcium, magnesium, and iron in soils.

Sheet silicates are represented by primary minerals such as mica and the secondary clay minerals. Both of these groups are composed of various combinations of three basic types of sheet structures: the silicon tetrahedral sheet (Figure 3.5), the aluminum hydroxide sheet (Figure 3.6), and the magnesium hydroxide sheet (Figure 3.7).

The silicon tetrahedral sheet is composed of silicon-oxygen tetrahedra linked in a hexagonal arrangement with the three basal oxygens in one plane and the apical oxygens in another. The aluminum hydroxide sheet is made up of aluminum-hydroxyl octahedra in which the hydroxyl ions are in two planes and the aluminum ions are

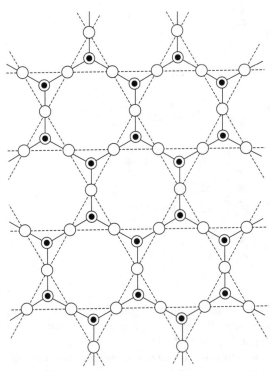

Figure 3.5 Diagram of the silicon–oxygen tetrahedral sheet. Open circles are oxygen and solid circles are silicon.

divalent, all of the sites in the middle plane are filled.

The micas are abundant and important soil-forming minerals that occur in a wide variety of igneous and metamorphic rocks. The basic structure is two silicon tetrahedral sheets with either an aluminum hydroxide (muscovite) sheet or a magnesium hydroxide (biotite) sheet between them (Figure 3.8).

One-quarter of the silicon ions have been replaced by aluminum ions, causing the sheet to have a net negative charge. This charge imbalance is satisfied by potassium, which bonds the sheets together. In biotite, about one-third of the magnesium in the middle sheet is replaced by ferrous iron. Mica is altered in soil to form hydrous mica (illite), kaolinite, chlorite, or even serpentine.

Serpentine is a secondary mineral that results from the alteration of olivine, hornblendes, biotite, and other magnesium minerals. Soils derived primarily from serpentine rock are infertile due to their high magnesium and low calcium content (Bohn *et al.*, 2001). They also often have peculiar physical properties and are sometimes referred to as serpentine barrens.

CLAY MINERALS ARE SECONDARY ALUMINO-SILICATES

Clay minerals are secondary alumino-silicates that are of great importance in soils. Six groups of these minerals are important in soils: hydrous mica,

sandwiched in a third plane between them. To satisfy all valences in this structure, only two-thirds of the positions in the middle plane are occupied by aluminum ions. The magnesium hydroxide sheet is similar in structure, but because magnesium is

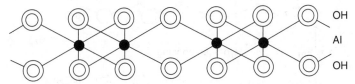

Figure 3.6 Diagram of an aluminum hydroxide sheet.

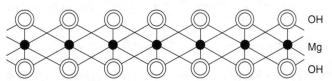

Figure 3.7 Diagram of a magnesium hydroxide sheet.

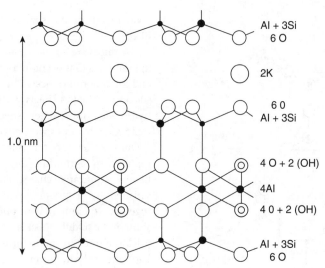

Figure 3.8 Diagram of muscovite mica.

kaolinite, vermiculite, montmorillonite, chlorite, and allophane. The first five are crystalline and are composed of silicon tetrahedral sheets and aluminum hydroxide and magnesium hydroxide sheets in various combinations. Allophane is an amorphous hydrous aluminum silicate or a mixture of aluminum hydroxide and silica gel with a variable ratio of silica to aluminum. It is prevalent in soils of volcanic origin but is also a common constituent of many other soils.

Hydrous mica, as its name implies, consists of hydrated muscovite and biotite particles of clay size. There are fewer potassium linkages in this mineral than in primary mica, but there are still sufficient linkages to prevent the lattice from expanding on hydration.

Kaolinite is the simplest of the true clay minerals. It is composed of one aluminum hydroxide sheet and one silicon tetrahedral sheet in which each apical oxygen replaces one hydroxyl group of the aluminum hydroxide sheet to form a 1 : 1-type clay structure (Figure 3.9).

Vermiculite is hydrated mica with the potassium ions replaced by calcium and magnesium and with increased isomorphic substitution of aluminum for silicon. It is made up of one hydroxyl sheet and two

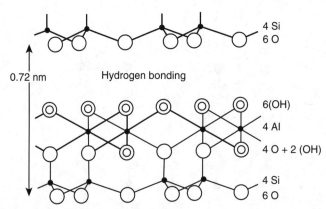

Figure 3.9 Diagram of the structure of kaolinite. Adjoining sheets of the 1 : 1 lattice are held rigidly together by hydrogen bonding. The intersheet spacing is a constant 0.72 nm.

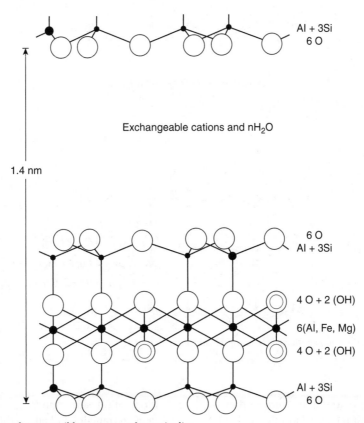

Al + 3Si
6 O

Exchangeable cations and nH_2O

1.4 nm

6 O
Al + 3Si

4 O + 2 (OH)

6(Al, Fe, Mg)

4 O + 2 (OH)

Al + 3Si
6 O

Figure 3.10 Diagram of one possible structure of vermiculite.

silicon tetrahedral sheets and is said to have a 2 : 1-type clay structure (Figure 3.10). This structure is capable of expansion on hydration because calcium and magnesium do not reform the broken potassium linkages. Therefore, the inter-sheet spacing varies with the degree of hydration. Montmorillonite is also a 2 : 1 clay (Figure 3.11).

The sheets that make up the structure exhibit considerable substitution of aluminum for silicon, as well as iron and magnesium for aluminum. The mineral has high affinity for water and expands and contracts a great deal in response to hydration and dehydration.

Chlorite exists as both a primary and a secondary mineral. In its primary form it is a 2 : 1 layer silicate with a magnesium hydroxide sheet between the mica layers in place of potassium. Secondary chlorite, as it is formed in soils, differs from the primary form by having an aluminum hydroxide interlayer sheet.

Clay minerals have several properties that are quite influential in soil behavior: their cation exchange capacity and ionic double layer, their ability to absorb water, and their ability to flocculate and disperse, to name but a few.

Minerals such as kaolinite, with its rigid structure and absence of isomorphic substitution, have low cation exchange capacity because their exchange sites are confined to the broken edges of crystals; where bonds are satisfied by hydrogen or other cations. Montmorillonite and vermiculite, on the other hand, have a great deal of isomorphic substitution of aluminum for silicon, resulting in a large number of exchange sites and a high cation exchange capacity. Allophane, with its intermediate structure, has a high but variable cation exchange capacity that is strongly influenced by external conditions, even the techniques used to measure it.

When clay particles are moist, each particle is covered with negative charges that are satisfied

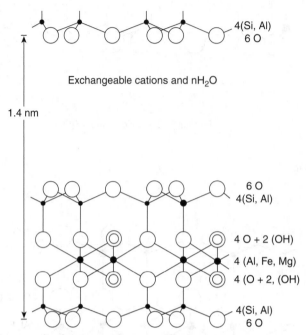

4(Si, Al)
6 O

Exchangeable cations and nH$_2$O

1.4 nm

6 O
4(Si, Al)

4 O + 2 (OH)

4 (Al, Fe, Mg)

4 (O + 2, (OH)

4(Si, Al)
6 O

Figure 3.11 Diagram of the generalized structure of montmorillonite.

by positive charges of ions in the surrounding solution. The innermost ions are most tightly held, but the outer ionic shell is also attached to some degree to the particle. This unit, the clay particle and its surrounding ionic double layer, is called a micelle.

Although clay minerals have a characteristic basal spacing – that is, the fixed distance from any point in one layer to the same point in the adjacent layer – many have the ability to expand and contract as they undergo hydration and dehydration. This ability, along with the small size of clay particles and the resulting large surface area, gives clays the capacity to retain a great deal of water against the force of gravity.

Flocculation and dispersion are also important properties of clays. Flocculation is the process of coagulation of individual particles into aggregates, and dispersion is the separation of the individual particles from one another. Flocculation in soils is brought about by a thin double layer such as the one that occurs when clays are calcium saturated. Dispersion results from a thick double layer such as the one that occurs when clays are sodium saturated.

Block silicates are minerals composed of silicon-oxygen tetrahedra linked at their corners into a continuous three-dimensional structure. The simplest of these is quartz, which is composed entirely of silicon-oxygen tetrahedra. Quartz, with its very regular and continuous structure, is extremely hard and resistant to weathering.

The feldspars are also block silicates, but they display a considerable amount of isomorphic substitution and contain a high proportion of basic cations. This substitution stresses even the three-dimensional structure and makes feldspars more easily weathered than quartz. There are three principal types of feldspar: sodium, potassium, and calcium. Members of the sodium to potassium series are known as orthoclase or alkali feldspars, and members of the sodium to calcium series are called plagioclase feldspars.

Stability series based on resistance to weathering have been developed, particularly for the silicate minerals. Figure 3.12 presents a comparison between stability series for sand- and silt-sized mineral particles and for clay-sized mineral particles.

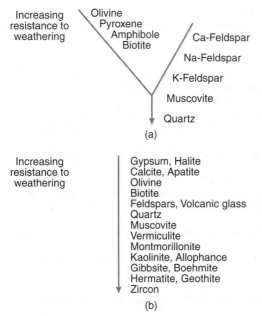

Increasing resistance to weathering

Olivine
Pyroxene
Amphibole
Biotite
Ca-Feldspar
Na-Feldspar
K-Feldspar
Muscovite
Quartz

(a)

Increasing resistance to weathering

Gypsum, Halite
Calcite, Apatite
Olivine
Biotite
Feldspars, Volcanic glass
Quartz
Muscovite
Vermiculite
Montmorillonite
Kaolinite, Allophance
Gibbsite, Boehmite
Hermatite, Geothite
Zircon

(b)

Figure 3.12 Mineral stability series for (a) sand and silt-sized particles, and (b) for clay-sized particles (adapted from Bohn *et al.*, 2001).

PARENT MATERIALS HAVE BOTH CHEMICAL AND STRUCTURAL FEATURES

Parent materials are made up of consolidated or unconsolidated mineral material that has undergone some degree of physical or chemical weathering. In order to organize our thinking, a simple system of classification based on parent material properties important in soil formation is useful.

A generalized classification of the rock types that commonly constitute consolidated soil parent materials is given in Table 3.2. Unconsolidated parent materials are usually classified according to the size and uniformity of their constituent particles, as well as their mode of origin. Thus, we often speak of alluvial deposits, loess, drift, till, marine clay, and so on. Such a system is not as accurate as that used for consolidated materials. Since a wide variety of materials can occur as alluvium or till, it is necessary to modify these terms not only with an indication of their texture and uniformity, but also, when

possible, with an indication of the type of rock from which they were derived.

Probably the most important characteristic of parent materials is their wide variability in composition within a short distance. Unconsolidated materials can vary greatly in texture, degree of sorting, and petrography in the space of a few meters, either horizontally or vertically. Even consolidated materials may have as much as a five-fold variation in one or more constituents within the space of a few centimeters. This is particularly true of the accessory constituents, which, although they make up only a small portion of the material, may markedly influence the chemical and physical properties of the rock.

CLIMATE IS A MAJOR DRIVER OF MINERAL WEATHERING

Climate is generally considered the most important single factor influencing soil formation (van Breemen and Buurman, 2002) and it tends to dominate the soil-forming processes in most forested areas. It directly affects weathering through the influence of temperature and moisture on the rate of physical and chemical processes.

Physical weathering plays an especially important role during the early stages of soil development, particularly in the degradation of parent materials. Changes in temperature result in rock disintegration by producing differential expansion and contraction. Changes in temperature also may convert water to ice. Since water reaches its maximum density at about 4°C, its volume increases as the temperature falls below that point. The maximum expansion is about 9%, and forces as great as 2.1×10^5 kPa may result. These forces can split rocks apart, wedge rocks upward in the soil, and heave and churn soil materials.

Chemical weathering can manifest itself in many ways. Initially, the principal agent is percolating rainwater charged with carbon dioxide dissolved from the atmosphere. Calcium and magnesium carbonates and other rock minerals such as feldspars and micas are affected. These latter minerals are hydrolyzed by the acid solution to produce clay minerals and to release cations, some of which

Table 3.2   Generalized classification of the major soil-forming rock types.

Rock Type	Petrography
	Ultrabasic
Peridotite	Coarse-grained igneous: olivine, augite. hornblende, plagioclase, biotite, magnetite
Serpentine	Coarse to medium-grained igneous: serpentine with accessory magnetite
	Basic
Basalt	Fine-grained, sometimes porphyritic. igneous: plagioclase, augite, olivine, magnetite, apatite
Dolerite	Medium-grained igneous: plagioclase (labradorite), augite, olivine, hornblende, biotite, magnetite, apatite, quartz
Gabbro	Medium- to coarse-grained igneous: plagioclase, augite. olivine, accessory hornblende, biotite, magnetite
	Intermediate
Andesite	Glassy to fine-grained igneous: plagioclase with feldspar phenocrysts, augite, hornblende, biotite
Diorite	Medium-grained igneous: plagioclase (labradorite). augite, olivine, hornblende, biotite, accessory magnetite, apatite, quartz
Syenite	Coarse- to medium grained igneous: orthoclase, hornblende, augite, biotite, accessory quartz, zircon, apatite, magnetite
Trachytes	Fine-grained igneous: orthoclase with phenocrysts of biotite, augile, hornblende, olivine, accessory apatite, magnetite, zircon
	Acidic
Rhyolite	Glassy to fine-grained igneous: orthoclase. quartz, orthoclase phenocrysts, accessory biotite, hornblende, apatite, zircon
Granite	Coarse-grained igneous: quartz, orthoclase, plagioclase, muscovite, biotile, accessory epidote, augite, apatite, zircon, magnetite
Aplite	Fine-grained granite variety
Quartz porphyry	Macrocrystalline igneous: quartz and feldspar with phenocrysts of quartz, feldspar, muscovite, biotite, augite
Pegmatite	Very-coarse-grained igneous: orthoclase (microcline). quartz, muscovite, garnet
Obsidian	Volcanic glass of granitic composition
Gneiss	Coarse-grained metamorphic: orthoclase, quartz, biotite, muscovite, hornblende
Arkose	Coarse-grained sedimentary: sandstone derived from granite and high in feldspar
Siliceous sandstone	Medium- to coarse-grained sedimentary: quartz, feldspar, accessory biotite, Muscovite, hornblende, magnetite, zircon
Acidic conglomerate	Very-coarse-grained sedimentary; quartz and feldspar in a siliceous matrix
	Extremely Acidic
Quartzite	Medium- to fine-grained metamorphic: quartz, accessory muscovite, magnetite
Ferruginous sandstone	Medium- to fine-grained sedimentary: quartz, accessory mica, hornblende, magnetite, zircon
Schist	Medium-grained metamorphic: quartz, feldspar, mica, garnet
Mica-schist	Medium- to fine-grained foliated metamorphic:biotite. quartz, muscovite, accessory epidote, hornblende, garnet
Shale	Fine-grained sedimentary: quartz, other highly variable constituents
Slate	Metamorphosed shale: quartz, accessory muscovite, chlorite
	Carbonaceous
Limestone	Crystalline, oolitic, or earthy sedimentary: calcium carbonate, other highly variable constituents
Marble	Metamorphosed limestone
Chalk	Soft white limestone of foraminifera remains
Dolomite	Medium to coarse crystalline sedimentary: equal parts calcium and magnesium carbonate
Calcareous sandstone	Medium- to fine-grained sedimentary: quartz, calcium carbonate, many accessories
Calcareous conglomerate	Very coarse sedimentary: quartz, feldspar, many others in a calcareous matrix

become attached to the clay particles. Resistant materials such as quartz are scarcely affected and tend to accumulate in the soil as sand and silt particles, while the water-soluble carbonates tend to be dissolved and removed.

Leaching of the readily soluble components continually depletes the surface soil of chlorides, sulfates, carbonates, and basic cations. On balance, this produces more and more acidic conditions. The degree of this impoverishment is controlled by the intensity of leaching from water passing through the soil, the speed of the return of materials to the soil surface via organic remains, and the weathering of rock fragments. In most forested regions there is a positive rainfall balance (after subtraction of transpiration, evaporation, and run-off losses), resulting in percolation, the formation of acid surface soils, and the absence of any accumulated carbonates in a subsoil horizon.

BIOLOGY ALSO INFLUENCES WEATHERING

Silicate minerals can be altered by biological systems. For example, tree roots and microorganisms can alter mica by causing the release of potassium and other ions from the mineral lattices (Boyle and Voigt, 1973). This process is a combination of the replacement of the interlayer K^+ and structural multivalent ions in the mica by H^+ ions from organic acids and the subsequent chelation of these released ions by organic acids (Boyle et al., 1974).

Many large organic molecules form polydentate ligands and are called chelates, from the Greek word for "claw." Such molecules absorb or chelate cations and are important not only in mineral weathering but also in the translocation of multivalent ions, including transition metals such as copper and iron, through the soil.

SOILS DEVELOP ACROSS TOPOGRAPHY OVER TIME

Topography can have a profound local influence on soil development, over-shadowing that of climate

(Daniels and Hammer, 1992). Soil relief affects development mainly through its influence on soil moisture and temperature regimes and on leaching and erosion. For example, in most coastal plains, small variations in elevation can have a pronounced effect on drainage. Water in soils with restricted drainage often becomes stagnant, and microorganisms and plant roots use dissolved oxygen faster than it can be restored. Under the ensuing anaerobic conditions, iron, aluminum, manganese, and some other heavy metals are reduced to a more soluble condition. Under reducing conditions, these metals move in the soil solution until they eventually oxidize and precipitate. Such precipitation produces gleyed conditions that are characterized by a gray to grayish-brown matrix sprinkled with yellow, brown, or red mottles or concretions in the B horizon. Subsoil colors are generally reliable guides to drainage conditions, ranging from dull bluish-gray for very poorly drained, bright yellows for reduced locations and reds for better-drained areas.

The genesis of a soil begins when a major event initiates a new cycle of soil development. The length of time required for a soil to develop, or to reach equilibrium with its environment, depends on its parent material, climatic conditions, living organisms, and topography. Weakly podzolized layers developed under hemlock within 100 years after fields in New England were removed from cultivation. These layers have also formed within relatively short periods in local depressions resulting from tree uprootings in hemlock forests. Recognizable mineral soil development under black spruce, and perhaps red spruce and balsam fir, can also be fairly rapid (Lyford, 1952). Buol et al. (2003) point out instances in which forest soils on glacial moraines in Alaska developed a litter layer in 15 years, a brownish A horizon in silt loam in 250 years, and a thick Spodosol (Podzol) profile in 1000 years.

Soil development from bare rock to muskeg took about 2000 years at one location in Alaska. The time required to develop modal Hapludalf (gray-brown podzolic) soil was estimated to be more than 1000 years in the northeastern United States. Development of distinctive profile characteristics can be extremely slow in arid regions and in some local areas where drainage or other conditions may slow the weathering processes. In short, there is no absolute timescale for soil development.

BIOLOGY ALSO DRIVES SOIL FORMATION

The roles of biota in soil development can hardly be overemphasized. Tree roots grow into fissures and aid in the breakdown of bedrock. They penetrate some compacted layers and improve aeration, soil structure, water infiltration and retention, and nutrient-supplying capacity. In addition to tree roots, a multitude of living organisms in the soil are responsible for organic matter transformations, translocations, decomposition, accretion of nitrogen, and structural stability. Bacteria and fungi are intermediaries in many of the chemical reactions in the soil, and many animals play a vital role by physically mixing soil constituents. Furthermore, the soil may be protected from erosion by a vegetative ground cover. A forest cover can significantly modify the temperature and moisture conditions of the soil below by its influence on the amounts of light and water that reach the soil surface, by a reduction in runoff and an increase in percolation, and by an increase in water loss as a result of evapotranspiration. Roots penetrate deep into the profile, extract bases, and return them to the surface in litter fall. The litter of many deciduous species that normally occur on base-rich soils decomposes rather rapidly, resulting in base-enriched humus. On the other hand, the litter of most conifers that normally occur on base-poor soils is rather resistant to decay and may encourage the process of podzolization. The composition of leachates produced beneath various plant covers can have a profound effect on soil formation. The presence of a thin, continuous Bh horizon below the E horizon on some wind-throw mounds but not on others in a disturbed forest probably is caused by the particular vegetation growing on or near those mounds (Lyford and MacLean, 1966).

Pyatt (1970) outlined examples of changes in soil processes that appear to have resulted from changes in vegetation brought about by human activity. One example is the increase in soil wetness resulting from deforestation during Bronze Age times in areas of Great Britain. Increased soil moisture apparently was responsible for the development of gleying within formerly well-drained soils on the North York Moors. A similar process commenced during medieval times with the spread of heath vegetation at the expense of woodland on some sites in south Wales.

Because of the close relationship between soils and plant communities and the influence of each on the formation of the other, an examination of the process of vegetation development should promote understanding of the influence of the soil on these processes.

VEGETATION AND SOILS DEVELOP TOGETHER, BUT AT DIFFERENT RATES

The Earth was formed more than 4.5 billion years ago, and as recently as 200 million years ago the major land masses were all joined in a single supercontinent. The supercontinent broke into fragments that largely define today's continents, and the fragments began slow, ponderous voyages across the planet (Matthews, 1973). Although North America and Europe are reported to be moving apart by a distance equal to a man's height during his lifetime, the two continents may have been joined in the north as recently as ~5 million years ago, with the Appalachian and Scandia mountains forming a continuous range.

Giant glaciers scoured the northern part of North America and the whole of northwest Europe as recently as 10 000 to 20 000 years ago. The vegetation of many of the warmer areas of the world has apparently had a rather constant composition for millennia. Revegetation of the abandoned ice fields and coastal areas inundated during intraglacial periods has taken place relatively recently.

Tree species that evolved and simultaneously colonized the European and North American continents while these two land masses were joined in the pre-Cretaceous period were largely eliminated by the repeated advances of ice sheets during the Pleistocene. It appears that while most species could continue to retreat to the south in advance of the North American glaciers, they were not so fortunate in Europe and Asia. Trapped between the Arctic ice mass in the north and the glaciers advancing from the several mountain ranges in the south, many species were apparently eliminated. These events help explain why the flora of much of Europe and Asia is simpler in species composition than that of North America and why Eurasia and North America now have few indigenous plant species in common.

Forests change over time, either so slowly we may not notice for decades, or so quickly that we're surprised. Some changes are relatively predictable, for example, the number of trees that fit into a hectare inevitably declines as average tree sizes get larger. Other changes depend on contingent events, such as droughts that combine with fires or beetles to restructure species composition across landscapes.

Parent material may dictate the plants that initially colonize an area, and it can affect the species composition of stable communities. This is quite evident in glaciated boreal regions. Finer-textured deposits are often imperfectly drained and are invaded by larch and alders. Over time, internal drainage improves as soil structure develops, and these areas may (or may not) become dominated by spruce and fir trees. In the same region, coarser deposits are initially occupied by open stands of aspen and birch or pine and shrubs. Eventually, all but the coarsest of these deposits develop a finer-textured B horizon and gradually become stable spruce–fir forests. However, the coarsest sand may remain in a series of pine forests.

Plant species differ in their effects on soils (see Chapter 11), so shifts in species composition over time can change soils (Fisher and Stone, 1969; Fisher and Eastburn, 1974). Abandoned agricultural fields have low levels of extractable nitrogen and phosphorus. Pines tolerate this poor nutritional status, while the new microflora and fauna associated with the conifers rapidly decompose soil organic matter and increase the levels of extractable nitrogen and phosphorus. The more demanding hardwoods then invade these more hospitable sites. Another change in microflora and fauna occurs, the temporarily degraded soils again accumulate organic matter, and the darkened surface soil thickens.

The study of succession has concentrated primarily on plant communities and only rarely on animal communities. It is clear that soil undergoes successive changes as great as, if not greater than, those that occur above ground, although these changes may occur on a somewhat different timescale. The changes are not as easily observed as those in vegetation and have received little attention. Adding these soil dynamics to our consideration of succession might help us to understand this complex phenomenon better and give new direction to the continuing debate on succession.

SOIL PROPERTIES INFLUENCE VEGETATION DEVELOPMENT

Five easily observable properties of soil (texture, structure, color, depth, and stoniness) can be used to infer a great deal about how a particular soil influences plant growth. Texture, whether the soil is a sandy loam or a clay loam for example, in large measure determines the moisture and nutrient-holding capacity of the soil. Structure, whether the soil is massive, single grain, or blocky, modifies the influence of texture on water-holding capacity and its corollary, oxygen availability. Color indicates the amount of organic matter in the soil, which is generally well correlated with fertility. Depth tells us how much water can be held and what volume of soil is exploitable for nutrients, and stoniness tells us how much non-soil objects dilute soil volume. If we know something about the climate, these properties can go a long way toward allowing us to determine the type and vigor of the plant community that any given soil might support. Many of these properties are strongly influenced by parent material but they change during soil development, and the ability of an area to support vegetation changes as the soil matures.

The type of parent material from which a soil is derived can influence its base status and nutrient level. Changes in vegetation type have been observed to coincide with changes in underlying parent rock. For example, soils derived from sandstone are generally acidic and coarse-textured, producing conditions under which pines and some other conifers have a great competitive advantage. On the other hand, limestone may weather under the same climatic conditions into finer-textured, fertile soils that support demanding hardwoods and such conifers as red cedar. In the Appalachian Mountains, it is not unusual to find a band of pine growing on sandstone-derived soils parallel to a strip of hardwoods growing on limestone soils. Although not all limestone-derived soils are productive, this parent material exerts an important influence on the distribution and growth of forest trees. Differential chemical weathering of the same parent material can also influence the distribution and growth of forest vegetation because of the changes in acidity, base status, and nutrient availability associated with the intensity of weathering.

Vegetation development on soils derived from transported materials may differ from that on adjacent soils derived from bedrock. In fact, the mode of transport may greatly influence the particle-size distribution (texture) of the soil. Wind-deposited soils are likely to be of finer texture than those deposited by running water. Outwash sands laid down by flowing water from melting glaciers are generally coarser-textured than the glacial till soils formed by the grinding action of advancing glaciers. The latter soils are generally more fertile and support hardwoods, while the former often support pine forests. Soils with high silt and clay content may offer more resistance to root penetration than coarser soils, thus excluding trees intolerant of drought or of low oxygen conditions in the root zone. On the other hand, sandy soils hold less water and nutrients but more air than clays. The rate of movement through a soil varies inversely with soil texture because the minute interstitial spaces in fine-textured soils offer resistance to water movement. While finer-textured soils hold more water per unit volume of soil, they also lose more water by direct evaporation than do sandy soils.

CHAPTER 4
Soil Organic Matter

OVERVIEW

Soil organic matter forms a small fraction of forest soils, but has profound impacts on the physical and chemical properties of the whole soil. Soil organic matter plays important roles in maintaining site productivity and is a major sink for atmospheric carbon; it is also the principal source and sink of plant nutrients in both natural and managed forests. The O horizon is the most visually distinctive feature of a forest soil, and it sometimes contains the major portion of the soil's organic matter. These O horizons have distinctive structure and are generally classified by their structure. They contribute significantly to the hydrologic and nutrient cycles of the site. Soil organic matter is the fuel that drives the biological engine that is at the heart of most soil processes. Although the O horizon is an important feature of forest soils, the organic portion of the mineral soil also influences most soil processes. It is a principal source of nutrients, is vitally important in determining soil structure, bulk density, and hydraulic conductivity, and contributes the majority of the ion exchange capacity in many forest soils. The amount of carbon stored below ground as soil organic matter may roughly match that stored above ground in the forest vegetation. The average turnover time for soil organic matter may be shorter than the lifespan of dominant trees in the forest, yet the average age of the carbon that remains behind in soils is usually much older than the trees. This seeming paradox results from the combination of a very large flow of carbon through the soil (relatively rapidly), and a very small proportion of the flow that accumulates in long-term, stabilized soil organic matter.

Soil organic matter plays fundamental roles in soils, comprising the energy source for most biogeochemical reactions, a binding agent that fosters soil aggregation and structure, and a reservoir for storing water and exchangeable ions. Carbon comprises about 45% of the mass of soil organic matter, and soils are one of the most important pools in the global C budget (see Chapter 15). Soil organic matter comprises the majority of the O horizon (forest floor) of soils, often constituting 1 to 15% of upper mineral soil (and B) horizons.

A focus on soil organic matter and its role in soil productivity is important because of the dynamics of its inputs, outputs, and net accumulation or loss in response to climate, topography, vegetation, and disturbance regime. We generally measure soil organic matter by a one-time determination of soil organic carbon, but this is akin to determining a person's wealth by looking at his assets and disregarding his liabilities. Ecosystems produce new organic matter and decompose old organic matter every year. When these inputs and outputs stabilize, soil organic matter also stabilizes, but if we decrease inputs or increase outputs through cultural activities, we alter the balance and soil organic matter may change dramatically. This could, in turn, decrease productive potential through a complex and poorly understood chain of events; the exact role of soil organic matter in soil productivity declines is unknown.

Soil organic matter consists of all of the carbon-containing substances in the soil, except inorganic carbonates. It is a mixture of plant and animal residues in various stages of decomposition, the bodies of living and dead microorganisms, and substances synthesized from breakdown products of all of these (Schnitzer, 1991). Soil organic matter occurs in solid, colloidal, and soluble states, in

Ecology and Management of Forest Soils, Fourth Edition. Dan Binkley and Richard F. Fisher.
© 2013 John Wiley & Sons, Ltd. Published 2013 by John Wiley & Sons, Ltd.

relatively distinct O horizons at the top of the soil and mixed intimately with mineral horizons.

THE HIGHEST SOIL ORGANIC MATTER MAY BE IN THE CANOPY

Rainforests around the world characteristically support large communities of epiphytic plants (and animals and microbes), and the accumulation of dead organic matter on large branches can proceed over centuries, leading to well-developed Histosols high above the soil surface (Nadkarni *et al.*, 2002). Canopy organic matter can develop to depths of a meter in some situations, with well-developed profiles grading from fibrous remnants of plant material to humified sapric material (Enloe *et al.*, 2006). Although most forest soils begin on the ground, it is fascinating to remember that soil-forming processes occur anywhere in which conditions are suitable.

FOREST SOIL PROFILES TYPICALLY BEGIN WITH AN O HORIZON

The O horizon is a distinctive feature of most forest soils, often accounting for 20% or more of total soil C. Typical forest O horizons have masses of 1 to $4 \, kg \, m^{-2}$ (10 to $40 \, Mg \, ha^{-1}$; see Figure 4.1), though

it's not unusual for O horizons to accumulate more than $5 \, kg \, m^{-2}$ in cold or wet locations. In general, O horizons are deepest where environments are cool and wet and where mixing by soil fauna is slight. Tropical forests (both plantations and natural forests) tend to accumulate 5 to $15 \, Mg \, ha^{-1}$, equal to about one to two years of aboveground litterfall inputs (reviewed by O'Connell and Sankaran, 1997). Temperate forests commonly accumulate 20 to $100 \, Mg \, ha^{-1}$ of O horizon. The O horizon may include substantial quantities of partially decomposed wood, ranging from less than $20 \, Mg \, ha^{-1}$ in tropical forests and areas subject to frequent fires to more than $500 \, Mg \, ha^{-1}$ for some cool rainforests with large, long-lived trees.

Grassland soils typically have minimal O horizons, and many ecologists simply refer to grassland O horizons as the litter layer. Wetland and tundra soils also tend to have large O horizons; indeed, many wetland and tundra soils are classified as organic soils, rather than simply possessing an upper O horizon like forests.

O horizons have had a bumpy time in the history of soil science and ecology. Much of the classification and nomenclature of soils was developed to provide insights into agricultural soils, where plowing prevents the normal development of O horizons. A 2006 FAO report (UNFAO/ISRIC, 2006) that aimed to foster communication and understanding about soils of the world essentially omitted

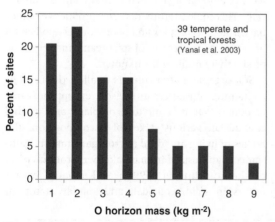
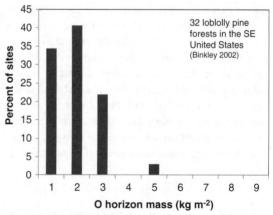

Figure 4.1 The mass of O horizons commonly varies from 1 to $4 \, kg \, m^{-2}$ (left). Even within a single forest type, O horizon mass varies several-fold (right) as a result of differences in climate, parent material, and soil animal activity.

consideration of the sorts of O horizons found in forests. When Berthrong *et al.* (2009) set out to examine the effects of afforestation on soil C accumulation, they found that more than 90% of published studies omitted the O horizon!

At the other extreme, some forest scientists have focused so heavily on O horizons that classification schemes may outstrip our understanding of the horizon's formation, structure, and function (see, for example, the "order, group, subgroup" approach of Klinka *et al.*, 1981 or the nine humus forms defined for French soils by Ponge *et al.*, 2011). Forest scientists sometimes refer to the O horizon as humus or the humus form. These terms are confusing because humus is generally defined as only a portion of the O horizon and also because humus or humic materials exist within the mineral soil (Stevenson, 1994; Piccolo, 1996). Similarly, the term "forest floor" has been used, but this has led many non-soil scientists to think of the O horizon as not being part of the soil.

The O horizon designates all organic matter that might pass beneath our feet as we walk through a forest, down to the depth where mineral material dominates (often defined as < 20% or 30% organic matter by weight). This includes freshly fallen plant litter (perhaps even tree boles), partially decomposed litter, and humic materials that were generated by microbes as by-products of decomposition. The concentration of mineral material in O horizons is typically low, except where physical mixing (by soil animals or treefall) is important. These organic matter layers and their characteristic microflora and fauna are a dynamic portion of the forest environment and the most important criterion distinguishing forest soils from agricultural soils.

The O horizon is a zone in which vast quantities of plant and animal remains are oxidized to CO_2, and humified into long-lasting humic compounds and structure. During the decomposition process, the recently added litter as well as more processed soil organic matter serves as a source of carbon for successive generations of organisms. The O horizon provides a source of food and a habitat for myriad microflora and fauna. Their activity is essential to the maintenance of nutrient cycles, particularly those of nitrogen, phosphorus, and sulfur.

The removal of forest litter for use as animal bedding in Germany prompted concern over site degradation and led to Ebermayer's (1876) classical work on litter production.

O HORIZONS HAVE A DISTINCTIVE STRUCTURE

Early forest scientists in Europe tended to see two or more major varieties of O horizons (Handley, 1954). Muller (1879) proposed the first generally adopted classification of O horizons based on field experiences in Denmark. Mor types had distinct O horizons resting atop a clear boundary with the A horizon (Figure 4.2). Mull types had O horizons that blended gradually into the upper mineral soil (A horizon), typically as a result of massive mixing of horizons by worms and other soil animals. An intermediate moder type had substantial signs of animal activity (including feces), but less mineral material mixed in than mull types.

Heiberg and Chandler (1941), Handley (1954), and Remezov and Pogrebnyak (1969) have given historical accounts of the evolution of terminology and classification of O horizon types, pointing out

Figure 4.2 In some cases, mor O horizons are woven so tightly together by tree roots and fungal mycelia that the entire horizon can be lifted intact from the soil, as in this northern hardwoods forest in northern Michigan, USA.

some of the variations in nomenclature used in various parts of the world. For example, the terms raw humus, peat, acid humus, duff, and mor have been used to describe one type of O horizon, while mild humus, leaf mold, and mull have all been used to designate another type.

Some general conclusions can be drawn concerning the origin of the distinctive features of mor and mull O horizons, but probably none of them holds for all soil conditions and plant communities. For example, mull O horizons are generally formed under hardwood forests and under forests growing on soils well supplied with base cations. Mor O horizons are most often found under coniferous forests and heath plants often growing on spodic (podzolic) soils. Mulls and mors, however, are by no means found exclusively under these forest types or on these soils (Romell, 1935). In fact, other factors in addition to the type of litter also influence the development of the O horizon. Climate has an influence through its effect on soil and vegetation development. The fragmentation, decomposition, and mixing brought about by organisms associated with the litter layer have a powerful effect on humus development. The invasion of exotic earthworms is converting the tightly woven mor layer in Figure 4.2 into a mull layer in just a decade or so.

Fossorial (burrowing) mammals are among Nature's other agents operating to alter O horizons, including gophers, moles, and shrews. These animals often pile soil around burrow entrances and move and mix soil when making tunnels (Troedsson and Lyford, 1973). Abandoned runways eventually collapse or become filled with soil material from above. Transport of mineral soil particles into the organic horizons may also occur as a result of the activity of ants, termites, earthworms, rodents, and other small animals. Where the mineral soil materials are deposited by these fauna, soil properties are likely to show sudden changes. The reverse process, transport of organic matter into the mineral soil, also results from animal activity.

Forest soil O horizons are often divided into three structural layers, progressing from less processed litter at the top to highly humified material at the bottom of the O (or in the transition from the O to the A horizon). These strata may not occur in all

soils (Hesselman, 1926; Green *et al.*, 1993). The upper layer was classically called the litter or L layer, and is currently referred to as the Oi horizon in the USA, or OL horizon in Europe (Zanella *et al.*, 2011) and other parts of the world (UNFAO, 1990). The Oi or OL horizon consists of unaltered dead remains of plants and animals whose origin can be readily identified. The middle layer is the "Formultning" (decay) layer with partly decomposed organic materials that are sufficiently well preserved to permit identification of their origin (Prescott *et al.*, 2000). This horizon is termed the Oe in the USA, and the OF in most other places. The lowermost layer contains material that is not easily identified by original litter structure; highly humified material gives the layer the name humus (though this sometimes is used to refer to entire O horizons), the H-layer, the Oa horizon in the USA, and the OH horizon in much of the rest of the world.

These structural differences correspond to a wide variety of difference in biotic activity and rates of biological processes that result in different rates of organic matter turnover. Franklin *et al.* (2003) used isotopic techniques to assess the mean age of O horizon layers in a Scots pine forest in northern Sweden. The OL had an average age of 5.2 years, the OF 13.0 years, and the OH 38.4 years. Summing up these layers and accounting for their masses led to an overall average age of the O horizon of 17.1 years.

MANY O HORIZON CLASSIFICATIONS HAVE BEEN DEVELOPED: ARE THEY USEFUL?

As with most natural systems, people have tried to develop classification systems to describe and account for differences in O horizons among forests. The variety of forms and processes in O horizons prevents any simple, unequivocal classification, so many approaches have been developed. It's not clear that classification efforts have led to clear insights that might, for example, provide an index of soil fertility or guide forest management. For example, Romell and Heiberg (1931), as revised by Heiberg and Chandler (1941), described humus layer types for northeastern United States upland

Figure 4.3 Thick mor humus in a spruce stand in central Norway. Note the gray, leached E layer over a spodic horizon.

soils. They defined mull as a humus layer consisting of mixed organic and mineral matter in which the transition to a lower horizon is not sharp. They divided mull humus into (1) coarse, (2) medium, (3) fine, (4) firm, and (5) twin types. They described mor humus as a "layer of unincorporated organic material usually matted or compacted or both, distinctly delimited from the mineral soil unless the latter has been blackened by the washing in of organic matter." They divided mor humus into (1) matted, (2) laminated, (3) granular, (4) greasy, and (5) fibrous types on the basis of morphological properties. Examples of mor humus are shown in Figures 4.2 and 4.3. However, while this classification was adequate for the northeastern United States, some O horizons in other regions were not clearly identified under this system.

Klinka *et al.* (1981) applied a species-taxonomy approach to classification of O horizons, defining orders of mor, moder, and mull, with 19 groups (such as hemimor and hemihumimor), and 69 subgroups (such as amphihemihumimor and microvelomoder). Green *et al.* (1993) proposed a comprehensive classification of organic and organic enriched mineral horizons. They defined three taxa: mor, moder, and mull at the order level based on the type of F horizon and the relative prominence of the Ah horizon. Below this order level are 16 taxa at the group level, reflecting differences in the nature and rate of the decomposition process. The latest foray into O horizon classification comes from Europe, with the proposition of six major types: Anmoor,

Mull, Moder, Mor, Amphi, and Tangel (Zanella *et al.*, 2011). This scheme accounts for variations that depend on soil features outside the O horizon. For example, the same sort of a mull horizon would be called a Terromull if it occurred on a well-drained soil; an Entimull if lying directly on bedrock; or a Paramull if plant roots were particularly dominant.

What practical insights might be supported by characterization of the structure of O horizons? One of the most important is the challenge presented in determining the mass (and element content) of the O horizon. Mor-type O horizons have a distinct lower boundary, and typical mass estimates would be reliable to about ± 20% (Yanai *et al.*, 2003b). O horizons with less distinct lower boundaries may be difficult to quantify with less than ± 50% uncertainty, and this may cloud discussions of rates of change over time (Yanai *et al.*, 2003a).

Another practical insight is that the acidity of O horizons tends to relate to the degree of incorporation of mineral material. Organic matter tends to be more strongly acidic than mineral surfaces, so mor-type O horizons may have lower pH (= more strongly acidic) than mull-type O horizons. Tree species and sites with acidic litter (and lower calcium) may have lower activity of soil animals (Reich *et al.*, 2005), leading to circular effects of O horizon chemistry, morphology, and mixing by animals.

Perhaps the most interesting aspect of O horizon structure would simply be the indication of the importance of certain types of soil organisms. Mull-type structures form as a result of mixing by soil animals (particularly worms), and macrofaunal biomasses may be three to five times greater than found in nearby soils with mor-type structure (cf. Staaf, 1987; Schaefer and Schauermann, 1990). Moder-form O horizons may result from activities of epigeic worms or other soil fauna such as millipedes which do not mix the organic material with mineral soil but rather ingest it (once it has become partially decayed and invaded by fungi). About 90% of the initial mass consumed is subsequently egested and may persist as a layer of fecal pellets, which have variable decay rates (C. Prescott, personal communication).

Large woody material is often a major pool of organic matter in forests. There seems to be no convention about whether large logs should be classified as part of the O horizon or as something distinct. The rate of organic matter input to the soil from fallen logs typically rivals the rate of leaf litter-fall (Sollins, 1982), though the influence of decaying logs on underlying soil organic matter appears limited (Spears *et al.*, 2003; Krzyszowska-Waitkus *et al.*, 2006). Woody material also contains large (slowly available) pools of nutrients; less than 5% of the annual soil N supply to trees likely comes from coarse woody material (Hart, 1999; Laiho and Prescott, 1999). In some cases, free-living biological nitrogen fixation is detectable in woody material (Jurgensen *et al.*, 1984; Hicks *et al.*, 2003), though the rates are very small compared to annual N flux in forests.

Much of the interest in O horizon morphology was driven implicitly or explicitly by hopes that structure would provide a key to important features such as tree nutrition and site productivity. This expectation has not been borne out; no universal pattern of O horizon morphology or decomposition has proven to be of much use in forest nutrition studies, probably because so many important interactions (of tree species, soil communities, and a suite of soil-forming factors) prevent simple patterns. Old expectations of mull O horizons (where mixing by soil animals comminutes the organic matter and buries it) being somehow richer than mor O horizons (where organic matter remains separate from mineral soil) have been supported only in the broadest terms: the richest forest soils have animals that mix litter into the mineral soil, and the poorest forest soils do not. Between these extremes, many examples have been found in which one O horizon type appears to be more fertile than another, with other cases showing the reverse pattern. For example, the supply of phosphorus to sugar maple trees is lower on mull soils than mor soils in Quebec (Figure 4.4), in contrast to classic expectations of higher fertility of mull soils. We suggest that after almost a century, these hopeful (but unsupported) ideas about litter decomposition rates and O horizon morphology should be set aside and the processes that determine the turnover of soil organic matter should be examined in both O horizons and mineral horizons, especially

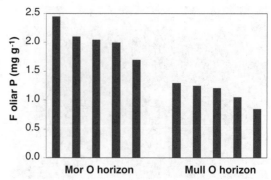

Figure 4.4 Sugar maple phosphorus deficiency develops in mull O horizons because mixing of organic phosphorus pools with mineral soil allows phosphorus to precipitate with iron and aluminum (data from Paré and Bernier, 1989a; 1989b).

recognizing potentially dominant influences of belowground sources of plant litter (cf. Pollierer *et al.*, 2007).

ORGANIC MATTER IS A VITAL PART OF MINERAL HORIZONS TOO

Identifiable pieces of plant material comprise only a portion of soil organic matter. The largest and most important fractions of soil organic matter are the colloidal and soluble fractions (McColl and Gressel, 1995). The majority of soil organic matter is insoluble and is bound as macromolecular complexes with calcium, iron, or aluminum or in organomineral complexes (Stevenson, 1994). Organic matter in mineral soils "knits" together the small mineral particles into larger soil aggregates (especially in soils with 2 : 1 clays; Six *et al.*, 2000), with far-ranging importance for soil porosity, aeration, and water infiltration, flow and retention.

The organic portion of these complexes is made up of diverse complex compounds, classically lumped into the term "humus." The huge variety of chemical compounds within humus has led to intensive efforts to fractionate the bulk pool into subpools that differ in chemical composition, reactivity, and turnover rates.

The most common characterization of humus in the twentieth century entailed separating fractions that were soluble in strong acids, weak acids, or neither. The defined humic acid fraction is soluble only in strong acid (pH < 2) conditions. Fulvic acids dissolve under weakly (or strongly!) acid conditions; and the acid-insoluble fraction is called humin (MacCarthy *et al.*, 1990). A greater proportion of the more soluble fulvic acids characterizes the humus of forest soils, whereas the humus of peats, grassland, and agricultural soils contains a greater proportion of the less soluble humic acids. The water (non-acid) soluble soil organic matter is a small proportion of the total soil organic matter. It is composed of amino, fatty, and nucleic acids, carbohydrates, polyphenols, and organic acids. In order of increasing resistance to decay, the organic constituents of the soil are carbohydrates (sugars, starches, hemicellulose, cellulose), proteins and amino acids, nucleic acids, lipids (fatty acids, waxes, oils), lignins, and humus. Lignins are complex, highly aromatic polymers that are difficult to describe and can form a substantial portion of humus. Along with lipids, they are insoluble in water and are resistant (but not impervious) to microbial decomposition (Tan, 1993). Some humus molecules remain in the soil for hundreds or thousands of years (Paul, 2007).

This classic chemical fractionation of soil organic matter based on solubility in strong acids and bases may not have been as good an idea as it once seemed. Some have argued that dissolving organic compounds in strong solutions can lead to the novel synthesis of huge molecular chains; essentially the large molecules of "humus" may be synthesized as a product of the laboratory techniques (e.g. Schmidt *et al.*, 2011). These perspectives have the potential to fundamentally change some of the core concepts of soil organic matter.

Another approach to fractionating soil organic matter is based on divisions of heavy and light fractions, or active, passive, and slow fractions. The active fraction consists of materials with relatively low carbon:nitrogen ratios and short half-lives. This fraction includes microbial biomass, some particulate organic matter, polysaccharides, and other labile organic compounds. The turnover time of the active fraction is rapid and can be measured in months.

The passive fraction of soil organic matter is made up of very stable compounds with extremely long half-lives. This fraction includes organomineral complexes, most of the humin, and much of the humic acid portion of the humus. This fraction accounts for 80–90% of the organic matter in most forest soils, lending stability to the soil and accounting for most of the cation exchange capacity and water-holding capacity role of soil organic matter.

The slow fraction of soil organic matter is intermediate between the active and passive fractions. This fraction includes some particulate organic matter with high carbon:nitrogen ratios and high levels of lignin or polyphenolics, as well as other chemically resistant compounds. The half-lives of the components in this fraction may be measured in decades, but this fraction is still an important source of substrate for soil microbes and mineralizable nitrogen.

These generalizations are the topic of ongoing research and modeling. In many cases it's difficult to assess how much a generalization reflects strong inference based on evidence, and how much it rests on interactions of assumptions. It may be useful to model the dynamics of soil organic matter by using fast, slow, and passive pools with unique turnover rates tuned for each application (for an interesting example, see Johnson *et al.*, 2009a), but a handy approach to modeling is not strong evidence that soil organic matter actually sits in relatively discrete pools, or that turnover rates have typical values (with low variances).

More recent approaches to examining the nature of humified materials may not separate the material; instead, they may simply characterize the relative frequencies of major types of C bonds. Fourier transform infrared (FTIR) spectroscopy uses infrared light to affect the vibrational frequencies of chemical bonds, and different types of bonds resonate differently. Oak litter from a calcareous site in Israel showed increasing absorbance with depth of soil sampling for phenolic OH^- groups, indicating greater humification of deeper material (McColl and Gressel, 1995).

Nuclear magnetic resonance (NMR) is a type of spectroscopy that depends on characteristic spin frequencies in an oscillating magnetic field. The

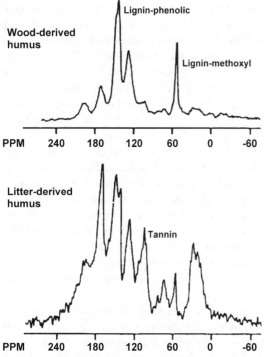

Figure 4.5 Dipolar-dephased ^{13}C CPMAS NMR spectra for humus (> 80% amorphous material) derived from wood and litter from a western red cedar–western hemlock forest. Differences in the origin of humic materials are evident from the high lignin-methoxyl peak for the wood-derived humus and for the tannin peak for the litter-derived humus (after deMontigny *et al.*, 1993).

terminology (and technology) is complex; the current state of the art is referred to as nondestructive solid state ^{13}CNMR with cross-polarization and magic-angle-spinning (CPMAS ^{13}CNMR), perhaps even with dipolar dephasing. The spectra produced by CPMAS ^{13}CNMR identify the major types of carbon bonds. For example, humus derived from woody litter was comprised of about 70% lignin-carbon, and 5% carbohydrate-carbon, and litter derived from leaf, twig, and root litter was 53% lignin-carbon and 13% carbohydrate-carbon (deMontigny *et al.*, 1993). The spectra for wood-derived humus lacked any peak for tannin-carbon, compared with a strong peak for tannin-carbon in the litter-derived humus (Figure 4.5).

How useful are these innovative approaches to characterizing organic matter in mineral soils? The answer to this question would be encouraging if either (a) soils showed clear, simple patterns in spectra as a function of some factor (like species or management), or (b) the spectra related clearly to articulated questions about forest soils (such as fertility gradients or responses to management). At this point the evidence typically seems to support the conclusion that these spectra indicate: "... a complicated interaction between physical protection and biochemical recalcitrance, making the land use and management induced changes of soil organic matter quality more complex" (He *et al.*, 2009). We know that organic matter in mineral soils is important, diverse in composition and dynamics, and not likely to be easily characterized in powerful ways.

CHARCOAL CAN ALSO BE IMPORTANT IN FOREST SOILS

The consumption of plant materials in fire involves complex chemistry and reactions, as the heat of the fire provides energy to break bonds in large molecules (pyrolysis), leading to volatilization of small molecules that are oxidized (burned) in the air. Unburned and partially burned particles are lofted into the air as smoke and embers, while partially burned materials are left on the soil surface or within the soil matrix. The blackened, partially burned material is called charcoal (or char), and the chemistry and fate of this material can be diverse (Knicker, 2006; Ohlson *et al.*, 2009; Nocentini *et al.*, 2010). Fresh charcoal has a high concentration of C (perhaps 80% or so), and over time the C concentration declines to values near 50%. Particle sizes range from large charred logs to fine pieces of twigs and needles. Charcoal may retain much of the chemistry of the original material, or may be almost completely reduced to pure charcoal. Some charcoal lasts for centuries or millennia, but some decomposes (as a result of physical, chemical, or biological processes). One study across boreal forests in Scandinavia found that charcoal accounted for an average of about $80\,g\,C\,m^{-2}$, with typical ages of centuries to millennia (Ohlson *et al.*, 2009). Sites with older charcoal tended to have less charcoal, indicating that charcoal does not last forever. Boreal forests in Alaska typically have 100 to $200\,g\,C\,m^{-2}$ as charcoal, with greater amounts on south-facing slopes compared to north-facing slopes. Forest soils in California, USA may have higher charcoal contents, ranging from 100 to $500\,g\,C\,m^{-2}$ (MacKenzie *et al.*, 2008). California forest soils with higher charcoal tended to have 10 to 20% less soil organic matter, consistent with some evidence which indicates that charcoal increases rates of decomposition of soil organic matter (for example, Wardle *et al.*, 2008). Charcoal accumulation can be even higher; two burned eucalyptus sites in Australia had 1 to $14\,kg\,C\,m^{-2}$ as charcoal, comprising 20 to 40% of total C in the soils (Hopmans *et al.*, 2005).

SOIL ORGANIC MATTER PERFORMS MANY FUNCTIONS

The life of the soil is carried out largely using soil organic matter as an energy source. The oxidation of organic matter fuels soil biogeochemistry, releasing carbon mostly as CO_2. The organic matter that remains behind, accumulating in the soil, is largely responsible for how "soils" differ from geochemical parent materials. The actions of soil fauna and flora enabled by soil organic matter are important in nutrient cycling, in maintenance of soil porosity, hydraulic conductivity, bulk density, and in soil detoxification processes.

The water-holding capacity of O horizons is impressive: a 5 cm thick O horizon in a subalpine forest in Alberta might have a mass of about $5\,kg\,m^{-2}$, and could retain about 10 L of water at field capacity ($-0.33\,kPa$; Golding and Stanton, 1972). In general, O horizons can absorb about 0.15–0.2 cm per centimeter of depth (Gessel and Balci, 1965; Remezov and Pogrebnyak, 1969; Wooldridge, 1970). Increasing the organic matter concentration of a sandy loam from 1% to 2% would be expected to increase the water-holding capacity by 10–15% (increasing from about $4\,L\,m^{-2}$ to 4.5 or $5.0\,L\,m^{-2}$ for the 0–20 cm mineral soil, based on trends from Rawls *et al.*, 2003). The retention of water in the O horizon results from water absorption by the organic material itself, whereas the effect of organic matter on water characteristics of mineral horizons depends primarily on enhancing soil structure (increasing aggregation and pore volume). This capacity is highest in the OH layer and lowest in the OL layer. The higher percentage of large pores in the OL and OF layers leads to increased aeration on wet sites, but it causes O horizons with these layers to dry quickly, and plant roots can frequently occupy the O horizon only temporarily. Mors generally have higher moisture-holding capacities than moders or mulls.

Soil organic matter also contains two major types of nutrient pools. Humic substances can have a net negative charge resulting from the dissociation of H^+ from hydroxyl ($-OH$), carboxylic ($-COOH$), or phenolic ($C_6H_{12}-OH$) groups. The ions held on cation exchange sites interchange readily with ions in soil solution. This dissociation is pH dependent (as these are essentially acid–base reactions), and cation exchange capacity of soil organic matter is commonly $500 - 3000\,mmol_c\,kg^{-1}$ (increasing with increasing soil pH). Silicate clays have cation exchange capacities of about 10 to $300\,mmol_c\,kg^{-1}$ for mineral soil. The relative importance of organic and mineral surfaces for cation exchange depends on both charge density ($mmol_c\,kg^{-1}$), and mass (or surface area); high mass of low-charge-density mineral surfaces indicates that mineral surfaces may account for more cation exchange capacity than organic compounds in horizons with moderate-to-low organic matter contents.

The second type of pool is simply the pool of nutrients bound in the organic molecules themselves. A typical O horizon might have 1% N, and if the rate of decomposition were $2\,kg\,m^{-2}$ annually, the rate of N release might be about $2\,g\,m^{-2}$ (a moderately large value). Dijkstra (2003) examined the rate of Ca release from organic matter decomposition in a hardwood forest in the northeastern USA, as well as the concentrations of Ca residing on exchange sites in the O horizon and 0–15 cm depth mineral soil. The annual release of Ca from decomposition added up to about $3\,g\,Ca\,m^{-2}$, compared with a static pool size of about $6\,g\,Ca\,m^{-2}$ on exchange sites in the O horizon, and $30\,g\,m^{-2}$ of exchangeable Ca in the upper mineral soil. The importance of decomposition of organic matter as a source of cation nutrients, as well as phosphate, should increase as the supply available in mineral pools declines (either across sites or over time).

Soil organic matter is important for the formation of soil aggregates (Tisdall and Oades, 1982; Kay, 1997; Six *et al.*, 2000). Organic compounds, along with fine clays and some amorphous and crystalline inorganic compounds, form the cement between mineral grains in soil aggregates. Soil organic matter contributes to aggregation and to aggregate stability. Soil organic matter is important in determining a soil's productive potential, in part through its influence on soil structure, which influences the soil's ability to receive, store, and transmit water, to support root growth, to diffuse gases, and to cycle nutrients.

O horizons physically insulate the mineral soil horizons from extremes of temperature and moisture content, offer mechanical protection from raindrop impact and erosional forces, and improve water infiltration rates. Mor and moder O horizons are also poor at conducting moisture upward and reduce evaporation from the mineral soil. Foresters and soil scientists have long been fascinated by differences in the vertical structure of O horizons, as well as differences that develop under the influence of various tree species, parent materials, and animals (Wollum, 1973).

THE CHEMISTRY OF O HORIZONS GOES BEYOND CARBON

As a general rule, mor O horizons are more acid (lower pH) than mull O horizons, but these two types often develop under wide ranges of tree species, soil types, and amount of mineral particle content. Wilde (1958) suggested that the pH of mull O horizons could vary from 3.0 to 8.0. Heyward and Barnette (1934) found the pH of the OF (or Oe) layer under longleaf pine to range from 3.4 to 5.0. These values were 0.25 to 1.0 pH units lower than those of the upper A horizon of the corresponding mineral soil. Lutz and Chandler (1946) reported average values of pH 4.3, 4.5, and 4.9 for the OL, OF, and OH layers (Oi, Oe, and Oa) layers, respectively, of mor O horizons under jack pine and 4.5, 5.9, and 6.5 for the

same layers in mull O horizons under a maple–basswood stand. The pH of OL, OF, and OH layers (Oi, Oe, and Oa) layers under spruce averaged pH 5.1, 4.9, and 4.7, respectively, while that of the corresponding layers under birch averaged 5.9, 5.7, and 5.7 in Russian forests (Remezov and Pogrebnyak, 1969). Nitrogen concentration in the O horizons varies commonly from 1.5 to 2.0% (Voigt, 1965).

The carbon:nitrogen ratio gives an indication of the availability of nitrogen in O horizons, as well as rates of decay. The carbon:nitrogen ratio of O horizons ranges from about 20 to 150. The ratio is greater in mor layers than in mull layers.

Relatively large quantities of nutrients are stored in the O horizon (Cole and Rapp, 1981). The amount and composition of the O horizon, which are influenced by the forest vegetation, climate, mineral soil, and the accumulation period following a major disturbance, dictate the total content of nutrients. In some forest soils, such as glacial outwash sands or sandy soils of coastal regions, the O horizon represents the major reserve of nutrients for tree growth. The total nutrient content of O horizons in warm regions may be only a fraction of that of O horizons in cooler areas, as the rates of decomposition and nutrient turnover are more rapid in the warm temperate forests than in cooler forests.

The chemical composition of the principal layers of the O horizon of some conifer and hardwood stands is given in Table 4.1. All of these stands had been protected from fire for extended periods. Note that the organic material under birch stands was generally higher in all nutrients than that from spruce stands in Russia. The nutrient concentration of humus from stands of northern coniferous species was generally higher than that of humus from three southern pine stands. The low concentration of nutrients in the southern pine O horizon probably reflects the low fertility of the soils of the lower coastal plain and may explain the relatively large accumulation of litter in these stands. There is a slower rate of decomposition than might be expected from the favorable temperature and moisture conditions in this area, apparently due to the low nutrient status of the litter.

The concentrations of potassium, calcium, and magnesium generally decreased from the surface

Table 4.1 Mass and element content of some O horizons.

Forest Type	Layer	Oven-Dry Mass (Mg/ha)	N	P	K	S	Ca	Mg	Al
						(kg/ha)			
Spruce	Ol	2.9	34	4	7	4	36	9	9
Russia[a]	Of	8.2	119	9	11	7	97	25	48
	Oh	10.1	133	9	9	8	97	30	110
Birch	Ol	1.1	15	2	3	1	15	4	5
Russia[a]	Of	6.1	101	13	11	13	84	20	29
	Oh	10.9	129	19	5	46	136	29	137
Longleaf-slash pine	Ol	10.2	53	5	6	—	45	12	—
USA[b]	Of	22.8	123	14	9	—	95	21	—
Slash pine	Ol	15.0	67	4	6	—	99	10	—
USA[c]	Of	24.5	137	4	9	—	282	12	—
Conifer	Ol	14.4	162	15	14	—	—	—	—
Mor	Of	22.4	313	24	24	—	—	—	—
USA[d]	Oh	121.0	1565	102	89	—	—	—	—
Conifer	Ol	13.6	171	15	15	—	—	—	—
Moder	Of	18.0	266	22	21	—	—	—	—
USA[d]	Oh	71.7	956	77	67	—	—	—	—
Red pine	Ol	5.3	33	6	7	—	29	3	—
USA[e]	Of	31.7	453	25	25	—	86	13	—
	Oh	31.2	353	25	25	—	44	22	—
Hemlock-maple	Ol	5.8	39	3	4	—	22	1	—
USA[e]	Of	45	669	45	45	—	94	13	—
	Oh	31.4	367	28	188	—	31	13	—
Southern pine	Ol	—	53	5	6	—	45	11	—
USA[b]	Of	—	123	13	8	—	96	20	—
Virginia pine	Ol	—	18	1	3	—	11	2	—
USA[f]	Of	—	217	13	10	—	77	9	—
Shortleaf pine	Ol	—	18	2	3	—	12	2	—
USA[f]	Of	—	217	13	10	—	77	9	—
Loblolly pine	Ol	—	28	2	4	—	17	4	—
USA[f]	Of	—	178	13	10	—	65	9	—
	Oh	—	59	4	4	—	20	3	—
Eastern white pine	Ol	—	23	2	3	—	28	3	—
USA[f]	Of	—	124	10	7	—	82	8	—
Douglas-fir Eastern Washington USA[d]	All	—	327	29	42	—	—	—	—
Douglas-fir Western Washington USA[d]	All	—	193	16	15	—	—	—	—
Black spruce Canada[g]	All	—	1214	213	382	—	102	430	—
Red spruce Canada[g]	All	—	1465	100	1952	—	253	154	—
Birch-spruce USA[h]	All	—	3187	152	91	—	617	—	—
Piñon-juniper USA[i]	All	—	80	12	21	—	216	41	—
Ponderosa pine USA[i]	All	25.1	191	15	80	—	432	83	—
White fir USA[i]	All	80.8	883	85	235	—	2674	339	—
Slash pine USA[c]	All	—	183	8	14	—	341	20	—

[a]Remezov and Pogrebnyak (1969).
[b]Heyward and Barnette (1936).
[c]Pritchett and Smith (1974).
[d]Gessel and Balci (1965).
[e]Lyford, Harvard Forest.
[f]Metz et al. (1970).
[g]Weetman and Webber (1972).
[h]McFee and Stone (1965).
[i]Wollum (1973).

litter to the lower layers of the O horizon under some stands, while aluminum concentrations increased with depth. On the other hand, the increases in aluminum concentration, as well as those of iron and manganese in the more decomposed layers, reflect a concentration of these elements and perhaps some contamination from the mineral soil.

SOIL ORGANIC MATTER IS MORE THAN THE BALANCE BETWEEN INPUTS AND OUTPUTS

As described in Chapter 15, the flow of carbon from atmospheric CO_2 into plants, into soils, and back to the atmosphere is a very important domain called carbon sequestration. The sizes of these flows determine whether a soil serves as a net "sink" for removing CO_2 from the atmosphere, or a net source that contributes to rising atmospheric CO_2.

The land has been a net source of carbon to the atmosphere since about 1860, when rapid expansion of agriculture began. Schlesinger (1995) estimated that the net annual release of carbon from agricultural lands was about $0.8\,Pg\,yr^{-1}$, or about 14% of fossil fuel emissions. Forests store about 15 to $40\,kg\,m^{-2}$ (150 to $400\,Mg\,ha^{-1}$) of carbon in the vegetation (MacKinley *et al.*, 2011), and about 15 to $40\,kg\,m^{-2}$ in the soil worldwide (Jobbágy and Jackson, 2000). Roughly 46% of the vegetative carbon and 33% of the soil carbon are stored in tropical rainforests. Currently, the deforestation in tropical areas leads formerly forested lands to be net sources of C to the atmosphere, though most of the added C comes from biomass rather than soil organic matter. Temperate forests are, in general, a net sink for carbon, especially where forests reclaim land from agricultural uses. However, the source–sink relationships of soil carbon are dynamic (Lal *et al.*, 1998a; 1998b).

At a more local scale, the dynamics of plant inputs and soil organic matter accumulation have strong, direct effects on soil fertility and ecosystem productivity. The amount of C added to soils by trees of course varies tremendously among forests, as does the fate of that C. One well-studied plantation of eucalyptus trees illustrates some general features (Giardina *et al.*, 2004). The trees added $0.4\,kg\,C\,m^{-2}\,yr^{-1}$ to the soil in the form of falling litter from the canopy, but aboveground litterfall accounted for less than 20% of the C added to the belowground portion of the ecosystem. The trees added another $2.0\,kg\,C\,m^{-2}\,yr^{-1}$ through their root systems, supporting the growth and respiration of root cells, the mycorrhizal community, and feeding the soil detrital system with dead roots. Of the $2.4\,kg\,C\,m^{-2}\,yr^{-1}$ added to the soil, $2.2\,kg\,C\,m^{-2}\,yr^{-1}$ returned to the atmosphere, so the net accumulation within the belowground portion of the ecosystem was just $0.2\,kg\,C\,m^{-2}\,yr^{-1}$. About 90% of the accumulating C was located in live (coarse-diameter) roots, and 10% in soil organic matter.

In this eucalyptus plantation, about 80% of the C sent below ground by trees (in litterfall and via roots) returned to the atmosphere within one year. This high "flow through" rate for the overall C budget showed little net change in soil organic matter. However, a pattern of little net change can result from either very little change, or from large offsetting gains and losses. The O horizon increased by $0.05\,kg\,C\,m^{-2}\,yr^{-1}$, but this small change resulted from very large offsetting inputs ($0.41\,kg\,C\,m^{-2}\,yr^{-1}$) and losses ($0.36\,kg\,C\,m^{-2}\,yr^{-1}$). The net change in mineral soil organic matter (to 30 cm depth) was a (non-significant) loss of $0.02\,kg\,C\,m^{-2}\,yr^{-1}$, which might seem to indicate a static system. On the contrary, stable ^{13}C isotope methods showed that the soil lost $0.24\,kg\,C\,m^{-2}\,yr^{-1}$ of old soil C (the soil C legacy prior to the plantation's establishment), at the same time it gained $0.22\,kg\,C\,m^{-2}\,yr^{-1}$ of the eucalyptus-derived C.

WHY DOES ORGANIC MATTER ACCUMULATE IN SOILS?

This seemingly simple question has been pondered for centuries, and new insights develop every decade. A simple overview answer might be:

Soil organic matter =
   Organic matter inputs − Organic matter outputs

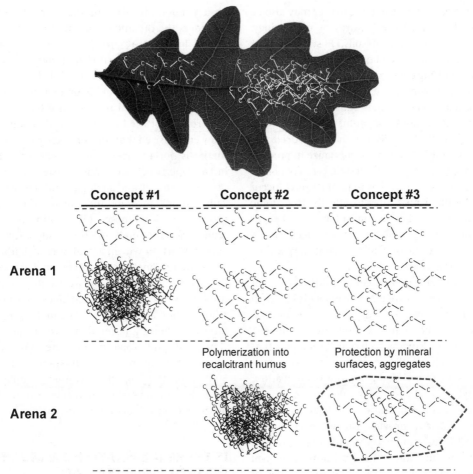

Figure 4.6 Three concepts have been used to explain why we find organic matter accumulating in soils. Concept 1 simply expects the small molecules in litter to be oxidized to CO_2 (in Arena 1), leaving the recalcitrant plant compounds to accumulate. Concept 2 has the complex plant compounds broken down into simpler compounds in Arena 1, with synthesis of new complex compounds in Arena 2. Concept 3 is similar to Concept 2, except the stabilization of organic matter happens through physical protection (or partial refuge locations) of otherwise decomposable compounds.

This simple equation must be true from a mass-balance perspective, but it provides only a black box level of clarity for *why* organic matter accumulates. Why doesn't the microbial community decompose all the soil organic matter? Three general concepts have been used to explain organic matter accumulation in soils (Figure 4.6). Fresh litter contains simple compounds that are readily used by microbes, and more complex compounds that are difficult to decompose. The first concept was simply that accumulated organic matter was mostly the left-over, difficult-to-decompose plant material. The second concept was that both the simple and complex components of plant litter break down, but a portion becomes transformed into decay-resistant humus. The third concept was that the complex soil organic compounds may be artifacts of analytical extraction methods, and that humus accumulates when otherwise decomposable organic matter becomes fixed to mineral particles or protected inside aggregates. Some organic matter may also be protected by some basic features of the biology, physics, and chemistry of soils. Ekschmitt *et al.* (2005) noted that a microbe faces a daunting challenge of producing exoenzymes that diffuse into the soil (into areas perhaps rich in suitable substrate, and "wasted" areas without substrate), with any products of decomposition diffusing both toward and away from the microbe that produced the enzyme. Microbes also tend to specialize in producing enzymes that target only certain types of bonds, so communities of microbes may need to co-occur in very localized microsites to effectively degrade organic matter. The net outcome of these challenges may be microsites that offer "partial refuges" where organic matter persists even though it may not be physically protected.

Why might some organic compounds take longer to break down than others? Concept #2 is based on the idea that chemical bonds contain free energy, and severing the bonds in soil organic matter releases energy. However, chemical reactions that have a net release of free energy may still have challenges in terms of the energy of activation; this is why wood is very stable under normal conditions, but breaks down (with great release of energy) if a flame provides the necessary energy of activation to launch the exothermic process. Enzymes are fundamental for breaking chemical bonds in soil organic matter, and some types of bonds are easily broken by enzymes (such as those binding chains of sucrose groups together into starch molecules), and others are more difficult (such as those binding C atoms in benzene rings). Soil organic matter may also be difficult to decompose for physical reasons; three-dimensional complexities in compound structure can limit enzyme access to bonds, and compounds adsorbed onto mineral surfaces or contained within small aggregates may be unavailable for enzymes. For example, Hagedorn *et al.* (2003) exposed young trees to elevated CO_2 for four years, and then looked at the stabilization of new C in the soil; acid loam soils retained almost twice as much C as calcareous sand soils.

This classic explanation for why organic matter accumulates in soils may, in fact, be missing very basic, important parts of the story. Concept #2 considers the organic matter accumulating in soils not as left-over, unprocessed plant material, but rather compounds synthesized by the microbial community following successful breakdown of plant material. However, Schmidt *et al.* (2011) concluded that chemical recalcitrance has only minor importance for the turnover of soil organic matter. They emphasized that the persistence of soil organic matter is not a property of the organic molecules, but a property of the ecological system that includes physicochemical and biological influences. Classically, complex compounds such as lignin were thought to decompose slowly, and some rather crude experimental methods supported these expectations. However, actual rates of lignin breakdown are reasonably fast, and lignin decomposition may be limited more by availability of readily used substrates than by chemical recalcitrance (Klotzbücher *et al.*, 2011). Lignin decomposition rates are typically faster than overall rates of soil C turnover. Bulk soil organic matter typically has a turnover time of about 50 years (ranging from 15 to 300 years), but lignin and other complex plant biomolecules have average turnover times in soils of 15 to 40 years (Schmidt *et al.*, 2011).

IS THERE A LIMIT TO HOW MUCH SOIL ORGANIC MATTER CAN ACCUMULATE IN FOREST SOILS?

It may seem logical that after many years of inputs of fresh organic materials and decomposition activity, soils should reach a balance where the pool of soil organic matter remains stable. The long-term stabilization of soil organic matter depends, in part, on the capacity of mineral soil surfaces to bind organic matter and form long-lasting soil aggregates, and this surface-binding capacity would be finite (Six *et al.*, 2002).

Surface O horizons might reach a stable level in cases where decomposition (or mixing by soil animals) is high. In other situations, O horizons may continue to accumulate for very long periods. McFee and Stone (1965) found that O horizons under mature yellow birch–red spruce stands growing on well-drained outwash sands in the Adirondack Mountains of New York continued to accumulate after a century, at a rate of about $0.05 \, \text{kg m}^{-2}$ annually over the next two centuries. Wardle et al. (2003) found that soil organic matter increased linearly for at least 6000 years in Norway spruce forests (in the absence of fire), at a rate of about $0.005 \, \text{kg m}^{-2} \, \text{yr}^{-1}$.

STUDIES OF LITTER DECOMPOSITION HAVE FASCINATED SCIENTISTS

Hundreds (or thousands?) of studies have examined the loss of mass from leaves and other plant materials, often by placing known amounts of material in mesh bags that reside on the soil surface for varying lengths of time (see Berg and Laskowski, 2006; Prescott, 2010). The loss of mass over time may (or may not) show a constant proportion of mass loss per unit of time, similar to the decay of radioactive isotopes. Nutrient patterns may be quite different: potassium often leaves litterbags faster than overall mass loss; phosphorus often remains within the litterbags until most of the matter is gone; and nitrogen may show an initial increase as mass declines. All these trends vary with the initial type of material, with varying patterns for live versus abscised leaves, pine needles versus spruce needles, twigs versus roots. The mass loss patterns also vary with geography, as a result of local soil communities (locally familiar substrates may decompose more rapidly than unusual substrates), and environmental factors such as temperature, moisture, and soil nutrient supply.

This fascinating range of variation has led to intensive studies of factors that may explain rates of mass loss, ranging from short-term studies by individual scientists with a single material at a single site, to decade-long studies with many scientists, many materials, and many locations. Very large studies include examinations of Scots pine decomposition across Europe (Berg and Laskowski, 2006), the Canadian Intersite Decomposition ExperimenT (CIDET; Moore et al., 2010), and the Long-term Intersite Decomposition Experiment (LIDET) in North and Central America (Currie et al., 2010).

Some studies recognize the local importance of macroinvertebrates, and vary the sizes of litterbag mesh to allow or exclude larger animals. Litterbag methods may lead to unrealistic patterns (such as exponential rates of mass loss), masking actual dynamics occurring in unconfined litter (such as asymptotic rates of mass loss; Kurz-Besson et al., 2005).

What questions do these studies seek to answer? Most of them deal with the rate of disappearance of mass or nutrients, rather than questions about what controls the rate of organic matter accumulation in soils, or the rate of nutrient supply to plants. Indeed, the majority of litterbag decomposition experiments last less than five years, and only a few continue beyond a decade, despite the fact that the relevance for soil organic matter accumulation (or loss) depends on the mass remaining after the initial years of rapid loss. For example, aspen leaves decompose (usually) faster than lodgepole pine needles, given lower ratios of lignin:N in aspen. Giardina et al. (2001) examined what this might mean for the stability of organic matter that actually accumulates under each species, and found the contrary result – that organic matter accumulated under aspen (with rapidly decomposing leaves) turned over more slowly than the organic matter under pine (with slowly decomposing leaves). The initial rates of mass loss from decaying litter may offer no insights into the accumulation or turnover rates of organic matter accumulated in the soils; any effect of tree species in the CIDET decomposition experiment disappeared within five years (Prescott, 2010). Other sites and comparisons have other results, but the key point is that we should not expect the initial rate of mass loss from litter to provide much insight into how organic matter accumulates in soils.

ORGANIC MATTER ENTERS SOILS FROM ABOVE AND FROM WITHIN

Forests drop massive amounts of dead leaves, twigs, and branches each year, generally adding one-half to one kg of material to each m^2 of ground area (Figure 4.7). The death of a tree leads to the input of

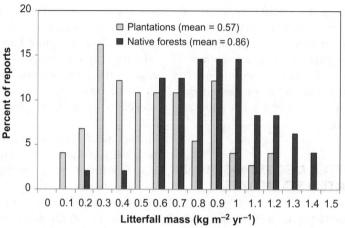

Figure 4.7 Tropical plantations add about 0.2 to 0.9 kg m^{-2} yr^{-1} of organic matter to soils via aboveground litterfall, compared with about 0.6 to 1.2 kg m^{-2} yr^{-1} for native forests (from data of Binkley *et al.*, 1997).

a very concentrated mass of organic matter onto a portion of the soil surface. Across decades and centuries, the episodic addition of organic matter in woody boles of dead trees may be similar to the more regular, annual inputs of smaller materials. As noted above, most of this material is relatively rapidly transformed into CO_2 and lost to the atmosphere, and some is transformed into longer-lasting soil organic matter. The transformed soil organic matter may reside within the O horizon, or may be transferred to the mineral horizons if animals mix the soil, or as soluble organic matter leaching out of the O horizon. Few studies have quantified the proportion of O horizon material that moves into the A horizon via percolating water; a classic study from a mountainous hardwood forest found between 15 and 20% of the original mass of aboveground litter entered the A horizon in solution (Qualls *et al.*, 1991).

The mass of aboveground litterfall generally shows modest response to gradients in site fertility; two stands may differ substantially in gross primary production and wood production, with much smaller differences in litterfall rates. Organic matter inputs in dead tree boles would generally be higher (over long time periods) in more fertile sites, and lower in managed forests where harvests remove boles.

A physicist recently claimed that " . . . roots add carbon to the soil, while stems and leaves mostly return carbon to the atmosphere" (Dyson, 2010). At first glance, this may seem likely, but the truth is that we really don't know. How much of the organic matter from a dead root becomes soil organic matter, and how much is oxidized to CO_2? How much of the organic matter of a dead leaf will become stabilized organic matter in the soil, and how will this be distributed between the O, A, and B horizons by soil-forming processes? We have lots of information that is relevant to these fundamental questions, but so far no clear answers.

The routine death of roots (and associated mycorrhizal hyphae) also adds large amounts of organic matter directly into mineral and O horizons. The location of fine roots within soil profiles ranges from almost all in the O horizon (see Figure 4.2) to almost all in the mineral soil. Belowground production generally shows less response across gradients and management activities (such as fertilization) than aboveground production; with increasing fertility, aboveground production increases and belowground production tends to remain constant or decrease. For example, Högberg *et al.* (2003) found that a fertility gradient with Scots pine and Norway spruce associated with a doubling of aboveground production showed no change in belowground production. At the other end of the spectrum of climate and management, Ryan *et al.* (2010) found that irrigation and fertilization of fast-growing eucalyptus plantations increased aboveground production by 25% with no change in belowground production.

Rates of belowground inputs are often estimated by a variety of methods, and comparisons across methods are not straightforward. As a general expectation, perhaps half of the organic matter input to soils comes from belowground fluxes (within the soil), and half from aboveground fluxes. The inputs may be roughly equal in mass, but they differ in almost all other respects. Dead roots exist in moist, moderate-temperature horizons, whereas leaves may fall onto dry, hot surfaces of O horizons. The organic chemistry (and food quality for soil microbes and animals) may differ substantially, along with physical sizes of litter materials.

If the mass of inputs of aboveground and belowground sources of organic matter were equal, would the contribution to long-term accumulation of soil C be equal between the sources? Not enough experimentation has been done to answer this intriguing question. Some insights are available from the Detritus Input, Removal and Transfer (DIRT) study, where collaborators modified the inputs of various sorts of materials and followed the soil responses. Trenching of plots removed the contribution of new roots and mycorrhizae, litter screens prevented addition of aboveground material, and additions of fine litter and wood supplemented background input rates. The removal of new root production seemed to reduce the development of new, longer-term soil organic matter in a deciduous forest in Pennsylvania, indicating that belowground inputs might be more important for stabilizing soil organic matter than aboveground inputs (Crow *et al.*, 2009). Aboveground litter seemed to be the major source for stabilized organic matter in the soil of an old-growth conifer forest in Oregon. Readily decomposed litter seemed to be an important "primer" of the detrital food web, accelerating decomposition of older soil organic matter. The contrasting stories from each of these forests highlight two features of forest soil research: we should not expect patterns from one site to extrapolate to another site, and we need a solid set of many sites to provide a good estimate of general tendencies, variations around those tendencies, and factors that might account for the variations.

SOIL ORGANIC MATTER SHOWS STRONG PATTERNS ACROSS LANDSCAPES

Forests change substantially in response to topographic factors. Increases in elevation often lead to cooler conditions, more precipitation, and less water limitation on production. Similar trends are associated with moving from south-facing slopes that receive high radiation loads to lower radiation, north-facing slopes (and vice versa in the Southern Hemisphere). Long-term soil development responds to these factors too, as do shorter-term rates of current processes.

Several topographic studies show the importance of variations in slopes, elevation, and topographic position. Griffiths *et al.* (2009) sampled 180 locations across 4000 ha of a mountainous landscape in Oregon; increasing elevation by about 500 m lowered soil temperature by 2.5°C, and increased the mass of mineral soil organic matter by about 25%. The same pattern was evident in moving from south-facing slopes to north-facing slopes: temperature decreased by 2.3°C, and mineral soil organic matter increased by 50%. A regional soil assessment in Vermont found that aspect had a larger effect on soil organic matter accumulation than elevation, and local slope position (influencing soil water supply and soil development) was also important (Johnson *et al.*, 2009b). Across more than 1400 locations in Sweden, the mass of soil organic C increased as the depth to the water table decreased; sites with water tables deeper than 2 m averaged $6.8 \, kg \, C \, m^{-2}$, rising to $8.0 \, kg \, C \, m^{-2}$ for 1–2 m deep water tables, and $9.8 \, kg \, C \, m^{-2}$ for water tables $< 0.5 \, m$ deep (Olsson *et al.*, 2009). Almost all the difference in these Swedish forests occurred in the O horizon rather than the mineral soil, and the mass of the O horizon increased with increasing site productivity.

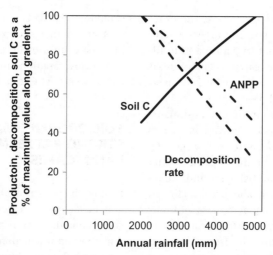

Figure 4.8 Increasing rainfall along a geographic gradient on the island of Maui leads to soil water saturation, low aeration, low aboveground net primary production (ANPP) and low decomposition. Taken together, these interacting factors lead to a doubling net increase in soil organic matter (C) storage as precipitation increases from 2000 to 4000 mm yr^{-1} (from data of Schuur et $al.$, 2001).

We may typically think of ecosystem productivity (and input of organic matter to soils) as increasing with increasing water supply. This is generally true, but too much water can reduce soil aeration, dramatically lowering rates of decomposition. On the island of Maui, too much water leads to soil saturation and low aeration, lowering both productivity and decomposition (Figure 4.8). The relative decrease in decomposition with increasing rainfall is greater than the relative decrease in production, leading to increasing soil organic matter. Excessive moisture also appears to drive large accumulations of organic matter in the upper soil of wet forests in the Pacific Northwest (Sajedi et $al.$, 2012), and of course very saturated sites are often peatlands with vast accumulations of organic matter.

Across a broad range of wet tropical lowland forests, Posada and Schuur (2011) found a weak relationship between mean annual precipitation and the turnover rate of soil C. Sites receiving 2500 mm yr^{-1} of rainfall had a soil C turnover rate of about four years, compared with six years for sites with more than 8000 mm yr^{-1}, but the effects of soil N and P supply had a stronger influence on C turnover than precipitation.

TROPICAL SOILS HAVE GREATER SOIL ORGANIC MATTER THAN TEMPERATE AND BOREAL SOILS

At the scale of a thousand kilometers, forests in warmer southern Sweden accumulate about twice as much soil organic matter as cooler forests in northern Sweden (Berggren Kleja et $al.$, 2010). But would this trend of increasing soil organic matter with increasing temperature apply at a global scale? One might expect that high year-round temperatures in the tropics would lead to rapid decomposition and lower accumulation of soil organic matter than in cooler temperate or cold boreal areas. Indeed, stories used to be told about the fragility of tropical soils as a result of the majority of organic matter and nutrients residing in aboveground biomass rather than the soil (for an interesting history, see Richter and Babbar, 1991). Speculation like this needs to be tested with data, and the empirical pattern from 900 locations clearly shows an opposite trend (Figure 4.9, Figure 2.2). Decomposition may be faster under warm (and moist) conditions in the tropics, but so is plant production; this

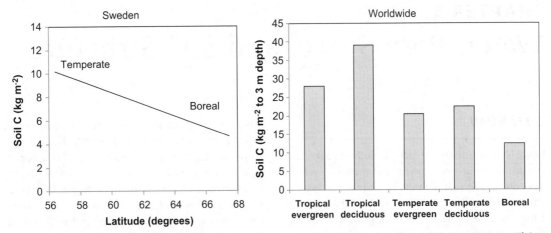

Figure 4.9 Forest soil C accumulation doubles from northern to southern Sweden (based on 1182 sites, Berggren Kleja *et al.*, 2010). Similarly, soil C accumulation increases from boreal to tropical forests at a global scale (based on 900 forest soils around the world, Jobbágy and Jackson, 2000). Large variation within each region ensures that broad generalizations are not locally informative (see also Figure 2.2 and Figure 15.9).

combination leads to much higher accumulations of soil organic matter in the tropics than in temperate forests, and in temperate forests than in boreal forests. The global scale pattern of temperature and soil organic matter accumulation matches the regional scale pattern in Sweden.

Tropical soils may be deeper than temperate or boreal soils, leading to greater volumes of soils for both the accumulation of organic matter and exploitation by plant roots (see Chapter 6, and Figure 16.1). Of course, soils low in organic matter are easy to find in the tropics, just as high organic matter soils are easy to find in boreal forests. As in all aspects of forest soils (and science in general), the value of general patterns may be high at one scale, and not relevant at another scale.

CHAPTER 5
Water, Pore Space, and Soil Structure

OVERVIEW

Water is the dominant site factor in determining forest composition and tree growth, and the soil is where trees obtain the vast majority of their water. Within any given climatic zone, the type of soil dramatically affects both vegetation type and growth potential. Trees root primarily in the vadose zone, the zone from the surface to the water table, and the physical properties of the soil in this zone dramatically affect the soil's ability to provide trees with water, oxygen, and nutrients. Physical properties of soil profoundly influence soil temperature, water relations, chemistry, and the life that depends on the soil. Differences in physical properties among soils, or within a soil over time, have major influences on tree growth. Soil texture refers to the size of mineral particles in the soil, and soil structure concerns the three-dimensional conglomerations of mineral particles and organic matter. The pore space within soils is important in influencing water infiltration, soil atmosphere composition, and ease of penetration by roots. Forest management operations may affect soil structure, especially by altering soil pore space and bulk density. Soil temperature regimes differ by regions and management activities, driving differences in the rates of soil processes, such as decomposition, and in the species composition and growth of forests. Water is held under negative potential within freely draining soils, and the characteristics of water retention and release as a function of water potential determine the availability of water to plants. Forest harvest reduces water losses through interception/evaporation and transpiration, increasing water yield from forest lands and the average soil water content.

Water is the dominant site factor in determining forest composition and tree growth. Of course, the major driving variable in water availability is precipitation, and we have discussed the role of climate in determining the location of forest biomes in Chapter 2. However, the soil is where trees obtain the vast majority of their water, and within a climatic zone the type of soil dramatically affects both vegetation type and growth potential.

The subsurface zone of soil containing fluid under pressure that is less than that of the atmosphere is termed the vadose zone (Selker *et al.*, 1999) Vadose (pronounced *vā'dōs'*) is a Latin word meaning surface. Pore spaces in the vadose zone are partly filled with water and partly filled with air. The vadose zone is bounded by the land surface above and by the water table below, and this well-oxygenated zone is where most plants have the majority of their roots.

This zone also includes the capillary fringe above the water table. The capillary fringe is the subsurface layer in which groundwater seeps up from a water table by capillary action to fill pores. Pores at the base of the capillary fringe are filled with water due to tension saturation. This saturated portion of the capillary fringe is less than total capillary rise because of the presence of a mix of pore sizes. If the pore size is small and relatively uniform, it is possible for soils to be completely saturated with water for several feet above the water table, and the fringe may have a very irregular upper boundary. Alternately, the saturated portion will extend only a few inches above the water table when pore size is large, and the fringe may be flat at the top.

The zone of soil or rock that contains the water table is termed the phreatic zone. Phreatic

Ecology and Management of Forest Soils, Fourth Edition. Dan Binkley and Richard F. Fisher.
© 2013 John Wiley & Sons, Ltd. Published 2013 by John Wiley & Sons, Ltd.

(pronounced *frē-ăt'ĭk*) is derived from the Greek word for well. In the phreatic zone, all pores and interstices are filled with fluid. Because of the weight of the overlying groundwater, the fluid pressure in the phreatic zone is greater than the atmospheric pressure. Some plants, termed phreatophytes, maintain roots in this oxygen-poor zone in order to obtain water.

The physical properties of soil, particularly in the vadose zone, profoundly influence the growth and distribution of trees through their effects on soil moisture regimes, aeration, temperature profiles, soil chemistry, and even the accumulation of organic matter. Some physical properties can be altered intentionally by management, including draining wet soils and plowing subsoil to break up hardpans. The most widespread result of management on the physical properties of soils may be inadvertent soil compaction during harvesting operations.

Soil water supply influences the growth of trees over short timescales, but it also has a slowly developing legacy over soil-development timescales. Water-driven erosion tends to limit the depth of soils on upper slopes, and deposition deepens them on lower slopes. In many cases, erosion over soil-development spans of time leads to shallow soils near ridges, and increasing soil depths toward toe slopes. These influences can have multiple effects on tree growth; lower-slope positions benefit from current water transport from upslope, and also have greater water storage capacity as a result of long-term soil formation. These hydrologic features can enhance overall biological activity as well, leading to lower-slope soils having greater supplies of both water and nutrients for trees (Figure 5.1).

This topographic pattern does not develop in all cases, especially where changes in slopes may relate to changes in parent material. For example, a four-year-old plantation of *Eucalyptus grandis* in Argentina had twice the volume on ridge-top sites as compared to lower-slope sites; in this case, erosion of the Oxisol (lateritic) soils had led to shallower soils lower on the slopes (Dalla-Tea and Marco, 1996).

Figure 5.1 Harvesting this *Eucalyptus* stand in Brazil revealed a very large gradient in soil fertility (particularly water supply) along the slope. The white and black bars are the same length, to highlight the shorter trees (with much lower volume) on the upper slope. Photo provided by J. L. Gonçalves and J. L. Stape, used with permission.

MINERALS ARE PLACED IN THREE SIZE CLASSES, COMPRISING SOIL TEXTURE

Soil can be conveniently divided into three phases: solid (including mineral and exchange phases), liquid, and gas. The solid phase makes up approximately 50% of the volume of most surface soils and consists of a mixture of inorganic and organic particles varying greatly in size and shape. The size distribution of the mineral particles determines the texture of a given soil. Schemes for classifying soil particle sizes have been developed in a number of countries. The classification used by the USDA, based on diameter limits in millimeters, is outlined in Table 5.1 (Soil Survey Staff, 1993).

Table 5.1 USDA classification of soil particle sizes.

Name of Particle	Diameter Limits (mm)
Sand	0.05–2.0
Very Coarse	1.0–2.0
Coarse	0.5–1.0
Medium	0.25–0.5
Fine	0.10–0.25
Very Fine	0.05–0.10
Silt	0.002–0.05
Clay	Less than 0.002

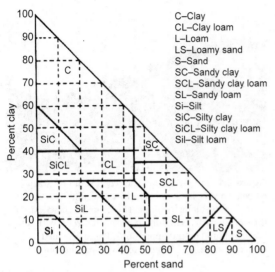

Figure 5.2 Soil textural triangle: percentage of clay and sand in the main textural classes of soils; the remainder of each class is silt.

Mineral soils are usually grouped into four broad textural classes – sands, silts, loams, and clays. Combinations of these class names are used to indicate intermediate textural classes (Figure 5.2).

The most important differences in soil texture relate to the surface areas of particles of different sizes. Medium-sized sand has a diameter of 0.25 to 0.50 mm, about 7000 grains per gram, and a surface area of $0.013\,m^2g^{-1}$ (Barber, 1995). Silt has a diameter of 0.002 to 0.050 mm, about 20 million particles per gram, and a surface area of $0.09\,m^2\,g^{-1}$. Clay particles are less than 0.002 mm in size, with 400 billion particles per gram, and a surface area of 1 to $10\,m^2g^{-1}$. The surface area per gram spans a range of three orders of magnitude, with dramatic effects on water potential, organic matter binding, cation exchange, and overall biotic activity.

Cobbles and gravels are fragments larger than 2 mm in diameter. They are not included in the particle-size designations because they normally play a minor role in agricultural soils. However, coarse particles are common in forest soils (up to 80% or more of some mountain soils). Major properties of coarse fragments are recognized by adjectives that indicate modification of the texture name (Table 5.2).

A moderate amount of rock in a fine-textured soil may favor tree growth. Coarse fragments may increase penetration of air and water, as well as the rate of soil warming in spring. Nevertheless, a coarse skeleton dilutes the soil and can be detrimental to tree growth if it occupies a large volume in sandy soils, because of low water-holding capacity and exchange capacity. Coarse-textured materials contribute little to plant nutrition.

TEXTURE INFLUENCES TREE GROWTH

Soil texture has major effects on forest growth, but these effects are indirect, manifested through the effect of texture on features such as water-holding capacity, aeration, exchange capacity, and organic matter retention. In general, within a climatic zone, coarser-textured soils support more xeric plant communities while finer-textured soils support more mesic plant communities. For example, deep, coarse, sandy soils often support low-productivity stands of pines, cedar, scrub oak, and other species that cope well with moisture and nutrient stress. However, some sandy soils have lenses of finer-textured materials within them. These variations in texture strongly influence soil properties such as water retention and cation exchange capacity and allow the soils to support more robust plant communities.

Table 5.2 USDA classification scheme for coarse fragments.

Shape and Kind of Fragment		Size and Name of Fragment		
Rounded fragments	2 mm to 8 cm in diameter	8–25 cm in diameter	>25 cm in diameter
	Gravelly	Cobbly	Stony or bouldery
Thin, flat fragments	2 mm to 15 cm in length	6–38 cm in length	>38 cm in length
	Channery	Flaggy	Stony or bouldery

SOIL STRUCTURE MODERATES THE EFFECTS OF SOIL TEXTURE

Soil structure refers to the aggregation of individual mineral particles and organic matter into larger, coarser units. This aggregation generally reduces bulk density (megagrams of soil per cubic meter), and alters the distribution of pore sizes in the soil, which alters water movement, water-holding capacity, and aeration. As structure develops in coarse-textured soils, water-holding capacity is increased. As structure develops in fine-textured soils, water-holding capacity decreases, but water movement and aeration increase.

Field descriptions of soil structure usually give the type or shape, class or size, and degree of distinctness in each horizon of the soil profile. The following types of structure are recognized by the U.S. Natural Resource Conservation Service: platy, prismatic, columnar, angular blocky, subangular blocky, granular, crumb, single grain, and massive. The size class of aggregates ranges from very fine to fine, medium, coarse, and very coarse (Soil Survey Staff, 1993).

Structural classes are determined by comparing a representative group of peds with a set of standardized diagrams. Grade is determined by the relative stability or durability of the aggregates and by the ease of separating one from another. Grade varies with moisture content and is usually determined on nearly dry soil and designated by the terms weak, moderate, and strong. The complete description of soil structure, therefore, consists of a combination of the three variables, in reverse order of that given above, to form a type, such as weak subangular blocky.

Important drivers of soil aggregation include mineral chemistry, salts, clay skins, oxide coatings, growth and decay of fungal hyphae and roots, freezing and thawing, wetting and drying, and the activity of soil organisms (especially earthworms). Soil texture has considerable influence on the development of aggregates. The absence of aggregation in sandy soils gives rise to a single-grain structure, whereas loams and clays exhibit a wide variety of structures. Soil animals, such as earthworms and millipedes, favor the formation of crumb

structure in the surface soil by ingesting mineral matter along with organic materials, producing casts that provide structure to the soil. The intermediate products of microbial synthesis and decay are effective stabilizers, and the cementing action of the more resistant humus components that form complexes with soil clays gives the highest stability.

Soil aggregation is also influenced by different tree species. In an experiment with 35-year-old plots with different tree species, the average size of aggregates ranged from 1.5 mm under white pine to 2.1 mm under Norway spruce (*Picea abies*) (Scott, 1998). Across the species, average aggregate size increased with increasing fungal mass ($r^2 = 0.66$) and declining bacterial biomass ($r^2 = 0.72$; bacterial biomass declined as fungal mass increased, $r^2 = 0.87$).

The influence of tree species on soil aggregation was also apparent from a lysimeter experiment in California in which 50 m^3 chambers were filled with soil and planted to pine or chaparral (Graham and Wood, 1991). After 40 years, soils under the influence of pine lacked earthworms, had developed a clay-depleted A horizon, and had accumulated enough clay in the B horizon to qualify as an argillic horizon. Soils under oak (*Quercus dumosa*) developed a 7-cm A horizon (90% of which was earthworm casts) enriched in humus and clay relative to the underlying C horizon. Thus, the plant species affected earthworm activity, which dominated structural development of the soil.

BULK DENSITY ACCOUNTS FOR THE COMPOSITION OF MINERALS, ORGANICS, AND PORE SPACE

Bulk density is the dry mass (of < 2 mm material) of a given volume of intact soil measured in Megagrams per cubic meter (which equals kilograms per liter). Well-developed soil structure increases pore volume and decreases bulk density. The particle density of most mineral soils varies between the narrow limits of 2.60 and 2.75, but the bulk density of forest soils varies from 0.2 in some organic layers to about 1.9 in coarse sands. Soils high in organic matter have lower bulk densities than soils low in this component. Soils that are loose and porous have low mass per unit volume (bulk density), while those that are compacted have high mass per unit volume. Bulk density can be increased by excessive trampling by grazing animals, inappropriate use of logging machinery, and intensive recreational use, particularly in fine-textured soils.

Increases in soil bulk density are generally harmful to tree growth for the same reasons that structure affects soil properties. Compacted soils have higher strength and can restrict penetration by roots. Reduced aeration in compacted soils can depress the activities of roots, aerobic microbes, and animals. Reductions in water infiltration rates are common when soils become compacted, and anaerobic conditions may develop in puddled areas.

The importance of differences in soil bulk density (and soil strength) for tree growth is illustrated by an extensive characterization of variation in soil bulk density across a stand of *Eucalyptus camaldulensis* on an Inceptisol in the savanna region of central Brazil (Figure 5.3). Where the bulk density of the 0 to 20 cm soil averaged 1.25 kg L^{-1}, stem volume of the four-year-old trees was 25 m^3 ha^{-1} (Gonçalves *et al.*, 1997). Where the bulk density averaged 1.06 kg L^{-1}, stem volume was more than three times greater (90 m^3 ha^{-1}).

Pore volume refers to that part of the soil volume filled by water or air. The proportions of water and air change over time, and soil water content drastically affects soil aeration. Gas molecules diffuse

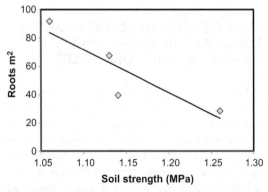

Figure 5.3 Variations across a stand in bulk density of an Inceptisol in the Brazilian savanna strongly affected the volume of wood accumulated by *Eucalyptus camaldulensis* at age 4 years (data from Pereira, 1990, cited in Gonçalves *et al.*, 1997).

about 10 000 times faster through air than through water. Coarse-textured soils have large pores, but their total pore space is less than that of fine-textured soils (although puddled clays may have even less porosity than sands). Because clay soils have greater total pore space than sands, they are normally lighter per unit volume (have lower bulk density), but differences in soil structure can override the basic influence of texture on pore volume and bulk density.

Pore volume is conveniently divided into capillary and noncapillary pores. Soils with a high proportion of capillary (small-diameter) pores generally have high moisture-holding capacity, slow infiltration of water, and perhaps a tendency to waterlog. By contrast, soils with a large proportion of noncapillary (large-diameter) pores generally are well aerated, and have rapid infiltration and low moisture-retaining capacity. The pore size distribution of a soil may be broad, with large portions of capillary and noncapillary pores, particularly where soil animals and old root cavities provide large-radius pores in clayey soils.

Sandy surface soils have a range in pore volume of approximately 35–50%, compared to 40–60% or higher for medium- to fine-textured soils. The amount and nature of soil organic matter and the activity of soil flora and fauna influence soil structure and thus pore volume. Pore space is reduced by compaction and generally varies with depth. Compact subsoil may have no more than 25–30% pore space. Tree species can also change the distribution of pore sizes in soils; soils under Norway spruce have been reported to have more pore space than adjacent soils under beech (Nihlgård, 1971).

The pore volume of forested soils is normally greater than that of similar soil used for agricultural purposes because continuous cropping results in a reduction in organic matter and macropore spaces (Figure 5.4). Porosity of most forest soils varies from 30–65%.

Soil aggregates are generally more stable under forested conditions than under cultivated conditions. Continued cultivation tends to reduce aggregation in most soils through mechanical rupturing of aggregates and by a reduction in organic matter content and associated cementing action of microbial exudates and fungal hyphae (Figure 5.5).

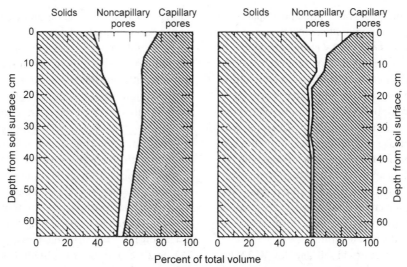

Figure 5.4 Porosity as measured in the surface 60 cm of Vance soil (Typic Hapludult) in the South Carolina Piedmont: (left) in an undisturbed mixed hardwood forest and (right) in abandoned farmland (Hoover, 1949).

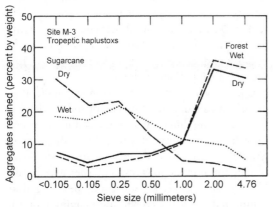

Figure 5.5 Stability of aggregates of an Oxisol used for forestry and for sugarcane in Hawaii (Wood, 1977). The forest soil has much larger aggregates that do not pass through a 1-mm sieve, whereas most of the sugarcane soil aggregates are much smaller. Reproduced from *Soil Science Society of America Journal*, **41**, no. l (1977):135, with permission of the Soil Science Society of America.

LIFE IN THE SOIL DEPENDS ON THE SOIL ATMOSPHERE

Soil air is important primarily as a source of oxygen for aerobic organisms, including tree roots. Soil air composition, like air volume, is constantly changing in a well-aerated soil. Oxygen is used by plant roots

and soil microorganisms, and carbon dioxide is liberated in root respiration and by aerobic decomposition of organic matter (Figure 5.6).

Gaseous exchange between the soil and the atmosphere above it takes place primarily through diffusion. Consumption of oxygen by respiration in the soil leads to a gradient from relatively high oxygen in the ambient air to the low-oxygen soil air. The oxygen content of air in well-drained surface soils seldom falls much below the 20% found in the atmosphere, but oxygen deficits are common in poorly drained, fine-textured soils. Under these conditions, gas exchange is very slow because of

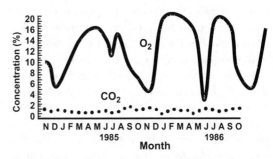

Figure 5.6 Trends in soil oxygen (solid line) and carbon dioxide (dotted line) concentrations at 20-cm depth in gleyic podzol (Spodosol) developed in glacial till in northern Sweden (data from Magnusson, 1992).

the high percentage of water-filled pore spaces. In very wet soils, carbon dioxide concentrations may rise to 5 or 6% and oxygen levels may drop to 1 or 2% by volume (Romell, 1922). However, oxygen is not necessarily deficient in all wet soils in spite of the absence of voids. If the soil water is moving, it may have a reasonably high content of oxygen brought in through mass flow. Soils saturated with stagnant water are low in oxygen, and they are very poor media for the growth of most higher plants. Soil air usually is much higher in water vapor than is atmospheric air, and it may also contain a higher concentration of such gases as methane and hydrogen sulfide, formed during organic matter decomposition.

Oxygen concentrations in soil air as low as 2% are generally not harmful to most trees for short periods. Some *Alnus, Taxodium, Nyssa, Salix* and *Picea* species can thrive at low levels of soil oxygen. Seedling root growth of many species is reduced at an oxygen content of less than 10%. Any restriction in gas exchange between the tree roots and the atmosphere will eventually result in the accumulation of carbon dioxide, anaerobic metabolic products, and root death.

SOIL STRUCTURE CAN BE HARMED BY MANAGEMENT ACTIVITIES

Compaction by heavy equipment or repeated passages of light equipment over the same track compresses the soil mass and breaks down surface aggregates, decreasing the macro pore volume and increasing the volume proportion of solids. Reductions in air diffusion and water infiltration often occur as soil strength increases in compacted soils. Compaction occurs more frequently on moist soils than on dry soils (because water lowers soil strength, "lubricating" soil particles) and more often on loamy and clayey soils than on sandy soils. Very wet soils are often so well lubricated, or soupy, that displacement, but little compaction, occurs. The effects of compaction may be less permanent in fine-textured soils (especially those containing considerable amounts of shrink–swell clays) than in some coarse-textured soils because of the density

reducing action of swelling and shrinking during wetting and drying cycles.

The use of large machines in forest harvesting is the primary driver of increases in soil bulk density and strength and decreases in pore volume. For example, ten passes with a rubber-tired skidder substantially increased soil strength in a harvesting unit in Australia (Figure 5.7), although in this case, the amount of soil compaction may not have been great enough to affect growth.

A greenhouse study by Simmons and Pope, 1988 illustrates the pieces of the soil compaction puzzle. When soils with a moisture content of -0.3 MPa were compacted to increase bulk density from 1.25 to 1.55 mg m^{-3}, the air-filled porosity declined from 36% to 20%, and soil strength increased from 2.3 to 4.1 MPa. When these soils were moister (with soil water potential near field capacity, -0.01 MPa), air-filled porosity declined from 22% in the 1.25 mg m^{-3} soil to 6% in the 1.55 mg m^{-3} soil. The wetter soils had much lower soil strength (0.4 to 0.8 MPa, too low to impede root growth), but the most compacted soil had poor root growth because of anaerobic conditions that resulted from the low air-filled pore volume.

Under severe compaction, soils may puddle. Puddling results from the dispersal of soil particles in water and the differential rate of settling that permits the orientation of clay particles so that they lie parallel to each other. The destruction of soil structure by this method may result in a dense crust that has the same effect on soil conditions as a thin, compacted layer. The crust is most common on soil surfaces where the litter has been removed by burning or by mechanical means. Reduced germination and increased mortality rates of loblolly and slash pine seedlings have been observed on soils compacted or puddled by logging equipment.

SOIL COLOR MAY BE OBVIOUS, INDICATING SUBTLE AND OBVIOUS SOIL CONDITIONS

Color is an obvious characteristic of soils that is used to differentiate soil horizons and classify soils. In many parts of the world, soils may be described as

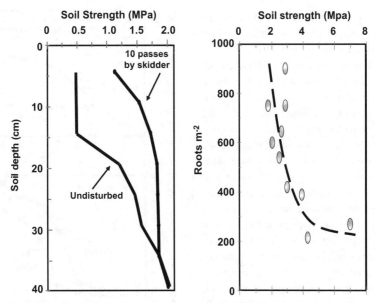

Figure 5.7 Soil strength increased substantially after ten passes with a rubber-tired skidder in Australia, and root penetration declined as root strength increased. However, the domain over which these effects occurred did not overlap; the skidder did not increase soil strength enough to reduce root growth (modified from Gracen and Sands, 1980).

Red and Yellow Podzolics, Brown Earths, Brown Forest Soils, and Chernozens (Black Earths).

Soil color depends on pedogenic processes and the parent material from which the soil was derived. Color is generally imparted to the soil by small amounts of colored materials, such as iron, manganese, and organic matter. Red colors generally indicate ferric (oxidized iron) compounds associated with well-aerated soils. Yellow colors may signify intermediate aeration. Ferrous (reduced iron) compounds of blue and gray colors are often found under reduced conditions associated with poorly aerated soils. Mottling (the interspersion of different colors) often indicates a zone of alternately good and poor aeration. Manganese compounds and organic matter produce dark colors in soils. Color intensity is often used to estimate organic matter content. It is not a foolproof system, however, because the pigmentation of humus is less intense in humid zones than in arid regions. Brown colors predominate in slightly decomposed plant materials, but more thoroughly decomposed amorphous material is nearly black.

Color itself is of no importance to tree growth, but color is correlated with several important characteristics of soil. These characteristics include geologic origin and degree of weathering of the soil material, degree of oxidation and reduction, content of organic material, and the leaching and accumulation of organic matter and calcium and iron, which may greatly influence soil quality.

Dark-colored surface soils absorb heat more readily than light-colored soils but because of their generally higher content of organic matter, they often have higher moisture content. Therefore, dark soils may warm less rapidly than well-drained, light-colored soils. Soil color influences the temperature of bare soils, but it has less effect on the temperature of soils beneath forest canopies. Soil color becomes important after fires, when removal of the canopy combines with blackening of the soil surface to increase temperature.

SOIL TEMPERATURE INFLUENCES BIOTIC AND ABIOTIC PROCESS RATES

Soil temperature is a balance between heat gains and losses. Solar radiation is the principal source of heat, and losses are due to radiation, conduction, convection, and the latent heat of volatilization consumed in the evaporation of water. When well-developed canopies are present, the temperature of upper soil layers varies more or less according to the temperature of the air immediately above it. Temperature fluctuations in deeper soil layers are moderated. In the absence of a forest canopy, topsoil temperature may be much higher than air temperature because of direct absorption of solar radiation.

Soil temperatures generally decrease with increases in elevation, although postharvest soil temperatures may be extreme at high elevations on sunny days. Soil temperature in winter is commonly warmer beneath the snowpack than above it. Conifer forests may also have cooler soils than hardwood deciduous species for several reasons. The higher leaf area of conifers may intercept more light and reduce solar heating of the soil, and may reduce convective heat losses as well. Conifer canopies also intercept more snow in winter, providing shallower snow layers on the ground to insulate the soil from frigid winter air. Temperatures also tend to decrease with latitude, and seasonal swings become more pronounced (Figure 5.8).

Aspect (direction of slope) influences soil temperature. Soils receive more solar radiation on south-facing slopes in the Northern Hemisphere and on north-facing slopes in the Southern Hemisphere. Higher temperatures typically lead to greater rates of evapotranspiration, so the heating effect on the soil may be amplified by drying.

The tree canopy and O horizon moderate extremes of mineral soil temperatures. They protect the soil from excessively high summer temperatures by intercepting solar radiation, and they reduce the rate of heat loss from the soil during the winter. Forest cover may influence the persistence of frost in soils in cold climates; however, freezing generally occurs earlier and penetrates deeper in bare soil than in soil under a forest cover.

The effect of forest canopies on reducing frosts has long been attributed to canopies' absorbing long-wave radiation emitted by the soil and then reradiating some of the energy back to the soil in a form of the greenhouse effect. Direct measurements of radiation budgets have not supported this story

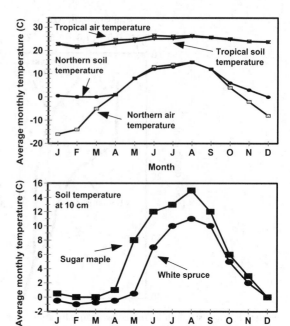

Figure 5.8 Seasonal traces of temperatures for air and upper soil at the Luquillo Experimental Forest in Puerto Rico (tropical site) and at Isle Royale, Michigan, for the northern forests of sugar maple and white spruce. The upper graph provides air temperature and soil temperature for the tropical, closed-canopy forest, which shows little seasonal variation and no difference between air and soil temperatures. The soil in the sugar maple stand remains warmer throughout the winter than the air because of the insulating snow layer. The lower graph contrasts soil temperatures under sugar maple and white spruce; spruce soils are colder because of less snow cover in winter and perhaps because of lower light penetration of the canopy (Luquillo data from X. Zou, personal communication; Isle Royale data from Stottlemyer *et al.*, 1998).

(Löfvenius, 1993); it appears that the beneficial effects of canopies on soil temperatures at night derive from the effects of canopies on air movement. Air at the surface of bare soils may become very cold as a result of radiative heat loss to the cold night sky. The presence of even a few trees per hectare can reduce the development of such cold air layers by intercepting some of the air currents higher above the ground and creating turbulent airflow that prevents stratification.

Part of the ameliorating effect of the forest cover on soil temperature is due to the O horizon

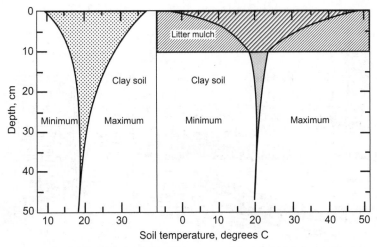

Figure 5.9 Daily temperature variations with depth for (left) an unmulched and (right) a mulched clay soil where the litter mulch has a lower thermal conductivity (Cochran, 1969).

(O horizons), as shown for clay soil in Figure 5.9. Organic layers have low thermal conductivity, and they lower maximum summer temperatures and raise minimum winter temperatures. Diurnal fluctuations are also dampened.

Snow also insulates soils because of its low thermal conductivity. A porous snow covering of about 45 cm was sufficient to prevent soil freezing in frigid northern Sweden (Beskow, 1935), and a covering of 20 to 30 cm was sufficient to prevent freezing in southern Sweden. Winter soil temperatures may often be higher at northern locations and high elevations, where the snow cover is thick. Sartz (1973) reported that aspect influences soil freezing in Wisconsin because the direction of the slope affects the amount of snow accumulation. Therefore, depending on the conditions, frost may be deeper on southern slopes than on northern slopes (Table 5.3).

Table 5.3 Snow and frost depths (in cm) on north and south slopes in Wisconsin (Sartz, 1973).

	Snow		Frost	
Year and Date	North	South	North	South
1970 (February 25)	25	0	8	11
1971 (February 25)	55	25	0	0
1972 (March 10)	45	25	16	23
1973 (March 01)	5	0	21	12

Specific heat and conductance of heat are inherent properties of soils that influence swings in daily temperature. Specific heat refers to the number of Joules needed to raise a unit mass of water by a defined number of degrees; raising the temperature of 1 kg (or 1 L) of water from 15°C to 16°C requires an input of 4.2 MJ of energy. Conductance refers to the movement or penetration of thermal energy into the soil profile. Both specific heat and conductance are influenced somewhat by texture, and especially by soil water content and organic matter content. Water has high specific heat and high conductance.

Wet soils are slower to change their temperature for two reasons: it takes a lot of energy to warm the water, and meanwhile the water transmits heat deeper into the soil. Organic matter has low conductance and impedes the movement of thermal energy.

Favorable soil temperature is essential for the germination of seeds and the survival and growth of seedlings. In some cold climates, a dense canopy may keep soil temperature so low that it delays germination and slows up seedling development. On the other hand, removal of the forest cover may result in increases in soil temperature to lethal levels for germinating seedlings. Soil surface temperatures vary among substrates, with high temperatures in dark residues (especially charred residues) and lower temperatures in moist, decayed logs (Figure 5.10).

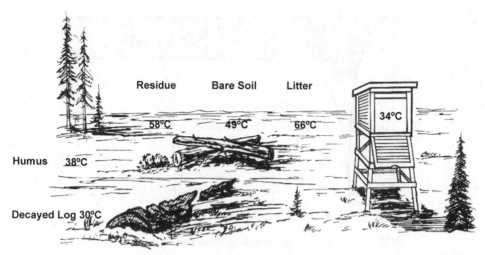

Figure 5.10 Surface temperatures of substrates in a clear-cut area in the mountains of Montana (Hungerford, 1980).

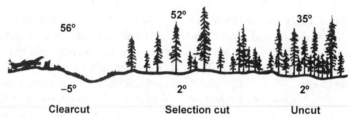

Figure 5.11 Average surface temperatures in uncut, partially cut, and clear-cut units in the mountains of Montana (Hungerford, 1980). Night temperatures in the clear-cut area fall below 0°C as a result of reduced turbulence without trees, which allows air at the soil surface to super cool; daytime temperatures reach extreme levels under high-altitude sunshine.

Clear-cut areas have different radiation and wind regimes, which tend to accentuate daily swings between low nighttime and high daytime temperatures (Figure 5.11).

Root growth is also affected by soil temperatures, as shown by Lyford and Wilson (1966). They found that day-to-day variations in growth rates of red maple root tips closely paralleled variations in surrounding soil temperatures.

SOIL WATER IS PART OF THE HYDROLOGICAL CYCLE

The supply of moisture in soils largely controls the types of tree that can grow, and the distribution of forests around the world relates to patterns in precipitation and soil moisture. Water is essential to the proper functioning of most soil and plant processes.

Carbon uptake by plants requires the moist interiors of leaves to be exposed to relatively dry air. Given the concentration gradient between wet leaves and dry air, about 200 to 500 molecules of water are lost for every molecule of carbon dioxide acquired. In addition to serving the metabolic needs of the plant, water is critical for many functions in the soil. Water is a solvent and a medium of transport of plant nutrients, a medium for the action of the exoenzymes produced by soil microbes, and has an important influence on soil temperature and aeration.

The retention and movement of water in soils and plants involves energy transfers. Water molecules are attracted to each other in a polymer-like grouping, forming an open tetrahedral lattice structure. This asymmetrical arrangement results from the dipolar nature of the water molecule. Although water molecules have no net charge, the hydrogen atoms sit to one side of the oxygen atom, giving the molecule a partial positive charge on one side and a partial negative charge on the other. As a result, the hydrogen atoms of one molecule attract the oxygen atom of an adjacent molecule. This hydrogen bonding (along with partial covalent bonding) accounts for the forces of adhesion, cohesion, and surface tension that largely regulate the retention and movement of water in soils. Adhesion refers to the attractive forces between soil surfaces and water molecules. At the water–air interface, surface tension may be the only force retaining water in soils. It results from the greater attraction of water molecules for each other (cohesion) than for the air.

Water in a zone of high free energy tends to move toward a zone of low free energy – from a wet soil to a dry soil and from the upper soil to the lower soil. The amount of movement depends on the differences in the energy states between the two zones; these differences are referred to as differences in potential. The potential of water in soil has three components. Matric potential is the attraction of water to soil surfaces; small pores hold water more tightly than larger pores do. Water that is low in dissolved ions tends to move into areas with higher ion concentrations, which represents the osmotic potential. Water also tends to move toward the center of the Earth because of the gravitational potential. The total soil water potential is the combined effects of matric, osmotic, and gravitational potentials. In physics, matter tends to move from zones of higher potential to zones of lower potential, and the unit for potential is the Pascal ($= 1$ Newton m^{-2}); other units that have commonly been used in the United States are bars or atmospheres, which equal about 100 kPa, or 0.1 MPa. Common soil water potentials range from near zero for very wet soils to -3.0 MPa or lower. The water potential at the bottom of a lake may have a positive potential because the mass of water above it has a positive pressure (or head).

Investigators have long attempted to devise useful equilibrium points or constants for describing soil moisture. Such terms as field capacity and permanent wilting point have found their way into soils literature over the years. Most of these terms deal with hypothetical concepts and do not apply equally

well to all soil conditions. Nevertheless, they are employed commonly and may be of use.

Field capacity describes the amount of water held in the soil after gravity has drained most of the water that is easily drained. Field capacity is difficult to determine accurately. Factors such as soil texture, structure, and organic matter content influence measurements. Soil layers of different pore size markedly influence water flow through a profile, greatly affecting field capacity. Field capacity may be defined as the water content and water potential after one or two days of draining following full rewetting of the soil, or 0.03 MPa for silt loam soils and 0.01 to 0.001 MPa for sandier soils.

The permanent wilting point is a classic term for the soil moisture potential at which plants remain permanently wilted, even when water is added to the soil. Just as field capacity has been widely used to refer to the upper limit of soil water storage for plant growth, the permanent wilting point is used to define the lower limit. The use of a test plant, such as a sunflower, is the most widely accepted method for determining this soil water potential, but it has no particular relevance to trees.

Soil water-holding capacity is a useful term that refers to the quantity of water held within soils between the freely drained level of field capacity and some arbitrary potential beyond which plant uptake becomes minimal. For example, the Commerce silt loam (a Fluvaquentic Endoaquept) holds about 45% moisture (water mass as a percentage of equivalent soil dry mass) at field capacity and about 10% moisture at -2 MPa, for a water-holding capacity of 35% of the soil's dry mass (Figure 5.12).

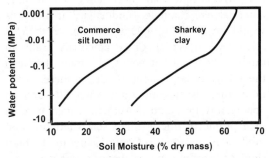

Figure 5.12 Relations of soil water potential to soil moisture content in Sharkey clay and Commerce silt loam (modified from Bonner, 1968).

Clays hold more water than silt loams, but they hold it more tightly. The Sharkey clay (a Chromic Epiaquert) holds about 65% water at field capacity and 35% at −2 MPa, for a water-holding capacity of 30% of the dry soil mass. At 35% moisture, roots obtain water easily from the silt loam, where the water potential is 0.01 MPa at that point, but not from the clay, where the water potential is −2.0 MPa at that point.

Soil water potential also affects the activities of microbes. A variety of studies have shown that carbon dioxide formation (an integrated measure of microbial activity) is reduced by half as soil water potential declines from saturated to −0.2 MPa (Sommers et al., 1980). Nitrogen mineralization also declines as soil moisture declines. One study in Kenya found that gross Nitrogen mineralization dropped from 2.2 to 0.08 $\mu g\,g^{-1}$ daily as soil water potential fell from −0.06 to −5.9 MPa (Pilbeam et al., 1993).

WATER FLOWS IN UNSATURATED SOILS

The availability of soil water to plants depends on its potential and on the hydraulic conductivity of the soil. In saturated soils, water uptake by trees is not limited by the rate of water movement through soils. As water drains from the soil, macropores empty and water is present only in capillary pores, which hold water with strong negative potential and also retard the flow of water. The rate at which water can move through a soil, hydraulic conductivity, is related to water-filled pore size and water potential. For example, conductivity at 0.02 MPa would be about 10 000 times greater than at −1.0 MPa. At very negative potentials, water in sands is held only at points of contact between the relatively large particles. Under these conditions, there is no continuous water film and no opportunity for liquid movement. Layers of sand in a profile of fine-textured soil, therefore, can inhibit downward movement of water in a fashion similar to that of compact clay or silt pans.

Layering or stratification of materials of different texture is common in soils as a result of differences in the original parent material or pedogenic development. These factors can result in silt or clay pans or lenses of sand or gravel. Layers of fine-textured materials over coarse sands and gravel, as well as the reverse situation, are common in glacial outwash sands in glaciated terrain. Many Ultisols (highly weathered soils with low base saturation) have sandy A horizons underlain by clay-rich B horizons. Changes in texture throughout a profile tend to slow water movement, whether the change down through the profile is from coarse to fine (the fine horizon has lower saturated conductivity) or from fine to coarse (water cannot leave the fine layer and enter the coarse layer except near saturation).

SEVERAL FACTORS AFFECT INFILTRATION AND LOSSES OF WATER

The term infiltration is generally applied to the entry of water into the soil from the surface. Infiltration rates depend mostly on the rate of water input to the soil surface, the initial soil water content, and internal characteristics of the soil (such as pore space, degree of swelling of soil colloids, and organic matter content). Overland flow happens only when the rate of rainfall exceeds the infiltration capacity of the soil. Because of the sponge-like action of most O horizons and the high infiltration rate of the mineral soil below, there is little surface runoff of water in mature forests. Overland flow is rarely a problem in undisturbed forests, even in steep mountain areas.

When rainfall does exceed infiltration capacity, the excess water accumulates on the surface and then flows overland toward stream channels. This surface flow concentrates in defined stream channels and causes greater peak flows in a shorter time than water that infiltrates and passes through the soil before reappearing as stream flow. High-velocity overland flow can erode soils. Soil compaction by harvesting equipment, disturbances by site preparation, and practices that reduce infiltration capacity and cause water to begin moving over the soil surface are of great concern to forest managers. Special care is warranted on shallow soils and those that have inherently low infiltration rates. Soils that are wet because of perched water tables or because of their position along drainage ways are unusually susceptible to reduction in porosity and permeability by compaction.

The O horizons beneath a forest cover is especially important in maintaining rapid infiltration rates. This layer not only absorbs several times its own mass of water, but also breaks the impact of raindrops, prevents agitation of the mineral soil particles, and discourages the formation of surface crusts.

The incorporation of organic matter into mineral soils, artificially or by natural means, increases their permeability to water as a result of increased porosity. Forest soils have a high percentage of macropores through which large quantities of water can move – sometimes without appreciable wetting of the soil mass. Most macropores develop from old root channels or from burrows and tunnels made by insects, worms, and other animals. Some macropores result from structural pores and cracks in the soil. As a consequence of the better structure, higher organic matter content, and presence of channels, infiltration rates in forest soils are considerably greater than in similar soils subjected to continued cultivation (Figure 5.13).

Wood (1977) found that water infiltration rates were higher on 14 of 15 forest sites than on adjacent sites used for pastures or for pineapple or sugarcane crops in Hawaii. In this study, lower bulk densities and greater porosities were found in forest-covered soils than in non-forested soils.

The presence of stones increases the rate of water infiltration in soils because the differences in expansion and contraction between stones and the soil produce channels and macropores. However, stones reduced the retention capacity of soils for 40 Oregon sites (Figure 5.14).

Burning of watersheds supporting certain types of vegetation may result in a well-defined hydrophobic (water-repellent) layer at the soil surface, which may increase erosion rates (see Chapter 13). This heat-induced water repellency results from the vaporization of organic hydrophobic substances at the soil surface during a fire, with subsequent condensation in the cooler soil below.

Evaporation from surface soils increases with increases in the percentage of fine-textured materials, moisture content, and soil temperature and decreases with increases in atmospheric humidity. Evaporation losses are also influenced by wind velocity. Heyward and Barnette (1934) found that the upper soil layers of an unburned longleaf pine forest contained significantly more moisture than did similar soils in a burned forest because of the mulching effect of the O horizon in the unburned areas.

Fine-textured soils have higher retention capacity for water than sands, and can store larger amounts of water following storm events. The effective depth of the soil (rooting depth) and the initial moisture content are factors that also influence the

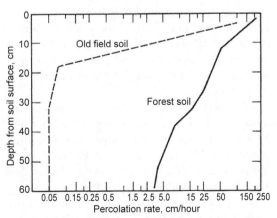

Figure 5.13 Comparative percolation rates by soil depth for a forest soil and an adjacent old-field soil in the South Carolina Piedmont (Hoover, 1949).

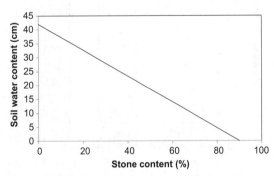

Figure 5.14 Retention storage capacity in the surface 120 cm of 40 soil profiles representing four soil series in Oregon as a function of average stone content (Dyrness, 1969).

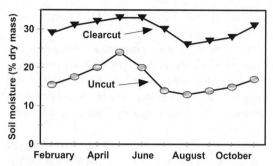

Figure 5.15 Seasonal course of soil moisture in a clear-cut and an uncut forest in the mountains of Montana (data from Newman and Schmidt, 1980).

amount of rainwater that can be retained in soils for later use by plants.

The condition of a forest has a strong influence on soil water as a result of differences in water use and water input. Clear-cutting a mixed conifer forest in Montana led to substantially wetter soils throughout the year (Figure 5.15). The difference was related to lower water use in the absence of trees but also to greater snow input (Figure 5.16).

Snow that accumulates in conifer canopies may sublimate without reaching the soil surface, lowering the input of water to the soil relative to

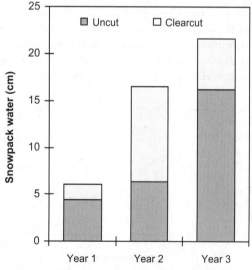

Figure 5.16 Water content of the snowpack across three years was substantially higher in clear-cut areas than in adjacent mature conifer forests in the mountains of Montana (data from Newman and Schmidt, 1980).

clear-cut areas, with less sublimation (or interception loss).

Differences in interception loss can also be important among tree species that support substantially different leaf areas. For example, Helvey and Patric (1988) contrasted the interception loss for a watershed dominated by a mixed-age, mixed-species hardwood forest with a watershed planted with a single-age stand of white pine. Average annual interception loss from the mixed-hardwood stand ($250 \, mm \, yr^{-1}$) was less than half of the loss from the white pine stand ($530 \, mm \, yr^{-1}$). Soil leachates were more concentrated under the pine stand, due in part to the reduced quantity of water leaching through the soils (data in Johnson and Lindberg, 1992).

TREES MAY GET WATER FROM THE CAPILLARY FRINGE

The upper surface of the zone of saturation in a soil is called the groundwater table. Extending upward from the water table is a zone of moist soil known as the capillary fringe. In fine-textured soils, this zone of moisture may approach a height of 1 m or more, but in sandy soils it seldom exceeds 25 to 30 cm. In some soils, the height of the water table fluctuates considerably between wet and dry periods. In forested soils in New England, water tables are highest during late autumn, winter, and early spring; gradually lower in late spring and early summer; and rise again in the autumn. In the flatwoods of the southeastern coastal plain of the United States, the water table may be above the soil surface during wet seasons and fall to a depth of 1 m or more during dry seasons.

Soil depth often reflects the volume of growing space for tree roots above some restricting layer, and in wet areas this restricting layer may be a high water table. The water table may be a temporary (perched) zone of saturation above an impervious layer following periods of high rainfall.

Large seasonal fluctuations in the depth to the water table are harmful to root development for many species. Roots that develop during dry periods in the portion of the profile that is later flooded may be killed under the induced anaerobic conditions. A reasonably high water table is not detrimental to tree growth as long as there is little fluctuation in its level. In an experiment in the southeastern coastal plain of the United States, slash pines grown for five years in plots where the water table was maintained at 45 cm by a system of subsurface irrigation and tile drains were 11% taller than trees in plots where the water table was maintained at 90 cm, but they were 60% taller than trees in plots where the water table was allowed to fluctuate normally (White and Pritchett, 1970).

The primary benefits derived from drainage of wet soils may result more from an increase in the nutrient supply than from an increase in the soil oxygen supply. The improved nutrition is brought about by an increase in the volume of soil available for root exploitation, and by faster mineralization of organic compounds. For example, fertilizer experiments in nutrient-poor wet soils of the coastal plain indicated that tree growth was increased as much by draining as by fertilization (Pritchett and Smith, 1974).

Trees obtain moisture from the water table or the capillary fringe within reach of their deep roots even when the water table is at a considerable depth. Thus, the water tables of some wet soils are lowered to a greater extent where trees are grown than where grasses and other shallow-rooted plants are grown. Wilde (1958) reviewed a number of reports of general rises in the soil water table following removal of a forest cover from level areas in temperate and cold regions. The increase in the water table level results from a reduction in water losses from evaporation on leaf surfaces (interception loss) and from transpiration of water from within leaves.

TREES REQUIRE GREAT QUANTITIES OF WATER

Tree roots absorb vast quantities of water to replace that lost by transpiration and that used in metabolic activities. Under favorable conditions, this loss can be as much as $6 \, \text{mm} \, \text{day}^{-1}$ ($60\,000 \, \text{L} \, \text{ha}^{-1}$) during summer. Trees use 500 to 1000 kg of water for each kilogram of wood produced. Tree roots can effectively exploit soil moisture even when the soil moisture content is relatively low. The mycelia of mycorrhizal fungi may play an important role in the extraction of soil, water, and nutrients.

In spite of the efficiency of most tree root systems, the distribution of trees is controlled to a great extent by the water supply. Wherever trees grow, their development is limited to some degree by either too little or too much water. Even in relatively humid areas of the temperate zone, forest soils may be recharged to field capacity only during the dormant season or for a brief time after very wet growing periods.

PHYSICS IN A LANDSCAPE CONTEXT CAN DETERMINE SOIL CHEMISTRY

The physical properties described above need to be placed in a landscape context. Most forest soils are affected by their context on landscapes. Ridge-top soils may be excessively well drained. Mid-slope sites receive water and nutrients from upslope sites but lose some water to downslope sites. Lower-slope positions may receive more water and nutrients than leach away, and in some cases may become saturated and experience periods of low oxidation–reduction potential. These landscape issues are sometimes lumped together under the term topography, and many generalizations may be offered. We stress that landscape position (or topography) is very important in forest soils but that broad generalizations may not be helpful. For example, the sequence of soil moisture supply, nutrient supply, and overall fertility commonly increases from ridges, to mid-slopes, to toe slopes in relatively young soils on the sides of hills, mountains, and valleys (Figure 5.1).

Rather than try to establish generalizations about the role of landscapes in soil physics, we present some of the biogeochemistry of a single landscape sequence (catena) for a boreal forest in Sweden. The same biogeochemical factors are integrated in different ways at other sites, but this case study illustrates the sorts of features that emerge in landscape contexts.

Table 5.4 Stand and soil characteristics along a 90-m transect of soils developed in sandy till in Sweden. *Source:* Data from Giesler *et al.* (1998).

Feature	Upper End	Lower End
Dominant tree	Scots pine	Norway spruce
Site index (m at 100 years)	17	28
Basal area (m^2/ha)	22	32
O horizon morphology	Thin Oa (L), indicating little soil faunal mixing Thick Oe (F) Thin Oi (H)	Thick Oi (H) only, indicating major soil fauna activity
O horizon solution		
pH	3.8	6.4
NO$_3$ (µmol/L)	10	160
PO$_4$ (µmol/L)	300	8
O horizon Extractable Fe + Al (mg/g)	2	35

The Betsele transect, located in northern Sweden in the valley of the Umea River, is 100 m long and descends a gentle (2%) slope of soils formed from sandy till deposits. At the upper end it is dominated by a 125-year-old stand of Scots pine with an understory of dwarf shrubs (mostly *Vaccinium*). The lower end is dominated by a Norway spruce stand of the same age, with a variety of tall herbs in the understory. The upper end is much less productive (site index of 17 m at 100 years) than the lower end (site index 28 m at 100 years) (see Table 5.4).

The key driver of the differences in soil chemistry and overall soil fertility is soil water. At the upper end the Betsele transect is excessively drained; water leaches downward out of the soil. The lower end is in a position to receive the "discharge water" from the upslope ecosystems, providing abundant water supply and changing nutrient cycles. The major biogeochemical implications are that the upper ecosystems acidify and are nitrogen limited,

and the lower ecosystems receive abundant alkalinity (acid-neutralizing capacity) from the discharge water and have high nitrogen availability. One surprising feature is the low (and limiting) concentrations of phosphorus at the lower slope position. Here, mixing of the organic litter with mineral soil brings phosphorus into contact with iron and aluminum, allowing sorption and removal of phosphorus from the soil solution. Alternatively, Giesler *et al.* (1998) have suggested that iron and aluminum enrichment of the forest floor at the toe slope may have resulted from dissolved inputs, particularly iron, which may have high solubility during anaerobic periods under saturated conditions. Overall, the authors conclude that the three-unit gradient in soil pH was driven primarily by the differences in base saturation, caused by the input of alkalinity into discharge water, rather than by differences in acid quantity or strength.

PART III

Life and Chemistry in Forest Soils

Part II

Life and Chemistry
in Forest Soils

CHAPTER 6
Life in Forest Soils

OVERVIEW

Soils are living systems, where tree roots connect with the microbial and animal communities in the soil. These living systems are very flexible and indeterminate, as feedback between processes and structures creates malleable "physiological" networks. Root systems penetrate deep into soils, typically several meters or more. Roots anchor trees to the soil, resisting the forces of wind on canopies, and roots also help anchor soils on hillslopes and mountainsides. The surface area of roots is augmented with very small diameter filaments in the form of mycorrhizal fungal hyphae; most of the uptake of water and nutrients comes through these hyphae rather than directly into roots. Trees fuel the soil food webs through the input of above-ground detritus, and especially through the direct supply of carbohydrates to roots (while living and after death) and the associated mycorrhizosphere community. Soil food webs are the locus of the great majority of genetic biodiversity in forests, exceeding the diversity of plants by orders of magnitude.

THE SINGLE MOST IMPORTANT THING TO COMMUNICATE ABOUT SOILS

Perhaps the most important insight to share with people who know little about soils is that soils are vibrant, living systems. Hans Jenny (1984) expressed this idea: "Many ecologists glibly designate soil as the abiotic environment of plants, a phrase that gives me the creeps." A non-living view of soils would expect that geochemical equilibria would explain soil chemistry, that change would be minimal, and that risks of damaging or enhancing soils would be small. A living view of soil recognizes the fundamental role of organic molecules in shaping soil chemistry, that soil processes are far removed from equilibrium as a result of free energy generated by living organisms, and that soils are very dynamic and subject to both harm and improvement.

Soils have often been an integral part of ecosystem-focused studies, and just as often have been ignored or placed in a "black box" where only the inputs and outputs warrant attention (reviewed by Binkley, 2006). Charles Darwin may have been the first scientist to begin to appreciate the biological nature of soil and to make systematic observations. A friend of Darwin's called his attention to the "sinking" of rocks placed on the surface of meadow soils, leading Darwin to ponder "On the Formation of Mould" (Darwin, 1883):

> "I was thus led to conclude that all the vegetable mould over the whole country has passed many times through, and will again pass many times through, the intestinal canals of worms. Hence the term 'animal mould' would be in some respects more appropriate than that commonly used of 'vegetable mould.'"

In modern terminology, we would use the term "humus" rather than "mould." Darwin continued to develop a very integrated, bio-centric focus on soil development:

> "Beneath large trees few castings can be found during certain seasons of the year, and this is apparently due to the moisture having been sucked out of the ground by the innumerable roots of the trees;

Ecology and Management of Forest Soils, Fourth Edition. Dan Binkley and Richard F. Fisher.
© 2013 John Wiley & Sons, Ltd. Published 2013 by John Wiley & Sons, Ltd.

for such places may be seen covered with castings after the heavy autumnal rains. Although most coppices and woods support many worms, yet in a forest of tall and ancient beech-trees in Knole Park, where the ground beneath was bare of all vegetation, not a single casting could be found over wide spaces, even during the autumn."

LIVING SOIL HAS THREE FUNDAMENTAL FEATURES

An inanimate system may be understood based on its physical structure and its passive interactions of energy and matter. Living systems have additional complications because feedback leads to elaboration of structure, alteration of processes, and the emergence of novel features. Small changes in inanimate systems, such as automobile engines, can lead to failure of the system; living systems routinely change without catastrophic failures, though some changes may be too severe for the system to persist.

This living view of the soil has three fundamental features:

- soils are "physiological" networks with intimate relationships at microscopic scales;
- feedback between processes and structure leads to changes in future processes and structures; and
- the state of living soils is flexible and indeterminate.

The processing of matter and energy in soils entails intricate interactions between the supplies of high-energy molecules (such as carbohydrates from plants), the generation and excretion of exoenzymes by microbes that digest large molecules into small, soluble molecules, the transport of these molecules through soils and across membranes of cells, and all the biochemistry that occurs within cells. The rates of these processes depend on the processes that influence the growth and death of plants (and plant tissues), the microenvironmental conditions that influence chemical and biological processes (such as soil moisture and oxygen supply), and the genetic potential of the suite of soil

organisms. This physiological nature of living soil leads to a sort of "uncertainty principle": the more an individual detail is isolated for study from the time and space context of the living soil, the less certain we can be about how this detail actually functions in a dynamic soil. Soil science depends largely on samples of soils that are cut from the living soil, placed in vials or on microscope slides, treated with chemicals and measured, but these acts of measurement, in part, determine what will be measured. Högberg and Read (2006) emphasized the need to take an integrated, physiological approach to soils:

> "The inherent complexity of the connected plant–soil system has long served to deter soil scientists from carrying out appropriately integrated studies of naturally intact systems. Although fragmentation of such systems is sometimes justified, we should recognize that soils are connected to plants almost as integrally as arteries are to veins in animals, and that our investigations need to embrace these connections rather than sever them . . . "

The current condition of any soil is a legacy of how previous processes altered the structure of the soil, with these changes in structure altering current rates of processes. The addition of organic matter alters the physical structure of soils, changing the three-dimensional movement of water, oxygen, and other chemicals over time. The processing of organic matter, and the nutrients and energy contained in organic matter, depends on organisms that are influenced by the supplies of water, oxygen, and other chemicals, leading to further changes in soil structure. A comparison of the properties of unaltered parent material and a well-developed soil reveals the cumulative history of feedback between chemical and biological processes and soil structure.

When a factory begins the assembly of a new automobile, there is no uncertainty in the outcome of the assembly process. The development of living soils is almost entirely unlike this process. The transformation of parent material into soil depends on the interactions of so many important factors, each of which is variable with some degree of uncertainty, that soil development is always somewhat indeterminate and only broadly describable. The influence of very dry or very wet environments is generally predictable as leading to indurate, calcified soils or saturated, peatland soils; but across smaller gradients in moisture regimes, the contingent influence of features (such as plant and animal species that may or may not be present) can largely determine soil development. The indeterminate nature of living soils means that soils change over time, and that management activities can substantially alter soils (for better or worse).

LIFE IN THE FOREST SOIL BEGINS WITH ROOTS

Hans Jenny also remarked that "trees and flowers excite poets and painter, but no one serenades the humble root, the hidden half of plants." Roots can be excavated from soils, but the act of removing them for study breaks the intimate living connections with mycorrhizal fungi, the rhizosphere microbe community, and the rapid chemical reactions that are, in part, driven by organic molecules from the roots (Figure 6.1).

Plant roots supply the connecting link between the plant and the soil, and studies of tree root systems are especially pertinent to forest soil science. Roots provide anchorage for the tree (Figure 6.2) and serve the vital functions of absorption and translocation of water and nutrients. They exert a significant influence on soil profile development by fueling most of the biogeochemistry within the soil, via the input of organic matter while roots are alive, when fine roots die over a period of months or years, and when the whole root system dies (McClaugherty *et al.*, 1982; Fahey *et al.*, 1988). It should not be surprising that the growth and distribution of roots are influenced by essentially the same environmental factors that affect growth of the aboveground portion of the tree. Not only do variations in the chemical, physical, and biological properties of the soil have profound effects on tree roots, but the influence of such factors as light intensity, air temperature, and wind may be

Figure 6.1 Roots connect the soil system with the sky and sunlight, bringing the organic materials that shape soil structure and fuel biogeochemistry. (Photo by Cristian Montes.)

reflected as much in root growth as in shoot growth. Around the world, trees typically have about 20% of their biomass hidden below ground (Cairns *et al.*, 1997).

ROOT SYSTEMS HAVE CHARACTERISTIC FORMS AND ENORMOUS EXTENTS

When grown under favorable soil conditions, tree species tend to develop distinctive root systems. Some send taproots deep into the soil; others tend to spread roots laterally in the upper soil; and some tree species do both (Figure 6.3). Large roots provide anchorage for the tree, and serve as core conducting members of the large number of feeder roots (Figure 6.4) that support the mycorrhizal network where absorption of water and nutrients occurs. The actual root system developed by individual trees is shaped, in part, by the microsite conditions of soil texture, strength, available moisture, impeding layers, and nutrition (Sutton,

Figure 6.2 The immensity of root systems and their influences on soils are easy to overlook under typical conditions. One of the most severe windstorms in four centuries toppled this ancient western hemlock in British Columbia, Canada, revealing the root system that fueled the biogeochemistry of the soil and shaped soil structure and biotic communities. The pile of soil raised by the falling tree, as well as the pit, will persist for centuries or longer (see review by Šamonil *et al.*, 2010).

1991). Factors such as stand density, competition among individuals, soil microrelief, and soil rock content have substantial effects on the extension of lateral roots. Under the most favorable conditions, it is not uncommon for trees to possess some lateral roots that extend two to three times the radius of the crown (Zimmerman and Brown, 1971; Stone and Kalisz, 1991).

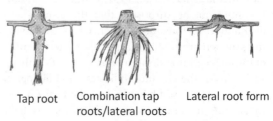

Tap root   Combination tap   Lateral root form
            roots/lateral roots

Figure 6.3 Diagrams of typical root structures.

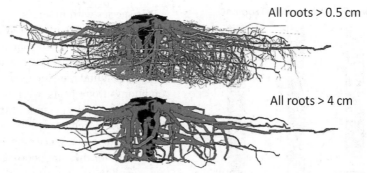

All roots > 0.5 cm

All roots > 4 cm

Figure 6.4 Root system of a mature maritime pine tree, showing the major coarse roots (1600 roots > 4 cm diameter) and all roots (6700 roots > 0.5 cm). With kind permission from Springer Science+Business Media: Plant and Soil, Assessing and analyzing 3D architecture of woody root systems, a review of methods and applications in tree and soil stability, resource acquisition and allocation, 303, 2007, page, Frédéric Danjon.

THE FORM OF ROOT SYSTEMS MAY BE RIGID OR FLEXIBLE

Root systems can be characterized on the basis of rooting habit (relating to the form, direction, and distribution of the larger framework roots) and root intensity (the form, distribution, and number of small roots). The habit, or form, of a root system is influenced by local site conditions, but also tends to be under some degree of genetic control. Taprooted trees are characterized by a strong, downward-growing main root, which may branch to some degree. Such root systems occur in species of *Carya, Juglans, Quercus, Pinus,* and *Abies.* Strong laterals from which vertical sinkers grow downward, as in *Populus, Fraxinus,* and some species of *Picea,* characterize the flat, lateral root habit. Root systems with strong roots radiating diagonally from the base of the tree without a strong taproot are characteristic of the combination form (sometimes called "heart root" form). Such systems are found in species of *Larix, Betula, Carpinus,* and *Tilia.* The three-dimensional "architecture" of root systems has been measured and modeled, revealing differences among species and responses to environmental factors (see Danjon and Reubens, 2008 for a review).

The rooting habit of a tree has considerable influence on the type of habitat in which it will thrive. Root form may determine whether or not a species is capable of fully exploiting a site and competing successfully with neighboring species. Root systems of species that are under strong genetic control, such as longleaf pine, tend to retain their rooting habit regardless of the soil conditions to which they are subjected; consequently, they grow well only in a limited range of site conditions. On the other hand, a few species, such as red maple, can adapt their juvenile root systems to a variety of environments, and the species can become established and grow on both wet and dry sites.

ROOTS PENETRATE MORE DEEPLY INTO SOILS THAN MOST INVESTIGATIONS SAMPLE

Schenk and Jackson (2002) summarized the patterns of rooting system depths around the world, and found that most investigations stop digging at less than half of the depth exploited by roots (Figure 6.5). In forests from the high latitudes to the tropics, the typical depth of the deepest tree roots is between two and three meters; much greater depths are not uncommon.

Trees that develop strong taproots are capable of penetrating the soil to great depths for support and moisture. For example, many pines, such as longleaf pine, have well-developed taproots and are capable of surviving on deep, relatively dry, sandy sites. This well-developed taproot form is also

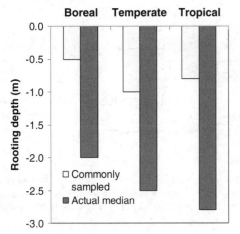

Figure 6.5 Common sampling for rooting depth stops at one or two meters for studies from boreal to tropical forests, but the typical depth of rooting is two to three meters (after Schenk and Jackson, 2002).

shared by a number of deciduous trees, exemplified by burr oak, black walnut, and many eucalyptus species (Figure 6.6). Many taprooted trees also have extensive laterals and sinkers that permit them to survive on shallow soils and soils with fluctuating water tables. Although roots do not usually persist in a zone of permanent water saturation, taproots of pitch and slash pine do develop below the depth of perched water tables (McQuilkin, 1935; Van Rees and Comerford, 1990). Cypress (*Taxodium*) and some other species common to swamps have special root adaptations that permit them to grow in saturated zones. If young taproots are injured, several descending woody roots may occur at the base of the tree in place of a single taproot. Taproots of slash pine are often fan-shaped from the capillary zone to below the mean water table. The fan-shaped roots were reported by Schultz (1972) to be two-ranked, profuse, and often dichotomously branched. Boggie (1972) reported that where the water table was maintained at the surface of deep peat, root development of lodgepole pine was confined to a multitude of short roots from the base of the stems, giving a brush-like appearance. Laterals developed only on trees where *Sphagnum* hummocks had raised the soil surface above the mean water level.

Tree species with inherently shallow or flat root systems and those with root systems under weak genetic control have particular advantages on shallow soils. Black spruce is an example of a species that can be found growing over a wide range of soil conditions, from peats with high or fluctuating water tables to deep sands. However, trees with shallow root systems may have a real disadvantage compared to other species on deep, easily penetrated soils, for they are poorly equipped to exploit these conditions and are more subject to windthrow than taprooted trees.

Species possessing a combination (heart root) form with a number of lateral and oblique roots arising from the root collar grow best on deep, permeable soils. However, they are also capable of exploiting fissures in fractured bedrock to a greater extent than other types of root systems.

SOIL CONDITIONS ALTER ROOT GROWTH

The physical, chemical, and biological properties of soils have profound effects on the rate of root growth and development, root habit, and intensity of root systems of established trees. Consequently, variations in root systems among individuals of a pliable species grown in different soils may be as great as those among different species grown on the same soil.

Soil bulk density strongly influences root growth because of limitations on the ability of roots to penetrate soils (soil strength), and because of influences on soil water supply and aeration. The bulk density that limits root penetration varies with tree species, soil moisture content, and soil texture (Sutton, 1991). The range of bulk densities that has been reported to limit root penetration is rather wide, $1.1\,kg\,L^{-1}$ in silty clay to about $2\,kg\,L^{-1}$ in clay loam. However, in general, roots grow well in soils with bulk densities of up to $1.4\,kg\,L^{-1}$, and significant root penetration begins to cease at bulk densities of around $1.7\,kg\,L^{-1}$. Basal tills and fragipans are examples of soil layers with high bulk density that often limit root growth.

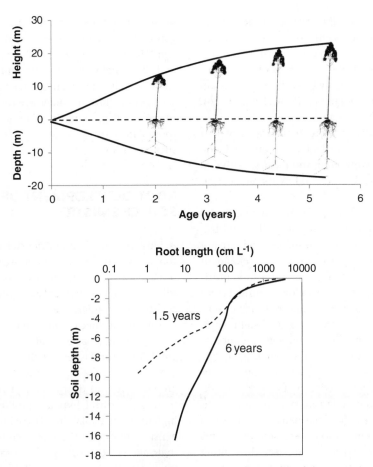

Figure 6.6 The penetration of roots into the soil equaled about 85% of the reach of the canopy into the sky for this eucalyptus stand. The proliferation of fine roots declined exponentially with depth, dropping from more than 1000 cm L^{-1} soil volume in the upper soil to 100 cm L^{-1} at 3 m (based on Christina *et al.*, 2011).

Soil strength is now often used to determine whether or not roots might penetrate a given soil layer. Soil strength is measured with a penetrometer. The values obtained with a penetrometer are determined not only by the bulk density and moisture content of the soil but also by the pore pattern of the soil and the shape and size of the penetrometer probe. Some roots are able to penetrate soil that has soil strength in excess of 3 MPa; however, the force that roots can actually exert is in the range of 0.5 to 1.5 MPa. Obviously, roots penetrate soil using mechanisms other than that of the penetrometer;

nonetheless, soil strength is a convenient guide to root growth limitation. In medium- and fine-textured soils, soil strength of 2.5 MPa, as measured by a penetrometer, curtails root extension. In coarse-textured soils, the pores permit the entry of the root tip into crevices in which wedging action permits the root to elongate. Thus, roots can extend into coarse-textured sand that has a soil strength of 5.0 MPa, as measured by a penetrometer.

Texture and structure may influence rooting through physical impedance and by their influence on soil aeration. A soil does not need to become anaerobic before root growth and function are harmed. Oxygen concentration needs to be approximately 20% to avoid interfering with respiration at the root tip; however, oxygen concentrations as low as 10% may be sufficient to sustain respiration of older root tissues (Waisel *et al.*, 1996). Trees differ in tolerance to reduced aeration. Loblolly pine apparently has a greater tolerance of poor aeration than does shortleaf pine, and red and ponderosa pines are among the most sensitive of all North American conifers to low-oxygen conditions in the root zone.

Moisture has a greater influence on root development and distribution than most other soil factors. Slow tree growth rates on shallow soils are not due directly to lack of space for root development but rather to the limitation of the water (and nutrient) supply associated with shallow rooting. Lyford and Wilson (1966) suggested that red maple roots grow rather well over a broad range of soil texture and fertility conditions if the soil is maintained at near-optimum moisture levels. Rooting habits change appreciably when maples are grown under very moist, poorly aerated conditions. Kozlowski (1968) reported that large root systems develop in tree seedlings grown in soils maintained close to field capacity, in contrast to the sparse root systems found in soils allowed to dry to near the wilting point before rewetting. Many investigators have noted that small roots die almost immediately in local dry areas (Sutton, 1991). Lorio *et al.* (1972) found that mature loblolly pine trees on wet, flat sites were nearly devoid of fine roots and mycorrhizal roots compared to neighboring trees on mounds. It is doubtful that any significant amount of root growth takes place when the soil moisture drops to near the wilting point (Waisel *et al.*, 1996), and growth may be exceedingly slow during the dry summer months that would otherwise be favorable for rapid root growth.

Optimum soil moisture content for root development depends on soil texture and temperature, among other factors, but soil moisture near field capacity is optimum for most species. Roots of most trees grow best in moist, well-aerated soils, and they generally proliferate in layers providing the greatest moisture supply only if well aerated. Water

saturation of the soil results in a deficiency of oxygen and an accumulation of carbon dioxide. Such conditions usually result in reduced root growth and, eventually, in root mortality. Some species, such as bald cypress and water tupelo and certain species of willow and alder, are capable of obtaining oxygen and growing in saturated soils; however, roots of most tree species will not survive long under such conditions.

ROOT DEVELOPMENT DEPENDS ON SOIL CHEMISTRY

Acidity and nutrient deficiencies can restrict plant root growth and development. Acid soil may inhibit roots because of the toxicity of aluminum, the solubility of which increases with increasing soil acidity. Root tolerance to acidity differs widely among plant species, and many tree species are very tolerant to acid conditions (Fisher and Juo, 1995; Tilki and Fisher, 1998).

Nutrient availability in the soil affects both the growth rate and distribution of roots. Roots (and their mycorrhizal hyphae) proliferate in zones of high nutrient availability. These increases in rooting density are due to the increased biological activity that takes place in these zones rather than to the tree's activity. Increases in site fertility, both along geographic gradients and in response to fertilization, typically lead to increases in aboveground tree growth but constant or declining belowground growth (Litton *et al.*, 2007; Högberg *et al.*, 2010).

ROOTS CAN SHAPE THE PHYSICAL PROPERTIES OF SOIL

Tree roots grow in a variety of forms and to varying degrees of intensity, governed to a large extent by genetic forces but modified by local conditions of soil and site. The influence of these latter factors on the nature and abundance of roots is very diverse and difficult to class into average conditions because of the challenges of studying roots under realistic field conditions. The fact that woody plant roots are perennial, penetrate the soil to great depth, and have the ability to concentrate in soil layers most favorable to growth makes deeper horizons much more important in forest soils than in agronomic soils. Other functions of roots that are often overlooked, but which are of tremendous importance, are those related to stability and development of the soil.

Tree roots provide a significant stabilizing force in mountainous areas where soil is subject to erosion or mass movement. Fine roots and fungal mycelia serve as binding agents for surface soil, and where larger roots penetrate below the surface horizons, they can anchor the soil mantle to the substrate (Swanston and Dyrness, 1973). The number of landslides after harvesting typically increases for three to five years because this is the time frame for roots to decompose enough to lose tensile strength (Figure 6.7). The harvesting of Japanese cedar forests in Japan led to a strong increase in landslide frequency shortly after harvesting as the strength of stump roots declined; recovery of root strength with regrowing trees dramatically lowered the frequency of landslides (Figure 6.8, Imaizumi *et al.*, 2008).

The value of living tree roots in anchoring shallow soils to the underlying subsoil is particularly important in small drainage areas, where winter storms can cause a sharp rise in groundwater level. Clear-felling further contributes to a decline in soil retention by roots through greater exposure of the soil surface and increased soil moisture levels following removal of the intercepting and transpiring tree canopy.

The form and extent of the tree root system also influence the amount of soil disturbance resulting from wind-throw. Soil mixing and the development of a pit and mound microrelief are brought about by the uprooting of individual trees during storm periods (Figure 6.2).

MYCORRHIZAL FUNGI CONNECT PLANT ROOTS WITH THE LIVING SOIL

Most plants (including all tree species) form symbiotic associations with mycorrhizal fungi; the intimate connection of roots and soils is largely a story of this symbiotic association (Brundrett *et al.*, 1996; Smith and Read, 2008). In 1885, Albert Bernhard Frank, a German forest pathologist, coined the term *mycorrhiza* (fungus root), to denote these particular associations of roots and fungi. The morphology of fine roots and mycorrhizae varies among plant and fungal species; individual tree species may form these associations with hundreds of species of fungi.

TWO CLASSES OF MYCORRHIZAE ARE PARTICULARLY IMPORTANT FOR TREES

Mycorrhizae are divided into two broad classes based on the spatial interrelation of thread-like fungal hyphae and root cells. Ectomycorrhizae are characterized by fungal hyphae that penetrate the intercellular spaces of the cortical cells and usually form a compact mantle around the short roots (Figure 6.9). Arbuscular mycorrhizae are characterized by hyphae

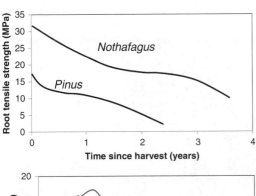

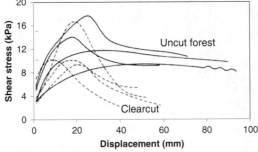

Figure 6.7 Tensile strength curves for decaying roots of *Nothofagus* and pine show the loss of strength over several years (upper; O'Loughlin and Watson, 1981), and the shear stress displacement curves from field tests of the upper 5 cm of soil in a *Nothofagus* forest (O'Loughlin *et al.*, 1982). Used with permission.

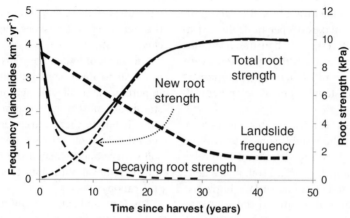

Figure 6.8 The frequency of landslides following harvesting of Japanese cedar (*Cryptomeria japonica*) forests is very high shortly after harvesting, as decay of stump roots weakens slopes. Slope failures decline as new roots increase the stability of the slope (modified from Imaizumi *et al.*, 2008).

that penetrate into the cells of the root cortex and do not form a fungal mantle (Figure 6.10). Two other types of mycorrhizae (ericoid, and those that associate with orchids) also occur in forests.

Ectomycorrhizae are restricted almost entirely to trees, and they occur naturally on many forest species. These fungi have been present for at least 50 million years based on fossil evidence, and

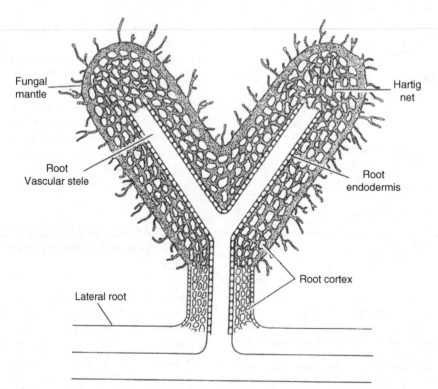

Figure 6.9 Graphic presentation (not to scale) of an ectomycorrhiza (after Ruehle and Marx, 1979).

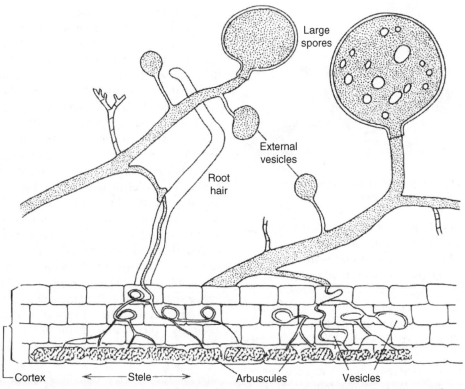

Large
spores

External
vesicles

Root
hair

Cortex ——— ← —— Stele —— → ——— Arbuscules ——— Vesicles

Figure 6.10 Graphic presentation (not to scale) of an arbuscular mycorrhiza. From T. H. Nicolson, Vesicular-Arbuscular Mycorrhiza – a Universal Plant Symbiosis. In: *Science Progress* 55 (1967): 561–581. Blackwell Scientific Publications Ltd., Oxford, UK. Used with permission.

probably more than 100 million years based on patterns of plant taxonomy and distribution (Smith and Read, 2008). More than 5000 species of ecto-mycorrhizal fungi are estimated to exist, and most of these are *Basidiomycetes*, but some also belong to the *Ascomycetes*. These fungi show a low level of specificity for host plants, and a single tree may host dozens of species of ectomycorrhizae.

The fruiting bodies of ectomycorrhizae produce spores that are readily and widely disseminated by wind, water, and animals. Ectomycorrhizae are characteristic of the families Pinaceae, Fagaceae, and Betulaceae. *Eucalyptus* and some tropical hard-wood species are also ectomycorrhizal, while such angiosperm families as Salicaceae, Juglandaceae, Tiliaceae, and Myrtaceae may be either ectomycor-rhizal or arbuscular mycorrhizal.

Ectomycorrhizal infection is initiated from spores or hyphae of the fungal symbiont in the rhizosphere

of feeder roots. Contact between hyphae of a mycorrhizal fungus and a compatible short root may originate from spores germinated in the vicin-ity of the roots, by extension through the soil of hyphae from either residual mycelia or established mycorrhizae, or by progression of hyphae through adjacent internal root tissue. Fungal mycelia usually grow over the feeder root surfaces and form an external mantle or sheath (Figure 6.11). Following mantle development, hyphae grow between cells without actually penetrating cell walls, forming a network of hyphae around root cortical cells. This "Hartig net" of hyphae may completely replace the middle lamellae between cortical cells, and is the major distinguishing feature of ectomycorrhizae (Kormanik *et al.*, 1977). Ectomycorrhizae may appear as simple, unforked feeder roots or as bifur-cated, multiforked coralloids, nodule-shaped modi-fications of feeder roots (Figure 6.12). Hyphae on

Figure 6.11 Ectomycorrhizae on slash pine. Note the fungal mantle on the short roots.

short roots radiate from the fungal mantle into the soil, greatly increasing the absorbing potential of the roots. Roots that are infected with ectomycorrhizae are comprised of about 60–90% of plant mass and 10 to 40% of fungal mass (Courty *et al.*, 2010). The biomass of ectomycorrhizal fungus might comprise about one-third of the total soil microbial biomass (Högberg and Högberg, 2002). About one-quarter of the CO_2 efflux from soils comes from respiration by ectomycorrhizae, accounting for 3–15% of total forest photosynthesis (Courty *et al.*, 2010).

Arbuscular mycorrhizae are the most widespread and important root symbionts, occurring throughout the world in both agricultural and forest soils. The fossil record shows these fungi can be traced back more than 400 million years; all species are obligate symbionts (none have the sort of saprophytic capacity that would allow them to subsist

Figure 6.12 Clumps of ectomycorrhizae on a conifer root, showing a bifurcated or coralloid short root.

independently). The number of species of arbuscular mycorrhizae is quite small, with perhaps 150 identified species for hundreds of thousands of species of plants (Smith and Read, 2008). Among the forest tree genera with these types of mycorrhizae are *Acer, Alnus, Fraxinus, Juglans, Liquidambar, Lirindendron, Platanus, Populus. Robinia, Salix,* and *Ulmus* (Gerdemann and Trappe, 1975).

Arbuscular mycorrhizae of trees are produced most frequently by fungi of the family Endogonaceae. They generally form an extensive network of hyphae on feeder roots and extend from the root into the soil, but they do not develop the dense fungal sheath typical of ectomycorrhizae. Arbuscular mycorrhizae form large, conspicuous, thick-walled spores in the rhizosphere, on the root surface, and sometimes in feeder root cortical tissue (Kormanik *et al.*, 1977). The fungal hyphae penetrate epidermal cell walls of the root and then grow into cortical cells, where they develop specialized absorbing structures called arbuscules in the cytoplasmic matrix. In some instances, the fungus completely colonizes the cortical region of the root, but it does not invade the endodermis, stele, or meristem (Gray, 1971). Thin-walled, spherical vesicles may also be produced in cortical cells, leading to the use of the term "vesicular-arbuscular" to denote this type of arbuscular mycorrhizae. Arbuscular mycorrhizal infection does not result in major morphological changes in roots; therefore, the unaided eye cannot detect it.

The fungi that form arbuscular mycorrhizae do not produce aboveground fruiting bodies or wind-disseminated spores, as do most ectomycorrhizal fungi. Because spores are produced below ground, the spread of arbuscular mycorrhizae is not as rapid as that of ectomycorrhizae, and they may be slow to invade fumigated soil and mine spoils.

SOIL FACTORS AFFECT MYCORRHIZAL DEVELOPMENT

Environmental factors influence mycorrhizal development by affecting both tree roots and fungal symbionts. After the formation of a receptive tree root, the main factors influencing susceptibility of the root to mycorrhizal infection appear to be photosynthetic potential and soil fertility (Marx, 1977). High light intensity and low-to-moderate soil fertility enhance mycorrhizal development, while the opposite conditions may reduce or even prevent mycorrhizal development.

The effects of soil fertility and fertilizer additions on the development of ectomycorrhizae appear to vary with the original fertility of the soil and the nutrient content of the host plant. Mikola (1973) reported that ectomycorrhizal formation on white pine seedlings is stimulated by applications of phosphate fertilizers to soils containing a low population of relatively inactive mycorrhizal fungi. It appears that growth suppression of pine following nitrogen applications to phosphorus-deficient soils results from a reduction in mycorrhizal development (and phosphorus absorption) caused by the high nitrogen: phosphorus ratio in host plant tissue (Figure 6.13).

Apparently, all mycorrhizal fungi are obligate aerobes, and mycelial growth likely decreases as oxygen availability decreases. The requirements of mycorrhizal fungi for nutrient elements are probably not very different from those of the host plants, although little research has been conducted on this point. The formation of ectomycorrhizae on tree

roots is often greatest under acidic conditions, though this may be due to low nutrient availability rather than to soil acidity. Richards (1961) concluded that poor mycorrhizal formation in alkaline soils was due to nitrate inhibition of mycorrhizal infection rather than alkalinity *per se*. However, Theodorou and Bowen (1969) reported that alkaline conditions in the rhizosphere severely inhibited the growth of some types of mycorrhizal fungi, apart from the possible effect of nitrate inhibition of infection. They further stated that nitrate inhibition of mycorrhiza formation under acid conditions is mainly due to a paucity of infection, not to poor fungal growth.

The presence of antagonistic soil microorganisms can influence survival of the mycorrhizal symbiont as well as root growth of the host plant. Fungicides used in plant disease control can inhibit mycorrhizal fungi under some conditions, or they may stimulate mycorrhizal development by reducing microbial competition. Eradicating ectomycorrhizal fungi from nursery soils by fumigation is generally not a problem because these fungi produce wind-disseminated spores that soon colonize the soil. The soil-borne spores of most arbuscular fungi are eliminated by fumigation of nursery soils, and artificial inoculation is essential if species requiring arbuscular mycorrhizae are to be grown successfully.

MYCORRHIZAE BENEFIT TREES LARGELY BY INCREASING SURFACE AREA FOR ABSORBING WATER AND NUTRIENTS

The dependence of most species of forest trees on mycorrhizae to initiate and support healthy growth has been most strikingly illustrated by the problem encountered in introducing trees in areas devoid of symbionts. Kessell (1927) failed to establish *Pinus radiata* in western Australian nurseries that lacked mycorrhizal fungi. The seedlings grew normally only after soil from a healthy pine stand was added to the beds and ectomycorrhizae formed. Similar results have been obtained where exotic species have been used in the Philippines, Zimbabwe (formerly Rhodesia), New Zealand, South America, and

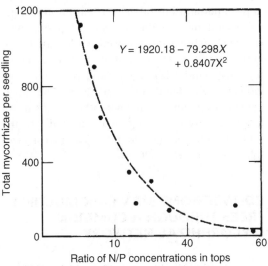

Figure 6.13 The relationship of the number of ectomycorrhizae to the nitrogen:phosphorus ratio in loblolly pine seedling tops. Reprinted from Pritchett (1972) with permission.

$$Y = 1920.18 - 79.298X + 0.8407X^2$$

Figure 6.14 After six months, radiata pine seedlings grew better in an infertile, sandy forest soil (C) than in a fertile prairie soil (A). Innoculating the fertile prairie soil with a small amount of the forest soil initiated the mycorrhizal association that allowed the best seedling development (B). (courtesy of S. A. Wilde).

Puerto Rico (Vosso, 1971). Other areas where ecto-mycorrhizal trees and their symbiotic fungi do not occur naturally include former agricultural soils of Poland, oak shelterbelts on the steppes of Russia, and the formerly treeless plains of North and South America (Marx, 1977).

The observed benefit of mycorrhizae (Figure 6.14) in the growth and development of trees largely results from the proliferation of absorbing surface area that provides trees with greater access to soil water and nutrients. Hyphae have much smaller diameters than roots, providing more than an order of magnitude increase in absorbing surface area per gram of mass; small-diameter hyphae are also more effective at absorbing ions from soil solutions (see Figure 8.11).

Ectomycorrhizae may also provide access for trees to organic forms of N dissolved in soil solutions. For example, Turnbull *et al.* (1995) showed that euca-lyptus seedlings without mycorrhizae could use ammonium, nitrate, and organic glutamine, but no other form of organic N. Inoculation with fungus allowed the seedlings to utilize a variety of amino acids and even soluble protein. The importance of the ability of ectomycorrhizae to use organic forms of soluble N would be determined by the availability of these compounds in soil solutions, and more research is needed.

Can decomposition be enhanced by mycorrhizae, gaining access to nutrients for trees? This question has two potential mechanisms. The provision of readily available C does seem to enhance the activity of decomposers (saprotrophs), and mycorrhizae-infested roots may support a host of microbes in the root "mycorrhizosphere." Some mycorrhizae play a direct role by increasing the proteolytic activity in decaying organic matter; indeed, mycorrhizae appear to span the entire gamut from obligate symbionts (receiving all their carbon from trees) to largely saprophytic (obtaining carbon from dead organic matter; Courty *et al.*, 2010).

Saprophytic fungi may be more effective in relatively fresh litter (with broad C:N compounds), whereas ectomycorrhizae may be more important in the processing of older, narrower C:N materials. For example, Lindahl *et al.* (2007) used DNA and isotope techniques to examine the spatial pattern of fungi in a Scots pine soil. Saprophytic fungi were found only in relatively fresh litter in the O horizon, where very few mycorrhizal hyphae developed. The lower O horizon and the mineral soil horizons were dominated solely by mycorrhizal fungi.

Mycorrhizae appear to be able to access both organic and inorganic pools of P in soils. The high-absorbing area of the fungal hyphae may be most important, but other mechanisms favoring P acquisition include the presence of phosphatase enzymes (to break down organic P), localized acid-ification, and excretion of anions such as oxalate that may bind with cations and release P from inorganic salts (Smith and Read, 2008). The apparent enhancement of mineral weathering and disso-lution of insoluble salts may depend on the community of bacteria found in the mycorrhizo-sphere (Courty *et al.*, 2010).

CONNECTIONS MAY LINK MULTIPLE TREES THROUGH A COMMON MYCORRHIZAL NETWORK

A single genetic individual (genet) of a mycorrhizal fungus may extend for several meters through a forest soil, linking with more than one tree. Indi-vidual trees may also be infected with more than a single fungus (or species of fungus), creating a common mycorrhizal–tree network at the scale of tens to hundreds of square meters (Courty *et al.*, 2010). Isotopic tracer studies have demonstrated flows of carbohydrate from one tree, through the common mycorrhizal network into another tree, and this sort of community may foster establishment of seedlings that can tap into the rich resource network. The actual magnitude of flows of water and nutrients between plants is difficult to determine, and issues of competition versus facilitation between plants need to be worked out for these networks.

ALMOST ALL THE DIVERSITY IN A FOREST IS FOUND IN THE SOIL

Soils are fundamental to forest ecosystems, as they form a reservoir for available water and nutrients, and anchorage for supporting massive canopies. Less apparent is the contribution of soils to the genetic biodiversity of forests; the diversity of plant species is barely a footnote in the complete community of organisms inhabiting the forest! A diverse tropical rainforest might contain more than 200 tree species in a hectare, plus a few thousand species of understory plants, epiphytes, arboreal arthropods, reptiles, amphibians, birds, and mammals. The biodiversity in the soil is almost inconceivably richer than this.

The concept of "species" applies well to some soil organisms, such as soil animals that pass genes through the generations by mating between males and females. Many small animals may be difficult to assign to traditional species, owing to very limited information about their biology and ecology. The species concept is even more difficult to apply to the microbial world, as genes can be transferred between unrelated types of cells. Modern DNA techniques can use a "barcode" approach, where operational taxonomic units (OTUs) are determined based on the level of shared versus unique segments of DNA between organisms. Fierer *et al.* (2007) used this DNA barcode/OTU approach to evaluate the biodiversity of a rainforest soil. They roughly estimated the soil community contained 20 000 OTUs of bacteria, 1000 OTUs of archaea, 2000 OTUs of fungi, and more than 10 million OTUs of viruses.

The incredible diversity of the soil community forms an intricate food web through which nutrients and energy pass. The study of this food web has lagged far behind the study of soil phenomena (DeAngelis, 1992; Hall and Raffaelli, 1993; Coleman and Crossley, 1996; Benckiser, 1997). The importance of the processes within this web to the maintenance of the soil's productive potential cannot be overestimated (Fisher, 1995), though the malleability of food webs may ensure sustained processing of matter and energy despite major impacts of disturbances or management. Food webs have been constructed for few soils, but it is clear that cultural practices applied to the soil can dramatically alter their structure.

The mass of the microbial community changes through seasons, influenced by temperature, moisture, and the supply of substrates and nutrients. A four-fold or greater range may be found in temperate forests, along with shifts in the mass of N contained in the microbes, and several-fold shifts in the C:N (Kaiser *et al.*, 2011). These shifts are driven in large part by current (and recent) photosynthesis by trees, and the rapid allocation of varying amounts of carbohydrates to the belowground portion of the ecosystem (see Högberg *et al.*, 2010). Huge changes in the biomass of the soil microbial community lead to remarkable variations in the structure of soil food webs, and the rates of processing of energy and matter.

BACTERIA AND ARCHAEA ARE SMALL, BUT WITH HUGE POPULATIONS

Although bacteria and archaea are small, typically between 0.1 and 10 μm in size, they are especially prominent in soils because of their large total biomass (1 to 10 Mg ha^{-1}, Brady and Weil, 2002) and numbers (on the order of 10^{18} ha^{-1}). The diversity of bacteria might be expected to show patterns in relation to latitude, temperature, or moisture regime; however, the best association seems to be with soil acidity (Figure 6.15). At a smaller scale and across the full suite of the soil microbial community, other approaches for characterizing microbial

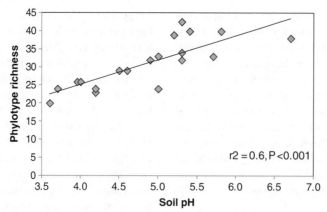

Figure 6.15 The diversity of soil bacterial communities (based on DNA techniques) in soils from humid forests around the world did not relate to latitude, temperature, or moisture regime, but did relate strongly with soil pH (data from Fierer and Jackson, 2006).

communities (phospholipid fatty acid profiles and enzymes) have shown notable differences among forests that relate to water regime and temperature (Brockett *et al.*, 2012).

Bacteria, archaea and fungi dominate in well-aerated soils, but bacteria and archaea alone account for most biological and chemical changes in anaerobic environments. In fact, the ability to grow in the absence of O_2 is the basis for further grouping of bacteria into three distinct categories: aerobes that live only in the presence of O_2; anaerobes that grow only in the absence of O_2; and facultative anaerobes that can develop in either the presence or the absence of O_2. The more meaningful grouping for understanding soil processes, however, is that based on energy and carbon source requirements.

Autotrophic bacteria and archaea are of two general types: photoautotrophs, whose energy is derived from sunlight, and chemoautotrophs, whose energy comes from the oxidation of inorganic materials. The chemoautotrophic bacteria using nitrogen compounds as energy sources include those that oxidize ammonium to nitrite (*Nitrosomonas* and *Nitrosococcus*) and those that oxidize nitrite to nitrate (*Nitrobacter*). The energy-yielding reactions involving these organisms are as follows:

$$2NH_4^+ + 3O_2 \rightarrow 2NO_2^- + 4H^+ + 2H_2O \ (\textit{Nitrosomonas})$$

$$2NO_2^- + O_2 \rightarrow 2NO_3^- \ (\textit{Nitrobacter})$$

Nitrification can also be performed by some heterotrophic bacteria (and fungi).

Of the chemoautotrophic mineral oxidizers, bacteria involved in the oxidation of inorganic sulfur are perhaps the most important in forest soils. Sulfur exists as sulfide in several primary minerals, and it is added to forest soils as plant and animal residues and to rainwater. Elemental sulfur is sometimes added to nursery soils to increase acidity and to control certain plant pathogens. The major part of the sulfur in soils exists in organic combinations and, like nitrogen, it must be mineralized to be useful to trees. Tree roots absorb sulfur largely as sulfate. While the initial decomposition of these organic materials and their conversion to inorganic sulfur compounds is accomplished by heterotrophic organisms, the oxidation of sulfides and elemental sulfur to sulfates can be carried out by both heterotrophs and chemoautotrophs.

The oxidation of elemental sulfur can result in the mobilization of some slowly soluble soil minerals as the result of the sulfuric acid formed. The solubility of phosphorus, potassium, calcium, and several micronutrients may be increased by the acidification resulting from this reaction.

Heterotrophic bacteria, which require preformed organic compounds as sources of energy and carbon, comprise the largest group of soil bacteria. This diverse group includes free-living and symbiotic nitrogen fixers and bacteria that decompose fats, proteins, cellulose, and other carbohydrates. They include both aerobic and anaerobic forms.

Biological nitrogen fixation is accomplished by free-living bacteria and by symbiotic associations composed of a microorganism and a higher plant or a fungus. It is primarily through the action of these organisms that part of the huge reservoir of atmospheric nitrogen is rendered available to higher plants. The free-living bacteria capable of utilizing nitrogen gas (N_2) are primarily aerobic species of *Azomonas, Azotobacter, Beijerinckia,* and *Azospirillum* and anaerobic species of *Clostridium* and *Desulfovibrio,* although strains of several other genera are also capable of the transformation under some conditions. Most are apparently not obligate because they can obtain nitrogen from organic and inorganic nitrogenous compounds as well as from the atmosphere. The amount of nitrogen fixed annually probably averages less than $2–5\,kg\,ha^{-1}\,yr^{-1}$ for non-symbiotic N fixation.

Organic matter-decomposing bacteria play a major role in the degradation of vast amounts of forest litter, plant roots, animal tissue, excretory products, and cells of other microorganisms. These materials are both physically and chemically heterogeneous and include such constituents as cellulose, hemicellulose, lignin, starch, waxes, fats, oils, resins, and proteins. With such a diversity of organic materials, it is not surprising that a complex population of heterotrophic bacteria, as well as fungi, is involved. The bacteria involved include both aerobic and anaerobic forms and the mechanisms of decomposition vary, depending on environmental conditions and the participating organisms. In all cases, organic matter provides the microflora with energy for growth and carbon for cell formation. The end products are carbon dioxide, methane, organic acids, and alcohols, in addition to bacterial cells and resistant organic materials.

In some wet soils with a pH near neutral, the activity of denitrifying organisms may be quite high. Under anaerobic conditions, certain bacteria derive their oxygen supply from the oxides of nitrogen (anaerobic respiration), reducing nitrates to nitrite and then to nitrous oxide or elemental nitrogen. True denitrification is largely limited to the genera *Pseudomonas, Achromobacter, Bacillus,* and *Micrococcus.* Because these organisms are facultative anaerobes, they can survive under a wide range of soil conditions, and the presence of a large population gives most soils a significant denitrifying potential. However, conditions must be favorable for the organisms to change from aerobic respiration to a denitrifying type of metabolism. This normally occurs in the presence of nitrates and a source of readily available carbohydrates when the demand for oxygen by the microflora exceeds the supply. Anaerobic conditions that favor denitrification are found in flooded soils and even in microsites on well-drained soils, but very little volatile loss of nitrogen is normally expected in acid forest soils because of the scarcity of nitrates in such soils.

FUNGI ARE PARTICULARLY IMPORTANT IN ACIDIC FOREST SOILS

The microbial biomass within the decomposing litter of forest soils is predominantly fungal, and fungi are probably the major agents of decay in all acidic environments. The large variety and number of mushrooms (fruiting bodies) seen in the forest during wet periods attests to the wide distribution of fungi. Fungi possess a filamentous network of hyphal strands in the soil. The fungal mycelium may be subdivided into individual cells by cross walls or septae, but many species are nonseptate. Fungal mycelia permeate most of the O horizon and are readily seen in mor and moder humus types. Taxonomically, most soil fungi are placed in one of two broad classes: *Hyphomycetes* and *Zygomycetes.* Species of Hyphomycetes produce spores only asexually; the mycelium is septate; and conidia of asexual spores are borne on special structures known as *conidiospores.* Zygomycetes and other fungi, in contrast, produce spores both sexually and asexually. At least six other classes of fungi are found less frequently in soils, but the biodiversity of soil fungi is very poorly understood (Coleman and Crossley, 1996). Fungi are so diverse that it is difficult to classify them on the basis of either morphology or source of carbon since the dominant soil genera can utilize a variety of carbonaceous substrates. They can be most rationally divided into functional groups important in forest soils:

1. Decomposition of cellulose and related compounds is one of the most important activities of fungi. They are active in the early stages of aerobic decomposition of wood and other organic debris in the O horizon. These materials include hemicelluloses, pectins, starches, fats, and the lignin compounds particularly resistant to bacterial attack. By degrading plant and animal remains, the fungi participate in the formation of humus from raw residues and aid nutrient cycling and soil aggregate stabilization.

2. Proteinaceous materials are utilized for both nitrogen and carbon by fungi; as a consequence of proteolysis, they are sources of ammonium and simple nitrogen compounds in the soil. Under some conditions, competition between fungi and higher plants for nitrate and ammonium leads to a reduction in available nitrogen for the higher plants.

3. Some fungi are predators on such soil fauna as protozoa, nematodes, and certain rhizopods and may thereby contribute to the microbiological balance in soil.

4. Pathogenic fungi can be either obligate or facultative parasites. Members of the genera *Rhioctonia*, *Pyritium*, and *Phytophthora* cause damping-off disease among nursery seedlings. *Fusarium* species may cause root rot in the nursery and in older plants. Representatives of the genera *Armillaria*, *Verticillium*, *Phymatotrichum*, and *Endoconidiophora*, among others, may also invade roots of older trees.

5. The mycorrhizal fungi were described earlier in this chapter.

6. Mycelia of fungi may account for the development of a hydrophobic property in some forest soils. These soils, mostly sands, are slow to wet once they become air-dry; consequently, their capacity for water retention is greatly impaired. The mechanism of this water repellency is not well understood.

THE MYCORRHIZOSPHERE IS A HOTSPOT OF MICROBIAL ACTIVITY

The roots of higher plants exert a profound influence on the development and activity of soil microorganisms (Lynch, 1990). Roots grow and die in the soil and supply soil fauna and microflora with food and energy. More important, live roots create a unique niche for soil microorganisms, resulting in a population distinctly different from the characteristic soil community. This difference is due primarily to the liberation by the root of organic and inorganic substances that are readily consumed by the organisms in its vicinity. The rhizosphere effect has been demonstrated with a wide variety of forest trees and other plants (Katznelson *et al.*, 1962). Many factors such as the kind and stage of development of the trees, the physical and chemical properties and moisture content of the soil, and other environmental conditions such as light and temperature influence the rhizosphere effect. These factors may act directly on the soil microflora or indirectly by influencing plant growth.

The most important influence of the growing plant on the rhizosphere flora results from the root excretion products and sloughed tissue which serve as sources of energy, nitrogen, or growth factors. The plant root absorbs inorganic nutrients from the rhizosphere, thereby lowering the concentration available for both plant and microbial development. On the other hand, root respiration may increase rhizosphere acidity and hasten the solubilization of less-soluble inorganic compounds. By this means, the amount of available phosphorus, potassium, magnesium, and calcium may increase (Youssef and Chino, 1987).

Root excretions affect the germination of the resting structures of several fungi, perhaps as a result of the energy sources in the rhizosphere. The stimulation can be deleterious to the host plant if the fungus is pathogenic. On the other hand, there is evidence that the rhizosphere flora protects the root from some soil-borne pathogens. Some antibiotic-producing microorganisms are found in abundance in the rhizosphere, but what effect these organisms have on pathogens is not known. It appears certain that exudates from some rhizosphere organisms form a kind of buffer zone, protecting roots against the attack of soil pathogens. An example is the protection afforded pine roots against *Phytophthora cinnamomi* by mycorrhizal fungi (Marx, 1977). Rhizosphere microflora also produce considerable amounts of growth substances such as indoleacetic acid, gibberellins, and cytokinins that may influence the growth of the host plant. Root exudates strongly influence microbial actions in the root zone, which in turn influence nutrient mobilization and uptake. Bowen and Rovira (1969) reviewed the early work in this area. Work by Grayson *et al.* (1997) and Dinkelaker *et al.* (1997) has increased our knowledge of these phenomena, but our understanding of the complex relationships in the rhizosphere remains vague.

SOIL FOOD WEBS ARE FUELED BY BOTH ABOVEGROUND LITTERFALL AND BELOWGROUND INPUTS

Soil animals, and the food webs they weave, have been thought to rely largely on aboveground detritus. The annual input of litterfall to the O horizon is very apparent, and many soil animals are concentrated in the O horizon. Mobile animals may frequent more than one location in soils, and isotopic tracer studies have shown that much of the soil animal community feeds on belowground roots and detritus within the mineral soil (Pollierer *et al.*, 2007).

SOIL ANIMALS COME IN ALL SIZES

Size divisions are useful for soil animals. Macrofauna (> 2 mm) are large enough to disrupt soil structure through their activity and include small mammals, earthworms, termites, ants, and other arthropods (Dindal, 1990; Stork and Eggleton, 1992; Berry, 1994). These organisms create macropores that influence infiltration rate and gas exchange. They are responsible for a great deal of soil mixing and can profoundly influence horizonation and rooting pattern. They often consume very high carbon:nitrogen ratio organic material and produce lower carbon:nitrogen detritus.

Mesofauna (0.1–2 mm) are large enough to escape the surface tension of water and move freely in the soil but small enough not to disrupt soil structure through their movement. They include some mollusks, springtails, mites, woodlice, and other arthropods. These organisms are extremely important in decomposition and mineralization processes (Crossley *et al.*, 1992).

Microfauna (< 0.1 mm) are usually confined to water films on soil particles. This group includes protozoa, turbellaria worms, rotifers, nematodes, and gastrotricha worms. These organisms are important predators on fungi and bacteria. They regulate microbial populations and are important in mineralization processes.

Since forest soils are seldom tilled or mixed by human beings, this great activity of a wide variety of decomposing and mixing organisms is of paramount importance. Soil biota are the tillers of uncultivated soils (Hole, 1981). The favorable environment of the O horizon encourages the proliferation of myriad organisms that perform many complex tasks relating to soil formation, slash and litter disposal, nutrient availability and recycling, and tree metabolism and growth.

Most soil animals make their homes on or near the surface litter or humus layers, where space and light conditions fit their particular needs. They are rather mobile, but because their distribution is dependent on an organic substrate, they seldom venture deeply into the mineral soil (Hole, 1981). Generally, the number of organisms is greatest in the O horizon and the rhizosphere, with a decreasing gradient with soil depth (Gray and Williams, 1971).

SOIL FAUNA PLAY VITAL ROLES

Essentially, all fauna that inhabit the forest environment influence soil properties in some way that eventually affects tree growth. Fauna range in size from wild beasts (and sometimes domestic animals) to simple one-celled protozoa. Their importance to the soil, however, is essentially inversely proportional to their size. The effects of large animals on forest soils are minimal except on overgrazed lands, where a reduction in ground-cover vegetation or an alteration in species composition, plus soil compaction and reduced water infiltration, result in soil erosion (Lavelle, 1997).

MICROFAUNA THRIVE WITHIN SOIL WATER FILMS

The soil microfauna include nematodes, rotifers, and many types of protozoa. Nematodes are nearly microscopic, non-segmented roundworms that commonly occur in mull O horizons and upper mineral soil. Densities of more than $100\,m^{-2}$ are not uncommon, and nematodes have a high reproductive capacity, and their populations can expand rapidly as soil conditions change. Only about a tenth of the 10 000 or so known nematodes are soil inhabitants. Although the populations of nematodes are always highest in the vicinity of plant roots, only a few appear to be root parasites. Most prey on bacteria, algae, fungi, protozoa, rotifers, or other nematodes.

Nematodes can be important as population regulators and nutrient concentrators in the soil ecosystem. The impact of their parasitism versus their role in decomposition is poorly understood (Benckiser, 1997). In a survey of three-year-old slash pine plantations in the lower coastal plains of the United States, plant parasitic nematodes were found in all 34 sampled sites. Spiral (*Helicotylenchus*) and ring (*Criconemoides*) nematodes were found most frequently, although 11 other plant parasitic genera were also identified. The actual damage to tree vigor and growth is not known, but several of the genera found in these plantations have been reported to cause damage to pine. In addition to the two genera mentioned above, others included sheath (*Hemicycliophora*), lance (*Hoplolaimus*), stunt (*Tylenchorhynchus*), and dagger (*Xiphinema*) nematodes.

Rather spectacular increases in growth have been reported for tree seedlings planted in some fumigated soils. This response is presumed to result from a reduction in the population of nematodes that attack pine roots, and tree responses have been particularly notable on dry, sandy sites. However, neither the actual causes of the increased growth nor the longevity of the effect of fumigation are well understood. The total concentration of nematodes in the soil may return to normal in a year or two following fumigation, but the spectrum of genera may be altered for longer periods.

Rotifers are aquatic animals that are active in soil when moisture is sufficient. At other times, they may form protective shells or simply enter a state of anabiosis. The soil rotifers are about equally divided between those that feed on organic debris and those that feed on algae or protozoa. They generally inhabit the litter and humus layers, and their populations can be quite high (more than $100\,m^{-2}$) on favorable sites.

Protozoa are the most abundant soil fauna; Waksman (1952) reported that their number varied from 1500 to 10 000 per grain of forest soil. These one-celled organisms may exist in either an active or a cyst stage, but they are generally aerobic and occur in the upper horizons. Their diet consists largely of decomposing organic materials and bacteria. Soil conditions that favor their development are similar to those that favor bacteria. They are found in soils supporting both hardwood and coniferous forests. In a strict sense, they are the only major group of soil fauna classified as microorganisms.

MESOFAUNA FRAGMENT LITTER AND PROMOTE SOIL STRUCTURE

Arthropods dominate the mesofauna of forest soils. Springtails, woodlice or sow bugs, spiders, ticks, and mites are abundant in essentially all forest soils and are particularly important in forest litter decomposition. The primary consumers chew and move plant parts on the surface and into the soil. Some, such as sow bugs, are active feeders on dead leaves and wood and are important in the disintegration of freshly fallen leaves.

Saprophagous mites constitute one of the most important groups of Arachnida, and by virtue of their great numbers (often exceeding one million m^{-2}) they play a major role in producing a crumb structure in some surface organic layers. They feed on decaying leaves, wood, fungal hyphae, and feces of other animals (Wallwork, 1970). Mites are important at several nodes in the soil food web (Coleman and Crossley, 1996; Benckiser, 1997). The springtails and bristletails are small, wingless, saprophagous insects that feed on decaying materials of the O horizon. Adults and larvae of many beetles and flies contribute to the breakdown of organic debris and improve the structure of surface soil (Kevan, 1962).

Fires generally reduce the number of arthropods in the O horizon, but the reduction is largely temporary and not all genera are reduced equally. Carabids are the most numerous arthropods in protected forests of northern Idaho, but *Acarina*, *Chilopoda*, *Thysanoptera*, *Protura*, and *Thysanura* are most numerous in areas that have received recent prescribed burns (Fellin and Kennedy, 1972).

MACROFAUNA ARE MOVERS AND SHAKERS

Vertebrates, consisting largely of four-legged animals, influence the soil through fertilization, trampling, scarification, and a form of cultivation. Many such animals burrow in the soil and aid in the breakdown of its organic constituents. They mix organic material with the mineral soil, hastening decomposition and fostering stabilization of organic matter. Rodents such as marmots, gophers, mice, shrews, and ground squirrels are particularly important in soil development. Moles are also especially active in European forests, fostering the development of mull humus layers; Lutz and Chandler (1946) speculated that the names "mole" and "mull" have common origin. While large animals may trample and compact the soil surface, penetration of water and air into the soil is greatly facilitated by the actions of smaller animals.

Mice, shrews, and other small animals are abundant in many forest soils. Hamilton and Cook (1940) found that in mixed hardwood forests of the northeastern United States there was an average of $5\,kg\,ha^{-1}$ of small mammals, or about twice the carrying capacity for white-tailed deer. Small mammals have a pronounced influence on microrelief and other soil properties. Their labyrinths of interconnecting tunnels allow for ready penetration of air and water, while their nests, dung, stored food,

and carcasses all enhance the organic matter content and fertility of the soil (Troedsson and Lyford, 1973). It has been suggested that the absence of an eluviated horizon in some Eutrochrepts of the Northeast may be caused largely by the mixing of surface and subsurface materials by fauna (Lyford, 1963). Burrowing animals have been reported to bring as much as several megagrams per hectare per year to the surface in forests (Paton *et al.*, 1995). Burrowing rodents may perform functions in arid and semi-arid areas similar to those of the earthworm in humid forests.

ARTHROPODS COMPRISE A BROAD GROUP OF SOIL ANIMALS

Arthropods have articulated bodies and limbs, and include insects and spiders. Macrofauna-sized arthropods are largely important as soil mixers; however, some are important decomposer organisms. Ants occur in essentially all forest soils; some are predatory carnivores, and others harvest tree leaves and farm fungus for food. Ants are very important soil mixers, transport large amounts of subsoil to the surface, and vice versa; ants may typically move $10\,Mg$ soil $ha^{-1}\,yr^{-1}$, and ten times that amount in some tropical and subtropical forests (Paton *et al.*, 1995).

Centipedes are predatory animals feeding on other soil fauna and play only a minor part in soil formation. Millipedes are largely saprophagous, feeding on dead organic matter. They are generally confined to soils with mull humus layers, especially those supporting stands of deciduous trees. Lyford (1943) noted that millipedes favored leaves of certain trees, particularly leaves with high calcium content. Millipedes are considered important in the formation of mull humus layers, although possibly not as important as earthworms.

Termites are important in tropical and subtropical wet and dry climates. They may be divided into three groups on the basis of food habits: those that feed on wood, those that consume litter and humus, and those that cultivate fungi for food. Termites do not produce their own cellulase; rather, they rely on a rich gut fauna of flagellates. Not all termites build mounds, but all tunnel in the soil, and they are important in transporting soil, decomposition, and mixing organic material with mineral soil.

WORMS MAY BE THE MOST IMPORTANT SOIL ANIMALS

The ordinary earthworm is probably the most important component of soil macrofauna (Edwards, 1998). Although there are many species of earthworms and earthworm taxonomy is in a state of flux, all large and middle-sized worms have been classified as *Lumbricidae*, while the smaller, light-colored "pot-worms" have been considered *Enchytraeidae* (Kevan, 1962).

The large, reddish *Lumbricus terrestris*, a species introduced from Europe, occurs widely in the Americas. A high population of earthworms is generally associated with mull humus formation, and this is particularly true of *L. terrestris*, which may make up as much as 80% of the total soil fauna by weight. These earthworms feed on fallen leaves and organic debris and pass it, together with fine mineral particles, through their bodies. Each year, earthworms have been estimated to pass more than $30\,Mg$ soil ha^{-1} through their bodies, where it is subjected to digestive enzymes and to a grinding action (Paton *et al.*, 1995). Earthworm casts are higher in total and nitrate nitrogen, available phosphorus, potassium, calcium, magnesium, and cation exchange capacity and lower in acidity than is the soil proper. Earthworms mix bits of organic materials into the mineral soils, and they promote good soil structure and aeration through their burrowing action. As a result of this transporting and mixing action, the upper layer of certain O horizons takes on the crumbly structure of the so-called earthworm mull. The concentration of earthworms in

forest soils has been estimated to be from 0.5 million to more than 2 million per hectare, the actual numbers depending on several climatic and soil factors. Although earthworms may flourish in acid soils, highly acid soils support fewer earthworms than less acid soils, with the optimum range being about pH 6.0 to 8.0. Sandy soils and soils that dry excessively are not favorable habitats for earthworms. *Lumbricus rubellus* and *L. festivus* appear to be more acid tolerant than *L. terrestris* and may be more common than the latter in coniferous forests and in mor humus.

Allobaphora species are approximately the same size as the *Lumbricus* species but are much lighter in color. They are active in forest soils of Europe and North America and are particularly important in the development of hardwood mull humus. The smaller worms, such as *Octolaseum* and *Dendrobaena* species, also devour organic debris, thereby improving the physical and chemical properties of the surface soil. Because of their smaller size, generally no more than a few centimeters long, they have not been considered as important as the other worms in forest soil formation. However, since these Enchytraeidae have less stringent environmental requirements than their larger relatives, they may be approximately as numerous in mor as in mull humus layers (Hendrix, 1995).

Earthworm invasions are transforming the soils of many northern hardwood forests across North America. Assemblages of European earthworms convert former mor O horizons (see Figure 4.3) into mull types. The exotic earthworm biomass can exceed the biomass of native soil animals, changing almost every aspect of soil structure and chemistry, even altering the vegetation (Fisk *et al.*, 2004; Szlavecz *et al.*, 2011). The growth of overstory trees may decline for two decades after invasions (Larson *et al.*, 2010).

CHAPTER 7

Forest Biogeochemistry

OVERVIEW

The productivity of forests depends on the supply of nutrients available within the soil for plant use; high rates of nutrient supply lead to rapid growth. The flow of chemical energy through the ecosystem involves oxidation and reduction reactions in which the flow of electrons is associated with the release and consumption of energy. Oxygen is the electron acceptor that provides the greatest amount of energy release for each electron. The nutrient cycles of forests involve pools of nutrient elements and flows of nutrients between these pools. The annual cycling of nutrients from the soil into trees is commonly greater than the annual input of nutrients into a forest, but this flow is only a small fraction of the total pool of nutrients in the soil. The annual decay of litter from leaves and roots provides most of the nutrients used by plants in a given year. However, the quantity of nutrients stored in the O horizon of most forests is either accumulating or in a steady state (until a fire or another disturbance creates drastic losses), and the increase in nutrient content of the accumulating forest biomass must come from annual inputs from the atmosphere, from mineral weathering, or from a declining pool of nutrients in slowly decomposing soil humus. The overall biogeochemical cycles differ among elements. Carbon, nitrogen, and sulfur have major oxidation and reduction reactions within ecosystems. Other elements, such as calcium and potassium, undergo no redox reactions but have strong interactions with geochemical pools (see Table 7.4 for a summary of the major elements). The accounting for pools and fluxes of elements in ecosystems includes three important rules:

1. pools do not represent fluxes (or supplies),
2. nutrient budgets need to balance, and
3. issues of scales in space and time need to be considered explicitly.

Forest productivity depends on supplies of water, light, and more than a dozen vital elements (nutrients). Soils with high supplies of water and nutrients are more productive than poorer soils, as evidenced by regional relationships between water supply and annual patterns of growth in wet and dry years and between nutrient supply and growth. The dependence of growth on nutrient supply is shown clearly in Figure 7.1.

Growth depends on the nitrogen supply for two reasons: the deployment of a large canopy to intercept and assimilate light requires a large supply of nitrogen, and the accumulating biomass of the forest contains large quantities of nitrogen. Gonçalves *et al.* (1997) plotted the relationship between aboveground biomass increment and nitrogen uptake for a variety of *Eucalyptus* species (from age 2–10 years) across a range of sites and silvicultural treatments. The patterns between biomass increment and uptake of calcium and potassium were as strong as those for nitrogen (Figure 7.2).

The dependence of growth on water and nutrient supplies is also apparent from experimental trials (Figure 7.3). The intersection of chemistry, geology, and life is the subject of biogeochemistry. This chapter expands the focus on the chemistry of soil surfaces and solutions in Chapter 8 to consider the broader sources and sinks for nutrients within forests, and the factors that influence these flows.

Ecology and Management of Forest Soils, Fourth Edition. Dan Binkley and Richard F. Fisher.
© 2013 John Wiley & Sons, Ltd. Published 2013 by John Wiley & Sons, Ltd.

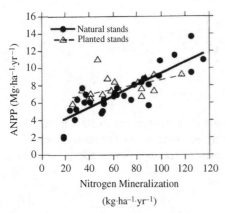

Figure 7.1 Aboveground net primary production (ANPP) of 50 forests in the north-central United States depends strongly on the nitrogen supply in the upper soil (Reich *et al.*, 1997).

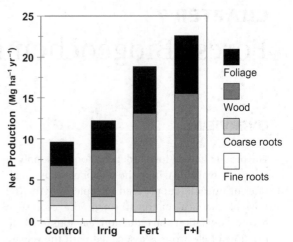

Figure 7.3 The productivity of an 8-year-old loblolly pine stand in North Carolina responded well to addition of water and very well to addition of fertilizer; addition of both water and nitrogen gave the strongest response (data from Albaugh *et al.*, 1998).

ENERGY FLOWS WITH ELECTRONS

The biogeochemistry of forests is dominated by flows of energy driven by biological processes such as photosynthesis and respiration. These flows of energy entail the flows of electrons from high-energy and low-energy states, commonly called oxidation and reduction reactions, or redox reactions. Photosynthesis utilizes radiant energy to boost electrons from low-energy bonds in CO_2 into high-energy bonds in carbohydrates. The oxidation state of carbon goes from 4+ to 0 as the electrons released in the splitting of water are transferred to the carbon:

$$2H_2O \rightarrow O_2 + 4e^- + 4H^+$$

$$4e^- + 4H^+ + CO_2 \rightarrow CH_2O + H_2O$$

These two equations represent half-reactions that generate electrons (the splitting of H_2O) and that accept electrons (reduction of CO_2). The sugar formed in photosynthesis provides for two fundamental aspects of plant growth: providing the carbon "backbone" that forms the structure of plant cells and tissues (plant tissues are 45–50% carbon), and providing a readily available source of energy to fuel chemical transformations throughout the plant.

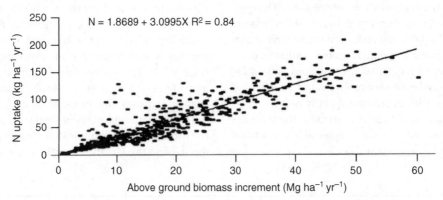

Figure 7.2 Aboveground increment related strongly to nitrogen uptake in over 200 stands of *Eucalyptus* species in Brazil (Gonçalves *et al.*, 1997).

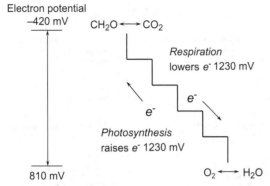

Figure 7.4 Photosynthesis uses sunlight to boost electrons from water (at 810 mV Eh at pH 7) to reduce CO_2 and form sugar (at Eh −420 mV). Respiration reverses this process, releasing energy as sugar is oxidized to CO_2 and oxygen is reduced to water.

Table 7.1 Reduction potentials for important half-reactions in soils. *Source*: Modified from Schlesinger (1997).

Half-reaction	Reduction Potential (Eh in mV at pH 7)	Change in Free Energy if Coupled to CH_2O Oxidation (kJ mol^{-1})
$CO_2 + e^- \rightarrow CH_2O$	−420	0
$H^+ + e^- \rightarrow H_2$	−410	−4
$CO_2 + e^- \rightarrow CH_4$	−244	−23
$SO_4^{2-} + e^- \rightarrow HS$	−220	−25
$Fe^{3+} + e^- \rightarrow Fe^{2+}$	−5	−42
$Mn^{3+} + e^- \rightarrow Mn^{2+}$	525	−97
$NO_3^- + e^- \rightarrow NO_2^-$	750	−118
$O_2 + e^- \rightarrow H_2O$	810	−125

The tendency of a chemical to release or accept electrons is gauged by the electrode potential. The electrode potential (abbreviated Eh; also called reduction potential and redox potential) can be pictured as a staircase, where substances higher on the stairs (more negative potential) tend to donate electrons to chemicals farther down (more positive potential; Figure 7.4). The difference in potential is 1230 mV between the half-reaction for oxygen to gain or release electrons and the half-reaction for CO_2 to gain or release electrons. Substantial energy is released when electrons cascade down the staircase (respiration), and energy is required to boost electrons up the staircase (photosynthesis).

Many compounds accept and donate electrons (Table 7.1). Where oxygen is present in soils, the reduction of oxygen to form water provides the greatest potential change for electrons from organic compounds. If all the oxygen is consumed in the soil (or at the microsite where the chemical reaction occurs), then alternative electron acceptors such as NO, Fe^{3+}, and SO_4^{2-} can be used as electron acceptors, releasing progressively less energy.

The redox potential of soils is commonly measured with a platinum electrode and may be expressed as pe:

$$pe = -log\, A_{e-}$$

where A_{e-} is the activity of e^- (activities are slightly lower than concentrations, and they approach concentrations at infinite dilutions). At dilute concentrations, pe = (Eh in V)/0.059. At pH 7, oxygen-rich soil has a pe of about 12. As oxygen becomes depleted, pe drops. Aerobic conditions pertain until pe drops below 10 as anaerobic conditions develop and may fall as low as 0 to −4 under very reducing conditions (Lindsay, 1979). The ability of a soil to buffer changes in pe is referred to as poise; a soil with a high supply of oxygen is well poised against changes in redox potential.

The sum of pe + pH tends to be constant for a given soil, which means that if soil oxygen is depleted and pe falls, then pH must rise. For example, Chen (1987) compared adjacent conifer stands on flat topography. A stand of loblolly pine was on a shelf about 1 m above a bottomland planted to bald cypress. She found that the pH was 6.05, with a pe of 10.22 in the cypress stand. The soil appeared to be more acidic under loblolly pine, with a pH of 5.02 and a pe of 11.25. In fact, the pine soil had a lower pH, but the reason was that the redox potential of the soil was higher. The pe + pH for both sites was about 16.25, indicating that the redox and acid systems of the two soils were essentially the same.

NUTRIENT CYCLES INVOLVE POOLS AND FLUXES

Forest nutrient cycles are commonly illustrated as pools and fluxes indicating major inputs, storage

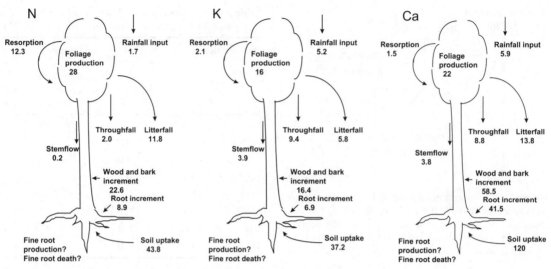

Figure 7.5 Nutrient cycles of a 16-year-old plantation of a eucalyptus hybrid in India. Major features include substantial resorption of nitrogen before litterfall, major throughfall loss of potassium, and large uptake of calcium for wood and bark increment, with very little resorption from leaves before litterfall. Nutrient requirements for the growth of fine roots and nutrient return in fine root death were not estimated (data from Negi, 1984).

pools, and outputs. Separate diagrams may be used for each element, and the relative importance of various pools and fluxes differs substantially among elements (Figure 7.5). A 16-year-old stand of eucalyptus in India took up about $44\,kg\,ha^{-1}\,yr^{-1}$ of nitrogen (not counting the unmeasured nitrogen requirement for fine-root production); this stand also had the benefit of another $12\,kg\,ha^{-1}\,yr^{-1}$ that was recycled from foliage prior to litterfall. The potassium cycle included very large losses of potassium from the canopy in throughfall and stemflow, which is typical for this ion that is not structurally bound within organic molecules. The uptake of calcium was very large, resulting primarily from the high calcium content of eucalyptus bark and wood; almost no calcium was resorbed prior to litterfall because calcium is bound in insoluble organic molecules.

ANNUAL NUTRIENT CYCLING IS GREATER THAN ANNUAL INPUTS

Nutrient inputs from the atmosphere and rock weathering are important to the long-term development of soils and ecosystems, but on an annual basis, nutrient recycling within ecosystems provides the major source of nutrients for plant use. Only a small portion of the total nutrient content of an ecosystem is cycled each year, so it is more valuable to know the rate at which nutrients are cycled each year than to know the total quantity of nutrients within an ecosystem.

At the decomposition stage of a nutrient cycle, nutrient atoms are released into the soil solution largely as by-products of microbial scavenging for energy. The rate of breakdown of organic matter by microbes depends on the:

- chemical quality of the material;
- availability of energy sources (such as readily oxidized carbon compounds) to fuel the microbes;
- activities of meso and microfauna; and
- environmental conditions.

Microbes such as bacteria and fungi excrete enzymes that digest organic molecules into smaller units that may then be absorbed by the microbes or plants. If the nutrient content of the decomposing litter is high, the microbes will find an abundance of nutrients relative to their energy needs and nutrient release will be rapid. If the litter is low in nutrients, the microbes may retain most of the nutrients to grow new cells, and availability to plants will be low (at least until the microbes turn over). This idea is used to explain why decomposition and nutrient release are relatively rapid for low carbon:nitrogen material (microbes have access to a good supply of nitrogen to decompose the carbon), and slow for high carbon:nitrogen material (insufficient nitrogen to allow decomposition of carbon, and the nitrogen that is released is immobilized by microbes to degrade more carbon). It's important to note that these carbon:nitrogen patterns may also simply index the types of carbon compounds present, and these types of compounds may simply differ in how easily they are broken down by enzymes. High carbon:nitrogen materials typically contain polycyclic aromatic compounds that are very difficult to degrade, and low carbon:nitrogen material contains higher-quality compounds (cellulose, proteins) that are readily used by microbes.

It may be good to keep in mind that the pool with the lowest carbon:nitrogen ratio in forest soils is the well-humified, recalcitrant fraction of soil organic matter. Hart (1999) compared the rates of net nitrogen mineralization in well-decayed wood and mineral soil humus in an old-growth Douglas fir forest. The woody material had a carbon:nitrogen ratio of 117 compared to just 26 for the mineral soil, but the nitrogen in the wood was released more than twice as fast as the nitrogen in the soil.

As with many issues of scale in forest soils, a generalization that works for part of the system (such as rapid decomposition of fresh litter with a low carbon:nitrogen ratio) does not extend to other parts (humus with a low carbon:nitrogen ratio decays very slowly; most of the carbon in woody materials with a high carbon:nitrogen ratio is not relevant to current microbial activity). Although microbes do not have absolute control of nutrient availability, they are strong competitors for available nutrients. Microbes are strong competitors with trees for available nitrogen (reviewed by Kaye and Hart, 1997), but the short life span of microbes provides better opportunities at a seasonal timescale for trees to compete effectively.

LITTER DECAYS LIKE RADIOACTIVE MATERIAL, BUT FOR DIFFERENT REASONS

A large number of litter decomposition studies have found that litter disappearance rates generally follow either a linear or an exponential decay trend. In an exponential pattern, the relative rate of weight loss is a constant proportion of the weight remaining at any given time, and the absolute weight loss is rapid in early stages but slows with time. This can be expressed by the equation:

$$B_{T+1} = B_T \bullet e^{-kt}$$

where B is the biomass remaining at time T (or one unit of time later, $T+1$), e is the base of natural logarithms, t is the time from the beginning of decomposition, and k is the rate constant. This curve can be shaped to fit the decomposition curve of most types of litter merely by changing the k value. The k value is useful for comparing the decomposability of various litter types and for estimating the time required for a given percentage of the litter to be decomposed. For example, 0.6931/k estimates the time needed for 50% disappearance, and 3/k estimates the 95% point.

Why should decomposition follow this pattern? If limiting nutrients accumulate in litter at the same time that biomass is respired, decomposition rates should *increase* with time rather than decrease. The answer to this puzzle is based on the chemistry of the organic molecules present in litter. Molecules that are readily respired, such as sugars, disappear quickly, whereas recalcitrant lignin and phenolic molecules are degraded more slowly (Figure 7.6). If litter were composed of a single compound, the classic exponential decay pattern probably would not apply. Some of the curve in decomposition studies may also be an artifact of litterbag methodologies; unconfined litter may follow more of an asymptotic pattern than an exponential decline (Kurz-Besson et al., 2005).

Measurements of the remaining litter biomass are net measurements and include the residue of the original litter, material synthesized by microbes during decomposition, and dead microbial cells.

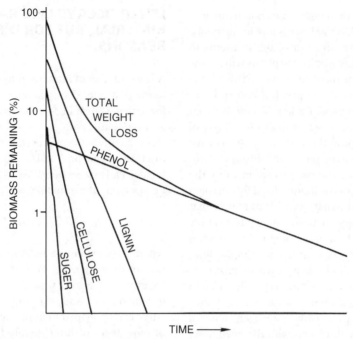

Figure 7.6 The common pattern of exponential decay of litter derives from rates of decomposition of different compounds within the litter rather than from a decrease in the rate of oxidation of particular compounds (after Minderman, 1968; used by permission of the *Journal of Ecology*).

As decomposition progresses, a large proportion of the remaining weight is composed of freshly synthesized materials. Paul and Clark (1996) estimated that by the time half of the mass of litter is gone, about half of the residual mass is actually composed of newly synthesized materials. Therefore, the actual rate of processing of the original litter molecules is faster than the rate of mass loss would indicate.

The overall rate of decomposition is influenced by the types of organic molecules and the nutrient content of the litter (as well as by environmental factors discussed later). One parameter of litter quality that relates best to decomposition and nutrient supply is the lignin:nitrogen ratio (Figure 7.7). Lignin represents a pool of complex compounds with high concentrations of phenolic rings and variable side chains. Lignin is more difficult to decompose than simpler compounds such as cellulose, so litter high in lignin-like compounds decomposes slowly. The nitrogen in litter is present in a wide range of compounds, many of which are relatively easy to decompose. The combination of the lignin and nitrogen concentrations in a single ratio provides a surprisingly strong prediction of rates of litter decomposition and even rates of nitrogen supply in soils. Relatively small variations in lignin:nitrogen within a tree species across sites do not explain the variations in nutrient supply, but the differences in lignin:nitrogen across species provide moderately strong relationships across environments.

These broad-scale patterns in nitrogen supply and lignin:nitrogen are typical of many patterns in forest ecosystems; the relationship across a very broad gradient in lignin:nitrogen (from 10 to 70) provides a strong correlation with the nitrogen supply ($r^2 = 0.74$). This broad insight into nutrient cycles may not be very useful in specific situations, where the range in the X variable (lignin:nitrogen) is narrow. For example, a site with a lignin:nitrogen ratio of 20 would average about 80 kg of nitrogen per hectare in annual net nitrogen mineralization compared with 40 kg of nitrogen per hectare for

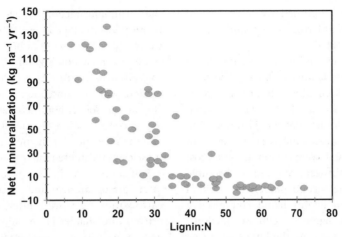

Figure 7.7 Net soil nitrogen mineralization as a function of the lignin:nitrogen ratio of aboveground litterfall from nine common garden experiments with temperate forest species (from Scott and Binkley, 1997; used by permission of Springer-Verlag).

a lignin:nitrogen ratio of 35. However, the ranges of the observations overlap very strongly. The range for a lignin:nitrogen ratio of 20 was 25 to 85 kg N ha^{-1} yr^{-1}, which overlaps substantially with the range of 5 to 65 kg N ha^{-1} yr^{-1} for a lignin:nitrogen ratio of 35.

Patterns of association between two biogeochemical features (like net N mineralization in soils and lignin:nitrogen of litter) need to be understood as correlations that do not provide unequivocal evidence for causal factors. Conceptually, net N mineralization might drive the lignin:nitrogen pattern, or lignin:nitrogen might drive net N mineralization, or both might be reflecting some third factor that actually drives the biogeochemistry. A good example comes from a "common garden" experiment in Wisconsin, USA where European larch, red pine, red oak, and Norway spruce were planted in replicated plots at a single location (Son and Gower, 1991; see Chapter 11 and Figure 11.11). The net N mineralization in the soil correlated well with the lignin:nitrogen of litterfall ($r^2 = 0.76$, $p = 0.05$). However, the correlation between net N mineralization and the biomass of fungi in the soils was stronger ($r^2 = 0.89$, $p = 0.02$). Linear thinking about cause and effect in biogeochemistry would usually be less helpful than thinking of mutually interacting factors that continue to influence many features over time.

Across species, litter with high nitrogen concentrations tends to decompose faster than litter with low nitrogen content. Would an increase in the nitrogen concentration of litter following fertilization lead to faster decomposition? Similarly, would addition of nitrogen directly to litter increase decomposition rates? Both of these expectations may be appealing, but empirical evidence does not support either one. This is a classic issue of logic: species that produce high-nitrogen litter show faster decomposition, but this correlation does not prove that high nitrogen concentrations cause rapid decomposition. Prescott (1996) noted that decomposition of jack pine litter was not increased with fertilization (over 1 Mg of nitrogen per hectare added!) and that litter from fertilized lodgepole pine trees decayed at the same rate as litter from unfertilized trees despite a five-fold increase in litter nitrogen concentration. She also found that birch leaves decomposed faster in microcosms with red alder O horizon material than in microcosms with Douglas fir material because of greater animal activity in the alder material, not because of the nitrogen supply (which was higher in the Douglas fir material). A lack of relationship between nitrogen supply and decomposition was also reported for red fir branches in California (McColl and Powers, 1998). After 17 years, branches (from a thinning operation) had lost about 40% of their original mass in unfertilized plots and in plots that received 300 kg N ha^{-1}.

Another error of logic and scale could develop if one expected the correlations with decomposition of fresh litter to work equally well for the decomposition of stabilized, humified organic matter. Melillo et al. (1989) found that the decomposition of red pine needles appeared to have two phases. In the first three years of decomposition, the litter lost mass, immobilized nitrogen, and lost acid-soluble carbohydrates. In the second phase, the residual material was high in lignin (lignin comprised a constant 70% of the lignin + cellulose fraction), and net nitrogen release from the litter began (see also Berg and Laskowski, 2006; Prescott, 2010 and Chapter 8).

General correlations between decomposition and litter nitrogen content can lead to mistaken inferences about the influence of the nitrogen supply on decomposition. Berg and Matzner (1997) reviewed this relationship and concluded that a high nitrogen supply might speed the decomposition of labile (readily decomposed) carbon compounds but that nitrogen actually *retards* the decomposition of lignified and humified compounds.

Microclimate also regulates litter decomposition and nutrient release. For example, the decay of aspen leaves increased with temperature up to 30 °C and with moisture contents up to three times the weight of the leaves (Bunnell et al., 1977). Of course, decomposition is impeded when soils are too hot and dry or wet. This strong response of fresh litter decomposition to temperature may not apply to the decomposition of old, well-humified soil organic matter.

The response of decomposition to temperature may differ among organic compounds, and depends on the extent of sorption onto mineral surfaces. Simple expectations about a decomposition/temperature response may be misleading (discussed by Conant et al., 2011).

The quantity of clay in soils may influence the accumulation of well-stabilized soil carbon that is resistant to decomposition. Giardina et al. (1999) reviewed laboratory incubation studies and concluded that when clays comprise less than 30% of the soil mineral fraction, the decomposition of carbon relates poorly to clay content. Where soil clay contents exceed 30%, the rates of carbon release are quite low (when expressed on a gram of carbon loss/gram of soil carbon basis). Torn et al. (1997) examined a four-million-year chronosequence in Hawaii and concluded that soil carbon increased for 150 000 years as the concentration of amorphous clays increased and then declined for the next four million years as these clays weathered into crystalline clays.

Soil animals can have a major impact on rates of litter disappearance and decomposition. When earthworms ingest fresh litter and soil, the litter is broken into smaller pieces and blended with mineral particles into moister conditions below the soil surface. Only a small fraction of the ingested organic matter is actually decomposed during passage through the earthworm guts, so the disappearance of litter from the soil surface does not represent immediate decomposition. However, the net effect of mixing by the worms may speed up decomposition. For example, the addition of earthworms to soils developed on mine spoils reduced litter accumulation from 6300 kg ha^{-1} to 1175 kg ha^{-1} over five years (Vimmerstedt and Finney, 1973).

Fungi in O horizons may "import" nitrogen from the mineral soil, providing a nitrogen supply for use in degrading material with a high carbon:nitrogen ratio in the O horizons. Hart and Firestone (1991) used ^{15}N techniques to estimate the rate of nitrogen transfer from the O horizons of an old-growth, mixed-conifer forest into the mineral soil, and the reverse flow. They found that leaching of organic and inorganic nitrogen from the O horizons transferred about 26 kg N ha^{-1} yr^{-1} into the mineral soil. The reverse flow, of nitrogen from the mineral soil into the O horizons via fungal hyphae, was estimated to be as large as 9 kg ha^{-1} yr^{-1}. The import of nitrogen into the O horizons offset about one-third of the release of soluble nitrogen from the O horizons.

Many forests have substantial accumulations of woody material on the soil surface, commonly ranging from 50 to 100 Mg ha^{-1} and much higher in some coastal rainforests of the Pacific Northwest. Does the large amount of this material indicate a major role of woody material in annual nutrient cycling? Over a period of 200 years, the nitrogen content of Douglas fir boles on the soil surface increases by about two-fold (Sollins *et al.*, 1987); if a site contained $100\,Mg\,ha^{-1}$ of downed Douglas fir logs with a nitrogen concentration of $1\,mg\,N\,g^{-1}$ wood, then the nitrogen content of the wood would increase from an initial $100\,kg\,ha^{-1}$ to $200\,kg\,ha^{-1}$ over two centuries, for an annual rate of increase of $0.5\,kg\,ha^{-1}\,yr^{-1}$. Subsequent decomposition in later centuries would release the nitrogen, with measured rates of release of 1.6 to $2.6\,kg\,ha^{-1}\,yr^{-1}$ leaching from well-decayed logs (Hart, 1999). In the absence of large amounts of woody material, the nitrogen would not be immobilized during initial stages of log decomposition or released during later stages. Overall, these rates of nitrogen removal and release are small relative to other fluxes in these forests, such as atmospheric deposition or even nitrogen uptake and release by the moss layer.

LITTERFALL AND ROOT DEATH ARE MAJOR PATHWAYS OF NUTRIENT RETURN TO SOILS

Nutrients taken up by trees face several fates: incorporation into accumulating biomass, recycling to the soil via litterfall or root death, leaching from leaves or roots, or recycling from leaves or roots for use the following year. These patterns vary with each nutrient, and general trends can be illustrated using a 14-year-old slash pine forest (Table 7.2). For nitrogen, the total annual uptake was $40\,kg\,ha^{-1}$; about 40% was incorporated into accumulating biomass, none was leached from leaves, and about 3% was resorbed into the tree before the remaining 60% was lost in litterfall. Internal recycling is often more

important in other forests; typically, about 20–35% of the nitrogen in leaves is recycled prior to leaf fall. Potassium in the slash pine forests exhibited a very different pattern. Of a total potassium uptake of $10.6\,kg\,ha^{-1}$, about 35% went into the biomass increment, 35% was returned to the soil in litterfall, and the remaining 30% was leached from the leaves by rainfall. None was resorbed from needles before litterfall.

These patterns change with stand development as forests progress through stages of relative nutrient abundance and scarcity. The role of internal recycling becomes particularly important in older slash pine ecosystems.

Rates of litterfall have been examined in scores of forests around the world. A review of studies from tropical forests found that plantations tend to have lower rates of nitrogen cycling in litterfall than natural forests do (Figure 7.8). Tropical plantations tend to cycle 10 to $80\,kg\,N\,ha^{-1}\,yr^{-1}$ in litterfall compared with 25 to $170\,kg\,N\,ha^{-1}\,yr^{-1}$ in natural forests. Any difference between plantations and natural forests is less clear for phosphorus, though low rates of phosphorus cycling in litterfall are more common for the plantations than for natural forests.

Of all the carbon that reaches soil over the life of a stand, how much comes from aboveground litterfall, root death, and stem death? These inputs should equal the average net production of each type of tree tissue over the life of the stand. In an average year, foliar production (and death) might be $6\,Mg\,ha^{-1}$, root production (and death) might be $6\,Mg\,ha^{-1}$, and wood production might be $8\,Mg\,ha^{-1}$. The wood tends to accumulate each year, and in the absence of forest harvest, all the wood production eventually reaches the O horizons. In this example, the leaves and roots would comprise

Table 7.2 Cycles of nitrogen and potassium differ substantially in slash pine plantations (foliage leaching is much more important for potassium), and change with age ($kg\,ha^{-1}$; from data in Gholz *et al.*, 1985).

Component	Age 14		Age 26	
	Nitrogen	Potassium	Nitrogen	Potassium
Stand uptake	40	10.6	30	10.0
Biomass accumulation	16	3.7	8	3
Foliage leaching	0	3.0	0	3
Retranslocation	1	0	16	2.5
Litterfall	23	3.9	22	4.0

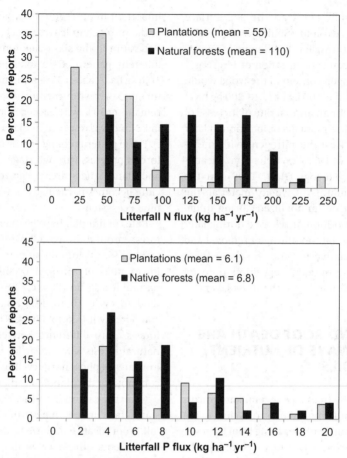

Figure 7.8 Patterns of nitrogen and phosphorus return to soils in aboveground litterfall in tropical plantations and natural forests (from a data compilation of Sankaran and O'Connell, in Binkley *et al.*, 1997; see also Figure 4.7).

about 60% of the carbon added to the soil and the wood would comprise 40%. But how much would each of these sources contribute to the carbon that ends up as well-humified, long-term soil organic matter? As noted in Chapter 4, this question remains unanswered; we simply don't know if soil organic matter tends to form more from one type of carbon input than from another.

TREES ADJUST TO NUTRIENT LIMITATIONS

The productivity of most forests is limited simultaneously by a variety of resources, including light,

water, and one or more nutrients. For a century, the idea of the "law of the minimum" (popularized by Justus von Liebig) dominated thinking about forest production. Droughty sites were assumed to be so limited by the lack of water that addition of nutrients would not increase growth. We now know that forests (and other ecosystems) may be limited by the lack of a single major resource or, more often, by the interacting effects of low supplies of several resources (Chapin *et al.*, 1987; Seastedt and Knapp, 1993; Binkley *et al.*, 2004a). The effects of increased supplies of multiple resources are often not linear. For example, nitrogen alone could not increase the growth of *Eucalyptus saligna* in New Zealand, but further addition of phosphorus

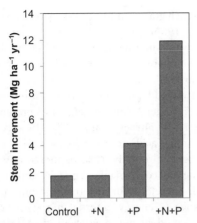

Figure 7.9 Growth of *Eucalyptus saligna* from age 1 year to age 2 years in New Zealand (extrapolated from data in Knight and Nicholas, 1996).

increased growth by much more than the sum of the individual nutrient responses (Figure 7.9). Simple linear thinking (X and only X controls Y) is probably not very useful in forest soils or ecology.

If the supply of nutrients increases on a nutrient-limited site, how do the trees respond? The most commonly examined response is wood production; fertilization on nitrogen- or phosphorus-limited sites can increase wood production by 20% to 50% or more. This increased production of wood may derive from increased net photosynthesis as a result of improved biochemistry in the leaves; from increased light interception by an expanding canopy; or from a shift in allocation away from roots and mycorrhizae to wood production. See Chapter 14 for detailed consideration of the ecology of the forest response to fertilization.

NUTRIENT TRANSPORT AND MOBILITY WITHIN PLANTS ARE IMPORTANT

Each nutrient serves unique functions in plants, and these roles affect their mobility. For example, potassium is an enzyme activator and regulator of osmotic potential, and in both these roles it remains a free, mobile cation. Therefore, some potassium can be recycled from senescent foliage before abscission. Calcium usually binds two organic molecules

together, remaining relatively immobile in plants, with little recycling before leaf fall. Nitrogen plays a wider variety of roles, and its mobility varies accordingly. If taken up as ammonium, it must be bound into amino acids (which then form proteins) to prevent ammonia toxicity. The simple organic nitrogen compounds then can be transported through the plant. If nitrogen is taken up as nitrate, it may be reduced (gain electrons) and converted to amino acids in the roots or it may simply be loaded into the xylem and transported to other parts of the plant. In either case, the nitrogen eventually is incorporated into various proteins and nucleic acids; some of these can be broken down within the plant, liberating the nitrogen for recycling, but other forms cannot be reused.

INTERNAL RECYCLING MAY INCREASE AS NUTRIENT AVAILABILITY INCREASES

It seems logical that nutrient conservation through internal plant recycling should be more important on low-nutrient sites, but this appealing idea appears to be wrong. Birk and Vitousek (1984) surveyed the literature from around the world for patterns in resorption of nitrogen and phosphorus as a function of leaf concentration (a measure of nutrient availability). They found no trend for nitrogen resorption when expressed as the percentage of nitrogen recovered from leaves. However, they found that leaves with higher concentrations of nitrogen showed greater resorption of nitrogen out of leaves when expressed as milligrams of nitrogen recovered per leaf. Fertilizer studies also show greater recovery of nitrogen from senescing needles in fertilized trees than from control trees. For example, unfertilized Scots pine trees resorbed 55% of needle N before abscission, compared with just 45% recovered from needles of heavily fertilized trees (Näsholm, 1994). However, accounting for the higher concentration of N in fertilized needles means that unfertilized trees removed $5.5\,mg\,N\,g^{-1}$ needle, whereas fertilized trees recovered $6.8\,mg\,N\,g^{-1}$ needle. A similar result came from a topographic sequence of sites in Japan, where the percentage of foliar N resorbed in Japanese black pine was higher for lower productivity sites, but there was no pattern in resorption when expressed as mg N g^{-1} needle (Enoki and Kawaguchi, 1999). Nambiar and Fife (1991) found the same pattern of greater nitrogen resorbed from radiata pine needles and went a step further to estimate the total resorption from the canopies. Fertilized trees had more nitrogen in their canopies, and total nitrogen recovered per tree was more than three times greater for fertilized trees than for control trees.

Why aren't nutrients resorbed more efficiently on poor sites? As noted earlier, the mobility of nutrients within plants depends on their functions and on the structure of the molecules they comprise. Nitrogen is present in a variety of compounds, some of which are mobile (or can be broken down into mobile parts) and some of which are insoluble. In the Scots pine example above, the fertilized trees had higher concentrations of arginine (a soluble amino acid form of nitrogen), which should be readily removed before senescence (Näsholm, 1994). The recycling of nitrogen may depend more on the form of the nitrogen compounds in the tissues than on the overall nutritional status of the plant or ecosystem. Plants growing under luxuriant nitrogen regimes typically accumulate more "mobile" nitrogen compounds, whereas nitrogen-stressed plants often have a larger proportion in structural, insoluble forms.

It is also important to keep in mind that nutrition is part of the overall physiology of the plant. Nutrient uptake, transformation, and recycling represent major energy costs, and if plants are energy limited, a complete evaluation of any nutrient-use strategy should consider these energy costs. This efficiency can be examined as the grams of glucose required to produce a gram of nitrogen contained in each type of molecule. For example, a root taking up ammonium expends about 1.5 g of glucose for every gram of nitrogen incorporated into glutamine (an amino acid). If the plant takes up nitrate, it must expend an additional 4.4 g of glucose to reduce 1 g of nitrate nitrogen to ammonium nitrogen. Processing the glutamine into more complex amino acids and proteins requires an additional 10 to 15 g of glucose, for a grand total of about 15 to 20 g of glucose to

process each gram of nitrogen taken up from the soil (Barnes, 1980). To complete the picture, some added cost should be included for the production and maintenance of roots to obtain the nitrogen from the soil.

What would be the energy cost of internal recycling of nitrogen? The conversion of protein containing 1 g of nitrogen into other forms (such as mobile amino acids or proteins) costs about 1 to 2 g of glucose, or roughly 10% of the cost (without including root costs) of taking up new nitrogen from the soil. Therefore, for any tree without an overabundance of energy, internal nitrogen recycling always would be more efficient than using soil nitrogen. Differences in internal recycling probably relate more to mobility of nitrogen compounds than to overall ecosystem fertility.

NUTRIENT INPUTS HAVE THREE MAJOR SOURCES

Nutrient inputs come from three major sources: deposition from the atmosphere, weathering of soil minerals, and, for nitrogen, biological nitrogen fixation. All three can be altered by vegetation. The "input" of nutrients from mineral weathering might be considered an internal ecosystem transfer. Weathering is conventionally called an input because the atoms are released from a pool that is unavailable to plants into pools that are readily available for uptake, leaching, and cycling within the forest.

The atmosphere contains a diverse soup of dust particles, ions, and gases, and the entrance of these into forest nutrient cycles has long-term importance for all forests. The dust particles (or aerosols) have high concentrations of nutrient cations, and sometimes nitrogen and sulfur in polluted environments. The ions dissolve in rainfall to provide substantial inputs of ammonium, nitrate, and sulfate to ecosystems. In some areas, high concentrations of ammonia, nitric acid, and sulfur dioxide may result in gaseous inputs of nitrogen and sulfur to forest nutrient cycles. A stand of loblolly pine in a moderately polluted area in Tennessee, USA had twice the nitrogen deposition of a Douglas fir stand in Washington, USA and the majority of the difference

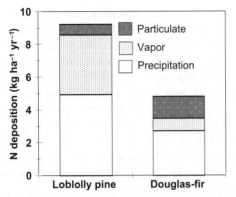

Figure 7.10 Estimated input of nitrogen to a loblolly pine stand in Tennessee and a Douglas fir stand in Washington (data from Johnson and Lindberg, 1992).

resulted from higher deposition of nitric acid vapor in the loblolly pine stand (Figure 7.10).

In the United States and Europe, networks of sampling sites are used to monitor the deposition of ions in precipitation. In the United States, the National Atmospheric Deposition Program (NADP) has documented the major patterns of deposition across the country, and has verified a very substantial decline in sulfur inputs to the eastern United States as a result of reduced emissions of sulfur dioxide from the industrial mid-1990s (Hovmand, 1999). At Whiteface Mountain, New York, sulfur deposition declined by half from 1984 to 2010 (Figure 7.11). The apparent decline in nitrogen

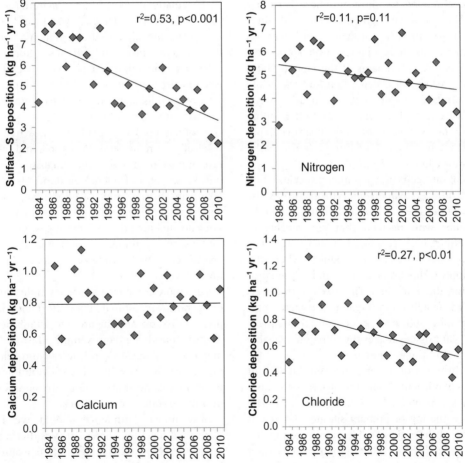

Figure 7.11 Trends in wet deposition for Whiteface Mountain, New York showed a 50% decline in sulfate-S deposition and an almost-significant decline of 20% in nitrogen deposition. Calcium deposition showed no trend, while chloride deposition dropped by 30% (data from NADP, 2012).

deposition was much less (averaging almost 20%), and was not statistically significant. Over the same period, calcium deposition showed no trend, and curiously, chloride deposition dropped by 30%.

In the eastern United States, nitrogen input to forests tends to average 5 to $10\,kg\,ha^{-1}\,yr^{-1}$ for low-elevation forests (including precipitation, vapor, and particulates) and 15 to $30\,kg\,ha^{-1}\,yr^{-1}$ for high-elevation forests (Johnson and Lindberg, 1992). Overall, less polluted areas commonly receive less than $5\,kg\,ha^{-1}\,yr^{-1}$ from the atmosphere, and more polluted areas receive more than $10\,kg\,ha^{-1}\,yr^{-1}$.

Vegetation can affect nutrient inputs from the atmosphere by altering the capture of particles, gases, and ions. The amount of surface area exposed by a forest canopy to the atmosphere can vary substantially, between high-leaf areas in conifers and lower-leaf areas in hardwoods and between evergreen and deciduous species. A red spruce forest in the Smoky Mountains of the United States had 20% greater sulfur deposition and three-fold greater nitrogen deposition than a nearby beech forest (Johnson and Lindberg, 1992). In the Solling area of Germany, a stand of Norway spruce experienced more than twice the sulfur input of a nearby beech forest (Ulrich, 1983).

Inputs of nitrogen can vary among species when symbiotic nitrogen fixation enters the picture. Some species of trees are capable of forming symbiotic relationships with bacteria that are capable of "fixing" atmospheric N_2 into ammonia (NH_3). The operation takes place in root nodules, where the plants supply carbohydrate energy and oxygen protection for the microbes. The fixed nitrogen is assimilated rapidly into organic nitrogen compounds and transported into the plant. Subsequent death of plant tissues makes the nitrogen available for recycling in the ecosystem.

The weathering of minerals transfers significant quantities of nutrients from mineral lattices to more available soil pools (Table 7.3; Figure 7.12). The rate of weathering of primary silicate minerals is a function of the ratio of silicon to oxygen in the crystals; minerals with higher ratios of Si:O weather faster.

A variety of approaches have been used to estimate the rates of mineral weathering and to examine the malleability of weathering under varying levels of acid inputs. The simplest approach is to estimate the rate of loss of cations from a watershed (the denudation rate), which should equal the rate of weathering minus the rate of atmospheric deposition and accumulation in vegetation or on the exchange complex. The classic watershed studies at Hubbard Brook in the 1960s used this approach (Likens et al., 1977). It was estimated that about $20\,kg\,Ca\,ha^{-1}\,yr^{-1}$ must come from mineral weathering to balance the calcium lost in streamwater or accumulated in vegetation.

A second approach to estimating long-term rates of weathering involves comparing the ratios of nutrient cations (such as calcium) to zircon in the parent materials. As mineral material weathers, calcium may be released, cycled, and lost over the years, but all the zircon remains because of its unreactive insolubility. If the A horizon has a calcium:zircon ratio that is 10% lower than the calcium:zircon ratio of the parent material, then 10% of the original calcium was lost from the parent material during weathering.

Another approach is to place a known amount of (typically) pure minerals into a soil, and follow the rate of dissolution over time. Augusto et al. (2000) put 3 g samples of a finely ground plagioclase in $20\,\mu m$ mesh bags, and placed the bags at various depths under five species of trees for nine years. They documented that weathering was twice as fast under spruce as under oak, and more than twice as fast in acidic, biologically active upper soil than in lower levels.

With more complex approaches, different weathering rates are estimated for the major minerals present in a soil, and the presence of minor constituents is considered. For example, Mast (1989) estimated that about 40% of the calcium released in mineral weathering in a Rocky Mountain alpine watershed came from minor intrusions of calcite into the granite/gneiss bedrock.

How much confidence should we have in watershed-level estimates of weathering rates? Klaminder et al. (2011) set out to examine this question by applying a suite of mineralogical and modeling approaches to estimating weather for a 50–100-year-old spruce, pine, birch forest in

Table 7.3 Estimated rates of mineral weathering (kg ha^{-1} yr^{-1}) for forest soils.

Site	Parent Material	P	K	Ca	Mg	Al	Method	Reference
Idaho, US	Feldspars	—	23	26	10		Watershed balance	Clayton, 1979
	Quartz monzonite		4	20	2.4		Watershed balance	Clayton, 1984
Colorado, US	Granite, gneiss, calcite intrusions		0.6	4.6	0.7		Watershed balance, mineralogy	Mast, 1989
New Hampshire, US	Quartz, feldspars	—	7	21	4		Watershed balance	Likens et al., 1977
Oregon, US	Andesitic tuff	0.2	2	12	7		Watershed balance	Sollins et al., 1980
California, US	Dolomite		4	86	52		Watershed balance	Marchand, 1971
	Quartz monzonite		8	17	2			
Maryland, US	Schist		2.3	1.3	1.7		Watershed balance	Cleaves et al., 1970
	Serpentine		<0.1	<0.1	34			Cleaves et al., 1974
New York, US	Glacial outwash		11	24	8		Lysimeter leaching	Woodwell and Whittaker, 1967
Virginia, US	Granite		1.3	8	2.6		Watershed balance	Pavich, 1986
Czech Republic	Biotite-muscovite gneiss		9.4	3.7	2.9		Watershed balance	Paces, 1986
	Biotite gneiss		23	19	14			
	Quartzitic gneiss		13	8.5	6.3			
British Columbia	Quartz diorite			34			Watershed balance	Zeman and Slaymaker, 1978
Scotland	Till		2.4	17	5.3		Watershed balance	Creasey et al., 1986
	Gabbro		1.2	20	5.9			
Luxembourg	Metashale		0.2	8.7	15.7		Watershed balance	Verstraten, 1977
Norråker, Sweden	Glacial till	0.092	0.8	5	7.92	6.03	Mineralogy relative to Zr	Bergholm et al., 1997,
Åseda, Sweden		0.054	3.6	2.2	3.6	8.01		Hallbäcken and Bergholm, 1997
Farabol, Sweden		0.005	1.2	0.6	0.72	1.8		
Loten, Norway		0.03	1.2	0.6	1.2	1.53		
Vardal, Norway		0.004	0.8	0	1.32	1.89		
Marnardal, Norway		1.13	28.4	18.2	12.12	73.08		
Birkenes, Norway		0.19	4	3.6	4.2	6.57		
Sodankylä, Finland		0.011	0.62	1.17	1.35			
Kemijärvi, Finland		0.012	0.71	1.21	1.16			
Heinola, Finland		0.03	1.49	1.39	1.32			
Australia	Dacite	3	21	19	8		Watershed balance	Feller, 1981
New Zealand	Sand	1.1	9.2	12	7.2		Modeled	Zabowski et al., 2007
	Tephra/pumice	0.5	5.4	14	5.0		Modeled	
	Tephra/pumice	0.2–0.5	6–12	15–20	4–4.7		Modeled/mass balance	
	Greywacke gravels	2.5	7.9	9.6	3.3		Modeled	
	Loess/gravel	0.2–1.5	3–15	3–12	4–10		Modeled/mass balance	
	Loess	2.2	31	24	6.6		Modeled	
Indonesia	Volcanic ash		12	18	25		Watershed balance	Bruijnzeel, 1982
Tropical, subtropical, 25 sites	Ultisols, Oxisols of various mineralogy		9–21	5–15	3–10		Watershed balance	Bruijnzeel, 1991

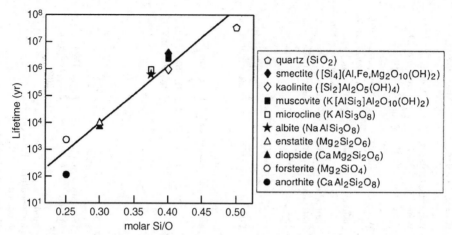

Figure 7.12 The time required to dissolve a 1-mm silicate mineral crystal depends strongly on the ratio of silicon to oxygen, increasing exponentially with linear increases in the ratio. Reprinted from Geoderma, 100, Oliver A. Chadwick, Jon Chorover, The chemistry of pedogenic thresholds, 321–353, 2001, with permission from Elsevier.

Sweden. The soils had developed in glacial till for several thousand years (after glacial retreat), with gneiss and schist as the primary mineral types. The average of all the approaches was $6.7 \, kg \, Ca \, ha^{-1} \, yr^{-1}$, and $3.9 \, kg \, K \, m^{-2} \, yr^{-1}$, which falls in the middle of estimates for many forest soils (Table 7.3; Figure 7.13). However, the variation in estimated rates was so high (standard deviations matching the

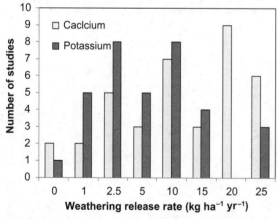

Figure 7.13 About half of the reported rates of mineral weathering in forest soils are $5 \, kg \, ha^{-1} \, yr^{-1}$ or less for potassium, whereas half of the estimates for calcium weathering are over $10 \, kg \, ha^{-1} \, yr^{-1}$. The full ranges for both elements are similar (from studies in Table 7.3).

means) that it would not be surprising if the "true" rates were near 0, or twice as high as the average estimate. The authors stress that such high uncertainty does not support high confidence in rotation-long nutrient budgets that evaluate whether forest harvest removals of cations are, or are not, balanced by weathering inputs.

How malleable are rates of mineral weathering? The concentration of H^+ in soil solutions clearly affects the rates of mineral weathering; greater acidity leads to greater rates of mineral weathering. In soils, the weathering story may depend very heavily on localized activities of roots and microbes that involve excretion of organic compounds that can directly degrade some types of minerals (see Chapter 3).

Direct assessments of rates of mineral weathering under different tree species are rare, but ecosystem budgets indicate substantial differences. For example, Eriksson (1996) examined the calcium contained in vegetation, O horizon, and mineral soil exchange pools under five species in a replicated, 23-year-old experimental plantation in northern Sweden. Norway spruce plots contained about $500 \, kg$ of calcium per hectare ($25 \, kmol_c \, ha^{-1}$) more than stands of Scots pine or lodgepole pine. Another species comparison by Bergkvist and Folkeson (1995) included estimates of deposition and leaching as well as internal pools, and the Norway spruce in this case needed about $1.6\,kmol_c\,ha^{-1}$ more weathering input of base cations to balance the nutrient budget. These replicated species trials do not provide conclusive evidence of differences in weathering rates, but until direct assessments are available, the null hypothesis should be that tree species do indeed affect rates of mineral weathering.

NUTRIENT OUTPUTS ARE GENERALLY LOW FOR FORESTS EXCEPT AFTER DISTURBANCES

Vegetation also affects nutrient output rates, but most undisturbed forests retain nutrients so efficiently that losses are small relative to the quantities cycled. Despite this general pattern, important exceptions of high nutrient losses from vigorous forests do occur. For example, Nadelhoffer *et al.* (1983) examined nitrate leaching losses from several forests in Wisconsin and found that a fast-growing stand of white pine was losing about $10\,kg$ $N\,ha^{-1}\,yr^{-1}$ in soil leaching. Another example involves high rates of nutrient leaching from vigorous stands of nitrogen-fixing red alder. Fixation of nitrogen increases both nitrogen capital and nitrogen mineralization rates while reducing the plant's need to take up nitrogen from the soil. Nitrate leaching losses from red alder ecosystems can exceed $50\,kg$ $N\,ha^{-1}\,yr^{-1}$ (Bigger and Cole, 1983; Van Miegroet and Cole, 1984; Binkley *et al.*, 1985). Some healthy forests in Bavaria are truly nitrogen saturated, with Norway spruce stands experiencing such high annual rates of N deposition that nitrate leaching may exceed $20\,kg\,N\,ha^{-1}\,yr^{-1}$ (Rothe *et al.*, 2002).

Nutrient losses usually increase after harvests, fires, and other disturbances because tree uptake is a major sink for available nutrient ions. Exposure of the O horizon and mineral soil to direct sunlight can increase temperatures, which combine with moister conditions in the absence of water uptake to accelerate decomposition and nutrient release. Thus, nutrient availability can be at a peak when plant uptake is minimal. Fortunately, microbial and geochemical immobilization are usually large enough to prevent large nutrient losses.

Is there a limit on how much nitrogen an ecosystem can retain? Intuitively, we might expect that forests will stop retaining added nitrogen when the inputs plus net mineralization exceed plant uptake. However, empirical studies have found very few systems (such as Norway spruce forests in Bavaria) that have reached a stage of nitrogen saturation. Fertilization with very heavy, repeated applications of nitrogen typically increases nitrogen leaching losses, but the losses remain lower than the input rates (see Chapter 14). Most European forests retain most of the nitrogen deposited from the atmosphere except under the highest deposition regimes. The mechanisms that account for nitrogen retention in forest ecosystems need more investigation.

CYCLES DIFFER GREATLY AMONG NUTRIENT ELEMENTS

The majority of the mass of plants is water. The dry mass of plants is composed primarily of oxygen and carbon (about 40–45% each), hydrogen (about 6%), and nitrogen (about 1.5% for non-woody tissues). The reliance of plants on these elements derives from their ability to form kinetically stable linear polymers (such as starch, cellulose, proteins, and nucleic acids) that can be readily controlled by catalysts (Williams and Frausto da Silva, 1996). About a dozen other elements comprise minor proportions of plant tissues, playing a variety of roles in modifying the structure and function of these polymers.

Each nutrient element is characterized by a unique biogeochemical cycle; the key features of each element are highlighted in Table 7.4. Potassium is the simplest, with only one form (K^+) that remains ionic throughout its entire cycle. Phosphorus cycling is more complicated, with the phosphate anion (PO_4^{3-}) playing several roles in organic compounds and with major biotic and geochemical controls on cycling. The nitrogen cycle adds the complexity of major oxidation and reduction steps. The sulfur cycle is perhaps the most convoluted, sharing some features of both the phosphorus and nitrogen cycles.

Table 7.4 Summary of features of the major nutrient cycles.

Element	Major Pool Used by Trees	Major Long-term Source for Tree Uptake	Key Biochemical Roles	Soil Chemistry	Limiting Situations
Carbon	Atmosphere	Atmosphere	In all organic molecules; fundamental to structure, energy flow, genetics	Complex organic C flows and storage, major role in overall soil fertility; bicarbonate and carbonate ions also important	Atmospheric concentrations may limit forest growth; limitation may decline with CO_2 enrichment of atmosphere
Oxygen	Atmosphere	Atmosphere	Oxidative phosphorylation, respiration, by-product of photosynthesis	Diffusion limited when soil water is high; major control of soil redox potential	Waterlogged soils
Hydrogen	Water	Water	In all organic compounds	Free H^+ influences many biological and chemical reactions	Extremely acidic or alkaline conditions
Nitrogen	Soluble nitrate, exchangeable ammonium; N_2 for N-fixing species	Soil organic matter; atmospheric N_2 for N-fixing species	Proteins, enzymes, nucleic acids	Availability dominated by microbe/enzymatic factors; most N in soils recalcitrant; important redox reactions, gaseous phases	Most temperate forests, many boreal forests, some tropical forests (esp. plantations also receiving P fertilizer)
Phosphorus	Soluble phosphate	Soil organic matter, adsorbed phosphate, mineral P	Nucleic acids, lipids, energy flow	Major organic and geochemical control; no gas phase, no redox reactions	Old soils high in iron and aluminum, common in semi-tropical and tropical situations
Potassium	Soluble K^+	Soil organic matter, exchange complex, mineral K	Enzyme cofactor, membrane regulation, ionic strength buffer	Remains in ionic form in plants, easily leached from leaves and soils	Miscellaneous situations, old soils, particularly if N and P fertilizers added
Calcium	Soluble Ca^{2+}	Soil organic matter, exchange complex, mineral Ca	Cell walls, also present as Ca-phosphate and Ca-oxalate	Dominates solution cations in non-acidic soils; precipitate with carbonate in dry alkaline soils	Rarely limiting; in some Oxisols; perhaps in some alpine spruce forests
Magnesium	Soluble Mg^{2+}	Soil organic matter, exchange complex, mineral Mg	Enzyme cofactor in chlorophyll, other enzymes	Similar to calcium	Rarely limiting, perhaps on some pumice soils

Element	Soluble form	Source	Biological function	Chemistry	Limiting status
Sulfur	Soluble sulfate, some SO_2 gas	Soil organic matter, atmospheric deposition, some mineral S	Three amino acids, sulf-hydryl bonds provide 3-D structure to proteins	Sulfate is a major anion in solutions, specifically adsorbed on oxides, important redox reactions in anaerobic situations	Rarely limiting
Manganese	Soluble Mn^{2+}		Component of 3 enzymes, cofactor in several dozen enzymes including one involved in photosynthesis	Redox reactions are important; Mn solubility increases as pH decreases	Rarely limiting; perhaps in some limestone-derived soils in New Zealand; South Africa
Iron	Soluble ferrous Fe^{2+}; chelated Fe^{3+}	Soil mineral phase	Critical in many electron transport enzymes	Major constituent of mineral soils; reduced form (Fe^{2+}) ten times more soluble than oxidized form (Fe^{3+}); organic chelates important for uptake	Limiting only in some alkaline situations for some species; addition with an organic chelate may be needed

CARBON FLOWS THROUGH FORESTS

Most nutrients in forests have the opportunity to cycle through the ecosystem; the same atom of nitrogen may be incorporated in tree biomass, return to the soil, and reenter the trees. Carbon cannot be cycled in the same way because when organisms use the carbon fixed by photosynthesis, it tends to be lost as carbon dioxide to the atmosphere. Of course, the carbon dioxide in the atmosphere can be refixed (the average residence time of carbon dioxide in the atmosphere is somewhere between a decade and a century), perhaps within the same ecosystem if the canopy leaves "catch" the carbon dioxide released from the soil. In addition, some autotrophic organisms such as nitrifying bacteria may fix carbon dioxide that is respired by tree roots. Overall, however, the vast majority of the carbon flowing through ecosystems does not reenter the same trophic level of the system without first passing through the vast atmospheric pool.

As the carbon flows through the ecosystem, it accumulates for shorter or longer periods of time in the biomass of plants and animals and in the soil. Most of the carbon fixed by photosynthesis leaves the trees the same year it was produced for deciduous trees, with a fraction (always less than half) accumulating for longer periods in wood. Evergreen trees may retain leaves as well as stemwood for more than a year, so the flow-through rate of carbon may be somewhat longer than for deciduous forests.

How long does carbon take to cycle through the soil and back to the atmosphere? An average turnover time for the whole soil carbon pool can be calculated by dividing the pool size by the annual input rate (and assuming that net changes in the pool size are small). This gives a surprisingly rapid turnover rate for most forests. The net primary carbon productivity of forests typically ranges from 5 to 25 $Mg\,ha^{-1}\,yr^{-1}$, and the carbon content of forest biomass plus soils is commonly on the order of 50 to 500 $Mg\,ha^{-1}$. The flux of carbon into forest systems, then, is on the order of 1 to 50% of the carbon stored in the system, for average turnover rates of a few decades to a century for all the carbon in the system. As noted above, decomposition of plant detritus typically follows an exponential decay pattern, so most of the carbon will be lost from the system rapidly, but small portions will persist for centuries or longer.

The chemistry of the carbon compounds that comprise soil organic matter is very important for microbes, soil fauna, and trees. These compounds differ in rates of turnover, and the nutrient supply to trees depends very heavily on these rates. If a pool of 50 000 kg of carbon per hectare contained 5000 kg of nitrogen per hectare and turned over every 100 years, the annual release of nitrogen might be on the order of 50 $kg\,ha^{-1}\,yr^{-1}$. If the same soil had an average carbon turnover rate of 50 years, the nitrogen availability might be more like 100 $kg\,ha^{-1}\,yr^{-1}$.

Soil carbon has a substantial effect on soil pH. Soil organic compounds are weak acids, and soils with high carbon contents have large capacities to (1) adsorb or release H^+ and (2) sorb nutrient cations. The acid strength (tendency to release H^+) of soil organic matter in forest soils can differ substantially among sites and under the influence of different species (see Chapter 4).

The flow of carbon through ecosystems has a major inorganic component. When carbon dioxide dissolves in water, carbonic acid is formed. This weak acid tends to dissociate to bicarbonate (HCO_3^-) and H^+. The bicarbonate typically leaches from soils in association with nutrient cations such as K^+, with the H^+ either replacing the nutrient cation on the exchange complex or driving the release of the nutrient cation by weathering of minerals. The respiration of roots and decomposition of organic matter lead to concentrations of carbon dioxide in the soil atmosphere that are orders of magnitude higher than those in the open atmosphere. These high concentrations lead to large production of carbonic acid, and to a large supply of H^+ for mineral weathering and soil development.

The carbon cycle is the primary focus of forest management for tree growth; accumulation of wood and fiber represents the rate of growth of carbon storage in this pool. The supply of carbon in the atmosphere is not amenable to management (at least not at a scale smaller than the globe), but the uptake of carbon through photosynthesis is very responsive to additions of nutrients and water. Each atom of nitrogen present in foliage can yield 50 to 100 atoms of photosynthetically fixed carbon each year. Addition of nitrogen in fertilizers involves substantial energy inputs to synthesize, transport, and apply the nitrogen, but despite these inputs, the net energy recovery in wood is still on the order of 15:1 (see Chapter 14).

Trees require far more water per unit of carbon fixed than they do of nitrogen; 1 kg of fixed carbon commonly "costs" about 300 to 600 kg (= 300 to 600 L) of water. For this reason, forest irrigation is currently practiced only in high-value situations, such as fast-growing plantations of eucalyptus or poplar to provide fiber to mix with conifer fibers in making high-quality paper.

Forest practices also manipulate the carbon cycle when soil carbon is increased or lost and when forest products are removed. The removal of carbon in harvested biomass is fairly straightforward to measure, but the effects of harvesting on later net losses or gains of carbon in the soil are less clear.

OXYGEN GIVES THE BIGGEST ENERGY RELEASE IN CHEMICAL REACTIONS

Oxygen is similar to carbon in that it tends to flow through ecosystems rather than cycle. Unlike carbon, the reservoir of oxygen in the atmosphere is vast, with a much longer turnover time (100 000 to a million years). Oxygen has one key role in soils with many facets: oxidation and reduction.

Oxidation and reduction reactions have to occur together, providing both a sink and a source of electrons, and molecular oxygen (O_2) has the highest potential of any electron acceptor in ecosystems (+810 mV at pH 7). Oxidation of sugar with oxygen gives five times the energy release of oxidizing sugar with carbon dioxide (forming methane).

Oxidation reactions with oxygen as the electron acceptor are also fundamentally important in soil development over time. Most mineral soils develop accumulations of iron and aluminum oxides; in some situations, iron oxides are present in the original mineral. In others, oxygen is engaged to oxidize reduced ferrous iron to oxidized ferric iron.

The diffusion of oxygen through soils depends heavily on soil aeration, as gases diffuse through air about 10 000 times faster than through water. The supply of oxygen in soils has two features: concentration (or partial pressure) and rate of resupply. Even small concentrations of oxygen in soils can maintain aerobic conditions with oxygen as the dominant electron acceptor in chemical reactions. However, the consumption of oxygen can deplete the supply unless the pore space remains filled with air (not water) or unless the rate of water flow is fast enough to bring in more oxygen. Oxygen diffusion into wet soils is typically too slow to prevent development of anaerobic conditions, leading to reduced microbial activity and gleying of the soil.

Soil oxygen supplies can be affected by management in two key ways. Soil compaction can reduce poor volume and restrict oxygen diffusion into the soil. The oxygen supply can also be increased intentionally in waterlogged areas through intensive systems of ditches and raised beds for planting seedlings. The improved oxygen supply then fosters microbial oxidation of organic matter (sometimes lowering the surface level of Histosols), net release of nutrients, and better root and top growth.

HYDROGEN ION BUDGETS INTEGRATE BIOGEOCHEMICAL CYCLES

Hydrogen is a critical nutrient in plants, but its ubiquitous availability in water prevents it from limiting plant growth. The more important feature of hydrogen and ecosystems is the availability of H^+ for chemical reactions. Many reactions in the soil solution depend on soil acidity, expressed as pH (negative logarithm of the H^+ activity). The pool of H^+ in soil solutions is typically small at any point in time, but this pool is strongly buffered by vast quantities of solid-phase soil acids. These solid-phase acids include the mineral and organic components of the exchange complex (see Chapter 8). Differences in pH between soils, or within a soil over time, depend, in part, on the equilibrium reactions between the soil solution and these large pools of solid-phase acids. Over the long term, the input and output of acids and H^+ determine the size and acid strength of the solid-phase acids.

The subject of H^+ budgets is complicated, but the basic features are clear. First, photosynthesis fixes carbon in the form of sugar. Next, some of the sugar is transformed into a variety of acids for biochemical use in cells, including acids such as malic or oxalic acid. These acids may dissociate, releasing H^+ and forming an acid anion such as malate or oxalate. The negative charge on the acid anions must be balanced by another cation if the H^+ is missing; candidates include K^+ and Ca^{2+}. When litter is produced that contains dissociated acids, these leaf acids decompose readily, consuming H^+ to return to CO_2 and H_2O, releasing the nutrient cation. The organic acids that accumulate in soil organic matter are produced through the incomplete oxidation of litter, which commonly results in the proliferation of carboxyl ($-COOH$) groups that behave as weak acids. Soil clays also act as weak acids, releasing and adsorbing H^+, depending on the concentration in the soil solution.

This solid-phase acid complex (or exchange complex) in the soil is then "titrated" by processes that add H^+ to or remove it from the system. The net sources of H^+ are:

1. Formation of carbonic acid; high concentrations of soil CO_2 form H_2CO_3, which dissociates at pH levels higher than about 4.5. The flux of bicarbonate leaching from the soil represents the net H^+ generation.
2. Net accumulation of base cations in vegetation; the uptake of calcium, magnesium, and potassium exceeds the uptake of phosphate, leaving a net uptake of cations that must be balanced by the excretion of H^+ into the soil. (Nitrate and sulfate are not counted in uptake because they are mostly reduced inside the plants, which consume any H^+ equivalent associated with uptake.)
3. Atmospheric deposition of H^+.
4. Net oxidation of reduced compounds; oxidation of ammonia to nitrate produces H^+. If the nitrate is assimilated or denitrified, H^+ is consumed, with no net effect. If the nitrate leaches (and therefore ammonia oxidation exceeds nitrate reduction), a net increase in H^+ occurs.

The major processes accounting for H^+ consumption are:

1. Release of base cations from decomposing biomass; this is the reverse of the biomass accumulation in the list of H^+ sources. Over the course of stand development, biomass accumulation of cations is a major source of acidification. When the biomass is burned or decomposes, this reverse process consumes the equivalent amount of H^+. Therefore, the biomass accumulation story is important at a timescale of years to centuries, but it has no net effect on long-term soil development unless biomass is removed.
2. Specific anion adsorption; the ligand exchange of sulfate or phosphate for OH^- groups in the coordination sphere of iron and aluminum consumes H^+ and protonates the OH^- to make H_2O.
3. Mineral weathering; as noted below, the weathering of silicates involves production of silicic acid, which consumes H^+.
4. Unbalanced reduction of oxidized compounds; nitrate deposited from the atmosphere may be taken up and used by microbes or plants or denitrified. In both cases, H^+ is consumed in the reduction of nitrate, so deposition of nitric acid rain can have no acidifying effect unless nitrate leaches from the soil.

The components of H^+ budgets differ with the perspective of the budget. At the level of plant cells, H^+ flows are associated with the synthesis of adenosine triphosphate (ATP) and the maintenance of electrical balances between cations and anions. These processes can be overlooked from an ecosystem perspective because they have no net effect at this higher level of resolution.

With a focus on the root–soil interface, H^+ excretion or absorption is used to maintain electrical balance between the uptake of cation and anion nutrients. In cases where nitrate is the major form of nitrogen, anion uptake usually exceeds cation uptake. With ammonium nutrition, this pattern is reversed. A balance is achieved with H^+ flows. If too many cations are taken up, plants synthesize organic acids that dissociate and supply H^+ to be secreted into the soil in exchange for the excess cations. When

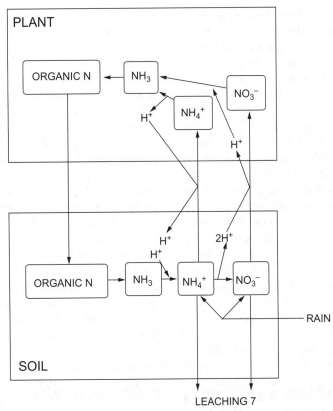

Figure 7.14 H^+ flux associated with transfers and transformation of organic and inorganic nitrogen. The nitrogen cycle has a net effect on H^+ only when it involves additions to or losses from the system, as the only form that accumulates in large quantities is organic nitrogen (for details, see Binkley and Richter, 1987).

anion uptake exceeds uptake of cations, H^+ can be absorbed from the soil to accompany the anions. In reality, plants may excrete OH^- (which combines with CO_2 and water to form HCO_3^-) rather than take up H^+. In either case, accounting for the flow of H^+ equivalents is sufficient for charge budgeting.

At a higher level of resolution, the source of the nutrient cations and anions must also be accounted for. Some surprising features emerge. For example, ammonium uptake requires excretion of H^+ into the soil, which increases the soil pool of H^+. However, if the ammonium came from mineralization of organic nitrogen, then it also involved the consumption of one H^+ ion when ammonia became ammonium. Plant uptake of ammonium merely replaces the H^+ consumed in the formation of ammonia. Similarly, nitrate uptake involves the uptake of one H^+ ion, but nitrification produces two

H^+ ions (Figure 7.14). One of these balances the consumption of the ammonia-to-ammonium step, and the other balances the H^+ taken up by the plant with the nitrate.

What about the plant? With ammonium uptake it loses one H^+, but when ammonium is converted to organic forms, one H^+ ion is released inside the plant. With nitrate uptake, one H^+ ion is taken up, but subsequent nitrate reduction consumes one H^+ ion, also leaving the plant in balance.

For these reasons, the transfer of soil organic nitrogen to plant organic nitrogen involves no net generation or consumption of H^+, regardless of the intervening transformations, because there was no net change in the redox state of the nitrogen atom.

If inorganic nitrogen is added directly to the ecosystem (in rain or fertilizer) rather than mineralized within the ecosystem, the H^+ budget may not balance so nicely. Added ammonium will not have consumed one H^+ ion previously, so H^+ excretion associated with ammonium uptake would represent a net increase in soil H^+. Similarly, nitrate added to a system would consume one H^+ ion from the soil on uptake, which would not be balanced by previous H^+ production.

This H^+ pattern has important implications for acid rain. Nitric acid (HNO_3) added to an ecosystem has no acidifying effect if the nitrate is used by the vegetation; one H^+ will be consumed for each NO taken up and utilized. If nitrate comes in as a salt (such as KNO_3) rather than as an acid, then use of the nitrate will consume H^+ from the soil. Ammonium enters as a salt (such as NH_4Cl), and plant use of the nitrogen results in production of H^+. In fact, if the ammonium is nitrified, two H^+ can be generated for each ammonium ion added. If the nitrate is later utilized, the net production drops back to one, but if the nitrate leaches from the ecosystem, the net production remains at two.

What about the other nutrient cycles? In general, the H^+ budget aspects of sulfur cycling resemble those of nitrogen cycling. Internal ecosystem cycling results in production and consumption of H^+, but these processes largely balance. Phosphorus cycling is a bit more complicated, but fortunately, the magnitude of H^+ flux in the phosphorus cycle is a very small part of the overall budget.

The accumulation of nutrient cations from inorganic pools in the soil into vegetation does represent a net flow of H^+ into the soil. Any flow of cations from organic pools in the soil into vegetation resembles the nitrogen cycle, with no net change in H^+. Over the course of a rotation, this can result in substantial H^+ production. When the biomass is decomposed or burned, though, release of cation nutrients is coupled with consumption of H^+, again completing the cycle, with no net change in the overall H^+. Forest harvesting prevents completion of the cycle and leaves the soil H^+ pool increased in proportion to the cation content of harvested biomass.

Harvesting may increase decomposition rates, and this pulse of H^+ consuming activity may decrease soil acidity by two processes. First, release of cations through the oxidation of organic matter

requires consumption of H^+ to form CO_2 and O_2. Second, some dissociated organic acids ($R-COO^-$) are oxidized, also consuming H^+ to form CO_2 and O_2. These processes can neutralize much of the acidity that was produced during the development of the previous stand. Indeed, if no biomass were removed, the magnitude of the neutralizing effect should be close to that of the acidifying effect. Little information is available from field studies on these dynamics, but limited work demonstrates the general tendency. For example, Nykvist and Rosén (1985) found that clear-cutting several Norway spruce stands increased the pH of the humus layer at several sites by about 0.5 units. They also compared exchangeable H^+ pools on plots with and without logging slash; decomposition of slash reduced the pool by about 30%. These trends should partly neutralize the acidity produced during the development of the previous stand and provide a more neutral starting point for the natural acidification that should begin with the development of the new stand. In contrast, harvesting a northern hardwoods stand in New Hampshire resulted in a decrease in pH in upper soil horizons following nitrification and nitrate leaching, along with an increase in pH in the B horizon (Johnson et al., 1991), perhaps as a result of nitrate assimilation.

Two case studies illustrate the major aspects of H^+ budgets in forests. A deciduous forest near Turkey Lakes, Ontario, experienced moderate inputs of nitrate in atmospheric deposition (about 0.35 $kmol_c$ ha^{-1} yr^{-1}; budget from Binkley, 1992), but the output of nitrate was large (1.75 $kmol_c$ ha^{-1} yr^{-1}). This net loss of nitrate indicates a net H^+ generation of 1.4 $kmol_c$ ha^{-1} yr^{-1} at this site as a result of the unbalanced oxidation of nitrate. Inputs and outputs of sulfate were similar, leaving little net H^+ generation or consumption. The average rate of cation accumulation in biomass was 0.65 $kmol_c$ ha^{-1} yr^{-1}, or about half of the acidification effect of the nitrate budget. Accounting for these and the net H^+ effects of other processes, this ecosystem was acidifying at a rate of 2.65 $kmol_c$ ha^{-1} yr^{-1}. This rate represents the quantity of H^+ available to "titrate" the exchange complex, and the rate of change in soil pH depends on the acid strength of the soil and the total quantity of acid stored in the soil. The net generation of H^+ is large relative to the low quantities of base cations in the exchange complex (about 41 $kmol_c ha^{-1}$), indicating that substantial soil acidification may be occurring.

The H^+ budget for a loblolly pine plantation in Tennessee showed a strong accumulation of nitrate; virtually all of the 0.5 $kmol_c ha^{-1} yr^{-1}$ input was retained, for a net consumption of 0.5 $kmol_c ha^{-1}$ of H^+. Mineral weathering (or other processes associated with dissociation of carbonic acid) consumed another 0.3 $kmol_c H^+ ha^{-1} yr^{-1}$. Sulfate output exceeded input, generating 0.4 $kmol_c ha^{-1} yr^{-1}$ of H^+. The rate of cation accumulation in biomass generated 0.65 $kmol_c ha^{-1}$ of H^+. The output of base cations in soil leachate exceeded inputs by about 1.5 $kmol_c ha^{-1}$, representing the largest net source of H^+ (the loss of base cations entails an increase in H^+, which is necessary to balance the lost base cations). Including other processes in this forest, the net rate of acidification was about 1.5 $kmol_c ha^{-1}$ of H^+. This rate of H^+ generation is very small relative to the large pools of exchangeable base cations (about 545 $kmol_c ha^{-1}$), so any changes in soil pH at this site are likely to be slight.

In summary, for H^+ budgets, most nutrient transfers and transformations either consume or generate H^+, but, fortunately, many of these changes are balanced later in the nutrient cycles and can be overlooked. For scientists interested in the intricacies of nutrient cycles, following H^+ cycles can ensure a clear understanding of the dynamics of all nutrients. For forest nutrition managers, it is important to realize that natural ecosystem processes generate and consume H^+, and evaluating the impacts of pollution or management treatments (such as fertilization) should be based on an understanding of the whole H^+ picture.

THE NITROGEN CYCLE DOMINATES FOREST NUTRITION

For a variety of reasons, nitrogen cycling has received more attention in forest research than any other nutrient. Nitrogen availability limits growth in more forests in more regions than any other nutrient, and it can be important even when it

is not limiting because substantial leaching of nitrate-nitrogen can occur when nitrogen availability exceeds plant uptake. Nitrate leaching is undesirable for several reasons. The nitrogen is lost from the site. The nitrate anion is also accompanied by cation nutrients, such as K^+ and Ca^{2+}, and by potentially toxic cations such as Al^{3+}. This sequence also generates H^+ and may acidify the soil. At high concentrations, nitrate may be toxic in drinking water.

Nitrogen is particularly important in plant nutrition as a component of amino acids, enzymes, proteins, and nucleic acids. In foliage, most of the nitrogen is present in the carboxylating enzyme RUBISCO; higher nitrogen in leaves leads to higher RUBISCO concentrations and higher rates of photosynthesis (depending on light and water supplies, of course).

The soil nitrogen cycle begins when nitrogen is added to the soil in organic form (from litterfall, root death, or throughfall) or inorganic form (throughfall; Figure 7.15). The cycling of nitrogen within the soil is very murky, with rapid and slow transfers between soluble and insoluble nitrogen, soil solution ammonium, and microbial nitrogen. The actual rates of these transfers are poorly known, as the best isotope techniques still provide net transfer information.

The release of ammonium from organic nitrogen pools is called gross mineralization, such as the oxidation of glycine to form CO_2, NH_3, and water:

$$CH_2NH_2COOH + 1.5O_2 \rightarrow 2CO_2 + NH_3 + H_2O$$

At the pH levels common in soils, the ammonia immediately absorbs one H^+ from the soil solution to become ammonium, NH_4^+. Much of the ammonium produced in soils is oxidized to nitrite and then to nitrate in an oxidation reaction called *nitrification*:

$$NH_4^+ + 2O_2 \rightarrow NO_3^- + 2H^+ + H_2O$$

Electrons are donated from the nitrogen atom to the oxygen molecule, releasing energy for use by the microbes. Both nitrate and ammonium may be used as nitrogen sources for protein formation. This equation represents autotrophic nitrification, a process carried out by microbes that utilize ammonium

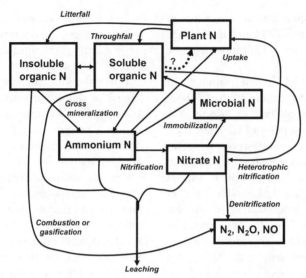

Figure 7.15 Basic features of the soil nitrogen cycle (the dotted line with "?" notes that some simple organic-N compounds may be taken up directly).

as an energy source. Another type of nitrification is performed by heterotrophic microbes, using ammonium or organic nitrogen as a substrate (Paul and Clark, 1996).

Both ammonium and nitrate are subject to leaching from the soil, but most soils lose more soluble organic nitrogen in leaching water than ammonium + nitrate (Figure 7.16). The concentration of nitrogen forms in small streams in forests in the United States often have similar levels of dissolved organic N and nitrate N (each averaging about $0.3\,mg\,N\,L^{-1}$

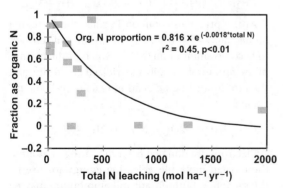

Figure 7.16 Fraction of total nitrogen leaching as organic nitrogen from 13 forests; low proportions of organic nitrogen leaching occur most often in high-leaching situations (data from Johnson and Lindberg, 1992).

across the country), with much lower concentrations of ammonium N (averaging $0.05\,mg\,N\,L^{-1}$; Binkley *et al.*, 2004b). Conifer stands are more likely to have higher concentrations of dissolved organic N and lower concentrations of nitrate than hardwood forests. Typically, only soils with very high rates of nitrogen leaching, such as those in areas with high nitrate deposition, lose more inorganic nitrogen than organic nitrogen in soil leachate.

Many compounds serve as electron acceptors in the absence of oxygen; oxygen is preferred merely because it gives the largest release of energy. An electron cascading down the redox staircase may stop at any level short of oxygen if soil aeration conditions are poor. In denitrification, an electron from some reduced carbon source is donated to nitrate, reducing it first to nitrite and then to nitric oxide (NO), nitrous oxide (N_2O), or nitrogen gas (N_2):

$$2HNO_3 + 10H^+ + 10e^- \rightarrow N_2 + 6H_2O$$

How far the reaction goes along this denitrification pathway varies primarily with the carbon (electron donating) status of the soil (Figure 7.17). Many studies characterize denitrification by "blocking" the production of N_2 (using acetylene) and measuring N_2O generation. The production of NO is more difficult to assess, but it is often an

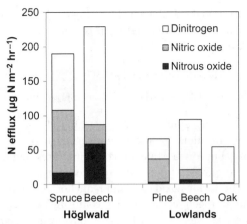

Figure 7.17 Denitrification in some German forests showed that dinitrogen (N_2) tended to be the major end product, followed by nitric oxide (NO) and then nitrous oxide (N_2O). The end products and the overall rates of denitrification appeared to differ among regions (Höglwald in Bavaria and the northeastern lowlands) and by species, but without replication (multiple locations for species within each region) these apparent patterns are confounded with other site factors Hourly rates would scale to annual loss rates of about 0.25 and 1.0 kg N ha^{-1} yr^{-1}. (from data of Papen *et al.*, 2005).

important component of denitrification (from 0 to 8 kg N ha^{-1} yr^{-1}; Butterbach-Bahl and Kiese, 2005). High carbon availability favors complete reduction to nitrogen gas. Denitrification losses are usually small in non-wetland forests (Figure 7.18). Rates of denitrification are notoriously variable in both space (at scales of 1 m or less) and time (hourly

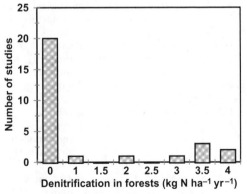

Figure 7.18 Loss of nitrogen through denitrification is typically low for forests, although a few sites may experience sizable losses (data from Davidson *et al.*, 1990).

to annually). For example, Kiese *et al.* (2003) found that denitrification losses from an Australian rainforest were low in a dry year (1 kg N ha^{-1} yr^{-1}) and quite high in a wet year (7.5 kg N ha^{-1} yr^{-1}).

In addition to denitrification leading to gaseous end products that are lost from the ecosystem, in some cases nitrate can be reduced to ammonium (called dissimilatory nitrate reduction to ammonium, or DNRA). Current evidence indicates that DNRA can be important in many cases, though more work will be needed to characterize the conditions and rates (Rütting *et al.*, 2011).

Aside from representing a loss of valuable nitrogen from a forest, denitrification is also a concern due to the production of nitrous oxide. This gas can rise into the stratosphere and react with ozone, reducing the quantity of ozone available for absorbing potentially harmful ultraviolet radiation. On balance, forests probably produce much less nitrous oxide than do agricultural soils. Nitrous oxide production by forests has also been fairly constant historically, while production from agricultural soils may have increased as a result of heavy use of fertilizers.

In some situations, plants appear capable of taking up small organic molecules that are dissolved in soil solutions (Kaye and Hart, 1997). The ability of trees to take up amino acids can be tested by applying ^{15}N-labeled molecules; of course, the amino acids could be broken down and the ^{15}N taken up as ammonium. For this reason, experiments also label some of the C atoms in the amino acids, and the appearance of both ^{15}N and ^{13}C in the tree roots indicates the whole amino acid was taken up. MacFarland *et al.* (2010) used this approach for a variety of tree species (at different locations) and found that when soils were injected with ammonium and with glycine, trees took up glycine with about 50–80% of their ability to take up ammonium. Trees with ectomycorrhizal associations appeared more effective at taking up glycine than trees with arbuscular mycorrhizae (see Chapter 6). The extent and importance of soluble organic nitrogen in tree nutrition should become clearer over the next decade.

One source of nitrogen that is important in some situations is nitrogen fixation, whereby plants supply energy in the form of carbohydrates to nitrogen-fixing prokaryotes housed in root nodules. Nitrogen fixation adds electrons to N_2 reducing it to ammonia, which is then added to proteins and used and recycled within the ecosystem:

$$N_2 + 8H^+ + 8e^- \rightarrow 2NH_3 + H_2$$

Once plants have taken up ammonium, it is aminated onto an organic molecule, such as amination of glutamate to form glutamine:

$$(CH_2)_2(COOH)_2CHNH_2 + NH_3$$
$$\rightarrow (CH)_2(COOH)_2CH(NH_2)_2 + H_2O$$

Various other nitrogen compounds are then produced by transamination. Nitrate uptake is followed by nitrate reduction. Some nitrate reduction may occur in roots, using carbohydrates from photosynthesis as an energy source.

Nitrate reduction also occurs in the leaves of some plants, where the reductant generated by the light reaction of photosynthesis can be used to reduce nitrate without consuming carbohydrates. This vector appears very important for *Rubus,* which in one study had four times the leaf capacity for nitrate reduction of competing pin cherry trees (Truax *et al.*, 1994).

As noted in Chapter 13, nitrogen combustion occurs when protein nitrogen is oxidized to N_2 or nitrogen oxides in fire:

$$4CHCH_2NH_2COOHSH + 15O_2$$
$$\rightarrow 8CO_2 + 14H_2O + 2N_2 + 4SO_2 + energy$$

On an annual basis, leaching losses and denitrification are the major vectors of nitrogen loss, but on a scale of centuries, fires remove more nitrogen from many types of forests.

The various oxidation and reduction reactions of the nitrogen cycle can be confusing, especially because terms such as nitrification and denitrification do not refer to precisely opposite reactions. Some of the confusion can be removed by identifying the role of each nitrogen compound as an energy source, or as an electron acceptor or donor (Table 7.5).

Nitrogen forms a major part of all plant tissues, but leaves (and fruits) always have the highest concentrations. The canopy often contains more than half of a tree's entire nitrogen content. As leaves develop throughout a growing season, high nitrogen concentrations at bud burst are rapidly reduced as the biomass of expanding leaves dilutes the nitrogen content. A plateau usually is

Table 7.5 Major oxidation and reduction reactions in the nitrogen cycle mediated by microbes.

Process	Microbe	Electron Donor	Electron Acceptor	By-products	Purpose
Nitrogen fixation	Bacteria, cyanobacteria, actinomycetes	Glucose or other high-energy C compound	N_2	NH_3, CO_2 (and sometimes H_2)	Provide N for use by microbe or plant
Nitrification, autotrophic	Bacteria, some fungi	NH_3	O_2	HNO_3	Obtain energy from NH_3 oxidation
Nitrification, heterotrophic	Bacteria and fungi	Organic-N	O_2	HNO_3	Obtain energy
Denitrification (dissimilatory nitrate reduction to form a gas)	Bacteria	Glucose or other high-energy C compound	NO_3^-	N_2 or N_2O	Obtain energy from C compounds in absence of O_2
Dissimilatory nitrate reduction to ammonium	Bacteria and fungi	Glucose or other high-energy C compound	NO_3^-	NH_4^+	Obtain energy from C compounds in absence of O_2

reached sometime after full leaf expansion; then leaf nitrogen gradually declines due to leaching from the leaves or resorption back into the twigs. Many trees recover substantial amounts of nitrogen from leaves before abscission. For evergreens, nitrogen concentrations in foliage gradually decrease with leaf age.

Ammonium is also sorbed onto cation exchange sites. These negatively charged sites arise from irregularities in the structure of clay minerals (such as broken edges and isomorphous substitution) and from dissociated organic acids. Forest soils commonly have 5 to 20 kg of nitrogen per hectare of ammonium sitting on exchange sites at any one time, but the flow through this very active (labile) pool is usually much greater than the average pool size would suggest. Some ammonium may also be "fixed" between the layers of clay particles, and this ammonium is only marginally available for plant uptake over long periods. Fortunately, ammonium fixation is generally not a large problem in forest soils.

The production of nitrate in soils was classically assumed to be the domain of autotrophic bacteria, but more recent work has shown that heterotrophic nitrification may be quite important in some forest soils. Heterotrophs may produce nitrate from organic nitrogen, particularly from oxidation of amines and amides (Paul and Clark, 1996). Both fungi and bacteria are capable of heterotrophic nitrification, and some of the nitrate produced in acid forest soils may be the work of heterotrophic fungi rather than autotrophic bacteria. Nitrate production by autotrophs is blocked during incubation in the presence of low concentrations of acetylene, but heterotrophic nitrification is not. Hart et al. (1997) used this acetylene-block technique to examine nitrification in conifer and alder–conifer stands in the northwestern United States and found that about two-thirds of the nitrate production in all sites appeared to be heterotrophic. More work on this basic feature of the soil nitrogen cycle is needed.

It was once assumed that low-nitrogen soils did not nitrify because trees were more effective competitors than microbes in obtaining ammonium when supplies were low. Net nitrification rates from incubations supported this idea; many low-nitrogen soils (particularly under conifers) showed no net accumulation of nitrate during 10 days or 30 days of incubation. However, this lack of nitrate accumulation (called net nitrification) commonly results from high rates of nitrate production coupled with high rates of nitrate immobilization by microbes. For example, incubations of soils from an old-growth Douglas fir site in Oregon showed no net accumulation of nitrate during incubations for the first month. The lack of nitrate accumulation resulted from a high rate of nitrate production that corresponded with a high rate of nitrate immobilization by microbes (Hart et al., 1994). The residence time for nitrate became much longer after the microbial community had depleted the supply of available C and reduced its "demand" for nitrogen.

Recent use of $^{15}NO_3$ has shown that most (although not all) soils produce respectably large amounts of nitrate, and the lack of positive net nitrification results from high rates of microbial uptake (immobilization) of nitrate. For example, Stark and Hart (1997) examined gross and net rates of nitrate production in soils from 11 forests in New Mexico and Oregon spanning a wide range of soil fertility. In the summer, 5 of the 11 forests showed net positive rates of nitrification, ranging from 0.3 to 1.0 kg of nitrogen per hectare daily. The sites with net nitrate immobilization had a nitrogen reduction of 0.4 to 1.2 kg per hectare daily. However, all of the sites showed strong rates of nitrate (gross) production, ranging from 0.3 to 2.0 kg per hectare daily.

The first product of autotrophic nitrification, nitrite (NO_2^-), is a toxic compound. Fortunately, conversion to nitrate releases additional energy, and microbial processing continues fairly nonstop, with little accumulation of nitrite.

Nitrate not taken up by plants is likely to leach as water passes through the soil because forest soils have only slight exchange capacities for anions. Most intact forests are very tight with respect to nitrogen cycling, and soil leaching removes less than 10% of the nitrogen cycled annually. Exceptions occur only on sites with very high rates of nitrogen mineralization, such as some hardwood forests, some areas receiving very high atmospheric deposition of nitrogen, and some ecosystems with nitrogen-fixing trees.

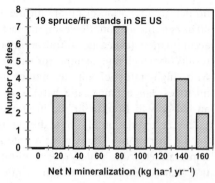
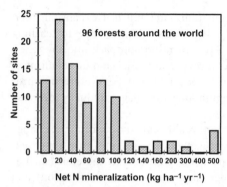

Figure 7.19 Patterns in net nitrogen mineralization from in-field incubations for a series of red spruce–Fraser fir stands in the southeastern United States (data from Strader *et al.*, 1989) and for forests around the world (data from Binkley and Hart, 1989).

Many studies have estimated in-field rates of net nitrogen mineralization on an annual basis. Within a forest type in a single region, the distribution of rates may follow a normal distribution curve (Figure 7.19). At a larger scale, including many forests around the world, the pattern is more of an F-distribution. The values above $400 \, kg \, N \, ha^{-1} \, yr^{-1}$ all came from forests in the humid tropics; only a few such studies are available.

The final process in the nitrogen cycle is the fixing of atmospheric nitrogen into ammonia. Only a few types of microbes can perform this reaction. Some, such as the bacterium *Azotobacter*, fix N_2 without assistance from higher plants. Others, such as *Rhizobium* and *Bradyrhizobium* bacteria (found in root nodules of legumes) and *Frankia* actinomycetes (found in actinorhizal nodules on alder and *Ceanothus)*, require a symbiotic relationship with plants. Symbiotic nitrogen fixation has the potential for higher rates than any free-living system for two reasons: the process is very energy expensive, and the presence of oxygen will break down the nitrogen-fixing enzyme (nitrogenase). Microbes in root nodules on legumes or other plants receive reduced carbon (energy) from the host plant, as well as protection from oxygen. Indeed, most host plants have evolved the ability to synthesize a form of hemoglobin to regulate the oxygen content of nodules.

Rates of nitrogen fixation in forest ecosystems range from near zero to several hundred kilograms per hectare annually. In the absence of symbiotic nitrogen fixers, rates are usually less than $1 \, kg \, ha^{-1} \, yr^{-1}$. Some symbiotic nitrogen fixers are much more reliable than others. For example, any member of the alder genus (*Alnus*) invariably has nodules, whereas members of the genus *Ceanothus* occasionally lack nodules.

PHOSPHORUS CYCLING IS CONTROLLED BY BOTH BIOTIC AND GEOCHEMICAL PROCESSES

The only form of phosphorus of importance in forest ecosystems is phosphate (PO_4^{3-}), and the phosphate cycle includes both biological and geochemical components (Figure 7.20). Phosphate enters into a wide variety of compounds, but the phosphorus atom remains joined with four oxygen atoms. Inside plants, phosphorus is found as inorganic phosphate, simple esters (C-O-phosphate, such as sugar phosphate), and pyrophosphate-bonded ATP (Marschner, 1995). When phosphorus supplies are limiting, leaves may allocate about 20% of their phosphorus to lipids, 40% to nucleic acids, 20% to simple esters, and 20% as free inorganic phosphate. As the phosphorus supply increases, the allocation to lipids and esters increases relative to that of nucleic acids, but the biggest increase is the pool of inorganic phosphorus, which may rise to about half of the total phosphorus in the leaves.

At the stage of soil organic matter, phosphorus is released primarily by direct action of phosphatase

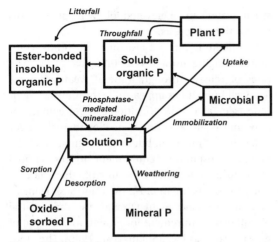

Figure 7.20 Soil phosphorus cycle.

enzymes. Some researchers believe that in the absence of phosphatases, it might take centuries for half of the organic phosphorus in soil to be released (Emsley, 1984). Organic P in soils also is present as mono- and di-ester forms; both forms are hydrolyzed by phosphatase enzymes, though mono-esters are more likely to be bound to mineral surfaces in ways that limit enzyme effectiveness (Fox *et al.*, 2011). Both microbes and plants can secrete various types of phosphatases into the soil. The relative importance of phosphatase types and sources in the phosphorus cycle of forests is not well known. Assays for phosphatase activity often reveal differences among soil types or vegetation types but usually show no correlation with plant-available phosphorus. This lack of correlation probably represents the difficulty of unraveling complex soil–plant–microbe interactions rather than the lack of importance of phosphatases. The complicated story of organic phosphorus mineralization is probably very important; Yanai (1992) estimated that about 60% of the phosphorus cycling in a northern hardwoods forest came from organic phosphorus pools.

Depending on the pH of the soil, dissolved phosphate will associate with one, two, or three H^+. At pH levels typical of forest soils, the $H_2PO_4^-$ form dominates. Phosphate in the soil solution faces four possible fates: uptake by plants or microbes; precipitation as barely soluble salts with calcium, iron, or aluminum; adsorption by anion exchange sites or sesquioxides; or leaching from the rooting zone of the forest.

Uptake of phosphorus by microbes or plants is followed by bonding one or two ends of the phosphate group to carbon chains. A single phosphate-carbon (P-O-C) bond is an ester, and two bonds are a di-ester. In these forms, phosphorus plays a pivotal role in plant energy transformations, protein and nucleotide synthesis, and cell replication. Much of the phosphorus in plants is bound to other phosphate groups through anhydride bonds. The energy of the phosphorus-phosphate bonds (anhydride linkage) is high, and is the source of energy when a phosphate group is removed from ATP to form ADP (adenosine diphosphate). Phosphorus transformations within plant cells are very dynamic. About 50% of the total phosphorus content of the plant is in the inorganic form. If cell pH is near 7, most of the free phosphate will be in the form HPO_4^{2-}. The leaves of many trees fall in the range of pH 5 to 6, so free phosphate would be found as $H_2PO_4^-$. Even in plants with severe phosphorus deficiencies, about 20% of the phosphorus content remains inorganic. Phosphorus compounds are fairly mobile, and a substantial portion of leaf phosphorus is resorbed prior to abscission.

In the soil, phosphate forms salts with calcium, iron, and aluminum that are only barely soluble (see Figure 8.4). Calcium phosphate salts are the most soluble but are important only in soils of near-neutral pH. Aluminum salts are the least soluble and dominate the low end of the pH spectrum. The solubility of iron phosphates depends on the redox state of the iron; reduced Fe^{2+} (ferrous iron) is ten times more soluble than oxidized Fe^{3+} (ferric iron).

If phosphate precipitates as an almost insoluble salt, do plants have access to these pools? Three mechanisms may help. First, simply by absorbing phosphorus, the plants create a disequilibrium that causes additional small amounts of potassium salt to dissolve. Second, many plants and microbes (including mycorrhizal fungi) also secrete large amounts of simple organic compounds such as oxalate into soils. Oxalate has a strong tendency to form salts (particularly with calcium) and may liberate phosphate by grabbing the salt cation.

Finally, the chemical status of the root zone (rhizosphere) may be substantially changed from that of the bulk soil; higher pH of the rhizosphere and other changes may favor the solubility of phosphorus salts.

Both the organic and the geochemical control of phosphorus cycling were important in explaining the potassium supply to sugar maple stands in a study from Quebec, Canada. Paré and Bernier (1989a; 1989b) found higher availability of phosphorus in stands with a mor-type O horizon, where the organic matter was not mixed by worms into the mineral soil, in comparison with stands with a mull-type O horizon (see Figure 4.4).

Given the high demand for phosphorus in forests and its low availability, losses of phosphorus are usually minimal. Even when the plant uptake component of the phosphorus cycle is removed, geochemical processes are often sufficient to retain all the phosphorus. For example, the phosphorus cycle of a northern hardwood forest in the northeastern United States was both dynamic and very tight. Uptake of phosphorus was about $12.5 \, kg \, ha^{-1} \, yr^{-1}$; $1.5 \, kg \, ha^{-1} \, yr^{-1}$ accumulated in biomass, and $11 \, kg \, ha^{-1} \, yr^{-1}$ returned to the soil in litter (Wood and Bormann, 1984). Only $0.007 \, kg \, ha^{-1} \, yr^{-1}$ of phosphorus leached from the ecosystem because phosphorus uptake by roots and microbes dominated in the upper soil, and aluminum and iron compounds adsorbed any phosphorus that reached the B horizon. When plant uptake of phosphorus was removed by deforestation, geochemical processes retained the extra phosphorus and prevented increases in streamwater concentrations.

Not all ecosystems retain phosphorus this effectively, especially following phosphorus fertilization. Humphreys and Pritchett (1971) examined the phosphorus content of soil profiles 7 to 11 years after fertilization with superphosphate or ground rock phosphate (see Chapter 14 for fertilizer formulations). They found that almost all of the superphosphate-P was retained in the top 20 cm of soils with low, moderate, or high phosphorus adsorption capacities, but that very little was retained in sandy soils with negligible phosphorus adsorption capacities. All soils retained the rock phosphorus. Chapter 14 notes that phosphorus concentrations in streamwater can be elevated following phosphorus fertilization of forests.

Phosphate inputs are also small in forests, commonly ranging from 0.1 to $0.5 \, kg \, ha^{-1} \, yr^{-1}$. The rates of release through mineral weathering are less well known but appear to be of similar magnitude (see Table 7.3). Nutritional requirements for phosphorus are also much smaller than those for potassium or nitrogen, but even so, the ratio of the annual phosphorus requirement to the annual phosphorus input is usually much higher than for other nutrients. In an old-growth forest of Douglas fir and other species in Oregon, the ratio of the phosphorus requirement for growth to the input of phosphorus was 18.2 (Sollins et al., 1980). The ratios were 6.5 for nitrogen, 3.0 for potassium, and 0.6 for calcium. This pattern emphasizes the importance of internal recycling and conservation in maintaining high phosphorus availability; it also illustrates the possible impacts of high phosphorus removals in harvesting or erosion.

The complex suite of phosphorus compounds in soils is often characterized by a sequence of extractions with bases and acid. The Hedley fractionations (Hedley et al., 1982; Cross and Schlesinger, 1995) first remove the phosphorus that is readily adsorbed by ion exchange resins. Next, the soil is extracted with bicarbonate and the extract is analyzed for inorganic and organic phosphate. Sodium hydroxide is used to remove less soluble phosphorus, which is analyzed in both inorganic and organic forms. The next step is addition of hydrochloric acid; the final, insoluble phosphorus fraction is termed residual phosphorus. These fractions probably represent a series of pools with varying turnover rates; bicarbonate-extractable phosphorus is probably more readily released for plant uptake than phosphorus from pools that are extracted by sodium hydroxide.

The long-term study of the loblolly pine stand at the Calhoun Experimental Forest documented changes in the Hedley P fractions across 45 years of forest growth (Richter et al., 2006). During this period, the vegetation (plus newly accreting O horizon) gained $82.5 \, kg \, P \, ha^{-1}$ with little apparent change in the labile pools of mineral soil P. Indeed, the bicarbonate-extractable pool of inorganic P increased by $22\,kg\,P\,ha^{-1}$. Tree uptake, and the increase in pools of labile P, resulted from reduction in the inorganic pool dissolvable by hydrochloric acid (mostly calcium-bound P, with some contribution from slowly mineralizable organic P and iron- and aluminum-associated P). These dynamics probably reflect the long-term legacy of agricultural fertilization prior to tree planting, and it will be fascinating to follow changes over the coming decades in the second rotation of pines at this site.

How well do these fractions represent the actual quantity of phosphorus that will be released and available for plant uptake? An elegant greenhouse experiment with *Brassica* plants examined the depletion of the phosphorus fractions within the rhizosphere (Gahoonia and Nielsen, 1992). After two weeks, the roots reduced the pH of the rhizosphere from 6.7 to 5.5. Bicarbonate-extractable inorganic phosphorus within the rhizosphere dropped by 34% (to a distance of 4 mm from the root surfaces), and the residual phosphorus (not extractable by bicarbonate or hydroxide) was also depleted by about 43% to a distance of 1 mm from the root surface. When a buffer solution was used to prevent acidification of the rhizosphere, plant uptake of phosphorus declined by about 17%, with less phosphorus obtained from the bicarbonate-extractable inorganic phosphorus and residual phosphorus pools. Overall, the plants obtained a little more than half of their phosphorus from the pool that would often be assumed to be the major source (bicarbonate-extractable inorganic phosphorus). This study underscores the point that rates of flow (or flux) into and out of pools cannot be determined simply by pool size.

POTASSIUM IS THE MOST MOBILE SOIL NUTRIENT

Entering the potassium cycle at the stage of O horizon and soil organic matter, decomposition (and simple leaching of soluble ions) releases K^+ to the soil solution. Potassium is present only in free, ionic form in plants, and its release from litter is usually faster than that of any other nutrient. In fact, about half of the potassium in leaves leaches prior to litterfall.

Potassium plays a wide variety of roles in plant biochemistry and ecophysiology. The pumping of K^+ into and out of the guard cells of stomata changes their osmotic potential, resulting in the opening and closing of the stomata. Free K^+ also balances the charges of major organic and inorganic anions within the cytoplasm, playing a major role in pH buffering. Potassium is also involved in enzyme activation, protein synthesis, and photosynthesis (Marschner, 1995).

Once in the soil solution, potassium has three possible fates. Plants and microbes may take up K^+, it may be retained on cation exchange sites, or it may leach from the rooting zone of the forest. Once taken up by plants, potassium remains unbound to any organic compounds and moves as a free cation through the plant to catalyze reactions and regulate osmotic potential. (Osmotic potential refers to the balance of water and dissolved compounds that help maintain the turgor, or rigidity, of plant cells.) If adsorbed on cation exchange sites, K^+ remains available for later use by vegetation. Some K^+ may become "fixed" within the mineral lattices of some types of clays and may be relatively unavailable for plant uptake. If not adsorbed, K^+ may be readily leached from the soil. Vigorous forests on relatively young soils tend to lose about 5 to $10\,kg\,K\,ha^{-1}\,yr^{-1}$ by leaching. Sandy soils and old soils are typically low in K^+, and leaching losses may be very small.

Potassium enters forest ecosystems from two sources: atmospheric deposition and mineral weathering. Precipitation involves a very dilute salt solution, and potassium inputs range from about 1 to $5\,kg\,ha^{-1}\,yr^{-1}$. Dust and aerosols also contain K, and such "dry deposition" is especially important near marine environments. Weathering of soil minerals typically adds another 5 to $10\,kg\,ha^{-1}\,yr^{-1}$ to young soils and less than $1\,kg\,ha^{-1}\,yr^{-1}$ to sandy soils and old, depleted soils (see Table 7.3).

In general, K^+ inputs exceed outputs. This element limits forest growth only in some very sandy soils, some organic soils, and some very old, weathered soils where millennia of leaching have depleted the K^+ supply.

Management of potassium in forests includes fertilization, often in association with additions of nitrogen and phosphorus (see Chapter 14).

CALCIUM AND MAGNESIUM HAVE SIMILAR BIOGEOCHEMISTRIES

Sources for plant uptake are the cations released by decomposition from organic matter and cations displaced from the cation exchange complex (Figure 7.21). Rates of weathering tend to be between 1 and 30 kg ha^{-1} yr^{-1}, depending primarily on mineralogy and perhaps secondarily on climate and biota. Controls on weathering are easy to list but hard to quantify, as discussed above.

Within plants, magnesium forms ionic bonds with nucleophilic ligands (such as phosphoryl groups in ATP), acting as a bridge to form complexes that vary in stability (Marschner, 1995). Magnesium is also a key, covalently bonded component of chlorophyll; the chlorophyll–magnesium pool accounts for 10–50% of the magnesium contained in leaves. Calcium within plants is found as exchangeable calcium at the surface of cell walls and membranes, bound within structures, and in vacuoles. The distributions tend to be about one-quarter water-soluble calcium, half pectate calcium (primarily in cell walls), about 15% bound with phosphate, and the rest consisting of calcium oxalate and miscellaneous other pools. In some cases, the oxalate calcium pool can be much larger, perhaps as high as 90% in Norway spruce (Fink, 1992).

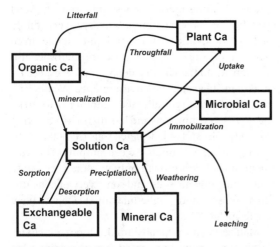

Figure 7.21 Soil calcium cycle; the cycles for other cations have a similar structure.

The biogeochemistry of calcium and magnesium includes large quantities bound in minerals in most soils (Oxisols are a major exception). This mineral pool is the long-term source for the moderate quantities adsorbed on cation exchange sites. Mass balance studies that characterize the quantity of exchangeable cations and the cations contained in tree biomass often show that the exchangeable pool is not likely to be large enough to account for the source of cations in biomass. For example, a 150-year-old beech forest in France had ten times more Ca in tree biomass than on the exchange complex, clearly indicating that mineral weathering was an important contributor to the net flux of cations into the trees (Uroz et al., 2009).

These "base cations" cannot consume H^+ or donate electrons, so they are not bases in the chemical sense. However, exchange complexes dominated by these cations maintain higher soil pH levels than those dominated by H^+ and Al^{3+}. Calcium and magnesium do not volatilize or oxidize in fires, so the primary fire-related losses occur if nutrient-rich ash is blown away by wind before rainfall leaches the cations into the soil.

SULFUR CYCLING IS MORE COMPLICATED THAN NITROGEN CYCLING

The sulfur cycle blends features of the phosphorus cycle with some from the nitrogen cycle (Figure 7.22). Like phosphate, sulfate anions can be specifically adsorbed by soils, resulting in high retention of sulfate in largely unavailable forms. Most sulfur in plants is in a reduced form (C-S-H), but some remains as free SO_4^{2-} and some is in compounds that are difficult to identify. Like the nitrogen cycle, the sulfur cycle involves oxidation and reduction processes.

Sulfur is found in the amino acids cysteine and methionine (and therefore in proteins), as well as in enzyme cofactors, sulfolipids, polysaccharides, and secondary chemicals (Marschner, 1995). Sulfur in amino acids and proteins is in the reduced form, and sulfhydryl bonds among amino acids are responsible, in part, for the three-dimensional structure of proteins. Oxidized sulfate is the form in sulfolipids and

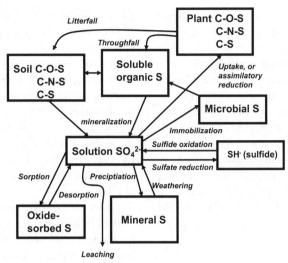

Figure 7.22 The soil sulfur cycle.

polysaccharides. Reduced sulfur can be oxidized to sulfate, a "safe" form for storage.

Sulfur is present in litter in three forms: as free sulfate (SO_4^{2-}), as reduced sulfur bound to amino acids, and in unidentified compounds, possibly including ester-bonded sulfate ($C-O-SO_3$). Free sulfate rapidly leaches from fresh litter; the nitrogen-bound reduced sulfur can be released as microbes scavenge for carbon energy sources. Ester sulfur is probably released primarily through the activity of sulfatase enzymes in soils.

Sulfate is the only free form of sulfur that is common in soils. Even when reduced sulfur is mineralized from nitrogen-bound compounds, microbes rapidly take advantage of sulfur's high-energy status to oxidize it to sulfate, much as nitrifying bacteria oxidize ammonium. Free sulfate in soil faces the same potential fates as phosphate: uptake by plants or microbes, precipitation as a salt, specific or nonspecific anion adsorption, or leaching from the rooting zone of the forest. Unlike phosphate, sulfate salts of calcium, iron, and aluminum are fairly soluble, so sulfate precipitation in forest soils is unimportant.

Sulfate is not toxic in plants, and plants with sulfur supplies in excess of current needs may accumulate sulfate in needles. Most organic sulfur compounds also contain nitrogen, and foliage generally contains about 1 sulfur atom for every 30 nitrogen atoms.

Turner *et al.* (1979) noted that some Douglas fir stands that failed to respond to nitrogen fertilization may have been limited by low sulfur availability. Nonresponsive stands had less than 400 mg of sulfate sulfur per gram of foliage, and responsive stands had up to 1000 mg per gram. These authors speculated that sulfur + nitrogen fertilization might produce larger growth responses than nitrogen fertilization alone, but careful, well-designed sulfate fertilization trials have not clearly supported this hypothesis (Blake *et al.*, 1990). Some more recent work offers support to the idea that S supply might limit response to N fertilization (Schmalz and Coleman, 2011).

Forests typically require only 5 to 10 kg S ha^{-1} yr^{-1}, and in unpolluted regions atmospheric inputs often range from 1 to 5 kg ha^{-1} yr^{-1}. Sulfur resorption prior to litterfall usually recycles about 20–30% of the sulfur in leaves. Sulfur dioxide (SO_2) pollution results in two sources of increased sulfur inputs: direct absorption of SO_2 gas and conversion of SO_2 to sulfuric acid ($H_2SO_4^{2-}$) in rainfall. Forests in industrialized regions often experience sulfur inputs of 20 to 50 kg ha^{-1} yr^{-1}. Where does the sulfur go? Only a minor proportion can be stored in plant biomass; the rest is either adsorbed in the soil or leached from the ecosystem. Leaching of sulfate sulfur is often in the range of 10 to 20 kg $ha^{-1} yr^{-1}$ in polluted regions.

Sulfate in unpolluted regions typically enters as a salt with sodium or calcium. In polluted areas, sulfuric acid is the dominant form. In both situations, sulfate tends to leave as a salt (in company with base cations). Sulfuric acid inputs coupled with sulfate salt outputs result in a net increase in ecosystem H^+. If sulfate leaches in company with H^+ or aluminum, acidification of streams and lakes may result. For this reason, the mobility of sulfate in forest soils is a major concern in the study of lake and stream acidification.

SMALL QUANTITIES OF MICRONUTRIENTS PLAY LARGE ROLES

The supply of micronutrients limits forest productivity in some notable cases, but overall these limitations are not common. See Marschner (1995) for more details on the features described in this chapter.

Iron is needed in plants for redox systems (cytochromes, peroxidases) and for some proteins. Key features of the iron cycle include redox reactions. The reduced form of iron (ferrous iron, Fe^{2+}) has a solubility of about $10^{-15} mol L^{-1}$ at neutral pH levels. Oxidized iron (ferric iron, Fe^{3+}) is only one-tenth as soluble as ferrous iron. This incredibly low solubility in water underscores the importance of organic chelates, which can increase iron solubility by several orders of magnitude (Lindsay, 1979). In acidic, waterlogged soils, some crops may experience iron toxicity owing to the increased solubility of ferrous iron at low pH; iron toxicity has not been reported in trees.

Manganese occurs in ecosystems in three oxidation states, with Mn^{2+} being most common. Manganese is a constituent of two plant enzymes: a manganese protein that splits water in photosystem II and superoxide dismutase. About three dozen enzymes are catalyzed by manganese, including RNA polymerase in chloroplasts. High levels of manganese in leaves of agricultural crops can be toxic (ranging from 200 to >5000 mg kg^{-1}, depending on the species), but manganese toxicity has not been reported in forests.

One of copper's primary functions in plants involves electron transport chains. The last step in electron transfer systems in cells occurs when copper enzymes perform the final step of transferring an electron to oxygen. About half of the copper in chloroplasts is bound in plastocyanin. Other roles for copper include a copper–zinc superoxide dismutase and phenol oxidases. Like manganese, copper at high levels can be toxic in agricultural crops, but no toxicities in forests have been reported.

The biochemical role of zinc in plants derives from its tendency to form tetrahedral complexes with nitrogen, oxygen, and sulfur. Many enzymes require zinc as a structural component, including carbonic anhydrase, alkaline phosphatase, and RNA polymerase. Some agricultural studies have shown that phosphorus fertilization may lead to zinc deficiencies, either by reducing zinc availability in the soil or by reducing zinc uptake and function in the plant.

Molybdenum is critical for two plant enzyme systems: nitrate reductase and, in nitrogen-fixing plants, nitrogenase. The oxidized state (molybdenum VI) is the most common state, with the majority of soil chemistry pertaining to molybdate (MoO_4^{2-}). Small quantities of molybdenum fertilizer (on the order of 10–20 g ha^{-1}) can substantially increase the productivity of clover pastures, but applications to nitrogen-fixing tree plantations have not been tested (because forest soils have not been molybdenum deficient). Some authors speculate that rates of nitrogen fixation in tropical forests would be far greater if molybdenum availability were slightly higher (Barron *et al.*, 2008). Molybdate chemistry is similar to phosphate chemistry, including substantial sorption on iron and aluminum sesquioxides.

Boron is the least understood of the elements required by plants. Plants that are deficient in boron show a variety of symptoms that may be related to cell wall thickness, lignification, membrane integrity, carbohydrate metabolism, and other features. The specific role of boron in supporting normal plant functioning remains unclear, even though deficiency symptoms are easy to induce in hydroponic culture, and forest deficiencies have been reported from New Zealand to Canada to Europe.

Nickel and chlorine are also regarded as necessary for normal plant functioning, but the amounts required are never limiting. Some other elements, including sodium, silicon, selenium, cobalt, and even aluminum, can be beneficial to some species even if an absolute requirement is difficult to demonstrate.

NUTRIENT USE AND NUTRIENT SUPPLY CHANGE AS STANDS DEVELOP

A classic study of an age sequence of loblolly pine stands illustrates a classic pattern: forest production rises to a peak early in stand development, followed by a substantial decline even though the forest remains vigorous and healthy. The pattern of nutrient use and accumulation tends to follow the same trend, but is this a cause, an effect, or a simple covariation? The decline in nutrient use in older forests may derive from a lower demand by the slower-growing forest. Alternatively, the decline in nitrogen use may come from a decline in nitrogen supply, which drives the decline in production. Not enough evidence is available to support broad generalizations, though it seems the decline in growth cannot be prevented by high rates of fertilization (see Ryan *et al.* 2004; 2010).

THREE RULES OF BIOGEOCHEMISTRY

The construction of nutrient budgets involves a wide range of issues, including the choice of processes to measure, experimental design, and accuracy of measurements. Some major mistakes can be avoided, or insights obtained, if three simple rules are used:

1. Pools do not represent fluxes (or supplies).
2. Nutrient budgets need to balance.
3. Attention to issues related to scales of space and time is important.

In nutrient cycling studies, it's often easier to measure the size of a pool at a given time than it

is to measure the flux of nutrients into and out of the pool. For example, the size of the cation exchange pool is easy to measure, but the flux of cations into this pool from the decomposition of organic matter or the weathering of soil minerals is very difficult to measure. The quantity of organic and inorganic phosphorus that can be extracted by sodium bicarbonate is easy to measure, but not the rate at which phosphorus enters or leaves either pool. We are often very interested in the rate at which nutrients become available for uptake, but it is important to remember that estimates of pool size are often poor indicators of availability. One cannot tell what the supply of nutrient cations is by measuring the pool that sits on the exchange complex at a particular time. If the supply of a nutrient has to be estimated by the size of a pool because direct measurement of fluxes is too difficult, one needs to retain healthy skepticism about the power of any interpretations. A good example comes from an agricultural study in Hawaii, where 15 rotations of a crop removed $4\,Mg\,K\,ha^{-1}$, but exchangeable pools of potassium declined from only 0.5 to 0.4 Mg ha^{-1} (Ayers *et al.*, 1947).

Two examples illustrate the second rule about balancing nutrient budgets. When only part of a nutrient budget is estimated, it may be useful to consider the implications of the measured components for the unmeasured components. A research project examined the loss of nitrogen from soils following harvesting. The authors sampled soil before harvesting and two years later. They reported that $1300\,kg\,ha^{-1}$ of nitrogen had disappeared from the soil (to a depth of 1 m). Only a few studies worldwide had ever reported rates of nitrogen loss even 10% as large as this one. If such a large loss had occurred, several implications for other parts of the nutrient budget would need to be considered. First, gaseous loss of this magnitude of nitrogen as N_2 or N_2O (denitrification) would be greater (by several orders of magnitude) than those found in studies that have estimated denitrification in such forests. If denitrification is unlikely to account for much of the loss, then nitrification and nitrate leaching would be the most likely causes. Leaching of dissolved organic nitrogen can be important where overall leaching rates are low, but no reports have suggested rates higher than $10 \, \text{kg ha}^{-1} \, \text{yr}^{-1}$ for leaching of organic nitrogen from forest soils. The generation of $1300 \, \text{kg ha}^{-1}$ of nitrogen as nitrate would involve a net production of over $100 \, \text{kmol}$ of H^+. The generation of so much acidity would have reduced soil pH substantially, and leaching that much nitrate would have entailed equivalent losses of base cations and aluminum. If this amount of nitrate were dissolved in $1 \, \text{m}$ of runoff from the site (a ballpark figure for precipitation minus evapotranspiration), the nitrate nitrogen concentration would have averaged about $150 \, \text{mg L}^{-1}$, 15 times the drinking water standard, and peak concentrations would probably have been much higher. Such large nitrogen losses are unlikely, and if the authors had high confidence in the world-record rate, readers would have benefited from the development of a balanced idea of the implications for nitrate leaching, soil acidification, and streamwater quality.

The value of looking for balanced accounting is also evident from a study that examined the cation nutrient budget for adjacent plantations of Norway spruce, beech, and birch. The authors used a model to estimate the rate of mineral weathering, and they concluded that the weathering release of base cations would be about $0.3 \, \text{kmol}_c \, \text{ha}^{-1} \, \text{yr}^{-1}$. How did this balance other parts of the nutrient budget? The spruce stands had greater amounts of base cations on the exchange complex, as well as greater base cation content in the larger accumulation of tree biomass. The leaching losses of base cations from the soil under spruce were more than twice the losses under the hardwoods. Summing the greater pools of base cations in soil and biomass with the greater loss rates showed that the spruce stand gained about $1.5 \, \text{kmol}_c \, \text{ha}^{-1} \, \text{yr}^{-1}$ more base cations than did the other stands. This balancing of the nutrient cation budget indicates either that weathering was far higher under spruce (at least five times greater than estimated in the model) or that one or more of the nutrient budget components was misestimated. In many studies, the degree of balance in nutrient budgets may not be known until the late stages of a project. In these cases, consideration of the accounting balance can help gauge the confidence that is warranted in the components of the budget, highlight areas of high uncertainty, and suggest areas for further research.

AS IN ALL ACCOUNTING, SCALE IS AN IMPORTANT ISSUE FOR NUTRIENT BUDGETS

The important components of nutrient budget calculations depend on the scale of interest, and extrapolation across scales needs careful attention. The first example of scale shows that important components of nutrient budgets differ across timescales. What is the rate of nitrogen loss from a lodgepole pine forest? At an annual scale, the average loss might be on the order of $0.5 \, \text{kg ha}^{-1} \, \text{yr}^{-1}$ from leaching, or $50 \, \text{kg ha}^{-1}$ in a century. But at the timescale of a century, the nitrogen loss is likely to be dominated by a single wildfire event that would remove over $100 \, \text{kg N ha}^{-1}$ in a few hours.

The second example of scale illustrates that patterns across stands may not match those that should be expected within stands. A pattern between soil nitrogen availability and net primary production across sites may be strong (Figures 7.1, 7.2). What would be the effect of increasing nitrogen availability within a single site? It may be tempting to expect that the pattern across sites would adequately represent the pattern within a single site, but such an expectation would need to be tested before much confidence was warranted.

The third scale issue is that the controls on a nutrient flux on one timescale may differ from the controls on another timescale. Short-term incubations show that carbon release often doubles when the incubation temperature increases by $10 \, °\text{C}$, and this relationship could be used to estimate the increase in carbon release from soils over long terms. However, the temperature responsiveness of readily oxidized carbon may be quite different from that of stabilized, humified carbon (Giardina et al., 1999).

Budgets of H^+ illustrate the importance of scales in both time and space. In most forests, one of the major components of the H^+ budget on an annual scale is the net accumulation of base cations in tree biomass. This acid production is commonly on the order of $0.5 \, \text{kmol}_c \, \text{ha}^{-1} \, \text{yr}^{-1}$, or about $50 \, \text{kmol}_c \, \text{ha}^{-1}$ in a century. Over the same time period, the H^+ generated from dissociation of carbonic acid (formed from the high concentration of soil carbon dioxide dissolved in soil water) might generate $25 \, \text{kmol}_c \, \text{ha}^{-1}$, giving the impression that cation accumulation in biomass has twice the acidifying effect of bicarbonate production and leaching. However, when the forest is consumed by fire or other disturbances, the base cations are returned to the soil and H^+ is consumed in the production of carbon dioxide and water, erasing the century-long legacy of acid production by biomass accumulation. The legacy of bicarbonate formation and leaching would not be erased when the forest was consumed. So, on an annual timescale, biomass accumulation was very important in the H^+ budget and soil acidification, but on the century scale, this process became a cycle with no net effect, and only the bicarbonate production needed to be accounted for to describe long-term soil acidification.

CHAPTER 8
Chemistry of Soil Surfaces and Solutions

OVERVIEW

The chemistry of soil solutions includes many different elements and inorganic compounds, as well as organic compounds and gases. The soil solution is the immediate source of most nutrients used by plants, and the composition and dynamics of the soil solution depend on interactions with the solid phases of the soil, as well as on the overall ecosystem biogeochemistry. Cation exchange reactions are important for elements such as calcium, potassium, and aluminum, and specific ion exchange is important for phosphate and sulfate. The total concentration of ions in solutions influences the acidity of the soil solution and the exchange reactions between the soil solution and the solid phases. Soil acidity is similar to the acid–base properties of any weak acids and bases, except that the anion portion of the soil acidity is solid. Soil pH differs across landscapes and under the influence of species and management. Differences in pH depend on differences in the types of acids present, the degree of dissociation of the soil acids (often referred to as base saturation), and the total salt concentration in the soil. Nutrient uptake by plants results in local depletion of the soil solution, and characteristics of the roots and mycorrhizae have reciprocal influences on the composition of the soil solution.

The chemistry of soils is fascinatingly complex, involving inorganic reactions between solid phases (including minerals, mineral surfaces, and organic matter), the liquid phase (near surfaces and in the bulk soil solution), and an incredible diversity of soil organisms. This chapter focuses on the chemistry of the soil surfaces and solutions as they influence chemistry relevant for plants. The soil solution is a dilute soup of dozens of chemicals, with the soup supplied with chemicals from atmospheric inputs and from the solid phases, and depleted by processes of plant uptake and movement into the solid phases (Wolt, 1994). The broader aspects of biogeochemistry determine the amount of flow (or flux) of chemicals through the soil solution on an annual scale (see Chapter 7). Shorter-term issues influence the composition of the soil solution, how the soil solution is buffered by relatively rapid interactions with the solid phases, and how the soil solution provides nutrients for plant uptake.

MAJOR SOIL ANIONS INCLUDE CHLORIDE, SULFATE, BICARBONATE, AND SOMETIMES NITRATE

Soil solutions contain dissolved chemicals, and many of these chemicals carry negative charges (anions) or positive charges (cations). The total charge of anions in forest soils commonly ranges between 100 and 500 $\mu mol_c\,L^{-1}$ (micromoles of charge per liter, or 10^{-6} moles of charge per liter), and maintenance of electroneutrality requires an equivalent concentration of cations. Figure 8.1 illustrates the anion composition of soil solutions from A horizons in six temperate forests. The Norway spruce forest in Norway had a soil solution dominated by chloride and sulfate. Chloride showed a moderate influence of inputs of salt from the ocean, and sulfate concentrations indicated substantial deposition of sulfur from polluted air. The loblolly pine soil in North Carolina, in the United States, had some chloride, a large

---

Ecology and Management of Forest Soils, Fourth Edition. Dan Binkley and Richard F. Fisher.
© 2013 John Wiley & Sons, Ltd. Published 2013 by John Wiley & Sons, Ltd.

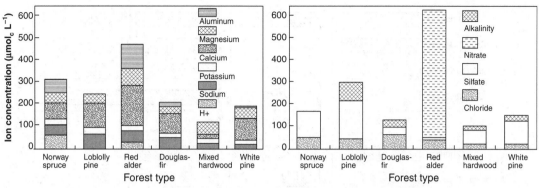

Figure 8.1 Concentrations of cations and anions in solutions collected by lysimeters placed in upper mineral soil from representative temperate forests. (data from Johnson and Lindberg, 1992)

amount of sulfate, and some bicarbonate (alkalinity). As described later, soils contain high partial pressures of CO_2 as a by-product of respiration, and the CO_2 dissolves into the soil solution to form carbonic acid. Depending on the pH of the soil solution, some of the carbonic acid may dissociate to form H^+ and bicarbonate (HCO_3^-). Both the Norway spruce and loblolly pine soils had carbonic acid in the soil solution, but only the pine soil had a pH level high enough for substantial dissociation to form HCO_3^- anions. Some of the alkalinity of soil solutions comes from soluble anions of organic acids rather than from inorganic HCO_3^-. The Douglas fir soil solution in Washington State, USA had almost no nitrate (NO_3^-), in contrast to extremely high concentrations of nitrate in the soil solution of an adjacent stand of nitrogen-fixing red alder. The mixed hardwood and white pine soil solutions came from watershed experiments in North Carolina, USA, and demonstrated greater deposition of sulfate in the larger conifer canopy. Other minor anions in soil solutions include phosphate and fluoride.

MAJOR SOIL CATIONS INCLUDE SODIUM, POTASSIUM, CALCIUM, MAGNESIUM, AND SOMETIMES ALUMINUM

The cations that balance the charge of the anions interact strongly with the solid phase of the soil, particularly with the cation exchange complex

(see later). Sodium is important in some soils, particularly near coastlines where inputs of marine salts are important. Potassium is a key nutrient for plants and is very mobile in soils; most soils have substantial quantities of potassium in solution, but potassium never dominates the total cation suite. Calcium is the dominant cation in most forest soil solutions; the major exceptions are found in soils with pH values below about 4.5, where dissolved aluminum ions become important (such as the Norway spruce soil in Figure 8.1) The increase in nitrate concentrations under red alder results in increases in concentrations of all cations, and aluminum responds more dramatically than calcium (see Figure 8.5). Magnesium concentrations are commonly 20–50% of the calcium concentrations, although some soil types may have more magnesium than calcium (such as the mixed hardwood site in Figure 8.1). Other minor cations in soil solutions include ammonium and a suite of metals including iron, manganese, zinc, and copper.

Soil solutions change in composition over time and with depth in the profile. The white pine soil from Figure 8.1 showed different patterns of change in solution concentrations with depth in the profile. Concentrations of sulfate reached a peak in the upper mineral soil (Figure 8.2), followed by strong reductions in the BC horizon after biologic uptake and adsorption of iron and aluminum sesquioxides (see later). Nitrate concentrations were high at the bottom of the O horizon but then dropped to near zero as this limiting nutrient was absorbed

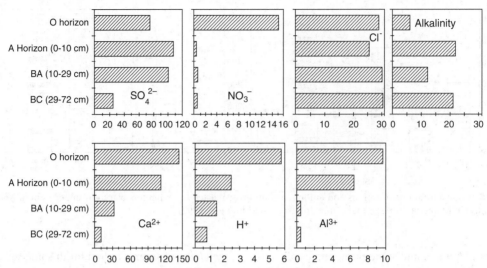

Figure 8.2 Soil solution chemistry profiles (all units are $\mu mol_c\,L^{-1}$) for a white pine soil at the Coweeta Hydrologic Laboratory in North Carolina. (data from Johnson and Lindberg, 1992)

from the solution by microbes and plants. Chloride concentrations showed little change with depth, demonstrating the low reactivity of chloride with minerals and biota. Some biological cycling of chloride is indicated by the match in upper soil and lower soil concentrations because the water percolating to the lower depth was reduced almost by half as a result of plant water uptake from the upper soil. Alkalinity showed the reverse pattern, generally increasing with depth. Soil solution cations generally decreased with depth, indicating that geochemical and biological processes remove these ions from the soil solution before they percolate very far into the soil.

SOIL SOLUTIONS ALSO CONTAIN SILICIC ACID, DISSOLVED ORGANIC CHEMICALS, AND GASES

Most forest soil solutions also contain non-ionized chemicals. Silicic acid (sometimes called silica) is a by-product of weathering of many primary minerals, and concentrations of silicic acid in forest soil solutions are commonly on the order of 1 to $10\,mg\,L^{-1}$. Silicic acid does not dissociate at normal pH levels in forest soils, and it is relatively soluble.

Therefore, silicic acid formed in soils mostly leaches out of the soil profile.

Dissolved organic compounds in soil solutions range from simple sugars to large, complex fulvic acids. The six soil solutions illustrated in Figure 8.1 ranged from about 300 to $2000\,\mu mol\,L^{-1}$ of dissolved organic carbon (assuming a carbon:nitrogen ratio of 20; Johnson and Lindberg, 1992). The organic carbon concentration of soil solutions is commonly about the same as that of the inorganic ions (on the order of 1 mole of carbon for 1 mole of charge of ions).

Soil solutions exchange gases with the soil atmosphere, and the soil solution commonly contains $<1\,mg\,L^{-1}$ of CO_2 and $10\,mg\,L^{-1}$ of oxygen. Depletion of oxygen leads to anaerobic conditions, changing many biogeochemical reactions of the soil system.

The importance of low-molecular-weight, soluble organic acids in soil solution chemistry has not been explored thoroughly. Plant roots and microbes excrete organic acids into soils, and organic acids leach from forest litter. Major acids include acetic, formic, oxalic, malic, citric, and shikimic acids (Fox, 1995; Kryszowska et al., 1996; Lundström et al., 2000). Some mat-forming mycorrhizae accumulate over $800\,kg\,ha^{-1}$ of calcium oxalate (Sollins et al., 1981). The concentration of malic, succinic, and lactic acids in soil solutions in an oak/pine forest in Alabama was 0.5 mmol L^{-1} (Hue *et al.*, 1986), and the concentration of oxalic acid in the rhizosphere of slash pine exceeded 1 g kg^{-1} soil (Fox and Comerford, 1990). Leachates from spruce O horizon in Maine averaged 10 μmol L^{-1} of acetic acid, 0.7 μmol L^{-1} of formic acid, and 3.3 μmol L^{-1} of oxalic acid (Kryszowska *et al.*, 1996). The sum of these three acids comprised a little more than 2% of the dissolved carbon in the soil solution. Lundström *et al.* (2000) found that low-molecular-weight organic acids comprised up to 5% of dissolved C, contributing up to 15% of the total acidity in soil solutions.

Some of these acids chelate metals, increasing the solubility of micronutrients such as iron and zinc and reducing the toxicity of monomeric Al^{3+}. The acids may also increase phosphorus availability by chelating the iron and aluminum that was bound to phosphorus (see later). The phosphorus sorption sites may also be reduced through the accumulation of low-molecular-weight organic acids.

Aluminum biogeochemistry is heavily influenced by soluble organic matter. The classic model of soil podzolization involves binding of aluminum (and iron) by soluble organic matter in the upper soil, followed by precipitation in the B horizon. For example, Driscoll *et al.* (1984) found that the concentration of monomeric aluminum in three Spodosols ranged from about 5 to 30 μmol L^{-1}, and 65–100% of the aluminum was bound with soluble organic matter. The deposition of metals in the B horizon may be a function of changes in soil pH and organo-ligand solubility, or of the degradation of the organic ions (Lundström *et al.*, 2000). This binding of aluminum with organic matter may be particularly important in studies that examine potential aluminum toxicity in relation to acid deposition; organically bound aluminum is non-toxic, and assays of total monomeric aluminum in soil solutions may greatly overestimate the concentration of toxic, inorganic, monomeric aluminum.

SOLUTION CHEMISTRY IS REGULATED BY INPUTS, OUTPUTS, AND REVERSIBLE REACTIONS

The variety of patterns in soil solution chemistry derives from chemical reactions and interactions between geochemical and biological components of soils (Figure 8.3). This chapter deals primarily with the reversible (sometimes called equilibrium) reactions that influence the soil solution (the bottom half of Figure 8.3). The fluxes into and out of the soil solution are covered primarily in Chapter 7.

THE SOLID PHASE OF THE SOIL HAS FOUR MAJOR COMPONENTS

The solid phase of the soil can be viewed as consisting of four functional components: the mineral

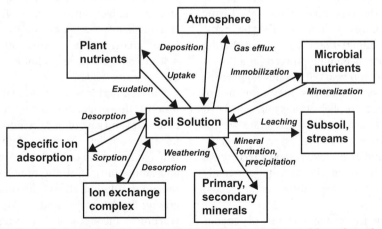

Figure 8.3 Major pools of soil chemicals and processes that transfer chemicals to and from the soil solution.

phase, the readily reversible exchange complex, the specific ion adsorption complex, and organic matter. The release of ions from primary or secondary minerals is called weathering, and this process accounts for the initial sources of most of the calcium, magnesium, potassium, and phosphate cycling in most forests.

Weathering processes are strongly influenced by biotic activity, including production of H^+, low-molecular-weight organic acids that complex with ions from soil minerals, and complex organic acids in leachates from decomposing organic matter. Rates of mineral weathering are discussed in Chapter 7; here we focus on the processes. The two primary processes that degrade soil minerals are:

1. Acid hydrolysis, in which H^+ displaces other ions from minerals, often forming silicic acid and bicarbonate salts with sodium, potassium, calcium, and magnesium.
2. Organic complexation, in which complexing anions (such as oxalate) have a very strong affinity for atoms that comprise the mineral (such as calcium).

Rock surfaces weather very slowly, on the order of 1 mm in 200 000 years for basalt (Colman and Dethier, 1986). In forest soils, interest in mineral weathering focuses on small particles of primary and secondary minerals that weather much faster than rocks. The dissolution and removal of soil minerals across the landscape of the U.S. Pacific Northwest averages about 0.3 $Mg\,ha^{-1}\,yr^{-1}$ (Coleman and Dethier, 1986). In situ weathering of minerals to form sapprolite proceeds at a rate of about 1 m per 30 000 years of soil development.

At the small scale of mineral surfaces, the suite of organisms in the rhizosphere (roots and associated mycorrhizae and bacteria) may accelerate weathering by production of acidity, especially organic acids (see Hoffland et al., 2004), and mycorrhizal fungi may proliferate in the vicinity of readily weatherable minerals (see Rosling et al., 2004; Rosling, 2009).

The weatherability of major classes of igneous minerals follows a reverse sequence of the Bowen reaction series of progressive crystallization as magma cools (Goldich, 1938). Olivine is one of the first minerals to crystallize as magma cools; it has a high energy of formation and weathers easily. Moving down the sequence of crystal formation in the direction of lower weatherability, the sequence progresses from olivine to augite to biotite mica to potassium feldspar, muscovite mica, and quartz. The sequence of decreasing weatherability for feldspars goes from calcium plagioclase to calcium-sodium plagioclase to sodium plagioclase to potassium orthoclase, mica, and quartz.

In addition to high energies of formation, the feldspars that crystallize at higher temperatures have tetrahedra that are half aluminum and half silicon. At temperatures on the Earth's surface, aluminum is more stable in octahedral structures; the aluminum charge in feldspar tetrahedra is balanced in part by Ca^{2+} ions between the tetrahedra, weakening the mineral structure (Bohn et al., 2001). More resistant feldspars contain only one-quarter aluminum, and the single available charge is met with monovalent Na^+ or K^+. Potassium feldspars are more stable than sodium feldspars because K^+ fits better between the tetrahedra. Finally, the degree of linkage of tetrahedra also influences weatherability; a high degree of linkage is more stable.

Another weathering sequence lists the minerals that characterize soils in various stages of weathering. The Jackson–Sherman stages start with gypsum, carbonates, olivine/pyroxene/amphibole, and feldspars in relatively unweathered soils (Sposito, 1989). Intermediate weathering stages characteristically have quartz, illite, vermiculite, or smectites. Advanced stages have kaolinite, gibbsite, and iron and titanium oxides.

The weathering of silica minerals includes alteration of the original structure (such as loss of potassium and gain of silica by muscovite to become mica) or complete dissolution and recrystallization (such as formation of kaolinite from dissolved silica and aluminum). If all the products of weathering remain in solution, the weathering reaction is congruent; if some of the products are not soluble, the reaction is incongruent. For example, the weathering of albite is congruous when aluminum ions and silicic acid are formed:

$$NaAlSi_3O_8 + 6H_2O + 2H^+ \rightarrow Na^+ + Al(OH)_2{}^+$$
$$+3Si(OH)_4 + 2H_2O$$

Alternatively, precipitation of gibbsite from the weathering of albite is an incongruent reaction:

$$NaAlSi_3O_8 + 6H_2O + H^+ \rightarrow +Al(OH)_3 + Na^+$$
$$+3Si(OH)_4 + H_2O$$

Biotic effects on mineral weathering are most apparent in rhizosphere soils, where organic complexation and simple reductions in soil pH may drive mineral dissolution. In a Norway spruce plantation in Sweden, rhizosphere soil was sampled by shaking off the soil that adhered to fine roots (Courchesne and Gobran, 1997). X-ray diffraction showed that this rhizosphere soil had substantially less amphibole and expandable phyllosilicates than were present in the bulk soil; no differences were found between the soil types for potassium feldspars. The rhizosphere soil also had higher concentrations of oxalate-extractable aluminum and iron. These differences in mineral weathering were associated with much greater organic matter concentrations in the rhizosphere soil and with a five-fold greater concentration of exchangeable base cations (Gobran and Clegg, 1996). A great deal of research is needed to provide a better understanding of the differences between bulk soils and those that are intimately affected by tree roots ad mycorrhizae.

The solubility of minerals may regulate the concentrations of ions in solution. For example, the concentration of phosphate in soil solutions depends on the solubility of calcium phosphates, iron phosphates, or aluminum phosphates. Above pH 6.5, calcium-phosphate compounds are the least soluble form in most soils (Figure 8.4), and phosphorus solubility increases with further increases in pH. At lower pH levels, aluminum-phosphate compounds dominate, and phosphorus solubility drops by one order of magnitude for each one unit decline in pH. If calcium phosphate (from apatite minerals or fertilizers) is present in low pH soils, phosphorus tends to be released from calcium into the soil solution, followed by precipitation (or specific adsorption; see below) by aluminum.

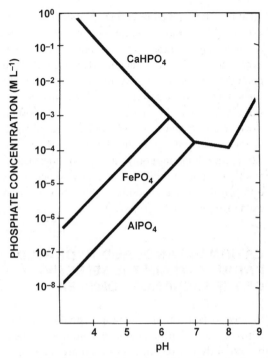

Figure 8.4 Solubility of phosphate depends on the availability of calcium, iron, aluminum, and soil pH. (after Lindsay and Vlek, 1977; reproduced from *Minerals in Soil Environments*, 1977, page 658; used by permission of the Soil Science Society of America)

CATION EXCHANGE IS A REVERSIBLE, ELECTROSTATIC SORPTION

Soil textbooks typically describe two types of positive soil charges that can retain cations: permanent charge that derives from isomorphic substitution of ions within minerals and pH-dependent charge that increases as pH rises. The key difference between these two types of soil charge results from the inability of soil clays to retain adsorbed H^+. If a clay is saturated with H^+ in a laboratory, most of the H^+ is rapidly replaced by aluminum ions in a matter of hours (Bohn *et al.*, 2001). Therefore, the ions on permanent charge sites can be displaced by concentrated salt solutions. Concentrated salt solutions cannot displace H^+ from soil organic matter, so the proportion of these sites available for cation exchange varies with soil pH.

Some positive charges can develop at low soil pH, providing some electrostatic anion exchange capacity for anions. If the pH is low enough, the size of the salt-exchangeable positive and negative charges may be equal. When the anion exchange capacity and cation exchange capacity are equal, the pH level is called the point of zero charge (PZC). At first glance, this might suggest that no ions will be retained on the soil surfaces because the charges cancel one another. However, the charges on soil colloids are not adjacent to each other and cannot dissolve and move, so the PZC point is important only for issues of flocculation (soils with a pH at the PZC flocculate and behave as uncharged colloids; Sposito, 1989), not for ion exchange properties.

CATION VALANCE AND HYDRATED RADIUS EXPLAIN THE SELECTIVITY OF THE EXCHANGE COMPLEX

In electrostatic cation exchange, smaller cations with higher valence (charge per atom) are held more tightly to exchange sites than are larger cations with lower valence. Accounting for hydrated radius and valence leads to the lyotropic series of cations in order of ease of replacement by ions farther along in the series (Bohn *et al.*, 2001):

$$Na^+ > K^+ \sim NH_4^+ > Mg^{2+} > Ca^{2+} > Al^{3+}$$

Given an exchange complex with a distribution of adsorbed cations, what determines the relative concentrations of ions supported in soil solutions? Two features are important. The total concentration of anions is matched by the total concentration of cations, so changes in the ionic strength of solutions lead to different quantities of cations in solution. The distribution of individual cation species is commonly described by selectivity coefficients that are analogous to the equilibrium constant of chemical reactions.

A variety of algebraic details have been used to characterize the selectivity coefficients of soil exchange reactions. These selectivity coefficients are at the heart of most simulation models that attempt to model the changes in soil solution as a function of the total ionic strength of the soil solution (see Reuss and Johnson, 1986, for a helpful overview).

Consider the replacement of Ca^{2+} on the exchange complex (represented by CaX) by Al^{3+}:

$$CaX + 2/3\, Al^{3+} \rightarrow 2/3\, AlX + Ca^{2+}$$

The equilibrium expression for this reaction (using the Kerr approach to selectivity coefficients) is:

$$K_s = \frac{[AlX]^{2/3}[Ca^{2+}]}{[CaX][Al^{3+}]^{2/3}}$$

The form of equilibrium constants is the concentration of the products multiplied in the numerator (with each species raised to the power of the number of molecules in the reaction) and the concentrations of the reactants multiplied (again raised to the power of the number of molecules) in the denominator.

This equation can be rearranged to describe the ratios of the concentrations on the exchanger and in solution:

$$\frac{[AlX]^2}{[CaX]^3} = K_s \frac{[Al^{3+}]^2}{[Ca^{2+}]^3}$$

(both sides of the equation were cubed to eliminate the fractional powers). The ratio of exchangeable aluminum to calcium will not change substantially over short periods of time, giving:

$$[Al^{3+}] = K_t[Ca^{2+}]^{3/2}$$

(where K_t is the square root of the right side of the equation above). The significance of these equations is that the changes in the concentration of Al^{3+} should be to the 3/2 power of the changes in concentration of Ca^{2+}.

How well does this description work in real soils? Seasonal changes in nitrate concentrations in soil solutions varied from 10 to 2500 $\mu mol_c\, mg\, L^{-1}$ in a red alder soil solution in Washington (Reuss, 1989). This wide range of nitrate concentrations led to huge variations in soil solution cations. The equations above predict that the change in the concentration of Al^{3+} would be the 3/2 power (= 1.5 power) of the change in the concentration of Ca^{2+}. The actual relationship is shown in Figure 8.5, where the Al^{3+} concentrations graphed as a

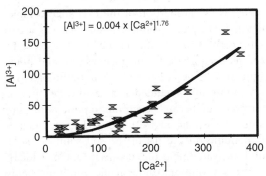

$[Al^{3+}] = 0.004 \times [Ca^{2+}]^{1.76}$

Figure 8.5 Changes in solution nitrate concentrations provide two-order-of-magnitude changes in concentrations of aluminum and calcium. Theory predicted that aluminum would increase to the 3/2 power of calcium; the actual relationship was a bit steeper (1.76, $r^2 = 0.83$; reworked from data of Reuss, 1989).

function of the Ca^{2+} concentrations showed that the best fit was with an exponent of 1.76, a bit steeper than the predicted 1.5.

THE RATIO OF ALUMINUM TO BASE CATIONS MAY (OR MAY NOT) BE IMPORTANT

Many studies have shown that some agricultural crops are sensitive not only to the total concentration of aluminum in soil solution, but also to the ratio of aluminum to base cations. Higher concentrations of aluminum may be tolerable if the concentrations of base cations are also high (Marschner, 1995). This concept has been extended to forest soils in an attempt to understand the possible impacts of soil acidification from acid deposition. Any value of the concept of aluminum:base cations has been discussed but not clearly established in forest soils, especially as much of the dissolved aluminum in forest soils is complexed with organic anions (removing any toxicity issues).

SOIL ACIDITY INVOLVES ACID–BASE REACTIONS BETWEEN SOLID AND SOLUTION PHASES

The dynamics of soil cations on the exchange complex and in the soil solution also depend

on the acidity of the soil. The hydrogen ions in solution influence exchange reactions, and the hydrogen ion concentrations, in turn, are influenced by the distribution of cations on the exchange complex.

Hydrogen ions are free, reactive protons that are involved in many chemical reactions, particularly hydrolysis (the gain or release of H^+) and redox reactions. Hydrogen ion concentrations span several orders of magnitude in soils and biological systems. A logarithmic scale is used to describe H^+ activities, expressed as pH:

$$pH = -\log AH^+$$

where AH^+ is the activity of H^+. Under oxidized soil conditions, common soil ranges for pH are 3.5 (very acidic) to 8 (mildly alkaline; Lindsay, 1979). Soils that resist changes in pH as H^+ is added or removed are referred to as well-buffered soils. In forest soils, pH tends to be < 4.0 only in organic horizons that lack aluminum (or iron). In soils where aluminum dominates the exchange complex, pH values commonly fall between 4.0 and 4.5. Exchange complexes dominated by so-called base cations (such as calcium, magnesium, and potassium) maintain pH levels of 5.0 to 6.5, and soils with high carbonate contents (such as those derived from limestones) may have pH values higher than 6.5.

Despite all the attention that soil pH has received in forest soils (including our discussion here), pH seems to have limited practical importance in relation to tree nutrition and growth. Some severe pH conditions can be very problematic for trees. Drainage of wetland soils can lead to oxidation of reduced sulfur compounds, driving the pH below 3.0 and making tree growth (and survival) low. High pH is almost never a problem in forests except on dry sites, where high pH and high salt concentrations combine to impair tree growth. Finally, some tree species are well adapted to soils with near neutral pH (pH 5.5 to 7.0); others grow well only in acidic soils (pH < 5.5), but most trees do well across the full range of common pH values. A decline in pH of 1 unit in response to natural processes or acid deposition has never been shown to affect tree growth. Increases in pH following forest harvesting or applications of lime commonly have no effect on growth, sometimes decreasing it and sometimes increasing it.

If pH is rarely of great relevance to tree growth, why has so much effort been invested in measuring and investigating it? Part of the answer comes from agricultural practices because food plants are often more sensitive to soil pH than in forests. Soil pH is also very easy to measure, and many researchers hope that an easy measure of a factor that is affected by a variety of soil processes might have more power of interpretation. Finally, some scientists expect that more insight can be obtained by understanding pH more thoroughly than at present.

THE PH OF SOIL SOLUTIONS IS BUFFERED BY THE SOLID PHASES

The pH of soils is important for a variety of reasons, including the solubility of aluminum (which is toxic to many plants and organisms), the weathering of minerals, and the distribution of cations on the exchange complex.

The concentration of H^+ in a soil with pH 5 would be about $10^{-5} mol L^{-1}$, or $10 \mu mol L^{-1}$. At pH 4 the concentration would increase to about $100 \mu mol L^{-1}$, and at pH 6 decline to just $1 \mu mol L^{-1}$. For comparison, the sum of the concentrations of all cations in forest soil solutions tends to be 200 to $1000 \mu mol$ of charge per liter ($\mu mol_c L^{-1}$). Hydrogen ions comprise only a small fraction of most forest soil solutions, but these low fractions are matched by strong effects.

What determines the pH of forest soils? Soil pH depends on the equilibrium between the soil solution and the solid phase of the soil. This buffering can be viewed as cation exchange, but the chemistry is more akin to acid–base reactions. The solid phase is a mosaic of weak acids, including clays (smectite, illite, kaolinite) and organic matter. Processes that remove H^+ from the soil, such as mineral weathering, essentially "titrate" the weak acids and make them more dissociated. Highly dissociated acids maintain higher pH values in solutions than undissociated acids. In soils terminology, the degree of acid dissociation is gauged as the base saturation,

which is the proportion of the exchange complex dominated by base cations. A soil that has a base saturation of 50% would have half of the exchange complex sorbing calcium, magnesium, and potassium and the other half sorbing aluminum and H^+.

Aluminum is included as an acid cation in soils terminology because each monomeric aluminum atom is associated with six water molecules. Depending on the pH of the soil solution, some of these water molecules dissociate and provide H^+ to the soil solution. Under acidic conditions below pH 5, all six water molecules retain their H^+. Between pH 5 and 9, one water molecule loses an H^+, and above pH 9 an additional H^+ is lost. This ability of hydrated aluminum ions to donate and accept H^+ would make it eligible to be an acid or a base, depending on the situation. Indeed, complexation of Al by soil organic matter may contribute to Al reacting as a base (H^+ consumer) in some situations (see Skyllberg *et al.*, 2001). A further reason may be that exchangeable aluminum is present in forest soils only under acidic conditions; at higher pH values, aluminum oxides are too insoluble to participate in the soil solution in any substantial way.

The total quantity of acid present in the soil solid phase is commonly on the order of $1000 kmol_c ha^{-1}$ or more, vastly exceeding the minute quantities present in soil solution at any one time (commonly $< 0.1 kmol_c ha^{-1}$). The balance between the solid and solution phases depends on:

1. The salt concentration of the soil solution.
2. The total quantity of acids present.
3. The degree of dissociation of the soil acids.
4. The acid strength of the solid phase.

The acid–base behavior of forest soils can be described in terms that are used to describe acid–base reactions in laboratories. It is important to note that the actual suite of reactions in soils encompasses a broad array of processes, including protonation/deprotonation of organic acids, titration of hydro-aluminum complexes, and even mineral weathering. The rates of some of these processes may be slow enough that equilibrium conditions do not apply. However, the general context of laboratory acid–base systems and forest soils is similar enough to provide a solid picture of how soil pH differs among soils and across time.

Factor 1: Increases in salt concentration decrease pH

If a soil solution had no ions in solution, the pH (7) would be determined simply by the tendency of water to dissociate into H^+ and OH^-. Forest soil solutions contain substantial amounts of ions, commonly 100 to 500 $\mu mol_c\, L^{-1}$. These ions equilibrate with readily exchangeable cations, and the final mixture includes H^+ (at least in acidic soils). If the total concentration of anions in the soil solution increases, the increase in cation concentration will include some extra H^+, and the pH will decrease. This effect of salt concentration is sometimes called the salt effect in the forest soils literature because additions of any salt (such as NaCI or KNO_3) will lower the pH of a soil solution even though the solid phase itself has not been altered substantially. The salt concentration effect is commonly on the order of 0.1 to 0.4 pH units, with the greatest variations occurring in areas where inputs of sea salt are variable, and in soils where accumulations of nitrate are large (and variable). For example, Richter *et al.* (1988) found that very dilute soil solutions (conductivity of soil solution of 2 mS m^{-1}) declined by about 0.3 units in pH when measured in a stronger salt solution (0.01 M $CaCl_2$), whereas soils with higher initial salt concentration (conductivity > 5 mS m^{-1}) showed no further effect of added salt on pH.

The same effect of salt concentration is apparent from field studies in areas where roofs have been built over small forest areas to block acidic deposition. The ambient deposition in the Soiling area of Germany contained about 500 $\mu mol_c\, L^{-1}$ of cations and of anions. Under the rainfall exclusion roof, very clean water (about 1 $\mu mol_c\, L^{-1}$) was used to replace rainwater (Bredemeier *et al.*, 1995). Within six months, the electrical conductivity of the soil solution (a measure of the total salt concentration) dropped from 26 mS m^{-1} to 6 mS m^{-1} and the pH of the soil rose from 3.7 to 4.1. The rise in pH resulted from the lower salt concentration in the soil under the roof, not from any major long-term change in the soil exchange complex.

Factor 2: Soil pH decreases as total soil acidity increases

The quantity of acid that constitutes the solid phase commonly differs by several-fold among soil types and may change by 10–50% within a soil as a result of changes in soil organic matter. Clays commonly have a charge of 50 mmol$_c$ kg^{-1} (kaolinite) to 1000 mmol$_c$ kg^{-1} (smectite) and even 3000 mmol$_c$ kg^{-1} (allophane). Soil organic matter contains 1000 to 5000 mmol$_c$ kg^{-1}. Given that soils also contain sand and silt with very little charge, the average total charge density in soils is commonly 20 to 100 mmol$_c$ kg^{-1}.

How does acid quantity affect the pH of soils? For soluble acids, such as acetic acid, the pH is directly determined by the equilibrium constant for the acid dissociation:

$$1.8 \times 10^{-5} = \frac{[H^+][CH_3COO^-]}{[CH_3COOH]}$$

When no other chemicals affect this equilibrium, the equilibrium concentration of H^+ will be 1.3×10^{-4} mol L^{-1}, which equals pH 3.9. If the total concentration of acetic acid were only 0.1 mmol L^{-1}, the pH would be 4.4. This simple picture does not apply to soils, where the acid anion is part of the solid phase rather than free to dissociate in solution. Therefore, the effect of acid quantity on soil pH derives from the combined effect of total acid quantity on influencing the balance between acid-neutralizing capacity (ANC) and base-neutralizing capacity (BNC), which is Factor 3 below.

Factor 3: The distribution of cations is often called base saturation

The degree of dissociation of a weak acid depends on the nature of the acid, but also on the extent to which it has been titrated by addition of H^+ (or OH^-). The pH maintained by a weak acid will

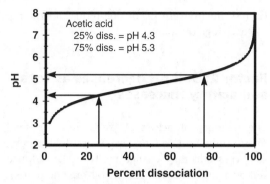

Figure 8.6 Titration of acetic acid with a base (such as OH^-) dissociates or deprotonates the acid, and the dissociated acid maintains a higher pH.

increase as the acid is titrated to become more dissociated (Figure 8.6). The equivalent of this phenomenon in soils is the degree to which the solid phase acid (the exchange complex) is protonated (acid saturated) or deprotonated (base saturated). An exchange complex of a soil is dominated by H^+ (or Al^{3+}), maintains a low soil pH, and progressive replacement of the H^+ by so-called base cations leads to the maintenance of a high pH.

In soil science, the degree of dissociation of the exchange complex is commonly referred to as base saturation:

$$\text{Base saturation} = \frac{[Ca^{2+}] + [Mg^{2+}] + [K^+]}{[H^+] + [Al^{X+}] + [Ca^{2+}] + [Mg^{2+}] + [K^+]}$$

This is roughly equivalent to the dissociation of the solid phase exchange complex. A methodological complication arises in equating base saturation with the degree of dissociation of the exchange complex. Much of the H^+ is not removed from the exchange complex when solutions of strong salts are added; these cations are removed only by titration of the soil with a strong base. Various methods and definitions have been developed to deal the salt-exchangeable acid:

1. Soil pH in water: H^+ in soil solution.
2. Soil pH in 0.01 M $CaCl_2$: H^+ in a soil solution with total salt concentration buffered at 0.01 M.
3. Exchangeable acidity: H^+ displaced from exchange complex in 1 M salt solution.

4. Titratable acidity: H^+ that can be consumed by addition of a strong base up to a defined endpoint (often pH 8.2 based on carbonate equilibrium). Titratable acidity is also called base-neutralizing capacity (BNC). Soils also have an acid-neutralizing capacity (ANC), where H^+ can replace so-called base cations of the exchange complex.

The classic use of base saturation deals with exchangeable acidity, which is only a subset of the total acidity of the solid phase. As noted above, the pH of a solution containing a weak acid depends on the degree of dissociation of the acid. Similarly, the pH of a soil is often related to the soil base saturation, as base saturation is a measure of acid dissociation. Soil pH in 16 stands (on a single soil type) in Hawaii ranged from 4.3 to 4.9 (Rhoades and Binkley, 1995). The relationship between base saturation and pH was strong, with an r^2 of 0.90. When the dissociation of the soil acids was expressed in terms of BNC (titratable acidity), the relationship improved significantly to 0.99.

The degree of base saturation of the soil acids can be reduced by removing base cations, which is a scenario for the effects of acid deposition on soil acidity. Added sulfate and nitrate may leach away base cations such as potassium and calcium, lowering the base saturation of the exchange complex and lowering the pH of the soil solution. A second approach to changing the degree of dissociation of soil acids involves adding (or removing) soil organic matter. In acidic soils, most of the charge capacity of organic matter is protonated. Addition of soil organic matter to a soil without commensurate additions of base cations will lead to an overall decrease in acid dissociation and a reduction in soil solution pH (as happened in the alder and Douglas fir case discussed below).

Factor 4: The strength of soil acids also influences pH

The strength of weak acids is characterized by equilibrium constants (K) representing the tendency of the acid to dissociate. The equilibrium constants are often converted to pK values (the negative of

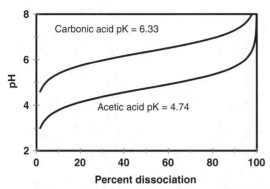

Figure 8.7 Carbonic acid is weaker than acetic acid, and a $1\,mmol\,L^{-1}$ solution maintains a pH about 1.5 units higher for the same acid quantity and degree of dissociation.

the logarithm of K). An acid that is half dissociated will have a pH that matches its pK. Acids that dissociate easily have a low pK and maintain low solution pH levels. For example, carbonic acid is weaker than acetic acid, maintaining lower solution pH for the same concentration and level of dissociation (Figure 8.7).

In soils, the acid strength varies greatly among the solid-phase acids, with kaolinite and vermiculite being weaker than smectites. In other words, a kaolinite clay that is half saturated with base cations and half with aluminum (with some H^{+}) will sustain a solution pH value that is higher than that maintained by a smectite clay under the same conditions.

Acid strength appears to be a key feature in determining the effects of some tree species on soil pH. In a replicated set of plantations of Norway spruce, white pine, and green ash, the pH was 3.8 under spruce, 4.2 under pine, and 4.6 under ash (Binkley and Valentine, 1991). The primary cause of the lower pH under spruce was the greater acid strength of the soil. Lower base saturation (22% under spruce, 42% under pine, 52% under ash) played a secondary role.

COMPARISONS OF TITRATION CURVES CAN EXPLAIN WHY PH DIFFERS AMONG SOILS

If soil pH differs among sites, or within a site over time, which factors account for the differences? An

empirical approach uses titrations of soils to determine the relative importance of factors that determine differences in soil pH among sites or over time. Comparisons of pH in water and in dilute salt (such as $0.01\,M\,CaCl_2$) identify any role of differences in the ionic strength of the soil solutions. Titrations are then used to characterize the relationship between pH and the degree of dissociation of the solid-phase acids. Additions of base identify the BNC (titratable acidity), and additions of strong acid identify the ANC. The titration curves may differ among soils as a result of greater acid quantity (ANC + BNC). The curves may show the same acid quantity, but the degree of dissociation [position along the curve, ANC/(ANC + BNC)] may differ. Finally, the quantity of acid and the degree of dissociation may be the same, but the strength of the solid-phase acids may differ, giving different pH levels in the soil suspension (for methodological details, see Valentine and Binkley, 1992). This empirical approach identifies the major factor(s) responsible for differences in pH, but the estimated magnitude of the contribution of each factor is only approximate. Variations in the titration (such as length of time and which ion saturates the exchange complex) alter the specific calculations.

The long-term experiment with soil change under loblolly pine at the Calhoun Experimental Forest in South Carolina, USA showed a striking acidification over two decades, with pH (in dilute salt) of the 0–7.5 cm mineral soil declining from 4.6 to 3.9 (Binkley et al., 1989). This decline was accompanied by a reduction of about $7.1\,kmol_c\,ha^{-1}$ of exchangeable bases, and a commensurate increase in exchangeable aluminum. The soil lost about $2.0\,kmol_c\,ha^{-1}$ of acid-neutralizing capacity (titratable alkalinity), and gained $6.6\,kmol_c\,ha^{-1}$ of base-neutralizing capacity (titratable acidity). How did these changes interact to account for the magnitude of the pH decline? The loss of acid-neutralizing capacity and gain in base-neutralizing capacity can be viewed in a classical sense as a reduction in "base saturation," and this was the primary driver of the pH decline. However, the empirical titration curves also showed a change in shape that indicated the acid strength had declined over time; if the acid strength had not become weaker, the observed decline in pH (driven by the decline in base saturation) would have been greater.

SOIL PH INCREASES FROM UPPER-SLOPE TO LOWER-SLOPE SITES

Most studies that have examined soil pH from ridges to toe slopes have found that pH increases. The major factor driving this pattern is the downslope movement of bicarbonate (HCO_3^-). High concentrations of CO_2 dissolve in the water in forest soils to form carbonic acid (H_2CO_3), which dissociates to H^+ and HCO_3^-. The H^+ tends to react within the profile, whereas the HCO_3^- leaches downslope in the company of cations such as Ca^{2+}, Mg^{2+}, K^+, and Na^+. The incoming HCO_3^- may then consume H^+ in the downslope soils, leading to higher soil pH levels.

PHOSPHATE AND SULFATE CONCENTRATIONS DEPEND ON SPECIFIC ADSORPTION

Phosphate and sulfate may be adsorbed electrostatically on positively charged sites that develop on the edges of some clay lattices or organic molecules. A more important type of sorption for these anions is called specific adsorption on iron and aluminum oxides. This ligand exchange involves substitution of phosphate or sulfate for one or more molecule of water (or OH^-; Figure 8.8).

The overall sorption of phosphate or sulfate by soils may be gauged from sorption isotherms. These curves are developed by equilibrating soil samples (for a specified time) with various levels of phosphate or sulfate in solution. The disappearance of the ion from solution is the amount adsorbed by the soil (Figure 8.9). Some of the phosphate or sulfate may actually be immobilized by the soil biota.

Sorption of phosphate and sulfate is often related positively to sesquioxide concentrations and negatively to organic matter concentrations. For example, in the 0 to 15 cm depth, 14 Oxisols in Queensland had a range of sesquioxides from 14–30% by soil weight, and phosphate sorption

Figure 8.8 Specific sorption of phosphate by iron sesquioxides (sesqui- means 3/2, referring to the number of oxygen atoms per iron atom) may release OH^- or H_2O to the soil solution.

capacity ranged from 0.05 to 0.2 mg g^{-1} (Oades et al., 1989). The organic matter concentration declined below the 15 cm depth. The increase in phosphorus sorption capacity with increasing sesquioxide concentration was much greater,

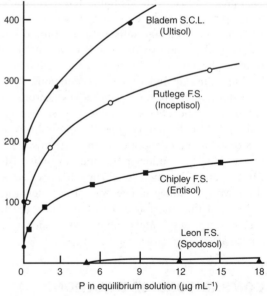

Figure 8.9 Phosphate adsorption isotherms for four soils from the southeastern Coastal Plain of the United States, representing four soil orders (S.C.L. = sandy clay loam; F.S. = fine sand). Reproduced from *Soil Science Society of America Proceedings*, **38** (2) (1974): 252, by permission of the Soil Science Society of America.

ranging from 0.05 to $1.2 \, mg \, g^{-1}$ soil for the same range of sesquioxide concentrations.

The release of phosphate from sorption with iron and aluminum may be facilitated by the activity of organic anions, including simple organic acids released by roots and microbes (such as oxalate and citrate). For example, Sato and Comerford (2006) found that addition of $10 \, \mu mol \, g^{-1}$ dry soil (from a rainforest Ultisol) of oxalate could release $1.2 \, \mu mol$ of iron g^{-1} dry soil, and thereby desorb $0.05 \, \mu mol \, P \, g^{-1}$ dry soil.

The interactions of phosphate with iron and aluminum compounds may be more than a two-dimensional issue of soil surfaces, as three-dimensional aggregation can play a large role in determining how "exposed" the phosphate may be to soil chemistry.

Sulfate adsorption also relates to soil properties such as sesquioxides and pH. Harrison and Johnson (1992) examined soils from ten temperate forests and found that sulfate sorption related well to oxalate-extractable aluminum in Spodosols and Inceptisols ($r^2 = 0.6$). Ultisols adsorbed large quantities of sulfate, regardless of the concentration of soil aluminum.

How reversible is the specific adsorption of phosphate and sulfate? This is a key question in studies of the availability of phosphate fertilizers and the changes in soil chemistry that follow increases or decreases in atmospheric deposition of sulfate. Some evidence of reversibility has been found in short-term laboratory studies. For example, Harrison and Johnson (1992) found that about two-thirds of the sulfate adsorbed in the soils from ten temperate forests was readily released in an extraction with phosphate, and one-third appeared to be irreversibly retained.

Soil organic matter may play a role in the capacity of soils to sorb phosphate and sulfate. Organic matter may sorb or coat the active surfaces of sesquioxides, reducing the available surfaces for phosphate and sulfate adsorption. Soil animals modify soil organic matter, and therefore may play an important role in modifying adsorption of phosphate and sulfate. In Quebec, the presence of earthworms in sugar maple forests greatly reduced the phosphorus supply to trees by mixing phosphorus from surficial organic

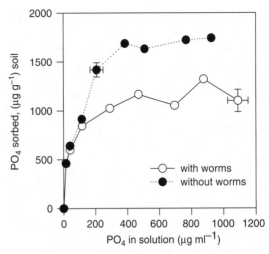

Figure 8.10 One-month laboratory incubation in microcosms with and without worms showed substantially lower phosphorus sorption capacity with worms than without worms (error bars = 1 standard deviation; Winsome, Horwath, and Powers, unpublished data), used by permission of Thais Winsome.

layers with mineral soil; aluminum bound the phosphorus (Paré and Bernier, 1989a; Paré and Bernier, 1989b). The activity of worms may have a somewhat opposite effect on phosphorus sorption within mineral horizons. In a microcosm experiment, pine needles were added to soils with and without the presence of earthworms. Where the worms had mixed the needles into the mineral soil, the soil sorption capacity for phosphorus was substantially reduced (Figure 8.10). Much more work is warranted on the overall effects of animals on soil chemistry and nutrient dynamics.

NUTRIENT SUPPLY AND UPTAKE

Nutrient supply can be viewed in three ways:

1. The mass of an element that is readily available in the soil solution.
2. The mass of an element in "labile" pools that may be accessed through the activity of microbes and plants.
3. The mass of an element actually acquired by plants.

The nutrients present in soil solution are typically a small fraction of the quantity cycled among the soil, microbes, and plants each year. The turnover rate for the nutrients contained in soil solution is typically on the order of hours to weeks; over these periods, the flux into and out of the soil solution will match the size of the pool contained in the soil solution. Soil solution concentrations are viewed as "intensity" factors, capable of driving chemical reactions.

The size of labile pools is difficult to assess, as they include a diverse array of compounds, and depend, in part, on the activity of the microbes and plants. The labile pool of soil nitrogen typically equals 1–3% of the total soil nitrogen pool, and includes the readily released atoms contained in microbial biomass, plant detritus, and humified soil organic matter. The labile pool of soil phosphorus is even more complicated, including major contributions from minerally and organically bound pools. The nutrients released from labile pools represent "capacity" or "buffer" factors, which determine the rate of resupply of the soil solution in response to nutrient uptake.

The actual quantity of nutrients available for use within plants depends on these intensity and capacity factors, but also on the ability of plants to access these pools. The investment by trees in the production of roots and mycorrhizal fungi is highly variable, both among sites and among species, and this belowground production can be a major determinant of nutrient acquisition by trees.

How large is the pool of nutrients dissolved in soil solution? The size of the soil solution pool differs across scales of time, both small and large. As a general illustration, consider a solution in a fertile soil with 1 mg of nitrate-nitrogen per liter, and a soil moisture level of 30% (water:soil by weight in a soil with a bulk density of 1.0) to a depth of 20 cm. The total quantity of soil water (to this depth) would be $60 \, L \, m^{-2}$, or $600 \, m^3 \, ha^{-1}$, giving $60 \, mg \, m^{-2}$ $(0.6 \, kg \, ha^{-1})$ of nitrogen. A typical forest might use $100 \, kg \, ha^{-1}$ of nitrogen annually, so the amount of nitrate in the soil solution at one time would be quite small relative to the annual uptake. Microbial processes replenish the soil solution pool of nitrate, and this buffering of the soil solution pool provides

the sustained source of nitrogen (and other elements) for tree growth.

The buffering of soil solution concentrations of nutrients depends on several processes, with some key differences among nutrients. Forest soils commonly contain 5 to $20 \, kg \, ha^{-1}$ of nitrogen as exchangeable ammonium, and the ammonium concentrations in soil solution may be partially buffered by sorption/desorption reactions. The overall picture of dissolved soil nitrogen is dominated by microbially mediated processes. Microbes commonly absorb more nitrogen daily than trees do; fortunately for the trees, microbial biomass turns over relatively rapidly (on a timescale of days to weeks), and the microbial pool of nitrogen may be tapped by trees. Laboratory studies with ^{15}N show that microbial uptake (immobilization) of nitrogen averages 75–95% of the nitrogen released through decomposition (gross mineralization), leaving only a fraction for tree uptake (net mineralization). The balance between microbial uptake and tree uptake remains unclear, although the importance of this competition has been well established (reviewed by Kaye and Hart, 1997). For example, Schimel et al. (1989) added ^{15}N to soils and found that 24 hours later, microbial biomass contained about half of the added nitrogen, and trees only 10%. The key question, of course, is what happens to the microbial nitrogen when the microbes die? How much goes into recalcitrant organic pools in soils and how much is obtained by trees? The interacting processes need much more attention before answers to this question can be generalized.

The soil solution concentrations of cation nutrients are well buffered by the soil exchange complex; removal of K^+ from the soil solution tends to lead to displacement of K^+ from the exchange complex. The balance between cations in soil solution and those adsorbed on exchange surfaces depends on the nature of the exchange complex, the relative distribution of ions in the exchange complex, and the ionic strength of the soil solution.

The concentrations of phosphate and sulfate in soil solutions are buffered both by microbially mediated processes (as for nitrogen), and by geochemical sorption and desorption (as noted above). Unlike the sorption and desorption of cations, phosphate and sulfate may be "specifically" adsorbed to sesqui-oxides, which leads to limited exchange between the solid and solution phases.

Nutrient uptake depends on soil chemistry and the absorbing surface area of tree roots and mycorrhizal fungi

The complex processes involved in nutrient acquisition by plants can be summarized by considering a few key features (after Barber, 1995): the concentration of the ion in solution, the buffering of the solution concentration (discussed above), the absorbing area of the roots and mycorrhizae, and the kinetics of uptake.

A key distinction between roots and mycorrhizae involves the surface area per mass of root or hyphae. A typical plant root hair might have a radius of 0.15 mm compared with a radius of 0.005 mm for hyphae. A meter of root hairs would have 30 times the surface area of a meter of hyphae, but a gram of hyphae would have about 900 times the surface area of a gram of roots. A typical forest soil (Rosling, 2009) might have 200 000 root tips in a square meter of soil, and 30 000 km of mycorrhizal hyphae!

The importance of diameter in roots versus hyphae is even greater than the relative surface area would indicate. Large-diameter roots act as large cylinders, with zones of depletion extending outward into the soil as a result of nutrient uptake; each square millimeter of root surface area draws on a zone of soil perpendicular to the root surface. Hyphae act as very tiny cylinders, and each square millimeter of hyphal area draws on a large area of "curved" soil around it. This geometrical effect is illustrated in Figure 8.11, where plant roots deplete the concentrations of phosphorus in soil solution to a distance of 0.8 cm. Fungal hyphae in the same situation can absorb more phosphorus (per unit of surface area) because the depletion zone is far smaller. In the simulation in Figure 8.11, the surface areas of the roots and hyphae are the same, but the root mass weighs 30 times more than the hyphae. Because of the smaller radius of the hyphae, the fungus can absorb twice as much phosphorus as the

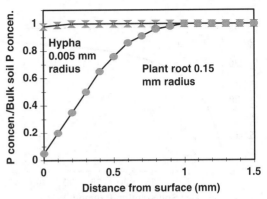

Figure 8.11 The phosphorus concentration near the surface of a root is much lower than that in the bulk soil solution, but the smaller diameter of mycorrhizal hyphae permits phosphorus uptake without local depletion of soil solution phosphorus (modified from Barber, 1995).

root hairs, even when the surface areas are equal. Nutrient uptake by plants has reciprocal effects on soil solution chemistry. Depletion of nutrient concentrations can promote rates of ion diffusion and mineral dissolution.

Nutrient uptake also depends on uptake kinetics

The absorption of nutrients by roots and mycorrhizae tends to show Michaelis–Menten-type kinetics for enzymatic reactions:

$$V = \frac{V_{max} * C}{K_m + C}$$

where V = the rate of uptake, V_{max} = the maximum rate of uptake when the nutrient concentration is not limiting, C = the concentration of the nutrient in the soil solution, and K_m is the concentration where $V = \frac{1}{2} V_{max}$. In the example in Figure 8.12, the maximum rate of uptake would be 40 nmol P m^{-2} of phosphorus absorbing area per second, and half of this value would be realized when the solution concentration of phosphorus was 5 μmol L^{-1}. As described below, trees that are limited by a particular element (such as phosphorus) tend to show higher V_{max} (when excised roots are

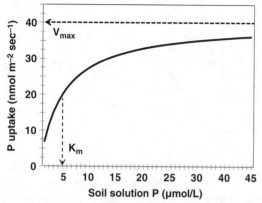

Figure 8.12 Hypothetical pattern of phosphorus uptake as a function of soil solution phosphorus concentration.

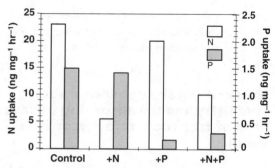

Figure 8.14 Bioassay of excised Norway spruce roots shows the root surface uptake kinetics in response to fertilization treatments. (data from Jones *et al.*, 1994)

exposed to saturating levels of phosphorus in solution) than trees with adequate nutrient supplies.

Putting all the factors together (soil chemistry and root characteristics), how sensitive is root uptake to changes in any single parameter? This is difficult to determine by direct measurement because of the number of variables involved, but Silberbush and Barber (1983) simulated the changes in phosphorus uptake when each factor was doubled (Figure 8.13). Doubling the phosphorus concentration in soil solution increased the simulated uptake by almost three-fold, Buffering of the phosphorus concentration and diffusivity of phosphorus in the soil showed linear effects on phosphorus uptake, but doubling of the V_{max} for uptake increased uptake

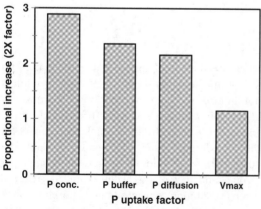

Figure 8.13 Simulated effect of changing factors that influence phosphorus uptake by two-fold (data from Barber, 1995).

only marginally. As mentioned above, substituting an equivalent mass of mycorrhizal hyphae for roots would double the rate of phosphorus uptake by reducing the absorbing radius.

Uptake kinetics at the root surface depend on the active uptake capability of the roots. Roots that are deficient in a particular nutrient tend to have greater capacity for taking up that nutrient. This uptake capacity is measured in excised roots that are exposed in laboratories to adequate concentrations of labeled nutrients (such as ^{32}P or ^{15}N).

Bioassays of excised roots from trees in control plots of Norway spruce showed high rates of uptake of both phosphorus and nitrogen, indicating possible deficiencies (Figure 8.14). In roots from plots that had been fertilized with nitrogen alone, nitrogen uptake rates dropped but phosphorus uptake rates remained high, indicating alleviation of nitrogen deficiency but not of phosphorus deficiency. Roots from plots fertilized with phosphorus alone showed low phosphorus uptake kinetics and high nitrogen uptake kinetics, and roots from plots fertilized with both nutrients showed reduced uptake kinetics for both nutrients (Jones *et al.*, 1994).

SOIL SOLUTIONS AND PLANT UPTAKE ARE LINKED TO ECOSYSTEM BIOGEOCHEMISTRY

These processes and patterns of soil solution chemistry and plant uptake interact with many other processes and patterns in forest ecosystems; this

Table 8.1 Rates of nutrient uptake ($kg\,ha^{-1}\,yr^{-1}$) to meet requirements for aboveground growth in relation to annual inputs from the atmosphere and total soil nutrient contents.

Forest Type		Element					
		Nitrogen	Phosphorus	Potassium	Calcium	Magnesium	Sulfur
Norway spruce	Uptake	27.6	4.0	21.8	21.3	2.9	1.9
	Uptake/input	2.5	—	8.7	9.8	0.40	0.09
	Uptake/soil total (to 50 cm)	0.007	—	0.0003	0.0007	0.0006	—
Loblolly pine	Uptake	30.2	7.3	12.9	26.4	5.6	4.3
	Uptake/input	2.2	240	2.6	3.8	6.2	0.12
	Uptake/soil total (to 80 cm)	0.02	0.006	0.0004	0.005	0.0005	0.002
Douglas-fir	Uptake	13.2	2.3	9.9	17.2	1.6	1.5
	Uptake/input	2.7	74	3.9	4.9	1.9	0.14
	Uptake/soil total	0.003	0.0006	0.0004	0.0004	0.00006	0.002
Red alder	Uptake	80.3	3.3	31.7	51.9	7.2	4.6
	Uptake/input	1.0	110	13	15	8.8	0.42
	Uptake/soil total	0.01	0.002	0.002	0.002	0.0003	0.007
Mixed hardwoods	Uptake	34.5	2.2	41.3	47.7	12.2	5.6
	Uptake/input	4.7	8.9	30	17	24	0.30
	Uptake/soil total	0.009	0.0008	0.0002	0.004	0.0001	0.004
White pine	Uptake	37.0	4.1	27.3	32.9	8.1	6.1
	Uptake/input	5.1	17	20	12	16	0.33
	Uptake/soil total	0.01	0.002	0.0002	0.009	0.0001	0.008

larger picture is called biogeochemistry. Biogeochemistry examines the interactions among geology, chemistry, and life, focusing on ecosystem productivity and nutrient cycling (Chapter 7). A key feature of biogeochemistry is the rates at which nutrients are taken up by trees, and how these rates compare to rates of nutrient inputs and the overall store of nutrients in soils. For the soils with solution data in Figure 8.1, the annual uptake of nutrients is higher than the annual inputs from the atmosphere for all elements except for sulfur in all forests, magnesium in Norway, and nitrogen in the nitrogen-fixing red alder (Table 8.1). In all cases, the annual uptake of nutrients represents a very small fraction (1% or less) of the total quantities of these elements stored in soils. The controls and interactions among these ecosystem processes are the subject of Chapter 7.

PART IV

Measuring Forest Soils

CHAPTER 9
Sampling Soils Across Space and Time

OVERVIEW

We measure soils, or more particularly their properties, for a variety of reasons. We may wish to better understand the genesis of a particular soil, how it has developed and how it may develop in future. We often wish to see how soils relate to vegetation type, or how they relate to vegetative productivity. We may wish to determine whether or not the addition of one or more chemical elements will enhance productivity. We may wish to create a "benchmark" so that we can determine how soils change over time, or with a change in vegetative cover. Measuring soils is a difficult and challenging task, but accurate information about soils is crucial to reliable soils research and sound ecosystem management. In the next two chapters we will discuss how we measure soils, and how the knowledge gained from those measurements can lead to a better understanding of how ecosystems function, and how we might better manage ecosystems either for preservation or production.

Recent advances in geographic information systems, global positioning systems, our understanding of the temporal and spatial scales of soil variation, and the development of more sophisticated taxonomies for the classification of ecosystem attributes have all contributed to our ability to better manage and preserve forests. By cleverly using these advances, we can better understand why a particular forest exists as it does. We now can also begin to understand why a change in ecosystem attributes, whether natural or human induced, affects the forest the way it does. Robust experimental designs and appropriate and adequate sampling are essential if we are to gain reliable knowledge. We have the tools – all we need to do is to use them wisely.

VARIATION OVER TIME AND SPACE IS FUNDAMENTAL IN FOREST SOILS

Soils on the landscape are complex physical, chemical, and biological systems that involve processes and interactions occurring within and between the atmosphere, hydrosphere, biosphere, and geosphere. These systems are dynamic in space and time and involve a constant interaction with environmental drivers. Scale issues are at the heart of many environmental problems because different processes may be dominant at different spatial and/or temporal scales. Few processes in nature operate in a linear fashion, but rather in a strong nonlinear manner. While a system may initially resist changes from a forcing factor, a relatively small change at some point may push the system across a threshold that leads to an abrupt change in the response (Hoosbeek, 1998; Goerres *et al.*, 1998; Bruckner *et al.*,1999; Lin *et al.*, 2005; Yavitt *et al.*, 2009).

To measure soils we must sample them, and to sample them effectively we must be cognizant of the fact that some soil properties vary rapidly over time while others are quite stable over time (Table 9.1).

The amount of ammonium and nitrate in a soil varies rapidly over time (Figure 9.1); however, the soil's total nitrogen content can be stable over a rather long period. The soil's ammonium and nitrate content also varies from point to point

---

Ecology and Management of Forest Soils, Fourth Edition. Dan Binkley and Richard F. Fisher.
© 2013 John Wiley & Sons, Ltd. Published 2013 by John Wiley & Sons, Ltd.

Table 9.1 Time and space scales in soil ecosystems.

Time Scales in Soil Ecosystems

	Time Scale in Years	Soil Process
10^4	10 millennia	Soil development
10^3	millennia	Calcium and iron eluviation
10^2	century	Organic matter response to disturbance
10^1	decade	Acidification and erosion
10^0	year	Mineralization
10^{-1}	month	Oxidation/reduction
10^{-2}	day	Leaching and diurnal temperature change
10^{-3}	hour	Nutrient uptake
10^{-4}	minute	Sorption
10^{-5}	second	Ion exchange

Spatial Scales in Soil Ecosystems

Area (m^2)	Spatial Scale
10^{14}	Global
10^{12}	Continent
10^{10}	Region/biome
10^8	Landscape
10^6	Small watershed
10^4	Hillslope, stand
10^2	Plot
10^0	Pedon
10^{-2}	Soil core
10^{-4}	Macro-aggregate
10^{-6}	Micro-aggregate
10^{-8}	Pore
10^{-10}	Colloids
10^{-12}	Molecular

over rather short distances (Figure 9.2); however, the soil's total nitrogen content may be similar over much longer distances.

In short, some soil properties vary from centimeter to centimeter, others vary from meter to meter, and others, for example parent material, may be constant for kilometers. It is important to consider both the temporal and spatial scales over which the property of interest varies if we are to sample it effectively.

SOILS ARE GEOGRAPHIC ENTITIES

A soil is a geographic entity and areas with similar soils are often defined as soil types. They tend to change as parent material and topography change.

In some regions, very small changes in topography can lead to a change in soil type. The sampling design needs to take into account all of the soil types in the area of interest.

When we sample the soil of an area, we also must remember that, in reality, we sample soil volumes. Each volume from which a sample is drawn is a population composed of many individuals that vary among themselves, both horizontally and vertically (Cline, 1944). Soil properties often vary greatly with depth from the surface. Of course, soil organic matter is a prime example. The variation in soil properties with depth can have profound influences on vegetation. For example, available phosphorus might be low in the A horizon of a particular soil, but high in the B horizon. In such cases, plants that root only in the surface soil would experience P

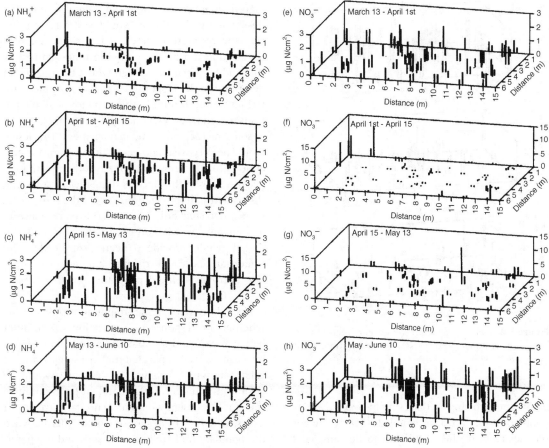

Figure 9.1 Spatial maps of NH_4^+ and NO_3^- availability over four time spans: (a and e) March 13–April 1; (b and f) April 1–15; (c and g) April 15–May 13; (d and h) May 13–June 10. (After Cain *et al.*, 1999).

deficiency, but trees that root more deeply would be able to acquire sufficient P.

POWERFUL SAMPLING DESIGNS ARE FUNDAMENTAL TO USEFUL SOIL MEASUREMENTS

Sampling soil involves the selection from the total population of a subset of individuals (or samples) upon which measurements will be made. These measurements are then used to estimate the properties of the total population. Sampling is essential if we wish to characterize the population, because measurement of the entire population would be impossible (Carter and Gregorich, 2008).

Sampling design involves the selection of the most efficient and effective method for choosing the samples that will be used to characterize the total population. The proper definition of the population is critical. It is important to clearly define the population of interest, and how the study findings will be extrapolated to that population (Binkley, 2008).

If one wishes to study how a particular plantation of loblolly pine growing on a Cecil soil (a fine, kaolinitic, thermic Typic Kanhapludult) in Georgia responds to the addition of fertilizer, then that plantation is the population that must be sampled. If one wishes to study how loblolly pine plantations growing on Cecil soils in Georgia respond to the addition of fertilizer, then all loblolly pine

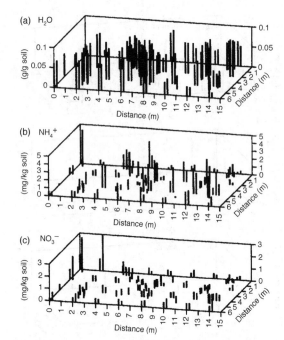

(a) H₂O

(b) NH₄⁺

(c) NO₃⁻

Figure 9.2 Spatial maps of (a) soil water content; (b) soil NH_4^+ concentration; and (c) NO_3^- concentration. (After Cain *et al.*, 1999).

plantations growing on Cecil soils in Georgia are the population that must be sampled. Clearly the correct selection of the population to be sampled is crucial if inferences made from the study results are to be valid.

The study of how a particular plantation of loblolly pine growing on a Cecil in Georgia responds to the addition of fertilizer might use four replicated blocks within the stand, with four levels of nitrogen fertilizer added to randomly assigned plots in each block. The aim of the study might focus on choosing which levels of fertilizer to add ($100 \, kg \, ha^{-1}$? or $200 \, kg \, ha^{-1}$?), and the measurements to use for characterizing the response to fertilization (diameter, height, volume, mortality?). In many cases, the post-treatment response might be more precise if pre-treatment differences among plots were considered (such as basal area, or growth rates within plots).

All of these questions need to be preceded by a clear definition of the population the researchers

plan to address with the experimental results. The design of this experiment would be strong for inferring the response of trees in this single stand to fertilization during the period in time spanned by the study. In fact, this design would be suitable for determining the response of all loblolly pine plantations to fertilization, if the response of all plantations to fertilization was not affected by factors such as soils, microclimate, and management. The design would be very weak, however, in providing insights about growth responses of any other plantation because these other factors dramatically confound the response to fertilization. Even if the researchers were interested only in the fertilizer response of loblolly pine growing on Cecil soils in Georgia, this design would provide 0 degrees of freedom for understanding the variability among sites within this narrowly defined population (with one site, $n - 1 = 0$).

A similarly poor design might aim to determine whether the response to fertilization would differ between well-drained soils and poorly drained soils by placing a well-replicated trial in a single stand on each soil type. Perhaps the response would be significantly higher in the stand on poorly drained soil, but this design would have no statistical power to test whether the response would differ between the population of well-drained soils and the population of poorly drained soils.

PSEUDOREPLICATION IS NO REPLICATION AT ALL FOR THE POPULATION OF INFERENCE

Pseudoreplication is a major problem with many field studies and often occurs when the target population has not been defined correctly. If the population for the loblolly pine study referenced above were defined as "this plantation", then the block design would indeed have four true replicates, and a researcher interested in just that plantation might conduct such a study. Frequently, the mistake is made of extrapolating the results of such a study to an entirely new population. If the population of interest was "loblolly pine plantations in Georgia," the design would have a single replicate (one plantation) with estimated responses based on four subsamples (actually, split subsamples) of the single replicate. A much better design for this latter objective would omit any replication of treatments within a single plantation, and apportion the experimental work across four independent plantations within the population of interest. The dispersed approach would have 3 degrees of freedom for interpreting the likely responsiveness of the whole population, despite having 0 degrees of freedom for interpreting the response within any single plantation. Replicate subsamples within a single plantation are only useful if they improve the representativeness of the response for that plantation, which might be important if, for example, basal area or soil type is variable within the plantation. Efforts invested in more true replicates are almost always more profitable than the same effort applied to obtaining more subsamples that contribute no degrees of freedom for extrapolating to the population (Stape *et al.*, 2006). Once we are certain that we have defined the population of interest correctly, the question becomes how to sample that population most accurately and efficiently.

PEDONS ARE SAMPLED FOR INFERENCES ABOUT LANDSCAPES

Sampling is often confined to a single pedon for the purposes of characterization; however, sampling of an area of land is more commonly required. Sampling of an area can either be haphazard or systematic. Sampling is haphazard when accessibility or convenience determines the location of samples rather than a conscious effort to ensure that the samples truly represent the range of variability in the population. Haphazard sampling yields data but no real information about the population, and consequently is of little real value.

Systematic sampling can either be based on judgment, termed purposeful sampling, or probability. Purposeful sampling involves using a sampling grid approach to spread sampling points over the area in question. Sometimes purposeful sampling is based

on knowledge held by the sampler. The reliability of data gained by sampler-driven purposeful sampling is only as good as the judgment of the sampler. Both of these sampling methods can produce reasonable estimates of population means, variances, and totals, but they provide only a weak measure of the adequacy of these estimates to truly characterize the area. Probability sampling selects sampling points randomly using one of a variety of specific sampling layouts. The probability of sample point selection can be calculated for any specific design, allowing an estimate to be made of the accuracy of parameter estimates, and a range of statistical analyses can be carried out based on the estimates of variability about the mean (Bélanger and Van Rees, 2008; Pennock, 2004; Pennock *et al.*, 2008).

THREE ADVANCES HAVE REVOLUTIONIZED SOIL GEOGRAPHY

In recent times, three major advances have taken place that have vastly improved our ability to measure soils, understand soils, and manage soils and landscapes: geographic information systems (GIS), geostatistics, and the global positioning system (GPS). Geographic information systems allow us to store information about an area of land in layers and then analyze the data to determine things like the productivity of different soil units or the response of different soil units to fertilizer addition. The global positioning system allows us to locate ourselves and to match points on maps to points on the ground accurately. Geostatistics allow us to extrapolate point data to area data and to calculate the reliability of those extrapolations.

Modern geographic information systems began in 1969 when Ian McHarg published his seminal work, *Design with Nature*. He proposed using transparent overlays to store geographic information in layers. These sheets, when placed over a map base, could allow one to interpret the landscape, combining the information from the various sheets or layers. Soon, computer-based applications were developed, and the McHarg technique became very popular in landscape architecture. Geologists and natural resource managers discovered the technique and geographic information systems were born. Several commercial systems and systems in the public domain are now available. These allow users to store geographically mapped information and tabular information in "layers" and then combine that information to make interpretations (Burrough and McDonnell, 1998; Bestelmeyer *et al.*, 2008; Krasilnikov *et al.*, 2008). GIS are now used by many natural resource managers to make informed management decisions.

The mapping of information such as soil type, slope, drainage class, vegetation type, stand type, parent material, etc. into a GIS is fairly straightforward. But what if we would like to map soil properties such as organic matter content, soil acidity, etc., which require us to sample the landscape and analyze the samples? Much of our soil information is in the form of an observation that relates to a particular location in space and time. Knowledge of an attribute value, say organic matter content or soil acidity, is thus of little interest unless location or time of measurement, or both, are known and accounted for in any analysis. Geostatistics provide a set of statistical tools for incorporating spatial and temporal coordinates of observations in data processing.

GEOSTATISTICS CHARACTERIZE SOIL PATTERNS

Geostatistics are a group of tools for studying and predicting the spatial structure of georeferenced variables. The use of geostatistical tools in soil science is diverse and extensive. They have been used for studying and predicting soil contamination in industrial areas, for building soil quality maps at the field level, and even to map physical and chemical soil properties at the landscape level (Goovaerts, 1997). Initially, geostatistics were used to describe spatial patterns by semivariograms (Figure 9.3) and to predict the values of soil attributes at unsampled locations by kriging (Trangmar *et al.*, 1985; Warrick *et al.*, 1986).

The theoretical variogram $2\gamma(x,y)$ is a function describing the degree of spatial dependence of a

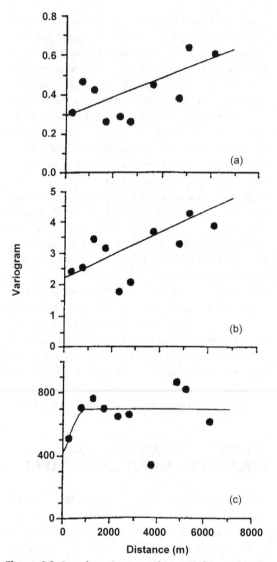

Figure 9.3 Sample variograms of (a) pH; (b) organic carbon content; and (c) solum coarse fragments with linear (a and b) and spherical (c) variogram models.

spatial random field or stochastic process $Z(x)$ defined as the expected squared increment of the values between locations x and y:

$$2\gamma(x,y) = E(|Z(x) - Z(y)|^2)$$

where $\gamma(x,y)$ itself is called the semivariogram. In the case of a stationary process, the variogram and semivariogram can be represented as a function $\gamma_s(h) = \gamma(0,0+h)$ of the difference $h = y - x$ between locations only, by the following relation:

$$\gamma(x,y) = \gamma_s(y-x)$$

If the process is furthermore isotropic, then the variogram and semivariogram can be represented by a function $\gamma_i(h) = \gamma_s(he_1)$ of the distance $h = \|y - x\|$ only:

$$\gamma(x,y) = \gamma_i(h)$$

The terms nugget, sill, and range are often used to describe variograms. The nugget is the height of the jump of the semivariogram at the discontinuity at the origin. The sill is the limit of the variogram tending to infinity of lag distances. The range is the distance in which the difference of the variogram from the sill becomes negligible. In models with a fixed sill, it is the distance at which this is first reached; for models with an asymptotic sill, it is conventionally taken to be the distance when the semivariance first reaches 95% of the sill (Cressie, 1993; Wackernagel, 2003).

KRIGING INTERPOLATES THE SPACE BETWEEN SAMPLES

The first step in ordinary kriging is to collect spatially defined samples, either at a constant spacing or, preferably, at a range of spacings (Figure 9.4). The values from the samples are used to construct a variogram from the scatter point set to be interpolated. A variogram consists of two parts: an experimental variogram and a model variogram. Suppose that the value to be interpolated is referred to as f. The experimental variogram is found by calculating the variance (g) of each point in the set with respect to each of the other points and plotting the variances versus distance (h) between the points. Several formulas can be used to compute the variance, but it is typically computed as one half of the difference in f squared.

Once the experimental variogram is computed, the next step is to define a model variogram. A model variogram is a simple mathematical function that models the trend in the experimental variogram (Figure 9.4). After a certain level of separation, the variance in the f values becomes somewhat random and the model variogram flattens out to a value corresponding to the average variance. Once the model variogram has been constructed, it is used to compute the weights used in kriging.

Kriging is a method of interpolation named after the South African mining engineer D. G. Krige, who developed the technique in an attempt to more accurately predict ore reserves. It is based on the assumption that the parameter being interpolated can be treated as a regionalized variable. A regionalized variable is intermediate between a truly random variable and a completely deterministic variable in that it varies in a continuous manner from one location to the next and therefore points that are near each other have a certain degree of spatial correlation, but points that are widely separated are statistically independent. Kriging is a set of linear regression routines which minimize estimation variance from a predefined covariance model. It allows us not only to establish values for unsampled points but also to create a contour map of values over a sampled area (Figure 9.5).

An important feature of kriging is that the variogram can be used to calculate the expected error of estimation at each interpolation point since the estimation error is a function of the distance to surrounding scatter points. When interpolating to an object using the kriging method, an estimation variance data set is always produced along with the interpolated data set. As a result, a contour or isosurface plot of estimation variance can be generated on the target mesh or grid.

New tools have now been developed to tackle advanced problems, such as the assessment of the uncertainty about soil quality or soil pollutant concentrations, the stochastic simulation of the spatial distribution of attribute values, and the modeling of space–time processes. Probably the most important contribution of geostatistics is the assessment of the uncertainty about unsampled values. This usually takes the form of a map of the probability of exceeding critical values, such as nutrient deficiency levels or criteria for soil quality.

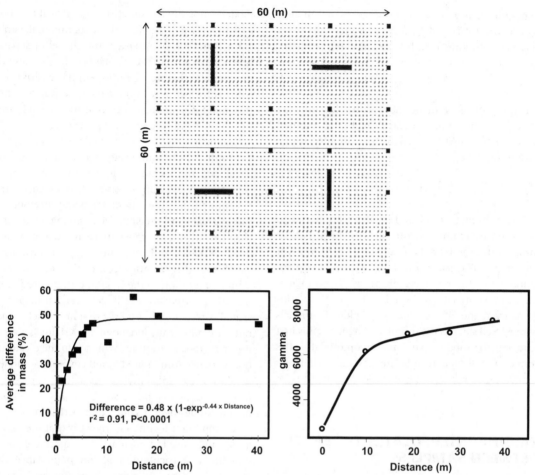

Figure 9.4 Intensive sampling of the O horizon in a 60 m × 60 m plot of white oak (*Quercus alba* L.) was designed to provide a range of distances between samples (upper graph). The black squares indicate 1 m² subplots that were sampled, at distances from 1 m (in adjacent cells) to 100 m (farthest diagonals). O horizon masses fell within a range of 25% for adjacent cells, but differed by about 50% for cells 8 m or more apart (lower graph left). The pattern of variation in O horizon mass can also be plotted as a semivariogram, which again showed spatial autocorrelation among samples was notable only below about 10 m (lower graph right). (Data analysis provided by Suzanne Boyden.)

This assessment of uncertainty can be used along with other knowledge in decision making in natural resource management. Stochastic simulation allows one to generate several models or images of the spatial distribution of soil attribute values, all of which are consistent with the information available. These scenarios can then be evaluated and the most appropriate policy or management action chosen.

THE GLOBAL POSITIONING SYSTEM (GPS) GUIDES SAMPLING, MAPPING, AND MANAGEMENT

The global positioning system (GPS) is a worldwide radio-navigation system formed from a constellation of 24 satellites and their ground stations. The system uses these "man-made stars" as reference points to calculate positions accurate to a matter of

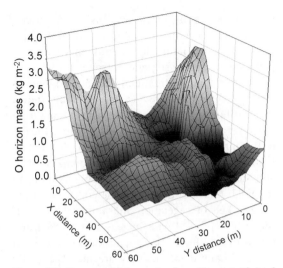

Figure 9.5 The spatial information from Figure 9.4 kriged to provide a map of the O horizon mass across the 60 m × 60 m plot in a white oak stand. The O horizon mass differed by up to four-fold over a distance of 20 m. (Data analysis provided by Suzanne Boyden.)

meters. In fact, with advanced forms of GPS you can make measurements to centimeter accuracy. The basis of GPS is a form of triangulation. A GPS receiver measures the distance to three or more satellites using the travel time of radio signals. By knowing the exact position of the satellites, the receiver can then calculate its own position and altitude. To achieve the greatest accuracy, corrections for various inaccuracies must be applied to standard GPS readings. For this purpose, differential GPS, or DGPS, has been developed. For most natural resource applications, standard GPS has proved quite satisfactory; however, to settle boundary disputes, DGPS may be required. For example, recent work using corrected GPS data has determined that "The Four Corners Monument" (the point where Arizona, Colorado, New Mexico, and Utah meet) is incorrectly located (Hoffmann-Wellenhof et al., 1994; Leick, 1995; Kaplan, 1996; Parkinson and Spilker, 1996).

To say that GPS has revolutionized natural resource management would be an understatement. Not only can we now determine our position on the landscape quite accurately, but we can also make very accurate maps of landscape features such as stand boundaries and soil types. The inclusion of a large amount of accurately georeferenced data in a GIS along with the development of knowledge-based interpretations allows us to achieve precision resource management.

SOILS ARE CLASSED INTO GROUPS ON LANDSCAPES FOR MANAGEMENT

Over the years, natural resource land segments within the overall landscape have been classified in a wide variety of ways, and many of these are still in use today. Taxonomies have been based on natural vegetative cover, land capability, climate, and soil characteristics. These taxonomies allow us to better understand and manage our natural resources. The earliest taxonomies appear to have been based on crop productivity. Pliny the Elder tells us that the Roman tax code of the first century A.D. taxed land based on the most profitable crop (e.g. grapes, grain, olives, grazing) that the land could support (Pliny 77–79). Similar taxonomies based on land capability were in use through the Middle Ages in Europe, in colonial America, and in share cropping in the southern United States after the mid-1800s.

Taxonomies based on the natural vegetation that the land supported developed first in Europe and then North America in the twentieth century (Braun-Blanquet, 1932; Wikum and Shanholtzer, 1978). Many of these vegetation or habitat type systems are still in use today. The most common form for these taxonomies is to use the climax or supposed climax overstory species as the initial classifier of an area and then to subdivide those types on the basis of the dominant understory species. Such taxonomies are easily applied, although the determination of the climax species is difficult and subjective in many cases. Their most common failing is that they do not account for the wide range of site productivity that may occur within a type.

A second type of taxonomy is that based on land capability or land use restrictions. The United States

Department of Agriculture land capability classification is a system of grouping soils primarily on the basis of their capability to produce common cultivated crops and pasture plants without deteriorating over a long period of time (Baldwin *et al.*, 1938; National Soil Survey Handbook Part 622, National Soil Survey Handbook,2011).

Class I soils have no or slight limitations that restrict their use. Class II soils have moderate limitations that reduce the choice of plants or require special conservation practices. Class III soils have severe limitations that reduce the choice of plants or require special conservation practices, or both. Class IV soils have very severe limitations that restrict the choice of plants or require very careful management, or both. Class V soils have little or no hazard of erosion but have other limitations, impractical to remove, that limit their use mainly to pasture, range, forestland, or wildlife food and cover. Class VI soils have severe limitations that make them generally unsuited to cultivation and that limit their use mainly to pasture, range, forestland, or wildlife food and cover. Class VII soils have very severe limitations that make them unsuited to cultivation and that restrict their use mainly to grazing, forestland, or wildlife. Class VIII soils and miscellaneous areas have limitations that preclude their use for commercial plant production and limit their use to recreation, wildlife, or water supply or for esthetic purposes. These classes are then divided into subclasses based on factors that limit land use. Subclass e is made up of soils for which the susceptibility to erosion is the dominant problem or hazard affecting their use. Erosion susceptibility and past erosion damage are the major soil factors that affect soils in this subclass. Subclass w is made up of soils for which excess water is the dominant hazard or limitation affecting their use. Poor soil drainage, wetness, a high water table, and overflow are the factors that affect soils in this subclass. Subclass s is made up of soils that have soil limitations within the rooting zone, such as shallowness of the rooting zone, stones, low moisture-holding capacity, low fertility that is difficult to correct, and salinity or sodium content. Subclass c is made up of soils for which the climate (the temperature or lack of moisture) is the major hazard or limitation affecting their

use. This taxonomy has proved useful in agriculture, but has been of little value to natural resource managers, since natural forests and range lands can occur on any, even all, land classes.

Another type of frequently used taxonomy is based on climate. In 1889, C. Hart Merriam studied the distribution patterns of plants and animals from the lower elevations of the Grand Canyon to the top of Humphreys Peak (elevation 3890 m) in the San Francisco Peaks near Flagstaff, Arizona (Merriam and Steineger, 1890). Based on his observations in the field, Merriam developed the concept of a life zone, a belt of vegetation and animal life that is similarly expressed with increases in altitude and increases in latitude. These zones can be delineated using climatic data, but they are more generally delineated based on vegetation.

To overcome imprecisions of Merriam's system, Holdridge developed the concept of life zones based on the relations between biotemperature (the mean of the annual temperature between 0 and 30 °C.), mean total annual precipitation and potential evapotranspiration ratio (Holdridge, 1967). This system was developed for use in the tropics, but it has been applied in many regions of the globe. Like Merriam's, system, life zones can be delineated using climatic data, but they are more generally delineated based on vegetation. As with the vegetation-based systems discussed earlier, these systems fail to account for the wide range of site productivity that occurs within any one zone.

The multifactor approaches to forestland classification differ from soil surveys mainly in their approach and their degree or intensity of mapping. The number of land features, stand characteristics, and soil factors used in delineating mapping units is great, and, as a consequence, maps based on these features provide flexibility for classifying on the basis of goals other than wood production. These schemes generally entail a combination of independent site variables, measured in the field or laboratory and superimposed as phasing elements on conventional soil series, to reflect variations within these units important to tree growth and land use. Such soil factors as soil depth, available water capacity, texture, organic matter, chemical composition, and aeration, as well as radiation, ground vegetation, and landform, have been studied individually and collectively.

Hills's (1952) total site classification of Ontario can be considered a multifactor approach to site evaluation. This holistic concept integrates the "complex of climate, relief, geological materials, soil profile, groundwater, and communities of plants, animals, and man." Physiographic features are used as the framework for integrating and rating climate, moisture, and nutrients. The integration of the various factors, of environment and vegetation at each level of this hierarchical classification scheme involves much subjective judgment and intuition (Carmean, 1975). Nevertheless, the system provides a good framework for stratifying large, inaccessible forest regions into broad subdivisions based on general features, of vegetation, climate, landform, and soil associations.

An alternative multifactor approach is the German site-type system, which was developed in Baden-Wurttemberg in the 1940s (Schlenker, 1964). Barnes *et al.* (1982) used this system successfully in Michigan. They found strong interrelationships among physiography, soils, and vegetation and used all of these factors simultaneously in the field to delineate site units. The reliance of this system on ecologists, who, in the field, simultaneously integrate ecosystem factors and are not bound to reconcile predetermined classes, leads to a high degree of subjectivity. Ecological site classification is common in Europe (Pyatt and Suarez, 1997; Wilson *et al.*, 2001; Farrelly *et al.*, 2009) and Canada (Klinka and Carter, 1990), and is popular with the United States Forest Service.

Foresters and ecologists have used site index as a tool to subdivide taxonomic units into areas of similar productivity (Skovsgaard and Vanclay, 2008). Site index is the term used to express the height of dominant and co-dominant trees of a stand projected to some particular standard age. This index or base age may be 25, 50, 100, or any other age appropriate to the growth rate and longevity of the species being considered. Site index is extremely important in site quality analysis in North America because it forms the standard against which all other forms of site evaluation are measured. In stands younger or older than the index age,

a family of height–age curves is required for projecting measured height to height at the index age (Beck, 1971). These curves are developed in a variety of ways. One common method is to measure the height and age of many stands at single points in time, fit an average curve of height-on-age to these data, and construct a series of higher or lower curves with the same shape as the guide.

There is evidence that these anamorphic curves often do not represent actual stand growth conditions accurately. For example, the guide curve is likely to be accurate only if the ranges of site indexes are equally represented at all ages. Unequal sampling may occur because of the timber harvesting and land abandonment trends in a particular region. It has been pointed out that trees reach merchantable size faster, and are often cut at a younger age, on high-quality sites. Consequently, a sample of stands selected at a particular time could result in a biased curve that would tend to underestimate the site index of stands younger than the index age and to overestimate the site index of older stands (Beck, 1971). It may also be important to apply the curves in a manner consistent with their construction. That is, if a set of curves is developed on the basis of the ten tallest trees per acre, it should be applied on other sites by using trees selected in a similar manner.

Furthermore, the assumption that the shape of the curve does not vary from site to site is generally false (Spurr, 1952; Beck and Trousdell, 1973). The degree of diversity in curve shape seems to vary with species and location, but the pattern of growth with change in site quality may be similar for many species. Instead of being proportional at all ages for all qualities of sites, as generally depicted by conventional curves, the rate of height growth rises rapidly on the best-quality sites and then becomes relatively gradual. On the other hand, growth rates on poorer sites increase slowly during the early years but may be maintained for a longer time. One might expect to find trees on sites with the poorest site indexes about equal to those on sites with the best indexes at some age older than the index age (Beck and Trousdell, 1973) (Figure 9.6(a)).

The bias introduced through the use of proportional rather than polymorphic curves is probably of

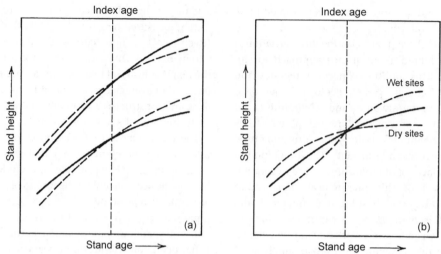

Figure 9.6 (a) Comparison of normal proportional or anamorphic curves (solid lines) with nonbiased polymorphic curves (dashed lines); (b) polymorphic curves (dashed lines) for wet and dry sites and the normal proportional curve (solid line) obtained by combining them.

little importance in the relatively short rotations of most intensively managed forest. Nonetheless, variations in growth patterns among sites due to differences in certain soil conditions can be quite striking in the early years of stand development. For example, slash pine growth during the first 10 to 20 years is often quite slow on the wet savanna soils of the coastal flats of the southeastern United States. The developing stand gradually draws down the mean groundwater table, thus increasing the effective rooting volume of the soil. The increase in soil volume provides for a faster tree growth rate through improved nutrition and aeration during the later stages of stand development. In contrast, slash pine planted on the well-drained sand hills of the coastal plain grows rather well during the first five to seven years after planting. However, growth tends to stagnate, as moisture becomes a limiting factor for good stand development (Figure 9.6(b)). That different sites can have height–age curves with several shapes, even though these sites may be of equal quality when measured at a common index age, has been reported for several species (Carmean, 1975).

It is apparent that the site index system of classification is highly empirical and subject to many different kinds of error. Nonetheless, we often

determine site index to the nearest 30–60 cm from site index curves that have confidence intervals of 1 m or more about each site index line. We then compare these "supposedly accurate" determinations of site quality to determinations made using other methods in order to gauge the accuracy of the method being tested. In so doing, we cannot hope to obtain a method of site quality determination that is more accurate than the site index method, which is fraught with error.

In many parts of the world productivity in $m^3 ha^{-1}$ is used to distinguish areas of differing productivity. This system has not been widely adopted in the U.S. As we move to the third, fourth and even fifth forest on many areas in the U.S., it would seem that we now have the knowledge and the data to move to this more precise indicator of site productivity.

Taxonomies based on the soil have been around for a long time, and in many cases they have proved very useful in natural resources management. V. V. Dokuchaev (Докучáев) is commonly regarded as the father of soil taxonomy. He introduced the idea that geographical variations in soil type could be explained in relation not only to parent material, but also to climatic and topographic factors, and the time available for soil formation processes to operate. Using these ideas as a basis, he created the first soil classification in 1883 (Glinka, 1927).

In 1938 the United States released its first official soil taxonomy (Baldwin *et al.*, 1938). Like the Russian system, the highest division in the system was based on climate. Subdivisions were then based on horizon properties. These taxonomies led the way for soil surveys and the production of detailed soil maps, which proved to be valuable in agriculture, but were, in general, deemed to be too imprecise to serve as a basis for site evaluation (Carmean, 1975; Grigal, 1984); however, they have proved to be quite valuable to forestry in many parts of the world (Valentine, 1986). Classification, mapping, evaluation, and interpretation of forestlands are pre-requisites to sound forest management. Collectively, they comprise an inventory program that is essential to proper intensive management programs and important for the development of effective, multiple-use land management plans. Classification and mapping are related but distinct steps in conducting a soil inventory. Classification is analogous to taxonomic systems used for plants and animals and consists of grouping land areas based on similarities and differences in properties. Mapping is an exercise in geography wherein homogeneous land areas are delineated on a map using the classification system.

The units within the classification are pure, although the properties that define the unit are allowed to vary over some prescribed range. Map units, the geographical areas delimited on the ground, are seldom pure. They almost always contain areas that belong to other units of the classification system or that remain undefined. Thus, the degree of purity of map units must be defined. The homogeneity of map units depends not only on the character of the area being mapped but also on the scale of mapping. At small scales, 1:50 000 or greater, map units must be large, 103–105 km^2, and of necessity quite heterogeneous; however, at large scales, 1:10 000 to 1:20 000, map units can be small, less than 100 km^2, and often quite homogeneous.

Inclusions are areas within a map unit that clearly belong to a different classification unit but that are too small to be delimited at the scale being used in the inventory. Often, the scale of mapping leads to map units composed rather equally of two or more soils. Such map units are commonly called complexes or associations. To be useful, such units must specify what proportion is contributed by each soil.

Evaluation is an assessment of the functional relationships within a land area. It can be carried out in the absence of formal classification and mapping. However, for evaluative information to be retained and communicated clearly, mapping is essential.

Interpretation is the development and presentation of interpretive management recommendations for classification of map units. It requires keen observation, statistical analysis of large volumes of data, and constant feedback to be effective. Yet, without interpretation, forestland inventory becomes a pedantic exercise of little value to land managers.

Soil and its associated environment can be thought of as the habitat or site of the forest trees and other organisms. Site includes the position in space as well as the associated environment (Barnes *et al.*, 1998). Site quality thus is defined as the sum total of all factors influencing the capacity of the forest to produce trees or other vegetation. The soil can be seen as an integrator and recorder of the environmental factors that influence a site. As such, "reading the soil" can yield information about the history as well as the current condition of the site. It is this ability to interpret difficult-to-observe site properties from easily observed soil properties that forms the basis of the soil survey systems used in agriculture the world over.

Since the early 1960s, the United States Department of Agriculture has been evolving a new soil taxonomy (Soil Survey Staff, 1999) that serves as the foundation of its national cooperative soil survey. This program of soil inventory and interpretation, which has parallels in Canada and many other countries, has the important advantage of employing well-established soil survey techniques based on an accepted system of soil classification. The system operates on the basic premise that soils can be identified as individual bodies and treated as integrated entities, so that knowledge of individual soil types can be interpreted in terms of their capacity to support various uses.

In the United States, detailed soil maps are produced at a scale of 1:24 000 and are published as county soil survey reports. The soil series of the classification system that are used to develop map units are correlated from one county to another so that similar soils are given a similar identifier in widely different places. Data on the management constraints associated with a particular soil series are collected widely. This allows interpretations of soil utility for a variety of uses, including forestry, to be made and published in the county survey report.

Several scientists have used soil survey information and maps successfully, by themselves or in combination with other observed properties such as drainage class, thickness of the surface horizon, and horizon depth, to predict site index (Coile, 1952; Stoeckeler, 1960; Broadfoot, 1969). Stephens (1965) reported that the soil taxonomic unit at the series level provided an accurate prediction of the Douglas fir site index on zonal soils of the Oregon Cascades.

In spite of the considerable potential of detailed maps from the cooperative soil surveys, they have been viewed with mixed emotions by most foresters. Some forest scientists consider them a valuable aid in management planning, but others find them of little value, particularly for predicting productivity. Whether or not they are useful to an individual apparently depends, to a large extent, on the landform and species under consideration. Van Lear and Hosner (1967) found little, if any, usable correlation between soil-mapping units and the site index of yellow poplar in Virginia. This conclusion was prompted by the wide variation in site indexes exhibited within each mapping unit. Other workers have encountered considerable frustration in attempting to group soils by series and type. Carmean (1961) pointed out the wide range of site values that may occur on single soil types. Several researchers in other areas (Coile, 1952; Broadfoot, 1969) have noted this same problem.

Many of the shortcomings of the soil survey maps derive from the fact that the classification system and survey methods were developed primarily for agricultural use. Some soil properties important to deep-rooted trees, such as water tables and subsoil textural changes at depths of 2 to 3 m, are not considered in delineating taxonomic or mapping units. By the time soil maps are made, a considerable amount of cooperation between soil scientists and agricultural users of soils information has gone into defining the principal soil series of interest to agriculture. Similar cooperation between foresters and soil scientists has not existed historically. At any rate, it is apparent that the standard USDA National Cooperative Soil Survey reports, as presently constructed, are not as effective as they could be for forestry purposes.

Industrial forest lands commonly have some type of detailed soil survey. In the absence of USDA soil surveys, or in an attempt to improve on the USDA method of mapping soils for purposes of forest management, several specialized forest soil survey methods have been developed. Coile (1952) developed a soil-based land inventory system that, in one form or another, is widely used in the southeastern United States, and the Weyerhaeuser Company (Steinbrenner, 1975) developed a system in the Pacific Northwest that has now been extended into the Mid-south and the Southeast.

Coile found soil-site maps desirable for both moderately intensive and intensively managed forests. He felt that soil maps should show the specific geographic extent of soil features and forest site classes useful in determining or making decisions on such things as (1) selection of species and spacing, (2) prediction of future yields, (3) drainage, (4) site preparation methods, (5) road construction, (6) definition of areas for seasonal logging, and (7) allowable costs for all phases of management based on expected returns.

Coile and his coworkers related tree growth to a limited number of easily measured or observed soil physical characteristics, such as texture of certain horizons, soil depth, consistency, and drainage characteristics, plus certain other selected features of topography, geology, and history of land use. He identified soil units with an alphanumeric code that characterized the drainage class, thickness and texture of the A horizon, depth to and texture of the B horizon, and other factors.

The value of these surveys depends, to a large extent, on the development of working relationships through mathematical trial-and-error testing of many combinations of variables (Mader, 1964). However, relatively little attention has been given to explaining the basic biophysiological relationships involved. They are based on the premise that a few factors will satisfactorily explain site differences over a wide range of conditions.

Coile (1952) stated that "the degree of success attained in demonstrating relationships between environmental factors and growth of trees is largely determined by the investigator's judgment in selecting the independent variables that are believed to be related to tree growth in various ways and in different combinations. How well the investigator samples the entire population of soil and other site factors determines the general applicability of the result." Coile limited his paired soil-forest stand observations to pure, even-aged pine stands over 20 years old that were fairly well stocked and growing in a relatively restricted area. He and his colleagues published information on growth and yield, stand structure, and soil-site relations of southern pines that dominated southern pine management for several decades (Schumacher and Coile, 1960; Coile and Schumacher, 1964).

In Weyerhaeuser's soil survey of mountainous terrain of the Pacific Northwest (Steinbrenner, 1975), topographic features are of paramount importance and the maps are based on a strong correlation between landform and soil series within a geologic unit.

The units mapped in this system are, in some cases, more narrowly defined than the National Cooperative Soil Survey units, but in some cases they are more broadly defined. In either event, they are given geographic place names much as are National Cooperative Soil Survey units. In the Weyerhaeuser system, topography, as evidenced by landform, is important to interpreting the survey for road construction, equipment use, and, in some cases, soil productivity. Mapping for productivity is a primary objective, and the interpretation for this purpose is developed through research. In addition, the maps are interpreted for land use, trafficability (for logging equipment), wind-throw hazard, thinning potential, and engineering characteristics for road construction. Productivity interpretations are the basis for determining allowable cut and for intensive forest operations, such as regeneration methods, stocking control, and thinning.

Mapping units provide the logical basis for delineating cuts in the logging plan, according to Steinbrenner (1975). The interpretations also indicate the type of equipment required and the timing of the harvesting operation, so that the impact on site quality is minimized. The wind-throw interpretation is utilized to minimize damage along harvest boundaries. Thinning potential is used to assign a priority to all lands for intensive forest practice. Engineers use the map in determining the best location for roads, drainage problems, and location and size of culverts needed. The soil survey provides information that is basic to sound forest management, and its usefulness increases as more interpretive detail is developed.

FOREST MANAGEMENT BEGINS WITH THE APPLICATION OF SOIL CLASSIFICATION TO LANDSCAPES

As forest management has intensified, soil maps have been constructed for a growing acreage of industrial forestland throughout the world. Systems similar to Coile's or Steinbrenner's have generally been utilized. Fisher (1980; 1984) proposed a system, variants of which are now commonly used in the southern United States and Latin America. These systems use many features of the USDA taxonomic and mapping systems, but they emphasize soil features that are closely related to site productivity in a particular area. For example, drainage class is very important in determining forest productivity in the southeast coastal plain, but has much less importance in the piedmont and mountains, where slope and stoniness are more important. Consequently, drainage class would be used to separate soil types at a higher level in a coastal plain soil survey, while slope and stoniness would be used to separate soil types at a higher level in a piedmont or mountain soil survey. This allows for regional taxonomies that can be refined so as to be of great value to the forest manager (Figure 9.7).

Our ability to discover the importance of various soil features or properties in determining site

Figure 9.7 Detailed soil maps are commonly imbedded into geographic information systems and used extensively in the management of industrial forest lands.

productivity has been greatly enhanced by GIS, GPS, geostatistics, and the fact that an increasing amount of forestland is growing its third, fourth, or fifth managed forest.

The global positioning system very accurately places the soil mapper as well as the soil map user in the landscape. We now have site productivity data for managed forest stands on many acres. These data give us a far better idea of site productivity than site index. Geographic information systems allow us to place the various diagnostic features of a soil type into different layers of the system, and then combine these features in different ways and test these for concordance with the site productivity data that have also been placed into the GIS. We can also use this methodology to discover the efficacy of management practices such as tillage, fertilization, and debris handling. This may lead to several different maps: one for species selection, one for productivity, one for trafficability, etc. In short, the utility of soil mapping for forest management is only limited by our imagination, intuition, ingenuity, and persistence.

CHAPTER 10

Common Approaches to Measuring Soils

OVERVIEW

The value of soil measurements is often limited by the choice of sampling designs, as highlighted in Chapter 9. The first step in measuring a soil is to be clear about the population that will be described. A robust sampling design is needed to address clear objectives. The variability of soil properties across space, and over time, needs explicit attention; only then can we know how many samples we will need to produce useful, reliable results. Too often, we reject or fail to confirm a hypothesis because our sample size is inadequate to address the variability of the property we are measuring. It is important to invest as much intellectual effort into our experimental and sampling designs as we invest physical effort into our sampling and analysis.

MOST OF WHAT MATTERS ABOUT SOILS IS HIDDEN FROM VIEW

One major difficulty in measuring soils is that they are seldom visible. They lie beneath our feet, are often covered with organic debris, and are generally unnoticed by the casual observer. To understand soils we must expose them, and that means digging or boring into them. Even when soils are exposed in a road cut or by natural erosion, we must remove the weathered surface of the cut to see the true soil. The exposed vertical section of a soil is called its profile, and there are conventions for describing it.

Soils develop in parent material from the time of its deposition under the influence of local climate, topography, and biota. Over time, a number of environmental forces act to create distinct layers, or horizons, parallel to the soil surface. This occurs through the differential downward movement of materials, such as Ca, Fe, organic matter, or clay particles. The movement and accumulation of materials at depth affects soil texture, structure, and/or color. These are the three principal properties that are used for distinguishing soil horizons (Soil Survey Staff, 1993; Schoeneberger et al., 2002; Buol et al., 2003). There are many soil classification systems in use around the world, but the nomenclature for describing soils is rather uniform. The nomenclature followed here is that of the United States Department of Agriculture's Natural Resources Conservation Service and the soil horizon designation nomenclature is outlined in Tables 10.1, 10.2, and 10.3 (Soil Survey Staff, 1993).

To describe a soil profile we begin by locating the preliminary horizon boundaries based upon differences in texture, structure, and color (see Chapter 5), and assigning each horizon a master horizon designation. If an O horizon is present, we measure its depth and determine its degree of decomposition. Remember that there may be more than one type of O horizon present (see Table 10.4 for an example of a soil profile description).

For each preliminary mineral horizon, we determine its texture (Figure 10.1). Texture classes can be accurately determined only by particle size analysis; however, in the field they are determined by feel.

Ecology and Management of Forest Soils, Fourth Edition. Dan Binkley and Richard F. Fisher.
© 2013 John Wiley & Sons, Ltd. Published 2013 by John Wiley & Sons, Ltd.

Table 10.1 Master, transitional and combination horizon nomenclature.

Horizon	Criteria
O	Organic soil materials
A	Mineral; Organic matter accumulation, loss of Fe, Al, clay
AB or AE	Dominantly A horizon characteristics but also contains some characteristics of B or E
A/B, A/E, A/C	Dominantly A horizon characteristics with discrete, intermingled bodies of B or E or C
AC	Dominantly A horizon characteristics but contains some characteristics of C horizon
E	Mineral; loss of Fe, Al, Organic matter, or clay
EA or EB	Dominantly E horizon characteristics but also contains some characteristics of A or B
E/A, E/B	Dominantly E horizon characteristics with discrete, intermingled bodies of A or B
E and Bt	Thin lamellae (Bt) within a dominantly E horizon
B	Subsurface accumulation of clay, Fe, Al, Si, humus, $CaCO_3$, $CaSO_4$; or loss of $CaCO_3$; or accumulation of sesquioxides; or subsurface soil structure
BA or BE or BC	Dominantly B horizon characteristics but also contains some characteristics of A or E or C
B/A or B/E or B/C	Dominantly B horizon characteristics with discrete, intermingled bodies of A or E or C
B and E	Thin lamellae (E) within a dominantly B horizon
C	Little or no pedogenic alteration; unconsolidated earthy material; soft bedrock
CB or CA	Dominantly C horizon characteristics but also contains some characteristics of A or B
C/B or C/A	Dominantly C horizon characteristics with discrete, intermingled bodies of A or B
L	Limnic material
R	Bedrock
W	A layer of liquid water or permanently frozen water (Wf) within the soil

Table 10.2 Horizon designation suffixes.

Horizon Suffixes	Criteria
a	Highly decomposed organic matter
b	Buried genetic horizon
c	Concretions or nodules
co	Coprogenous earth (used only with L)
d	Densic layer (physically root restrictive)
di	Diatomaceous earth (used only with L)
e	Moderately decomposed organic matter
f	Permanently frozen soil or ice (permafrost); continuous, subsurface ice; not seasonal ice
ff	Permanently frozen soil; no continuous ice; not seasonal ice
g	Strong gley
h	Illuvial organic matter accumulation
i	Slightly decomposed organic matter
j	Jarosite accumulation
jj	Evidence of cryoturbation
k	Pedogenic carbonate accumulation
m	Strong cementation
ma	Marl (used only with L)
n	Pedogenic, exchangeable sodium accumulation

The ability to do this accurately comes only with practice, and field determination should always be followed up with particle size analysis. It is important to also note the presence of coarse fragments, e.g. gravel, channers, flags, or stones (see Table 5.2), that might be present either on the surface or within

Table 10.3 Other horizon modifiers.

Numerical prefixes	Used to denote lithologic discontinuities. By convention, 1 is understood and is not written; e.g., A, Bt, 2Bt, 2C, 3C
Numerical suffixes	Used to denote subdivisions within a horizon; e.g., A1, A2, Bt1, Bt2, Bs1, Bs2
The prime	Used to indicate the second or subsequent occurrence of an identical horizon descriptor in a profile or pedon; e.g., A, E, Bt, E′, Btx, C. Double and triple primes are used to indicate subsequent identical horizon descriptors in a profile or pedon; e.g., A, E, Bt, E′, Btx, E″, C

Table 10.4 An example of a profile description.

Horizon	Description
Oi	10 to 7 cm; intact needles and leaves; abrupt smooth boundary.
Oe	7 to 4 cm; moderately decomposed needles and leaves; indistinct wavy boundary.
Oa	4 to 0 cm; highly decomposed organic matter; indistinct wavy boundary.
A	0 to 20 cm; very dark gray (10YR 3/1) fine sand; weak fine granular structure; clear irregular boundary. (10 to 25 cm thick)
Bh	20 to 36 cm; dark brown (7.5YR 3/2) fine sand; weak fine granular structure; friable; gradual irregular boundary. (1 to 30 cm thick)
Eg	36 to 94 cm; light gray (10YR 7/2) fine sand; many coarse distinct grayish brown (10YR 5/2) mottles; few medium prominent (7.5YR 3/2) Bh bodies; few fine prominent strong brown (7.5YR 5/8) streaks along root channels; single grained loose; clear wavy boundary.
Btg	94 to 145 cm; light brownish gray (10YR 6/2) fine sandy loam; many coarse prominent yellowish brown (10YR 5/8) and few fine prominent red (2.5YR 4/6) soft masses of iron accumulation; weak coarse subangular blocky structure; wavy boundary. (10 to 48 inches thick)
Cg	145 to 203 cm; reticulately mottled gray (10YR 6/1), strong brown (7.5YR 5/8), and red (2.5YR 4/6, 10R 4/6) fine sandy loam; massive.

a horizon. We also need to estimate the percentage of the soil volume that the coarse fragments occupy, and to apply the appropriate modifier to the texture designation (e.g. gravelly sandy loam).

Next, we break out a handful of structural aggregates and determine the horizon's structure. Structure is described in terms of shape, size, and grade.

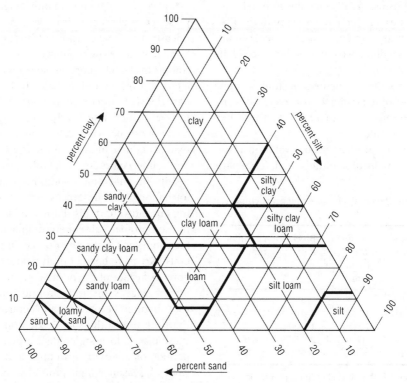

Figure 10.1 The texture triangle (see also Figure 5.2).

Shape:

  Granular: the units are approximately spherical or polyhedral and are bounded by curved or irregular faces that are not casts of adjoining peds.

  Blocky: the units are block-like or polyhedral and are bounded by flat or slightly rounded surfaces that are casts of adjoining peds. The structure is described as angular blocky if the faces intersect at sharp angles and as subangular blocky if the faces are a mixture of rounded and planar faces and the corners are mostly rounded.

  Platy: the units are flat and plate-like. They are generally oriented horizontally.

  Prismatic: the units are bounded by flat to rounded vertical faces and are distinctly longer vertically. The faces are typically casts of adjoining units.

  Columnar: the units are similar to prisms and are bounded by flat or only slightly rounded vertical faces. The tops of columns, in contrast to those of prisms, are distinct and rounded.

Size: five classes are employed: very fine, fine, medium, coarse, and very coarse. The size limits of the classes differ according to the shape of the units (Table 10.5).

Grade: describes the distinctness of units. Criteria are the ease of separation into discrete units and the proportion of units that hold together when the soil is handled.

  Weak: the units are barely observable in place. When gently disturbed, the soil material parts into a mixture of whole and broken units and much material that exhibits no planes of weakness.

  Moderate: the units are well formed and evident in undisturbed soil. When disturbed, the soil material parts into a mixture of mostly whole units, some broken units, and material that is not in units.

  Strong: the units are distinct in undisturbed soil. They separate cleanly when the soil is disturbed.

Then, we determine the color or colors present in each preliminary horizon using the Munsell Soil Color Charts (Munsell, 1990; Soil Survey Staff, 1993). Like texture by feel, the ability to determine soil color accurately is only gained through practice. It is important to note whether or not there is any mixing of different colors (mottling) in the horizon, since this is an indication of the soil's drainage class; that is, how often and for how long the soil is saturated with water.

Soil colors are predominantly associated with the presence or absence of iron. Weathering causes the release of mineral constituents to the environment. Soluble weathering products are removed from the soil profile, while more stable compounds will precipitate. Iron migrating through the soil coats soil particles with thin oxide coatings. Well-aerated soils in areas of moderate weathering are typically brown to yellowish-brown in color from these ferric (Fe^{3-}) oxide coatings. In areas of intense weathering, red and yellow colors develop. Prolonged soil saturation results in anaerobic conditions that lead to the formation of mobile ferrous (Fe^{2-}) iron. The migrating groundwater redistributes the iron throughout the soil profile. Subsequent drainage restores aerobic conditions, but some iron coatings on the minerals may have been entirely removed, leaving the grayish surface of the mineral grains exposed. During drainage, some areas around pores, cracks, and root channels become dry and aerated more quickly than the rest of the soil. Ferric iron precipitates in these places, forming reddish-brown spots. During periods of alternating wetting and drying cycles, such as seasonal high groundwater, ferrous iron does not transfer out of the soil profile entirely

Table 10.5   Size class limits for describing structural shapes.

Size Class	Platy[1] mm	Prismatic and Columnar mm	Blocky mm	Granular mm
Very Fine	< 1	< 10	< 5	< 1
Fine	1–2	10–20	5–10	1–2
Medium	2–5	20–50	10–20	2–5
Coarse	5–10	50–100	20–50	5–10
Very Coarse	> 10	> 100	> 50	> 10

[1]Plates are described as thin instead of fine and thick instead of coarse.

but moves over short distances only and precipitates during the drying phase. Such conditions are characterized by blotches of gray and reddish-brown soil colors occurring at the same depth. The longer the saturation period, the more pronounced the reduction process, and the grayer the soil becomes. This pattern of spots or blotches of different color or a shade of color, interspersed with the dominant color, is called soil mottling. A quick look at the sample profile description will reveal that it is a somewhat poorly drained soil.

Now, if we are satisfied that the preliminary horizon delineations are correct, we can apply the master and subordinate horizon designations based upon our soil description. We then need to measure and record each horizon's thickness, which may vary from point to point along the exposed face, and describe the boundary between horizons (e.g. is it abrupt or gradual, and is it smooth, wavy, or irregular?).

If we are interested in the soil only at the location of the soil description, we are finished; however, if we are interested in the soil in a particular stand, we need to determine whether or not the soil we have described is the only one present in the stand. We can estimate this soil's extent in the landscape by traversing the stand, taking soil cores, and comparing horizons in those cores to our profile description. In general, but not always, changes in the soil profile occur at changes in topography, and this serves as a key to where we must check for potential profile changes.

INFERENCES ARE DEVELOPED FROM PROFILE DESCRIPTIONS

Having a profile description allows us to infer a great deal about a site. Texture is an important variable in determining site quality for tree growth. Fine particles provide exchange sites for nutrients that are in cationic form, and they produce a large proportion of small soil pore sizes that retain water in the soil, while coarse particles produce large pore sizes and increase soil aeration. Coarse fragments neither provide nutrient exchange sites nor do they aid in water retention,

but they do produce many macro-pores that increase infiltration rate and reduce erosion risk. Therefore, we can infer something about soil nutrient and moisture status, trafficability, and erosion hazard from soil texture.

Structure tends to modify texture properties. Sandy soils with well-developed structure retain more water than those with little or no structure, while clayey soils with well-developed structure are better aerated and have higher infiltration rates than clayey soils with little or no structure. The clay skins that develop on the peds in soils with well-developed structure also increase the number of cationic nutrient exchange sites. Soils with well-developed structure also have greater soil strength than those with weak or no structure. If dense soil horizons are present within the profile, they may impede root growth and reduce site quality for trees. Therefore, we gain knowledge about soil nutrient and moisture status, and trafficability from soil structure.

The color that develops in soil horizons is another guide to soil nutrient status and serves as a guide to soil drainage class, the amount of time that the soil is saturated with water. The dark colors of the A horizon are related to soil organic matter content and nutrient status, particularly nitrogen content, and drainage class is an important factor in determining a site's suitability for a particular tree species, its trafficability and resistance to compaction, and site productivity. By coupling this knowledge gained from soil texture, structure, and color with knowledge of horizon depth and thickness, we can gain a great deal of understanding about a site, either in the field or by studying the profile description in the office.

SOIL WATER CONTENT SHOWS IMPORTANT PATTERNS

This is another soil property that we frequently wish to measure. Soil moisture content is defined as the mass of water relative to the mass of dry soil particles, and is often referred to as the gravimetric water content. Wetness depends largely on the porosity of a soil, and for that reason clayey soils, which have a high porosity generally, have higher water content than do sandy soils. Also, if a soil has been compacted, the result will be that the water content will decrease, as the inter-basal spacing between the pores will have been decreased. The traditional method of measuring gravimetric moisture involves the removal of a soil core and the subsequent determination of the moist and dry weights. The moist weight is determined by weighing the sample as it is at the time of sampling, and the dry weight is obtained after drying the sample at 105 °C in an oven. The mass of water lost per mass of oven-dry soil is commonly known as *wetness* (Hillel, 1982).

Core samples are fairly impractical, as they are so time consuming. Taking samples, sealing these to prevent loss of initial moisture, transporting from the field to an oven, many hours in the oven, and repeated weighing all take too much time to consider using this as a realistic, regular method. The standard method of drying in itself is arbitrary, as some clayey soils may still contain perceivable amounts of water even at 105 °C. On the other hand, some organic matter may oxidize and decompose at this temperature, so that the weight loss may not be entirely due to evaporation of water. The errors in this method may be reduced by increasing the number of samples, but this then has to be balanced with the destruction and disturbance that taking so many cores can cause the site.

Another way of expressing the moisture content of a soil is the volumetric water content, which is defined as the volume of water relative to the total volume of soil. The use of volumetric water content rather than wetness is often more convenient because it is more directly adaptable to the computation of fluxes and water quantities in the soil. Volumetric water content also represents the depth of water per depth of soil.

Volumetric soil water content can be measured using the neutron probe, time domain reflectometry (TDR), and capacitance techniques, e.g. gypsum blocks. The neutron probe method utilizes the fact that hydrogen slows down fast neutrons much more effectively than other substances. In the soil, the vast majority of hydrogen is in water molecules; therefore, changes in hydrogen content are closely related to changes in water content. The TDR and capacitance techniques exploit the strong dependence of soil dielectric properties on soil water content (Gardner *et al.*, 2000).

Soil water content classes can be estimated in the field. The feel and appearance of the soil indicates soil moisture status. Use a soil tube, soil auger, or shovel to sample soils to determine moisture content. Take soil samples at intervals throughout the depth of the active root zone. Make an estimate of soil moisture status by firmly squeezing a handful of soil and comparing the results with Table 10.6.

Table 10.6 Soil moisture interpretation chart.

Soil Moisture Deficiency	Moderately Coarse Texture	Medium Texture	Fine and Very Fine Texture
0% (field capacity)	Upon squeezing, no free water appears on soil but wet outline of ball is left on hand.		
0–25%	Forms weak ball, breaks easily when bounced in hand.*	Forms ball, very pliable, slicks readily.*	Easily ribbons out between thumb and forefinger.*
25–50%	Will form ball, but falls apart when bounced in hand.*	Forms ball, slicks under pressure.*	Forms ball, will ribbon out between thumb and forefinger.*
50–75%	Appears dry, will not form ball with pressure.*	Crumbly, holds together from pressure.*	Somewhat pliable, will ball under pressure.*
75–100%	Dry, loose, flows through fingers.	Powdery, crumbles easily.	Hard, difficult to break into powder.

*Squeeze a handful of soil firmly to make ball test.

WATER RELEASE CURVES RELATE SOIL WATER CONTENT WITH WATER POTENTIAL

The water release or retention curve is the relationship between the water content θ (theta) and the soil water potential ψ (psi). This curve is characteristic for different types of soil, and is also called the soil moisture characteristic. It is used to predict soil water storage, water supply to the plants, and soil aggregate stability (Figure 10.2).

The general features of water retention and release curves can be seen in the figure, in which the volume water content, often labeled θ, is plotted against the water tension or matric potential, often labeled ψ. At potentials close to zero, a soil is close to saturation, and water is held in the soil primarily by capillary forces. As θ decreases, binding of the water becomes stronger, and at small potentials (more negative), water is strongly bound in the smallest of pores, at contact points between grains and as films bound by adsorptive forces around particles.

Sandy soils will involve mainly capillary binding, and will therefore release most of the water at higher potentials, while clayey soils, with adhesive and osmotic binding, will release water at lower (more negative) potentials. At any given potential, peaty soils will usually display much higher

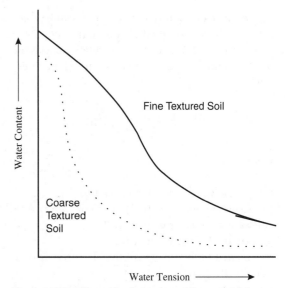

Figure 10.3 Effects of soil texture on the water release curve.

moisture content than clayey soils, which would be expected to hold more water than sandy soils. The water-holding capacity of any soil is due to the porosity and the nature of the bonding in the soil (Figure 10.3). The degree of soil aggregation, particularly in finer-textured soils, can have a pronounced effect on the water release curve (Figure 10.4).

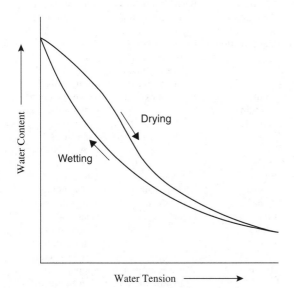

Figure 10.2 Water retention and release curves.

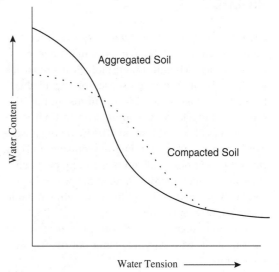

Figure 10.4 Effects of soil aggregation on the water release curve.

The shape of water retention curves can be characterized by several models (Cornelis *et al.*, 2005; Khlosi *et al.*, 2008). The van Genuchten model is the most commonly used model (van Genuchten, 1980):

$$\theta(\psi) = \theta_r + \frac{\theta_s - \theta_r}{[1 + (\alpha|\psi|)^n]^{1-1/n}}$$

where

$\theta(\psi)$ is the water retention curve [L^3L^{-3}];
$|\psi|$ is suction pressure ([L^{-1}] or cm of water);
θ_s is saturated water content [L^3L^{-3}];
θ_r is residual water content [L^3L^{-3}];
α is related to the inverse of the air entry suction, $\alpha > 0$ ([L^{-1}], or cm^{-1}); and,
n is a measure of the pore-size distribution, $n > 1$ (dimensionless).

This model is not perfect and other models have been developed to overcome some of its specific weaknesses; however, no model has, as yet, replaced it in general use.

SOIL ORGANIC MATTER IS AT THE HEART OF SOIL STRUCTURE AND BIOLOGY

Organic matter is widely regarded as a vital component of a healthy soil. It has important influences on soil's physical, chemical, and biological properties (Chapter 4). In its broadest sense, soil organic matter comprises all living soil organisms and all the remains of previous living organisms in their various degrees of decomposition. The living organisms can be animals, plants, or micro-organisms, and can range in size from small animals to single-cell bacteria only a few microns long. When we measure soil organic matter, it is the non-living organic matter that we measure.

Non-living organic matter can be considered to exist in pools with varying degrees of overlap and separation:

- Dissolved organic matter in soil water.

- Particulate organic matter ranging from recently added plant and animal debris to partially decomposed material less than 50 microns in size, but all with an identifiable cell structure. Particulate organic matter can constitute from a few percent up to 25% of the total organic matter in a soil.
- Humus, which comprises both organic molecules of identifiable structure, like proteins and cellulose, and molecules with no identifiable structure (humic and fulvic acids and humin) but which have reactive regions which allow the molecule to bond with other mineral and organic soil components. These molecules are moderate to large in size (molecular weights of 20 000–100 000). Humus usually represents the largest pool of soil organic matter, comprising over 50% of the total.
- Inert organic matter or charcoal derived from the burning of plants. Can be up to 10% of the total soil organic matter.

An alternative way of classifying these pools is shown in Table 10.7.

Table 10.7 Organic matter pools.

Organic Matter Pool	Constituents	Role
Microbial biomass	Bacteria and fungi (i.e. the living part)	Decomposes the organic matter
Light fraction	Organic matter that has recently been incorporated	Food for microbes, releases soil nutrients
Soluble organic matter	E.g. root exudates	Moves through the soil profile – binds soil particles, available for plant uptake
Protected organic matter	Protected chemically or physically	Can't be decomposed by microbes
Inert organic matter	E.g. charcoal	Does not breakdown, contributes to soil structure
Humus	Well-decomposed organic matter	Supplies nutrients

Table 10.8   Comparison of field instrumentation for determination of soil C (modified from Gehl and Rice, 2007).

Method[1]	Principle	Approximate Detection Limit	Sampling Time	Sampling Attributes
LIBS	Intense laser pulse is focused on in-tact sample core, microplasm emission spectrally resolved	300 mg total C kg^{-1}	< 60 s	Invasive
INS	Inelastic scattering of neutrons from C nuclei emits gamma rays that are detected and analyzed for C peak intensity	0.018 g C cm^{-3}, or 1.2% C for 1.5 g cm^{-3} bulk density sample	-	Non-invasive
NIRS	Diffuse reflectance properties in the 700 to 2500 nm spectrum are correlated to soil C content	0.6% organic C	200 ms spectrum^{-1}	Invasive

[1]LIBS – laser-induced breakdown spectroscopy; INS – inelastic neutron scattering; NIRS – near-infrared spectroscopy.

The most common methods for measuring soil organic matter in current use actually measure the amount of carbon in the soil (Sparks, 1996). This is done by oxidizing the carbon and measuring either the amount of oxidant used (wet oxidation, usually using dichromate), the CO_2 given off in the process (combustion method with specific detection), or the weight loss on ignition. Soil organic matter is derived by multiplying the amount of soil carbon by 1.72. However, this conversion factor is not the same for all soils, and it is more precise to report soil carbon rather than organic matter. Soil organic carbon varies greatly with soil depth, but it is usually rather constant at a given depth over fairly large horizontal distances. However, it can vary greatly with the changes in micro-topography that are common in forest soils. Samples for laboratory analysis must be taken using a well-thought-out sampling design. Subsamples from several locations are generally composited for analysis. The samples must be handled carefully after sampling.

During the collection and handling of the samples, losses of organic compounds may occur due to: microbial degradation, sample drying, oxidation, volatilization, and sample processing biases (e.g., selective removal of carbon-bearing components). Soil samples can be collected by numerous different tools but, once collected, the samples should be kept cool ($\sim 4\,°C$) and should be analyzed within 28 days. While microbial degradation is greatly reduced at $4\,°C$, it is not completely stopped, leading to some potential loss of organic materials. Prior to analysis, some methods may require or recommend drying (either air drying or oven drying) of the sample. Samples that have been in an anaerobic environment will undergo some loss of organic compounds when exposed to the atmosphere; these losses are generally small (probably < 1% of the TOC).

The laboratory-based measurements of soil C or soil organic matter involve intensive, time-consuming, and costly methodology that limits applicability for large land areas. Recently, research efforts have focused on measuring soil C *in situ* using a variety of methods (Gehl and Rice, 2007). These methods include laser-induced breakdown spectroscopy (LIBS), inelastic neutron scattering (INS), near-infrared spectroscopy (NIRS), and remote sensing (Table 10.8). However, none of these techniques is, as yet, in common usage.

pH INDEXES THE ACIDITY AND ALKALINITY OF SOILS

Soil pH is a measure of the acidity and alkalinity in soils. pH is defined as the negative logarithm (base 10) of the activity of hydrogen ions (H^+) in solution (Chapter 8). Along with soil oxidation/reduction potential, soil pH is a master variable that influences many chemical processes. It specifically affects plant nutrient availability by controlling the chemical forms of the nutrient; therefore, we frequently measure soil pH.

Acidity in soils comes from H^+ and Al^{3+} ions in the soil solution and sorbed to soil surfaces. While pH is the measure of H^+ in solution, Al^{3+} is important in acid soils because between pH 4 and 6, Al^{3+} reacts with water (H_2O) forming $AlOH^{2+}$, and $Al(OH)_2^+$, releasing extra H^+ ions. Every Al^{3+} ion can create 3 H^+ ions. Many other processes contribute to the formation of acid soils, including rainfall, fertilizer use, plant root activity, and the weathering of primary and secondary soil minerals. Acid soils can also be caused by pollutants such as acid rain and mine spoils.

Alkaline soils have a high saturation of base cations (K^+, Ca^{2+}, Mg^{2+}, and Na^+). This is due to an accumulation of soluble salts, and these soils are classified as saline soil, sodic soil, saline-sodic soil or alkaline soil. All saline and sodic soils have high salt concentrations, with saline soils being dominated by Ca and Mg salts and sodic soils being dominated by Na. Alkaline soils are characterized by the presence of carbonates.

Most forest soils are acidic, and many drier rangeland and savanna soils are basic. Methods of determining pH range from simple observation to laboratory analysis:

- Observation of soil profile. Certain profile characteristics can be indicators of acid, saline, or sodic conditions. Strongly acidic soils often have poor incorporation of the organic surface layer with the underlying mineral layer. The mineral horizons are distinctively layered in many cases, with a pale eluvial (E) horizon beneath the organic surface; this E is underlain by a darker B horizon in a classic spodic horizon sequence. This is a very rough gauge of acidity, as there is no correlation between thickness of the E and soil pH. The presence of a caliche layer indicates the presence of calcium carbonates, which are present in alkaline conditions. Also, columnar structure can be an indicator of sodic conditions.
- Observation of predominant flora. Some plant species, e.g., Ericaceae species, *Pinus*, and *Ulex*, prefer an acidic soil, while others, e.g., *Cornus*, *Juniperus*, and *Lonicera* prefer basic soil.
- Use of a field pH testing kit.
- Use of litmus paper.
- Laboratory testing with a pH meter.

In the laboratory, pH is generally measured in a 1:1 mixture of soil:water with a pH electrode (Sparks, 1996). Even when the soil's acidity is unchanged, the pH may vary from year to year or season to season. This fluctuation occurs because of variations in the salt content of the soil. The pH electrode requires an electrolyte (acid/salt) to work. The amount of salt in the soil affects the pH reading, and since the amount of salt in the soil varies over time, the pH varies over time. Therefore, pH is often measured in a dilute salt solution. A 0.01 M $CaCl_2$ solution is commonly used, although other concentrations and salts are sometimes used. pH measured in dilute $CaCl_2$ solution is commonly about half a unit lower than that measured in water, but it is much more constant from year to year and season to season.

Soil pH and soil organic matter vary greatly with soil depth, but are rather similar at any one depth over larger horizontal distances. They can, however, vary substantially with the micro-topography common in forest soils. Careful soil sampling is necessary if we are accurately to determine the average pH of a plot or site. Bulking of subsamples from similar topographic positions can help to reduce the amount of sampling necessary to obtain an accurate idea of soil pH.

SOIL NITROGEN IS TIGHTLY RELATED TO SOIL ORGANIC MATTER

Total soil nitrogen content is closely related to soil organic matter content and can be measured by wet or dry oxidation methods, just as total soil carbon content is; however, total soil nitrogen content is only weakly related to site productive capacity, so we seldom wish to measure it. Plants take up nitrogen in the form of ammonium (NH_4^+), nitrate (NO_3^-), and in some cases low-molecular-weight organic N (Chapter 7). The quantity of these compounds in soil solutions is small at any single time, so the rate at which these pools are replenished from slower, less-available pools is important. Nitrogen mineralization to produce NH_4^+ and NO_3^- from the soil's organic nitrogen pool is an ongoing process controlled by many biotic and environmental variables.

We can measure the amount of NH_4^+ and NO_3^- in the soil by taking samples and extracting and measuring the NH_4^+ and NO_3^- present, but the residence time of NO_3^- in the soil ranges from hours to one to two days (Stark and Hart, 1997). Ammonium may be sorbed onto cation exchange sites, increasing its residence time, but we still have no idea of how fast the NH_4^+ we find in our sample would be replaced if that NH_4^+ were taken up by plants. At any given sampling date, several kilograms per hectare of ammonium nitrogen and nitrate nitrogen are present in forest soils, although some soils may not have measurable nitrate. This pool of "mineral" or inorganic nitrogen is incredibly dynamic. Although some researchers have inferred nitrogen availability from these ephemeral pools, this approach is simplistic, as these pools are only a small fraction of the nitrogen that passes through them on an annual basis. The average residence time of nitrate in soils is on the order of hours to days. This reasoning is similar to estimating a person's income based on the amount in a checking account at one point in time. Even repeated samplings throughout an annual cycle may not reflect adequately the total flow of ammonium and nitrate through the "clearinghouse" of available pools.

Consequently, we generally wish to measure nitrogen mineralization, or nitrogen supplying power. There are many ways to assess nitrogen mineralization. For example, we could use anaerobic or aerobic incubation either *in situ* or *ex situ*, lysimeters, ^{15}N isotopic pool depletion, or buried resin bags (see Binkley and Hart, 1989 for a review of methods of assessing nitrogen availability in soils). These methods require us to pay close attention to experimental design and sampling protocols. A frequent error in measuring nitrogen mineralization is to have too few samples inadequately distributed over the plot, site, or stand of interest.

PHOSPHORUS SUPPLY RATES ARE MORE COMPLICATED THAN NITROGEN

The total phosphorus content of a soil sample may be determined following digestion in strong perchloric acid ($HClO_4$), but this number provides little information for site assessment because most of the total phosphorus pool is unavailable to plants. Phosphorus availability in soils is not measured easily, but a number of extraction procedures have been developed that correlate well with plant-available phosphorus under various conditions (Olsen and Sommers, 1982).

Bray and Kurtz (1945) suggested a combination of HCl and NH_4F to remove easily acid-soluble P forms, largely Al- and Fe-phosphates. In 1953, Mehlich introduced a combination of HCl and H_2SO_4 acids (Mehlich 1) to extract P from soils in the north-central region of the U.S. Sulfate ions in this acid solution can dissolve Al- and Fe-phosphates in addition to P adsorbed on colloidal surfaces in soils. In the early 1980s, Mehlich modified his initial soil test and developed a multi-element extractant (Mehlich 3) which is suitable for removing P and other elements in acid and neutral soils. Mehlich 3 extractant (Mehlich, 1984) is a combination of acids (acetic [HOAc] and nitric [HNO_3]), salts (ammonium fluoride [NH_4F] and ammonium nitrate [NH_4NO_3]), and the chelating agent ethylenediaminetetraacetic acid (EDTA).

The so-called double acid method (Mehlich 1) has been popular for assaying phosphorus availability in pine plantations in the southeastern United States. In a comparison of these methods with the loblolly pine height growth response to phosphorus fertilization, Wells *et al.* (1973) found a fairly close correlation between the methods ($r^2 = 0.56$ to 0.67). Nevertheless, the methods accounted for only 25–40% of the observed variation in fertilizer response.

Olsen *et al.* (1954) introduced 0.5 M sodium bicarbonate ($NaHCO_3$) solution at a pH of 8.5 to extract P from calcareous, alkaline, and neutral soils. This extractant decreases calcium in solution (through precipitation of calcium carbonate), and this decrease enhances the dissolution of Ca-phosphates. Moreover, this extracting solution removes dissolved and adsorbed P on calcium carbonate and Fe-oxide surfaces. This method is common for phosphorus assays of neutral to alkaline pH soils, as well as for soils high in calcium carbonate (limestone) and many tropical soils. The bicarbonate anion causes precipitation of calcium, freeing phosphorus from calcium salts. Most forest soils are acid, so this technique has not proven generally useful in forestry. For example, Kadeba and Boyle (1978) found that a sodium bicarbonate extractant dissolved about the same amount of phosphorus from acid soils as acid extractants did, but the phosphorus availability estimates were unrelated to the phosphorus taken up by pine seedlings. In this experiment, the acid extractants did not perform impressively either, but an anion exchange resin method worked fairly well.

Ion exchange resins were first used in soil analysis to assess phosphorus availability in laboratory assays (Amer *et al.*, 1955). This method merely mixes anion exchange resins with water and soil and allows the resin to adsorb phosphorus from the soil for several hours. The resins are then removed from the soil and extracted to determine phosphorus adsorption. Of all the methods tested by Kadeba and Boyle (1978), resin phosphorus accounted for the most variation (66%) in pine seedling phosphorus uptake.

Ion exchange resin bags were developed to assess the supply of nutrient ions as a function of the mineralization of ions and their transport to the resin bags (Binkley and Matson, 1983). Hart *et al.* (1986) found that IER bag phosphorus correlated weakly with current loblolly pine growth in the coastal plain of North Carolina ($r^2 = 0.31$), as did double-acid extracts ($r^2 = 0.26$). Neither method alone predicted the response to nitrogen + phosphorus fertilization, but combining either method with pre-fertilization growth accounted for 80–85% of the variation in growth response.

It is important to note here that a thorough knowledge of the degree of correlation between the phosphorus assay method to be used and plant growth response on a particular type of soil is essential if a measure of soil P is to be useful. Available soil P varies greatly with soil depth, but is rather similar at any one depth over larger horizontal distances. Surface soil available P can vary substantially with the micro-topography common in forest soils. Since trees often root deeply into the soil profile, it is often necessary to sample subsurface horizons in order to understand a site's ability to meet tree P uptake requirements. A robust experimental design and careful soil sampling is necessary if we are accurately to determine the average available soil P of a plot or site. Bulking of subsamples from similar topographic positions can help to reduce the amount of sampling necessary to obtain an accurate idea of available soil P.

CATION SUPPLIES MAY OR MAY NOT RELATE TO EXCHANGEABLE POOLS

We frequently measure soil cations; however, the true availability of cations is difficult to assess. Although the cation nutrients held on exchange sites form a readily available pool, they do not represent the cation-supplying ability of the soil. Cations removed from exchange sites often are replenished rapidly from other sources, such as organic matter decomposition, mineral weathering, or release of ions "fixed" within the layers of clay minerals. For example, 15 years of a grass crop removed a total of 4000 kg ha^{-1} of potassium without substantially depleting the exchangeable potassium pool in a Hawaiian soil (Ayers *et al.*, 1947). Exchangeable cations may well represent the total available pool in old, highly weathered soils, but the exchangeable pool is only part of the available pool in most forest soils.

Exchangeable cations can be extracted from soil using a variety of extracting solutions (Sparks, 1996). The effectiveness of extracting solutions varies depending on soil type; therefore, it is critical to choose the correct solution for the soil of interest. The extract can then be analyzed by a variety of methods to determine the content of the cation or cations in question. Extraction procedures for estimating soil cation-supplying power have also been developed (MacLean, 1961; Sparks, 1996); however, knowledge of how these supplying power estimates correlate with plant growth in any particular soil–plant system is necessary if they are to be useful.

Cation exchange capacity (CEC) is the maximum quantity of total cations, of any class, that a soil is capable of holding. CEC is highly dependent upon soil texture and organic matter content. In general, more clay and organic matter in the soil will produce a higher CEC. It is usually measured by saturating the soil sample with a single cation, e.g., potassium or ammonium, and extracting and measuring that cation (Sparks, 1996). It is a measure of potential soil fertility.

Base saturation (BS) is the amount of the CEC that contains basic cations, e.g., Ca, Mg, K, or Na. If we have measured those cations and the CEC, we can calculate the BS. The availability of nutrient cations such as Ca, Mg, and K to plants increases with increasing BS; therefore, it is a measure of soil fertility.

Exchangeable cations, CEC, and BS vary greatly with soil depth, but they are rather constant over horizontal distances. Consequently, extensive soil sampling is seldom used to determine these variables, and frequently samples from a single pedon are used to establish their values.

THE BIOTA OF SOILS IS EVEN RICHER THAN THE CHEMISTRY

Forest soils contain a vast array of biota that have a profound effect on soil quality (Doran and Jones, 1997; Battigelli and Berch, 2002; Chapter 6). We often wish to quantify the soil biota. For our purposes, it is simplest to divide the biota into the two classic kingdoms – animals and plants (Coleman *et al.*, 2004).

Macrofauna (> 2 mm) are species large enough to disrupt soil structure through their activity and include small mammals, earthworms, termites, ants, and other arthropods (Dindal, 1990; Stork and Eggleton, 1992; Berry, 1994). Mesofauna (0.1–2 mm) are those soil-dwelling organisms large enough to escape the surface tension of water and move freely in the soil but small enough not to disrupt soil structure through their movement. They include some mollusks, springtails, mites, woodlice, and other arthropods (Crossley *et al.*, 1992). Microfauna (< 0.1 mm) are usually confined to water films on soil particles. This group includes protozoa, turbellaria worms, rotifers, nematodes, and gastrotricha worms (Sohlenius, 1982).

Macrofauna are sampled either by trapping or using reagents to drive them to the surface of the soil and then capturing them (Edwards, 1991; Ruiz *et al.*, 2008).

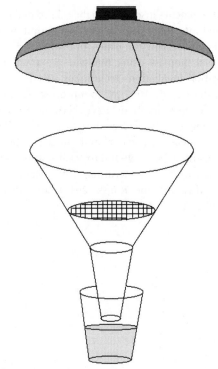

Figure 10.5 A Berlese funnel set-up for sampling soil fauna.

Mesofauna are commonly sampled by using heat to drive them from a soil sample (Figure 10.5). They can then be classified and counted using a dissecting microscope or automated techniques (Cole *et al.*, 2004; Chamblin *et al.*, 2011).

Microfauna can be sampled by heat or solution extraction from a soil sample. They can then be classified and counted using a compound microscope (Qualset and Collins, 1999; Forge *et al.*, 2003).

In all cases, we must remember that soil fauna populations vary greatly diurnally, seasonally, and with depth in the soil and micro-topographic relief (Weaver *et al.*, 1994; Coleman, 2001). Measuring soil faunal populations is challenging, and we should be confident that we have sound experimental and sampling designs before undertaking it.

Archaea, bacteria (including actinomycetes) and fungi play very important roles in the soil, and therefore we often wish to enumerate them or measure their activity in the soil (Weaver *et al.*, 1994). Traditionally, these organisms have been measured using CO_2 production, the decomposition of cotton patches inserted into the soil, coated plates inserted into the soil for a specified period of time and then observing the plates, or by measuring soil microbial biomass using fumigation-extraction or fumigation-incubation techniques (Kieft *et al.*, 1987; Beck *et al.*, 1997; Nachimuthu *et al.*, 2007). Recently, new methods have become available. Phospholipid fatty acid analysis (PFLA) has been used to measure bacteria and fungi in the soil (Frostegård *et al.*, 1993; Frostegård and Bååth, 1996; Bardgett and McAlister, 1999; Kaur *et al.*, 2005; Strickland and Rousk, 2010), and appears, as of now, the way of the future.

Enumeration and speciation of soil microbes has been done using microscopic analysis; however, this has limitations. rRNA analysis has been used to measure the genetic diversity of bacteria, archaea, fungi, and viruses in the soil (Nannipieri *et al.*, 2003; Fierer *et al.*, 2007), and rDNA fingerprinting (Fierer and Jackson, 2006) has been used to enumerate soil microbes. These methods both produce results superior to microscopic enumeration. Measuring soil microbial populations is time consuming and equipment intensive. We need to be confident that we have sound experimental and sampling designs before undertaking it.

PART V
Dynamics of Forest Soils

CHAPTER 11
Influence of Tree Species

OVERVIEW

The vast diversity among forest soils connects to tree species in two important ways: the success of various species depends on soil properties, and soil properties may be strongly influenced by the species of trees growing at a site. Tree species differ in their influence on soil water dynamics and soil temperature, with cascading effects on soil biology and chemistry. The quantity and chemistry of organic matter added to the soil by trees influences the species and activities of soil animals and microbes, which in turn influence nutrient turnover rates and tree nutrition. Common garden experiments allow the effects of tree species to be separated from other soil-forming factors, and these experiments show that the supply of soil N commonly differs by more than 50% under the influence of tree species. The patterns of overall influences of trees on soils are relatively simple to measure, but accounting for the direct processes that create these differences will remain a fertile area of research. Tree species that fix nitrogen tend to have the strongest impacts on soils as well as ecosystem productivity. Several approaches are available for estimating rates of nitrogen fixation in forests, including long-term accretion of nitrogen in the ecosystem and short-term assays of nitrogen fixation activity using acetylene. Nitrogen-fixing trees typically add 15 to 20 g of stabilized organic C for each gram of N accumulated, and this gain in soil C likely results from a combination of stabilization of recently added C and greater retention of older soil C.

" . . . different vegetation gives rise to different soil. Local inhabitants have long noted that land formerly under forest has different qualities according to the kind of forest which it bore." (Vasily Dokuchaev 1890, cited in Remezov and Pogrebnyak 1965).

Did these obvious differences develop because of the presence of various tree species, or were the tree species successful on various sites because other soil-forming factors had developed soils where certain species survived particularly well? Any conclusion about cause and effect needs to carefully weigh the evidence with the web of mutual influences between trees and soils.

More than a century of speculation and research have focused on notions of how trees shape the chemistry and fertility of soils (Handley, 1954; Remezov and Pogrebnyak, 1965). Early concepts acknowledged that certain tree species generally occur on specific types of soils, and that such soils may indeed be developed by the dominant species. For many decades, European foresters and soil scientists commonly thought that conversion of beech forests to spruce led to soil degradation. Most early comparisons, however, examined stands where the observed differences in the development of soil profiles occurred over centuries rather than within the tenure of the current stand. Some monocultures of spruce showed declining productivity, but this was often due to litter raking and not to the species planted (Baule and Fricker, 1970). Stone (1975) summarized the situation and concluded that the effects of different species on soils had not been well demonstrated. Research in subsequent years has supported some ideas and refuted others.

The challenge of understanding the influences of various tree species on soils can be summarized by considering Figure 11.1 and Figure 3.1. In a set of old-growth, natural forests with varying species composition in Austria, the concentration of carbon in the A horizon differed by a factor of 4; N concentrations differed by a factor of 3; and pH differed by up to 3.4 units. These old forests developed on various parent

---

Ecology and Management of Forest Soils, Fourth Edition. Dan Binkley and Richard F. Fisher.
© 2013 John Wiley & Sons, Ltd. Published 2013 by John Wiley & Sons, Ltd.

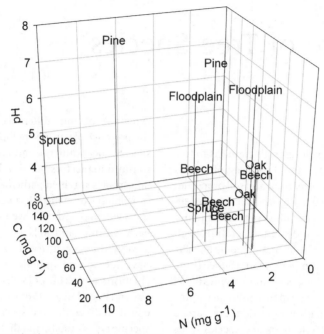

Figure 11.1 Combinations of soil carbon, nitrogen, and pH for 12 old-growth forests in Austria resulting from a variety of influences on soil formation, including elevation (temperature), precipitation, parent materials, and species (from data in Zechmeister-Boltenstern *et al.*, 2005). The success of individual species depends in part on these same factors, and tree species may shape soils differently, posing a major puzzle for soil-formation studies.

materials, over a span of 800 m in elevation, and a three-fold range in precipitation (Zechmeister-Boltenstern *et al.*, 2005). Several major factors of soil formation (Figure 3.1) likely influenced the differences among these soils, and it would not be sensible to attribute the differences solely to the influence of beech, pine, or spruce species.

The diversity of soil-forming factors can be partially constrained by restricting sites to a single region, at a single topographic position (such as flat, well-drained slopes), and similar soil textures. In one study, this approach revealed that we might expect soil N supply to differ by almost two-fold under the influence of tree species (Figure 11.2). Even though this approach removes some of the potential confounding factors of soil-forming factors, the possibility remains that somewhat subtle, pre-existing differences among microsites led to the pattern of successful establishment of each species, leaving the effect of tree species potentially confounded with some other soil factors.

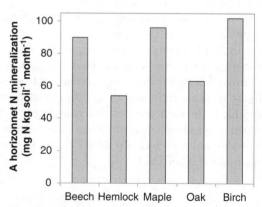

Figure 11.2 The availability of nitrogen (gauged as the net mineralization of N during laboratory incubations) differed by up to 90% within a set of similar sites in the Catskill Mountains of New York, USA, sampled with six plots dominated by each species. Hemlock has significantly lower rates of N supply than beech or birch (from data in Lovett *et al.*, 2004).

A similar approach was used by Dijkstra *et al.* (2003) to examine soils in a single watershed containing several species of large, old trees. The soils developed over the past 10 000 years or so, following glacial melting. Mineralogy differed substantially between soils collected under sugar maple trees and hemlock trees. Plagioclase feldspar comprised about 16% of the mineral particle surfaces (50–500 μm in size) in soils from beneath maple trees, and just 7% under hemlock. Similarly, biotite accounted for 9% of the very small mineral surfaces (10–50 μm in size) under maple, but just 0.5% under hemlock. This analysis revealed significant differences in the soils under each species, but were the minerals altered by the trees (in the past century or two), or did the suite of soil-forming factors over ten millennia lead to differences in soil properties that favored the successful establishment of maple on some sites, and hemlock on others? The answer might be "yes" to both. Frelich *et al.* (1993) found that sugar maple and hemlock tended to reinforce differences in soils (at a scale of 10–50 meters) in a mosaic where both species are dominant; over a 3000-year period, soils changed under the influence of each species in ways that favored future generations of the same species.

TREE SPECIES DIFFER SUBSTANTIALLY IN GROWTH RATES AND ALLOCATION ABOVE GROUND AND BELOW GROUND

The accumulation of organic matter in soils may relate, in part, to the rate at which trees fix carbon from the atmosphere. How different are the growth rates of species? Planting species in pure plots at single locations can gauge typical magnitudes. A case study experiment with five species was planted in Costa Rica (Figure 11.3; Fisher, 1995a; Russell *et al.*, 2010). The accumulation of biomass (above ground and below ground) and the current rate of production show differences among species on the order of 50% to more than two-fold. The long-term influence of tree species on soils would also be mediated by a range of factors, including the chemistry of the organic materials created by the trees, the influence of quantity and chemistry of plant materials on soil animals and microbes, and other factors that influence the long-term stabilization of C in soils (as discussed in Chapter 4).

Another common garden experiment in Ontario, Canada showed that 30 years of soil development in adjacent plantations of 30-year-old white pine and red oak showed a much greater accumulation of organic matter in the forest floor (O horizon)

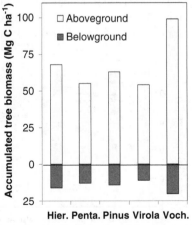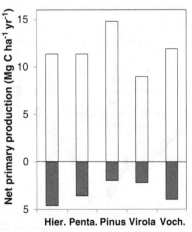

Figure 11.3 Plantations of replicated plots of five species (*Hieronyma alchorneoides, Pentaclethra macroloba, Pinus patula, Virola koschnyi,* and *Vochysia guatemalensis*) in Costa Rica showed that, after 17 years, the range in aboveground tree biomass differed by up to about two-thirds among species, compared with about two-fold ranges in belowground biomass, and the current rates of net primary production (based on Russell *et al.*, 2010).

Figure 11.4 After 30 years in a common garden at the University of Toronto Observatory, white pine plots accumulated much larger O horizons than red oak plots, and earthworm-mixing of soils led to greater humus accumulation in the A horizon under oak.

under pine than under red oak, but much more accumulation of humus in the A horizon under oak (Figure 11.4).

Measurements of soil properties in common garden experiments show how much soil characteristics have diverged under the influence of different species, but a challenge remains for understanding (1) which processes drove the observed changes; (2) would these same processes occur under the same species on other types of soils; and (3) would the overall outcomes be the same?

Tree species might alter soils by modifying the amount of precipitation reaching the soil, or how long water remains in a soil. The leaf area of conifer canopies is often higher (year-round) than broad-leaved canopies, which might lead to differences in interception loss (with more rain or snow evaporating from the canopy surfaces of conifers, leaving soils drier) or transpiration loss (if transpiration is proportional to leaf area and duration). Augusto *et al.* (2002) summarized the European evidence for the percentage of bulk precipitation that is lost to canopy interception, and values were about 20–25% for broadleaved species and 35–40% for conifers. A comparison of tree species in a common garden in Denmark revealed this same influence: soil moisture was lower beneath Norway spruce throughout the year than under broadleaved species (Figure 11.5).

Leaf area duration may have a large effect on soil temperature in areas with cold winters, as illustrated for white spruce and sugar maple in

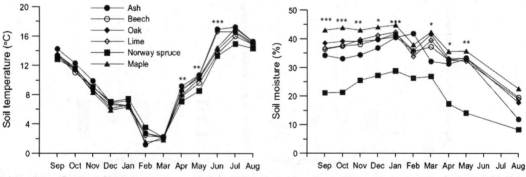

Figure 11.5 Influence of six tree species on soil temperature (left) and moisture (right) from six sites in Denmark. Reprinted from Forest Ecology and Management, 264, Lars Vesterdal, Bo Elberling, Jesper Riis Christiansen, Ingeborg Callesen, Inger Kappel Schmidt, Soil respiration and rates of soil carbon turnover differ among six common European tree species, 185 - 196, Copyright 2012, with permission from Elsevier.

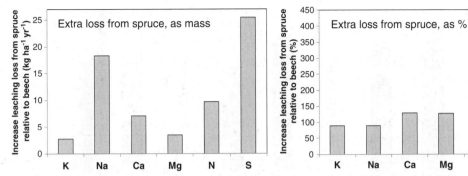

Figure 11.6 A comparison of five sites in France, Sweden, and Germany showed higher losses of all elements from Norway spruce stands compared with beech stands (from data in Augusto *et al.*, 2002), illustrated as the extra mass lost (left) and as a percentage of the loss rate in beech.

Figure 5.8. Where winter temperatures are less severe, as in Denmark, the influence of tree species on soil temperature may be slight (Figure 11.5).

Elemental losses in leachate and streamwater can differ substantially under the influence of different species, as illustrated by comparison of Norway spruce and beech (Figure 11.6). Losses from spruce forests tended to be two to four times the loss rates under beech forests, as a result of greater atmospheric deposition on large leaf-area canopies of spruce, and perhaps increased rates of weathering (perhaps indicated by higher leaching of sodium, which is not an essential tree nutrient).

COMMON GARDEN EXPERIMENTS SHARE SOME COMMON LIMITATIONS

Common garden plantations have some important limitations, as might be expected whenever complex ecological interactions are controlled and simplified to examine individual factors. These plantations usually represent artificial conditions, sometimes incorporating the effects of former agricultural practices or the introduction of exotic species. The rates of divergence seen in common garden experiments are probably near the upper bound of the rates of change that would occur under normal forest development, where changes in rates of growth and nutrient cycling would take longer to develop.

Another important limitation of common gardens is their small population of inference. When all replicates of an experiment are nested within a single site, the statistical population of inference is just a single site (Binkley, 2008). The demonstrated effects of species may or may not be consistent across the variety of soil-forming factors that apply at other locations. This limitation can be overcome by spreading replicate plots across larger geographic gradients, providing the chance to see if species influences are strong enough to rise above other sources of variation.

An experiment in Michigan, USA sought to determine the effects of planted red pine trees versus naturally regenerated maple and oak trees on soil C accumulation (Gahagan *et al.*, 2012). Adjacent plots of each type were chosen, but rather than nest all pairs within a single stand, the sampling plots were chosen from stands spread across 600 ha. Pines accumulated much larger O horizons, but C accumulation in the 0–50 cm mineral soil did not differ by species. Combining both horizons gave an average difference of about $1 \, kg \, C \, m^{-2}$, or a rate of divergence over 60 years of $17 \, g \, C \, m^{-2} \, yr^{-1}$. This difference in C content was too small to warrant strong confidence; either there is not a notable effect of species on C accumulation across the 600 ha population of inference, or more samples would be needed to detect the difference with confidence.

A similar approach was used over two larger areas in Austria (Berger *et al.*, 2002; Figure 11.7). The

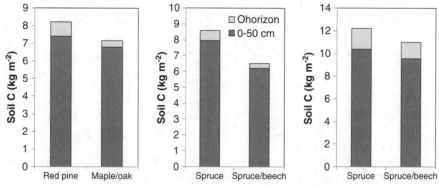

Figure 11.7 Experimental designs that spread replicates across landscapes give confidence that soil influence of species is robust relative to other soil-forming factors. The comparison of red pine with maple and oak is from eight pairs of plots across 600 ha in Michigan, USA (Gahagan *et al.*, 2012). The middle graph of spruce and spruce/beech is from nine pairs of stands on sandstone/marl parent materials across more than 10 000 ha in Austria (Berger *et al.*, 2002). The graph on the right is from the same study, but with nine pairs of plots on sandy sediments from a population of inference of a few thousand ha.

spruce monocultures had 2.1 kg C m^{-2} more than spruce–beech mixtures when growing on sandstone/marl parent materials (a divergence rate of about 28 g C m^{-2} yr^{-1}). The same comparison on soils derived in sandy sediments showed spruce soils had 1.2 kg C m^{-2} more than spruce–beech soils (a divergence rate of about 16 g C m^{-2} yr^{-1}). How confident should we be that more C accumulated under spruce monocultures? The authors did not test the overall species effect in this study, though they did apply statistical tests to the effect of species within each parent material type. The species effect was highly significant for the sandstone/marl subset of sites, but not for the sandy sediment subset. The consistent trend across both soil types might indicate an overall increase in soil C under spruce, but a statistical calculation with the original data would be needed to test for the overall species effect.

Common garden experiments have supported some of the classic generalizations (such as soil acidification under spruce relative to hardwoods), but not others (such as greater N availability under hardwoods than under conifers; Table 11.1).

Tree species may differ in their effects on soils by a suite of mechanisms (Table 11.2). How substantial would the combined effects of these processes be? It's very common for tree species to differ by 50% in the mass of litterfall added to the O horizon annually, and in about one-third of common

garden studies, the difference is more than two-fold (Figure 11.8). The N content of litterfall typically differs by up to 40% between species (not including N-fixing species). The pattern among species for O horizon mass mirrors the pattern for litterfall, with 50% differences being common and some cases showing several-fold differences. Net N mineralization shows an interesting pattern of gradually declining frequency of larger differences; about one-third of the species effects are 50% or less, and about one-third are greater than two-fold. Given the prevalence of N limitation on tree growth, this tree-species influence on N supply may be quite important. Differences in soil pH of 0.3 units are very common, and differences > 0.5 units are not rare. Curiously, base saturation usually differs by 50% or more among species, accounting in part for the patterns in soil pH.

Somehow soil acidity has been linked with soil fertility in much of the literature. This expectation probably came from agricultural soils, where amelioration of acidity often increases crop production. However, most tree species that occur on acidic soils are well adapted to acidic soils, and the relationship from agriculture may not apply to forestry (except where acid-intolerant species are planted on acidic soils). Do changes in forest soil pH affect changes in net N mineralization? The common garden data do not support this idea. Norway spruce almost always

Table 11.1 Some proposed generalizations regarding tree–soil interactions, and the extent of supporting evidence from common garden plantations (modified from Binkley, 1995; Binkley and Giardina, 1998).

Generalization	Evidence
Norway spruce acidifies soils by accumulating strongly acidic organic matter	Almost a dozen common garden studies support part or all of this generalization.
Norway spruce degrades soils, particularly in comparison with beech	No evidence supports this view. Aboveground net primary production is higher in spruce forests than in beech; studies of adjacent stands typically show equal or greater net N mineralization and porosity under spruce.
White pine may increase soil N availability	Three common garden experiments found notably higher net N mineralization under white pine.
N-fixing trees increase soil carbon, and rates of cycling of all nutrients	More than a dozen studies documented higher soil C and rates of nutrient cycling in litterfall under N fixers than under non-N-fixers.
Hardwoods promote soil N availability relative to conifers	Mixed evidence; at least some cases show equal or higher net N mineralization under conifers such as white pine (relative to green ash), larch, or Norway spruce (relative to beech).
"Mull" O horizons indicate more fertile conditions than "mor" O horizons	Probably supported regionally (mor O horizons are found on the poorest sites), but not locally (mull O horizons may indicate lower availability of P, for example).
Increases in soil C represent increased soil fertility	Some evidence, though untrue for Histosols. Common garden experiments do not show greater growth for species that develop greater soil C; but no "second generation" reciprocal plantings have assessed the legacy of increased C for soil fertility for more productive species.
Lignin:N of litterfall is a good indicator of net N mineralization in soil	Moderately strong evidence.

develops the lowest pH among species in common gardens, yet net N mineralization under Norway spruce tends to match or exceed rates beneath other species. The landscape-scale experiment illustrated in Figure 11.2 showed no correspondence between differences in pH beneath each species and soil N supply. Similarly, Prescott *et al.* (1995) examined differences in soil N availability and pH in a common garden with 11 tree species. The trend between these soil properties was far from significant, and the data would support *declining* N availability with increasing pH more than they would any increase in N availability (Figure 11.9).

Chapter 8 described how the pH of two soils may differ, as a result of differences in the ionic strength of soil solutions, the degree of dissociation of the soil

Table 11.2 Mechanisms by which species differ in their effects on soils (modified from Binkley and Giardina, 1998).

Process	Common Size of Effect
Atmospheric deposition	Up to doubling of deposition of S, N, and H^+ in polluted areas
Nitrogen fixation	Input of 50 to 150 kg N ha^{-1} yr^{-1} for symbiotic N-fixing species
Mineral weathering by exudates, microflora	Mostly unknown, but some evidence of 0.3 to 1.5 kmol$_c$ ha^{-1} yr^{-1} greater input of base cations; higher P supply
Quantity, quality of C compounds added to soil	30%+ difference in quantity, large differences in quality, such as two-fold range in lignin:N
Soil communities	Populations, ratios of functional groups, etc., commonly differ by several-fold
Physical properties (temperature, water content, structure)	Moderate to large differences
Pedogenesis (such as podzolization)	Some notable case studies of horizon development (especially E and Bs horizons)

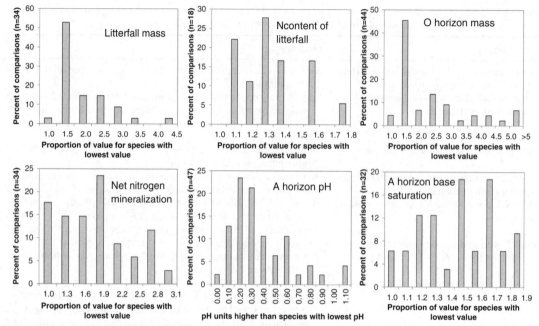

Figure 11.8 Lowest values reported within common gardens were set at 1.0, and values for other species calculated as proportions of this lowest value. For pH, species effects were tabulated in pH units above the lowest value found for any species within the common garden (modified from Binkley and Giardina, 1998; with additional comparisons from Menyailo *et al.*, 2002; Hagen-Thorn *et al.*, 2004; Reich *et al.*, 2005; Hansen *et al.*, 2009; Hansson *et al.*, 2011).

acid complex (~ base saturation), and the quantity and strength of soil acids. The relative importance of these factors has been evaluated in a few common garden studies (Table 11.3). The strength of soil acids and the degree of dissociation seem to be the most important factors, and in some cases opposing changes in these factors can actually prevent a difference in pH between tree species.

Conifers generally are thought to produce more acidic soils than hardwoods on similar sites. This is probably a bad generalization, and it may be more productive to characterize the factors that influence soil acidity and discuss how these differ among species and sites. About half of the studies of nitrogen-fixing species have shown significant soil pH declines, and about half have not.

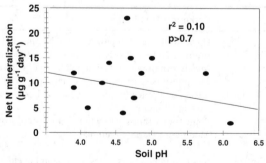

Figure 11.9 Soil net N mineralization from 11 species in a common garden species trail at Clonsast Bog, Ireland showed no relationship with the soil pH that developed under the influence of each species (modified from Prescott *et al.*, 1995).

Table 11.3 Explanations for how tree species differ in soil pH.

Comparison	Difference in pH	Relative Ranking of Factors	Reference
50 yr Norway spruce vs. white pine	Spruce pH 3.8 Pine pH 4.2	Strength > Dissociation > Quantity	Binkley and Valentine, 1991
50 yr Norway spruce vs. green ash	Spruce pH 3.8 Ash pH 4.6	Strength > Dissociation > Quantity	
50 yr white pine vs. green ash	Pine pH 4.2 Ash pH 4.6	Dissociation > Strength > Quantity	
8 yr *Eucalyptus* vs. *Falcataria*	*Eucalyptus* pH 4.5 *Falcataria* pH 4.3	Dissociation > Strength > Quantity	Rhoades and Binkley, 1995
55 yr old conifer/alder vs. conifer-rich site	Conifer pH 3.9 Alder/conifer pH 3.6	Dissociation > Strength > Quantity	Binkley and Sollins, 1990
55 yr old Douglas fir/alder vs. Douglas fir	Douglas fir pH 4.0 Alder/Douglas fir pH 4.0	Increased dissociation under alder offset by increased acid strength	
50 yr old Douglas fir vs. red alder	Douglas fir pH 5.2 Red alder pH 5.0	Quantity > Strength > Dissociation	Calculated from Van Miegroet *et al.*, 1989

The processes that influence soil pH under nitrogen-fixing trees can be illustrated with some details from a comparison of 55-year-old stands of red alders and conifers in Washington State. The pH of soil 0–15 cm in depth (measured in distilled water suspension) under the alder–conifer stand at the less fertile site was 5.1, compared with 5.4 in the pure conifer stand (Binkley and Sollins, 1990). The same comparison in dilute calcium chloride suspension showed no differences between species. The cause of the difference in water, but not in salt, was the higher concentrations of soluble ions in the mixed stand. The levels of soluble anions were almost twice as high in the soil from the alder–conifer stand, and of this extra anion charge, about 3% of the balancing cation charge came from H^+; this modest increase lowered soil pH (measured in water) by 0.3 units. Measurement in dilute salt suspensions removed this effect of ionic strength and showed no difference between stands. This analysis showed no effect of species on soil pH, but that doesn't mean that the species did not differ in their effects on the factors that influence soil pH (see Chapter 8). The base saturation was higher in the 0 to 15 cm soil depth in the alder–conifer stand than in the conifer stand, and higher base saturation indicates that the exchange complex is "more dissociated" as an acid and should maintain a higher pH in the soil solution. To have a higher base saturation without a higher pH, some other factor must have worked to lower the pH. In this case, the acid of the organic matter in the mixed stand was stronger than that of the conifer stand.

In another case study, a 50-year-old stand of red alder had a pH in the A horizon of 5.0, in contrast to a pH of 5.2 for an adjacent plantation of Douglas fir (Van Miegroet *et al.*, 1989). Titrations of the soils showed the same ANC (about the same quantity of base cations) but much larger BNC (acid quantity) under alder (Figure 11.10). The dissociation of the acid complex was 41% in Douglas fir soil but only 29% in red alder soil. If the alder soil had the same total quantity of acid as the Douglas fir soil, its pH would have been higher (5.2). If the acid strength of the alder soil were adjusted to match that of the Douglas fir soil, the pH would also have risen to about 5.2. If the acid quantity in the Douglas fir soil were increased to match that of the alder soil, the pH would have dropped from 5.2 to about 4.8. If the acid quantity in the Douglas fir remained the same

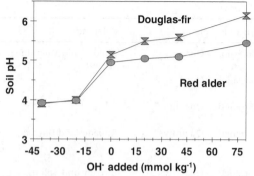

Figure 11.10 Titration curves for soils from adjacent stands of Douglas fir and red alder (data from Van Miegroet *et al.*, 1989). Both species had similar ANC to pH 4, representing the same level of exchangeable base cations. Alder soil had twice the BNC to pH 5.5, indicating much larger pools of soil acids.

the biogeochemical mechanisms by which species alter soils. A common garden experiment in Wisconsin, USA showed net N mineralization was more than twice as high under larch as under other species (Figure 11.11). The difference corresponded to a very low ratio of lignin:N in aboveground litterfall, which might explain rapid decomposition and N turnover. However, the pattern was just as strong (or stronger) for fungal biomass as a driver of net N mineralization. A strong (inverse) covariance between fungal and bacterial biomass indicated that bacterial biomass would be equally related to the pattern of species influence on N mineralization. The multiple covarying factors still prevent clear insights about direct causes for any observed effects.

but the strength matched that of the red alder soil, the pH would have dropped to 5.0. These calculations indicate that the most important difference between these soils was the quantity of acid in the soil (which altered the degree of dissociation), with the greater acid strength playing an additional role.

Determining the pattern of tree species effects on soils is a first step that can be followed by examining

TREE SPECIES INFLUENCE SOIL ANIMAL COMMUNITIES

Soil animal communities are complex webs that respond very strongly to tree species and the organic matter the trees add to the soil. An example of the sorts or patterns that might be found comes from an afforestation experiment in Siberia, where several tree species were planted in a former agricultural

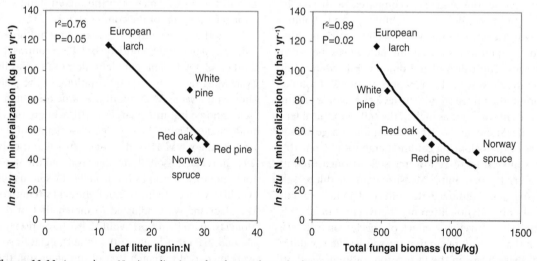

Figure 11.11 Annual net N mineralization related strongly to the lignin:N of aboveground litterfall, but also to fungal biomass. These sorts of correlations can form an interesting basis for new hypotheses, but they cannot establish cause and effect. (from data of Son and Gower, 1991; Scott, 1998).

field. After ten years, the density of large inverte-brates (worms, beetles, ants, fly larvae, and others) ranged from about 60 to $130\,m^{-2}$ (Figure 11.12; Bezkorovaynaya, 2005) compared with $40\,m^{-2}$ in the adjacent agricultural field. Twenty years later, the population density more than doubled for some species, and dropped by half in another. A common garden experiment in Poland also found huge dif-ferences in the density of small invertebrates (enchytreids and collembola) among tree species.

Some soil properties (such as clay content and % C) are very stable through a growing season, but biological properties such as animal populations differ substantially as a result of changes in temper-ature, moisture, and overall biological activity of the food web. Properties that are sensitive to environ-mental changes through seasons may need to be sampled multiple times within a year before we can be confident that apparent differences between years are much larger than any potential bias from sampling during different seasons (or recent weather conditions). Do the apparent changes in large invertebrate populations in the Siberian experiment indicate true changes at a decade scale, or might the season/environment conditions at the time of sampling mistakenly confound seasonal and decadal trends? The Polish study, for example, found that populations of enchytreids and collem-bola fluctuated by several-fold between samplings

in spring and autumn. Another important source of variation is the depth of sampling. The Siberian study found that 80% of the large invertebrates were found in the 0–20 cm mineral soil, with the O horizon and the deeper mineral soil accounting for the remaining 20%. Soil animals move up and down in soil profiles in response to temperature and moisture, so this distribution may shift seasonally.

Differences in animal populations can be meas-ured in common gardens, but the biogeochemical significance can remain murky. For example, an experimental plantation in Hawaii examined the effects of eucalyptus and an N-fixing leguminous tree (see later in this chapter). The N-fixing species enriched the total N capital of the soil and dramati-cally increased N availability while decreasing P availability (Garcia-Montiel and Binkley, 1998; Kaye *et al.*, 2000). At one sampling period, the density of small earthworms was about $90\,m^{-2}$ under eucalyptus, compared with $470\,m^{-2}$ under the N-fixing trees (Zou, 1993). As the worms feed on litter, we may reasonably expect that the N-fixer's litter was higher quality worm food. What might the effect of such a large population of worms be on the turnover of N and P? Would the apparent effects of the tree species be different if the exotic worm species had not been present?

The difficulty of interpreting the importance of animal populations is evident in the Polish common

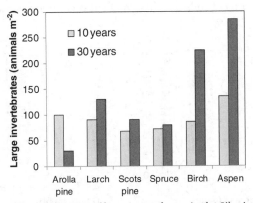 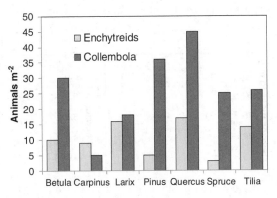

Figure 11.12 Density of large invertebrates in the Siberian afforestation experiment (left) differed by more than two-fold across the species, and also over stand development (data from Bezkorovaynaya, 2005). The density of small enchytreids and collembolla in a common garden experiment in Poland showed several-fold difference among tree species (Rożen *et al.*, 2010).

garden study mentioned above. The biomass of the O horizon declined (across tree species) as earthworm densities increased (Figure 11.13; Reich *et al.*, 2005). Given that earthworms are major mixing agents of soils, this pattern would be expected if trees differed as suitable providers of earthworm food. Curiously, the pH of the O horizon increased with increasing earthworm density. Does this reflect a change in acidity for material passing through earthworm guts, an effect of mixing mineral particles with organic material, or perhaps the original chemistry of tree litter that also happens to influence both decomposition and pH in the O horizon? To complete the circular story, the pH of the O horizon also correlated with O horizon mass. Unavoidable feedback in ecosystems leads to circles of cause, effect, and covariation.

THE INFLUENCE OF TREE SPECIES MIGHT BE EXAMINED AS A FUNCTION OF LITTER CHEMISTRY RATHER THAN SPECIES IDENTITY

A major step in developing generalizable insights about species effects on soils may come from studies that intentionally vary the chemistry of litter inputs by selecting and arranging species in pure and mixed plantings. Litterfall inputs (as well as root inputs) in the middle of a monoculture plot would be influenced by a single tree species, but at the boundaries between plots, the influence of

neighboring species blends in. Mixed-species plots have the same sort of variation in species influence, but at a smaller (single-tree) scale. A combination of spatially explicit sampling with a litter quality characterization (perhaps including the balance among nutrient elements, or "stoichiometry") may be promising (see Hättenschwiler *et al.*, 2008; Vivanco and Austin, 2008; Marichal *et al.*, 2011). A particular challenge to this approach might be the divergence of chemistry and decomposition between leaf litter and fine roots within the same species. Hobbie *et al.* (2010) examined these aspects in the Polish common garden experiment, and found that leaves and roots may decay at the same rate for some species, or leaves may decompose at twice (or half) the rate of roots of the same species.

CHANGES IN SOILS DO NOT LAST FOREVER

A major remaining question about the effects of different species on soils is how long these effects will last after the original species is replaced by another. Will these changes have major legacies, or will the species effects evaporate because they're ephemeral features that disappear quickly as soils adjust to new species? Would a species that doubles the rate of soil N supply leave a legacy of high N for the next generation of trees or would the legacy be very short lived? Laboratory incubations of soils from the Wisconsin common garden

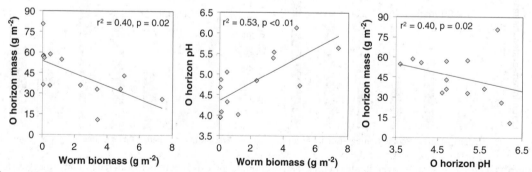

Figure 11.13 Covarying patterns of earthworm (dry matter) mass, O horizon mass, and O horizon pH in the Polish common garden experiment illustrate why circular feedback makes it difficult to assign simple cause and effect in the study of soil ecology. (from data of Reich *et al.*, 2005).

trial showed a five-fold difference in net N mineralization among species after 30 days (Scott, 1996). After 300 days, the differences among species had shrunk to less than two-fold. The laboratory incubations suggest that the apparent effect of species may be most pronounced on the relatively small, rapidly cycling pools of N, with less influence on the larger, more stable pools. Field tests are clearly needed to determine the robustness and longevity of tree species effects.

NITROGEN-FIXING SPECIES REDUCE N FROM THE ATMOSPHERE AND ENRICH SOILS

The nitrogen-fixing enzyme, nitrogenase, reduces atmospheric nitrogen to ammonia through addition of electrons and hydrogen ions:

$$N_2 + 8e^- + 8H^+ \rightarrow 2NH_3 + H_2$$

Because electrons are added to N_2, this is a reduction process, requiring energy input. The production of hydrogen gas represents an unproductive energy drain, and some N-fixing organisms have a hydrogenase system that oxidizes hydrogen to form water and release energy. The theoretical energy requirement is about $35\,kJ\,mol^{-1}$ of N fixed, but it is difficult to assess the actual cost under field conditions. A good approximation of the actual cost might be about 15 to 30 g of carbohydrates (6 to 12 g C) for every gram of ammonia N produced (Schubert, 1982; Marshner, 1995). About half the carbohydrate cost is for the N-fixation process, and the other half is for the growth and maintenance of nodules and for assimilating and exporting the fixed ammonium. The nitrogenase enzyme is also very sensitive to oxygen, and the high energy requirements coupled with the need for oxygen protection set limits on N-fixing systems. Interestingly, a substantial amount (25–30%) of the CO_2 released during the N-fixation process can be "refixed" inside the root nodules by PEP carboxylase.

A wide range of tree species have evolved an association with nitrogen-fixing bacteria, including many legume family genera (such as *Acacia, Robinia, Falcataria, Albizia*), and "actinorhizal" species of *Alnus, Eleagnus, Casuarina,* and others. The bacteria infect roots, causing formation of nodules where bacteria proliferate, consume carbohydrates (while protected by low oxygen partial pressures), and fix nitrogen. The fixed nitrogen is exported to the trees as amino acids, and used in all growing tissues. When tree tissues die, the N enters the decomposition process, and may be taken up by the roots and mycorrhizae of all plants in the soil.

Rates of nitrogen fixation vary among species, sites, and over time, and the local details of these factors need to be considered for each case. Overall, the major N-fixing species that have been studied intensively tend to fix about $75\,kg\,N\,ha^{-1}\,yr^{-1}$, though rates half this large, or twice this large, are not uncommon (Figure 11.14). To put these rates in context, $75\,kg\,N\,ha^{-1}\,yr^{-1}$ would equal 10 to 25 times the annual rate of input from atmospheric deposition, and about the same rate at which N circulates from soil into trees through the course of one year.

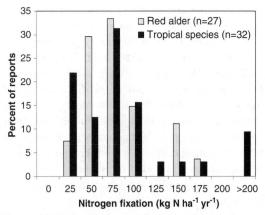

Figure 11.14 Reported rates of nitrogen fixation for forests dominated by red alder in the Pacific Northwest (data from Binkley *et al.*, 1994), and for a wide variety of tropical species (data from Binkley and Giardina, 1997, and Garcia-Montiel and Binkley, 1998; Pearson and Vitousek, 2001; Forrester *et al.*, 2007; and Bouillet *et al.*, 2008 – including original data and summarized tropical forest cases).

NITROGEN FIXATION ACCELERATES NITROGEN CYCLING

What happens to the nitrogen once fixed? The ammonia is rapidly aminated to become part of organic amino acids and proteins. In free-living systems, the bacteria and cyanobacteria simply use the N as any other microbe would – to grow and synthesize new cells. In symbiotic systems, the fixed N is shipped from the nodules into the plant roots, where it performs the usual functions. Soil enrichment from N fixation comes only after microbe and plant tissues have died and decomposed to release the extra N.

Increases in N availability may have a positive feedback that stimulates N mineralization. This is one of the major benefits of mixing N-fixing species with crop trees, and can be illustrated in mixed stands of alders and conifers. Portions of a Douglas fir plantation on Mt. Benson in British Columbia, Canada, contained shrubby Sitka alder, red alder, or no alder (Binkley, 1983; 1984). The site index in the absence of alder was low, about 24 m at 50 years. Sitka alder fixed only about $20 \, kg \, N \, ha^{-1} \, yr^{-1}$, but after 23 years the N content of litterfall was $110 \, kg \, N \, ha^{-1} \, yr^{-1}$ with Sitka alder, compared to only $16 \, kg \, N \, ha^{-1} \, yr^{-1}$ without alder. The N fixation rate of red alder was about twice that of Sitka alder, and the N content of litterfall was about $130 \, kg \, N \, ha^{-1} \, yr^{-1}$. What accounted for this increase in N cycling? The N content of alder leaves (and alder litter) was very high, about 3% of the dry weight, as compared with less than 1% for Douglas fir needles. Most litter immobilizes N from the surrounding soil in the initial stages of decomposition, but N-rich litter begins mineralizing N immediately. Ecosystems with N-fixing species enjoy both greater inputs of N and produce litter rich in N; these two factors combine to greatly accelerate N cycling and availability.

NITROGEN FIXATION RATES ARE DIFFICULT TO MEASURE

Nitrogen fixation rates can be estimated by a variety of methods, with four approaches used commonly in forest research: ^{15}N methods, N accretion, chronosequences, and acetylene reduction assays. The addition of highly enriched ^{15}N to soils may provide the best estimates of N fixation, but the cost of sufficient ^{15}N to label soils in the field has limited the use of this method for trees. Some processes in N cycling alter the ratio between naturally occurring ^{15}N and ^{14}N, and if certain key assumptions are met, this "natural abundance" approach can provide inexpensive estimates of N fixation. Unfortunately, the key assumptions are probably not met in many (if any) forests. Nitrogen accretion is also a simple approach, involving measuring the total N content of an ecosystem at two points in time. The difference between samplings represents the addition of N from fixation, plus any precipitation inputs. This accretion method assumes outputs are negligible, and underestimates true N fixation rates on sites that experience high nitrate leaching or denitrification. The chronosequence approach is really a variation of the accretion approach but with more assumptions (see also Chapter 16). Several sites of different ages are used to represent trends to be expected within one site over time. If all site factors are constant except for stand age, then spatial patterns can be interpreted to represent temporal patterns. Finally, the acetylene reduction technique takes advantage of the nitrogenase (N-fixing) enzyme's preference for reducing acetylene to ethylene (both of which are easily measured) rather than dinitrogen to ammonia (not measurable without ^{15}N methods).

^{15}N LABELING SHOWS HIGH RATES OF N FIXATION FOR *LEUCAENA* AND *CASUARINA*

Parrotta *et al.* (1996) used the ^{15}N addition method to estimate the rate of nitrogen fixation in planted pure and mixed replicated plots of *Eucalyptus*, *Leucaena*, and *Casuarina*. Small subplots were labeled with ^{15}N, so that N derived from the soil would be substantially enriched in ^{15}N relative to N derived from the atmosphere. The ratio of $^{15}N:^{14}N$ in *Eucalyptus* trees represented the isotope ratio of N derived from the soil, and the lower enrichment in the N-fixing trees allowed separation of the N-fixers' reliance on the atmosphere and soil. This approach showed that *Leucaena* fixed about 90% of its N supply through age 2, and about 60% of its N supply by age 4; *Casuarina* derived about 50 to 60% of its N from the atmosphere for both periods. This decline in N derived from the atmosphere for *Leucaena* was only a relative decline; the quantity of N fixed remained steady at about 70 to $75\,kg\,N\,ha^{-1}$ yr^{-1}, but the uptake of N from the soil increased as the tree demand for N increased.

^{15}N NATURAL ABUNDANCE IS AFFECTED BY N FIXATION

In the atmosphere, ^{15}N comprises about 0.3664% of the N_2, and ^{14}N comprises 99.6336%. The isotope ratio in soils is typically enriched in ^{15}N because the minor difference in mass of the N atoms favors the reaction, mobility, and loss of lighter ^{14}N. If a soil is sufficiently enriched in ^{15}N relative to the atmosphere, it might be possible to use the natural abundance of N isotopes to estimate the input of N from fixation.

Natural abundance levels of ^{15}N are often gauged as delta ^{15}N (or δ^{15}N), which equals the parts per thousand deviation of a sample from the percentage of ^{15}N in the atmosphere. A soil sample that contained 0.3710% ^{15}N would have a δ^{15}N of 4.6 ((0.3710 − 0.3664) * 1000). As a first approach, the contribution of the atmosphere to the N economy of an N-fixing tree could be estimated by comparing the δ^{15}N of foliage with the δ^{15}N of total soil N. Reality is actually messier: the δ^{15}N within trees may differ by several per mil from the roots to the leaves, and the δ^{15}N of soil total N often increases by several per mil with depth. In addition, the δ^{15}N of soil ammonium and nitrate commonly do not match that of the total soil N, or each other. One approach to circumventing these problems is to use a non-N-fixing reference tree to gauge the δ^{15}N of available soil N. For example, total soil N in a pure Douglas fir plantation had a δ^{15}N of 5.8, significantly higher than the 2.3 δ^{15}N for a mixed stand of red alder and Douglas fir (Binkley *et al.*, 1985). Accounting for the higher

%N in the mixed stand soil, these data suggest that the alder had contributed about 60% of the N present in the 0–15 cm soil depth. The confidence in this pattern had to be tempered, however, with data from two other sites where the δ^{15}N of total soil N was too close to the atmospheric level to be useful, and by a fourth site where the δ^{15}N in the mixed alder–Douglas fir stand was *more* enriched in ^{15}N. Högberg (1997) suggested that differences in δ^{15}N might need to be greater than 5 per mil before interpretations about N fixation could be made with confidence.

A meta-analysis of nitrogen-fixation studies that used ^{15}N natural abundance indicated that leguminous trees show fairly high variation in the proportion of N obtained from the air (via fixation) and that obtained from the soil (Andrews *et al.*, 2011). Actinorhizal species were notably more consistent in obtaining the major portion of their N from the air. Curiously, legumes were more likely to shift to reliance on soil N on more fertile sites, whereas actinorhizal trees continued to rely heavily on N fixation regardless of soil fertility.

Bouillet *et al.* (2008) investigated rates of N fixation by *Acacia mangium* in pure and mixed-species plots, using several approaches. Based on dilution of labeled ^{15}N added to the soil early in the plantation, they estimated the acacia fixed about $30\,kg\,N\,ha^{-1}\,yr^{-1}$ when mixed 1:2 with eucalyptus trees. The natural abundance patterns suggested that nitrogen fixation was double this rate, indicating again that natural abundance approaches are too imprecise for reliable quantitative estimates.

NITROGEN ACCRETION AND CHRONOSEQUENCES CAN DETECT MODERATE AND LARGE N FIXATION RATES

These two approaches can be illustrated by several research projects on a nitrogen-fixing shrub, snow-brush (*Ceanothus velutinus*). Youngberg and associates intensively characterized the N pools of a recently logged site, and resampled for N accretion several times over the next 20 years. Zavitkovski and Newton (1968) chose a chronosequence approach in hopes that differences among sites of varying ages would track the N accretion occurring within each site, providing a rapid estimate of N fixation. Youngberg *et al.* (1976; 1979) found that N accumulated in vegetation + soil (to 23 cm) at a rate of about 110 kg N ha^{-1} yr^{-1} for the first ten years, and 40 kg N ha^{-1} yr^{-1} for the next five years. Zavitkovski and Newton (1968) found a very similar pattern (in N content of vegetation + soil to 60 cm depth), despite high variability among sites (Figure 11.15). A regression equation of ecosystem N content with age shows a significant slope in the chronosequence series of sites of about 80 kg N ha^{-1} yr^{-1}. A 95% confidence interval around this relationship ranges from 20 to 140 kg N ha^{-1} yr^{-1}. However, Zavitkovski and Newton found few nodules on any plants younger than five years, and concluded that the three youngest stands were probably unrepresentative of the rest of the sequence. Deleting these stands, they concluded that N fixation was negligible in the chronosequence.

The chronosequence method can be converted into the accretion method if sites are resampled at a later date. Hopefully, each ecosystem would progress along the trajectory established for the chronosequence. Using this approach, six of the original 13 sites examined by Zavitkovski and Newton were resampled 15 years later. Five sites showed no significant change in soil (0 to 15 cm depth) N content, but one showed a very significant increase of 280 kg N ha^{-1} (17 kg N ha^{-1} yr^{-1}; Figure 11.16). Some N accretion may have occurred in other ecosystem pools, but it is unlikely that high rates of N fixation took place without evidence of increased N in the upper soil. This resampling showed that the apparent trend in the original chronosequence was not, in fact, a reliable estimate of N fixation for other sites, and that the sites in the chronosequence probably differed fundamentally from the single site sampled by Youngberg. As with most features of forest soils, site-specific details were important in addressing the rates of N fixation by snowbrush.

ACETYLENE REDUCTION ASSAYS ESTIMATE CURRENT RATES OF N FIXATION

The fourth approach to estimating rates of N fixation also involves large sources of variation. If the average rate of N fixation per gram of nodule (activity rate) were known, an estimate of the annual N fixation rate could be obtained by multiplying the rate by the nodule biomass per hectare. Nodule activity usually is assessed with an acetylene

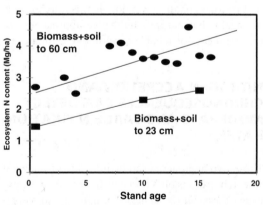

Figure 11.15 N-fixation estimates for snowbrush based on a chronosequence (ovals, based on Zavitkovski and Newton, 1968), and at a nearby site where accretion was followed within a single site over time (squares, based on Youngberg *et al.*, 1979).

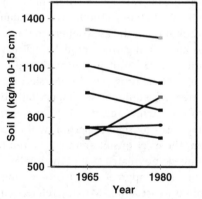

Figure 11.16 Fifteen-year changes in total soil N (0–15 cm depth) in six stands sampled by Zavitkovski and Newton (1968), resampled in 1980 (Binkley and Newton, unpublished data). Only one of the six sites showed a significant accretion of N; overall, the average annual rate of accretion across all sites was −0.7 kg N ha^{-1} yr^{-1}.

reduction assay, where the nitrogenase enzyme reduces acetylene to ethylene rather than reducing nitrogen gas to ammonia:

$$C_2H_2 + 2H^+ + 2e^- \rightarrow C_2H_4$$

Acetylene and ethylene are easier to measure than the nitrogen compounds. Recalling that N fixation requires the addition of 6 e^- for every mole of N_2 consumed, it would appear that the reduction of 3 moles of acetylene would be equivalent to 1 mole of nitrogen. However, H_2 gas is an unavoidable drain on the N-fixing system when N_2 is reduced, but no generation occurs when acetylene is reduced. Another potential artifact in the acetylene reduction method is called nitrogenase derepression. Normally, the product of N fixation (ammonia) has a negative feedback on further synthesis of the nitrogenase enzyme. This feedback balances the rate of N fixation with the plant's ability to process ammonia into proteins. With acetylene reduction, no ammonia is produced, no feedback develops, and the synthesis of extra nitrogenase is uninhibited. For these reasons, the conversion factor for acetylene to nitrogen is not always constant, but usually averages between 3 and 10 moles C_2H_2 per mole of N_2 under carefully controlled incubation conditions.

Nodule activity also varies on a diurnal cycle; rates usually decline at night (Figure 11.17). Seasonal cycles are also important, as are patterns in water availability. It is very difficult to account for all these sources of variability; most studies simply try to perform assays on many nodules at many points in the growing season in hopes of hitting a reasonable approximation of average rates. In most cases, it would be fortunate to achieve a confidence interval of ±50% on a seasonal estimate of average nodule activity.

Nodule biomass per plant is easy to determine on small plants by excavating entire root systems, but large plants present mammoth problems. Ecosystems with large plants usually are sampled for nodule biomass by digging pits or taking soil cores at random across the site. The soil is then sieved to collect all nodules. The variability is again quite high, and a standard error is likely to be 25% to 50% of the mean.

Because high variability is unavoidable in the acetylene reduction method, these estimates should be viewed as very rough approximations. Acetylene reduction assays can be useful in gauging the general magnitude of N fixation (such as less than 20 kg ha^{-1}, or more than 100 kg ha^{-1}). They are probably most valuable for assessing physiological aspects of N fixation, such as response to light and water regimes.

How well do rates of N fixation from acetylene reduction approaches compare with N accretion estimates? Adjacent 55-year-old stands of Douglas fir and Douglas fir mixed with red alder were compared by the accretion method, and the current

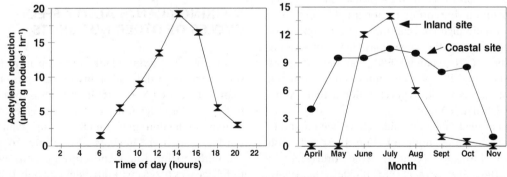

Figure 11.17 The rate of nitrogen fixation by nodules of Himalayan alder begins to rise after canopy photosynthesis begins in the mornings, and tapers off as evening approaches (left, from data of Sharma, 1988). Nitrogen fixation by nodules on red alder trees occurred for a longer period at a coastal site in Oregon with a mild, wet climate compared to red alder trees at a higher elevation, inland site in Washington (right, from data of Binkley *et al.*, 1992b).

rates of N fixation were estimated by the acetylene reduction method (Wind River site from Binkley *et al.*, 1992b). At the less fertile site, the biomass and soil (to 0.9 m) in the mixed stand contained about 2800 kg ha^{-1} more N than the pure Douglas fir stand, for an average rate of accretion of about 54 kg N ha^{-1} yr^{-1}. Leaching losses were about 20 kg N ha^{-1} yr^{-1} higher from the mixed stand, raising the estimated rate of N fixation to 74 kg N ha^{-1} yr^{-1}. The seasonal average rate of acetylene reduction was 8.3 μmol g^{-1} hr^{-1}, and the nodule biomass was 325 kg ha^{-1}. The rate of acetylene reduction at a hectare scale would be: 8.3 μmol g^{-1} hr^{-1} * 325 000 g nodules ha^{-1} * 2880 hr season^{-1}, divided by 3 mols C$_2$H$_2$ per mol N$_2$, times 28 g N mol^{-1}, which totals 73 kg N ha^{-1} for the season. This almost perfect match between methods is coincidental in part. A similar comparison at another site yielded an accretion-based estimate of 102 kg N ha^{-1} yr^{-1}, and a current acetylene-reduction based rate of 85 kg N ha^{-1} yr^{-1} (Binkley *et al.*, 1992b).

DOES N FIXATION DECLINE AS SOIL N INCREASES?

Although N fixation by agricultural legumes does tend to decrease as availability of soil N increases, no feedback inhibition has been shown for actinorhizal species under field conditions. Several greenhouse studies showed that high concentrations of ammonium or nitrate can decrease N fixation, but the concentrations used in these studies far exceeded soil solution concentrations under field conditions. Abnormally high concentrations of ammonium or nitrate around the roots may have misleading effects on N fixation. For example, Ingestad (1981) found that a high supply rate of nitrogen to alder roots at a low concentration actually stimulated N fixation rates.

Under field conditions, alders have been found to fix substantial quantities of N on forest soils with some of the highest N contents in the world (Franklin *et al.*, 1968; Sharma *et al.*, 1985; Binkley *et al.*, 1994). The high rate of nitrate leaching from many alder forests (sometimes in excess of 50 kg N ha^{-1} yr^{-1}) provides a final piece of evidence for lack of strong feedback regulation by N availability on N fixation. Nitrogen fixation by falcataria trees in Hawaii was greater on soils that had less N (Garcia-Montiel and Binkley, 1998). In some cases, nitrogen fixation does decline as trees age, such as an example with *Acacia koa* in Hawaii (Pearson and Vitousek, 2001). However, the koa decline in N fixation was not driven by increasing soil N supply, as soil N supply actually declined along with N fixation.

Why would plants expend energy to fix N on sites where soil N was abundantly available? No conclusive evidence is available, but some speculation can be advanced. Nitrogen fixation requires more energy than assimilation of ammonium, so ammonium uptake might appear to be more efficient than N fixation. Nitrate assimilation costs may be similar to those of N fixation. The picture is actually more complex, as metabolic costs are not the only energy costs involved. From a whole-plant perspective, the cost of developing and maintaining root systems to obtain ammonium and nitrate must also be included. Ammonium is not very mobile in soils, so a larger root system may be needed to obtain soil ammonium than for nitrate. Nitrogen gas is far more mobile than either ammonium or nitrate. Indeed, nodulated (N-fixing) alders have been shown to develop smaller root systems than non-nodulated (non-N-fixing) plants. From a whole-plant energy perspective, N fixation may not be more costly than assimilation of soil N, but this speculation needs further testing.

N-FIXING PLANTS ALSO AFFECT CYCLES OF OTHER NUTRIENTS

A mixture of N-fixing plants and crop trees may show changes in the availability of nutrients other than N. Increased N availability may directly affect the cycling of other nutrients, and increased ecosystem production may rapidly tie up available nutrients in tree biomass. Returning to the Mt. Benson study, the P concentration of Douglas fir foliage was 0.22% without alder and only 0.12% with Sitka alder. Sitka alder leaves contained a healthy level of 0.30% P. Red alder trees in the same plantation further reduced Douglas fir foliage

P concentrations to 0.09%, and red alder leaves contained only 0.14% P. Why did the alder impair Douglas fir P nutrition? The rate of biomass accumulation increased by 75% with Sitka alder, and by 260% with red alder. The P contained in the ecosystem's biomass totaled 45 kg ha^{-1} without alder, 105 with Sitka alder, and 150 kg ha^{-1} with red alder. Thus, the decrease in soil P availability may simply be due to accumulation of P in rapidly growing biomass. As the stand without alder develops, the accumulation of P in biomass may also reach a point where P availability in the soil declines. Acceleration of stand production increases demands for all nutrients, and the benefits of N-fixing species may be limited by the availability of other nutrients on very poor soils. This pattern would also be expected if N fertilizer were used to greatly accelerate stand growth.

Limitations of nutrients other than N will affect N fixation as well as the performance of crop species. Other studies with red alder have found both substantial declines in P availability (Compton and Cole, 1998, assuming the large site difference in total P did not confound species effect on available P) and substantial increases (Giardina *et al.*, 1995). A strong pattern of reduced P supply was also apparent in 16-year-old plantations of *Eucalyptus saligna* and N-fixing *Falcataria moluccana*. Nitrogen availability in pure *Falcataria* plots was four times higher than in pure *Eucalyptus* plots, but pure *Eucalyptus* plots had five times more available P than pure *Falcataria* plots (Figure 11.18). Phosphorus availability under N-fixing species appears to be increased in some cases and depressed in others; this is another situation where simple generalizations are not helpful.

NITROGEN-FIXING SPECIES ALTER SOIL CARBON IN TWO WAYS

Soil organic matter often increases under nitrogen-fixing species, and also changes the turnover rate of organic matter. Across a range of N-fixing case studies, the accumulation of soil N was accompanied by large increases in soil organic matter (soil C). Typical rates of C accretion range from 0.5 to 1.25 Mg ha^{-1} yr^{-1} (Figure 11.19). The size of a pool

in a soil could increase because of increased inputs to the pool, or reduced losses from the pool (or both). Nitrogen-fixing species often increase the total productivity of a forest, so it may seem reasonable that C accretion results from larger inputs of organic matter in aboveground litter and belowground tissues. Nitrogen-rich tissues may decay faster than low-N materials, and faster decomposition might be expected to lead to less accumulation of soil organic matter. However, as noted in Chapter 4, litter decay rates should not be assumed to relate well to rates of organic matter accumulation, because these related processes are subject to many other influences.

The *Eucalyptus/Falcataria* study described above was planted on a site following several decades of sugarcane agriculture on a high-organic matter Andisol soil. Sugarcane is a C4 plant; the different photosynthetic pathway of C4 plants leads to a difference in the ratio of the stable isotopes ^{13}C and ^{12}C. After 17 years of tree growth, this means the soil had three pools of C: very old C3-derived material from millennia of forests prior to clearing for agriculture; moderately old C4-derived material from sugarcane agriculture; and recent C3-derived material from the current crop of trees. The stable-isotope approach cannot separate the pool of very old C3 material from the recently added C3 material, but it may be reasonable to expect very little change in an old pool that lasted through decades of agricultural plowing. Any difference in the C3-derived soil organic matter between *Eucalyptus* and *Falcataria* should represent the accumulation of new material during the current rotation. Given that the input of C4 material stopped with the last crop of sugarcane, any difference in the remaining C4-derived material would result from different rates of loss of soil C between the species. As with other N-fixation experiments, soil C accumulated more under N-fixing *Falcataria* relative to *Eucalyptus*. The divergence between the species was a linear function of the percentage of *Falcataria* trees in a plot, with the monocultures diverging by about 1.2 Mg C ha^{-1} yr^{-1}. Intriguingly, the stable-isotope evaluation revealed that about two-thirds of the difference resulted from the accumulation of more new material under the N-fixing trees, and

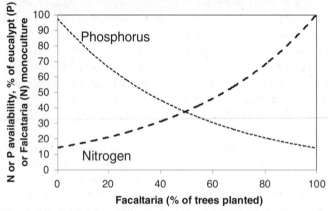

Figure 11.18 A study with 18-year-old monocultures and mixtures of *Eucalyptus saligna* (straight trees, with dark bark) and *Falcataria moluccana* (mottled bark) showed exponential increases in soil N availability (with resin bags) with increasing percentage of N-fixing *Falcataria* in the plot ($r^2 = 0.95$). Conversely, the availability of P increased exponentially with increasing percentage of *Eucalyptus* ($r^2 = 0.89$; from data of Kaye *et al.*, 2000).

one-third from the reduced loss of older (sugarcane) C (Figure 11.20).

This pattern of greater addition of C and greater retention of C was clear for this single experiment, but would the same pattern apply to other cases with N-fixing trees? Resh *et al.* (2002) performed a similar analysis for three other experiments, including another Andisol in Hawaii, and a Vertisol and an Inceptisol in Puerto Rico; the patterns all supported the generalization that higher N-fixing trees raise soil C by adding more C and reducing rates of loss of old C.

Further support for the importance of reduced loss rates of older soil C under N-fixing plants came from one of the Puerto Rico sites. This site was labeled with ^{15}N as part of an assessment of N-fixation rates (Parrotta *et al.*, 1996, described earlier in this chapter). Kaye *et al.* (2002) collected soils seven years after the ^{15}N label had been applied, and incubated the soil samples under moist, warm conditions for over a year in the laboratory. About 45% of the label N was still present in the soil after seven years, and the intensive activity of microbes during the incubation managed to release almost

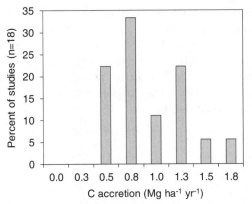

Figure 11.19 Stands with nitrogen-fixing trees increase both soil N and soil C. Carbon accretion averages about 0.8 to 0.9 Mg ha^{-1} yr^{-1}, or about 17 g C for each g of N accretion in the soil. (from data of Binkley 2005).

40% of the ^{15}N from the *Eucalyptus* soil, but only 25% of the ^{15}N from the soil from the N-fixing *Albizia* and *Casuarina*. The soil organic matter that accumulated under the N fixers was more recalcitrant to further decomposition than the soil under the *Eucalyptus*, consistent with the C3–C4 findings in the same experiment.

What mechanism might account for the better stabilization of soil C under N-fixing species? The first possibility would be a mechanism where higher N supply helps stabilize humified C pools.

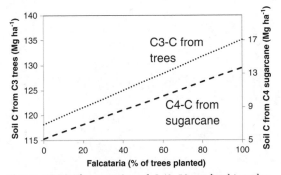

Figure 11.20 The accretion of C (0–50 cm depth) under N-fixing *Falcataria* resulted from a 2/3 contribution of the addition of new C derived from trees, and 1/3 by the greater retention of older C that was added under the previous land use of sugarcane agriculture. (from data of Kaye *et al.*, 2000; see also Figure 11.18)

This idea was tested on an Andisol in Hawaii (five kilometers away from the *Eucalyptus/Falcataria*) by adding over 1 Mg N ha^{-1} as inorganic fertilizer in a *Eucalyptus* plantation over a six-year period (Binkley *et al.*, 2004c). High uniformity in soil properties gave a very precise estimate of the effect of added inorganic N on the accumulation of new soil C and loss of old soil C: exactly zero. The soil changes engendered by the N-fixing *Falcataria* were *not* directly related to increasing soil N supply. Other possibilities include the enhanced populations (and activities) of earthworms, a lower ratio of fungi/bacterial biomass, and other biological factors rather than simple chemistry.

WHEN DOES FIXED NITROGEN SHOW UP IN OTHER VEGETATION?

Trees growing near N-fixing trees may benefit from increased soil N supplies, and secondary benefits from this enhanced N nutrition (for a review, see Richards *et al.*, 2010). Only a few studies have examined how long it takes for a nitrogen-fixing plant to provide significant amounts of nitrogen for associated species. The major way that fixed nitrogen becomes available to associated trees is through decomposition; the annual litter produced by the nitrogen fixer decomposes, and the released nitrogen is taken up by trees. Minor amounts of nitrogen may pass directly from the root system of the nitrogen fixer to associated trees through mycorrhizal connections (Ekblad and Huss-Danell, 1995), but the rates of exchange among roots are probably too low to be important. About 1–2% of the fixed N may leak from roots into the soil (Uselman *et al.*, 1999).

The Puerto Rico ^{15}N labeling study described above found that the *Eucalyptus* trees began taking up nitrogen fixed by *Casuarina* or *Leucaena* sometime between 2 and 3.5 years of age (Parrotta *et al.*, 1996). A study using natural abundance levels of ^{15}N concluded that a *Leucaena* plantation derived about 70% of its nitrogen from the atmosphere and that by age 6 years, the understory plants were as reliant on nitrogen fixed recently by *Leucaena* as was the *Leucaena* itself (van Kessel *et al.*, 1994).

In temperate forests, three studies have examined mixtures of poplars and alders for short-rotation biomass production (DeBell and Radwan, 1979; Hanson and Dawson, 1982; Cote and Camire, 1984). All found that poplars neighboring on alders grew better in the first two to four years than poplars neighboring on poplars.

This increased growth could have resulted from soil nitrogen enrichment by alder (DeBell and Radwan, 1979) or from the size differences between the species. Because the poplars were larger than the alders, poplars with alders probably experienced less competition than poplars with poplars (Cote and Camire, 1984).

CHAPTER 12

Soil Management – Harvesting, Site Preparation, Conversion, and Drainage

OVERVIEW

Until recently we did not think of managing forest soils, but the advent of intensive plantation management for wood production opened new perspectives and opportunities for working with soils to increase tree growth. Forest plantations are established and maintained through site manipulation and silvicultural treatments. Decision support systems allow a forest manager to model the results of various soil management activities on wood production, and choose the most cost-effective series of activities. Forest soils in plantation forests are drained, tilled, fertilized, limed, and even irrigated. Considerable knowledge has accumulated to support forest soil management in intensive plantation forestry, but even more knowledge will need to be generated to meet future wood needs from intensively managed plantations in a sustainable, environmentally sound manner. Some forest soils present unique management problems. These include dry, sandy soils that are prone to wind erosion, seasonally flooded soils, organic soils or peatlands, Gelisols (soils that contain permafrost), mining spoils, and soils that have had other intensive uses. Many diverse techniques for managing forest growth on such soils have been, and continue to be, developed. Forest tree nursery soils are not really forest soils, but their management is essential to plantation forestry. The management of these soils employs many agronomic techniques and has become a highly developed subdiscipline within forest soils.

With the advent of wood production plantation forestry, soil management has become widespread and both economically and ecologically important. Good soil management calls for an intimate knowledge of the soil's properties and the soil's response to management activities. In clonal plantation forestry, accurate matching of clones to soil properties and management interventions is critical for maximizing yield and controlling negative impacts of management on the environment. Our ability to maintain detailed long-term records on the performance of plantation forests and to use those data to refine soil management practices allows us to increase forest productivity while minimizing environmental impact.

SOIL MANAGEMENT PLAYS A MAJOR ROLE IN INTENSIVE PLANTATION FORESTRY

The millions of hectares of the world's forests managed as plantations include a wide range of soil and climatic conditions and a variety of tree species. In general, tree plantations are established for one of three reasons: (1) afforestation of open sites that have not supported forests for some period of time, (2) reforestation of sites that have recently been harvested, or (3) conversion of one type of forest to another. Examples of open lands that have been planted to forests are former agricultural lands in England, southwestern France, central Europe, and the southern United States; sand dunes in New Zealand and Libya; arid regions in Israel and other Middle Eastern countries and parts of sub-Saharan

Ecology and Management of Forest Soils, Fourth Edition. Dan Binkley and Richard F. Fisher.
© 2013 John Wiley & Sons, Ltd. Published 2013 by John Wiley & Sons, Ltd.

Africa; steep slopes in Europe and Japan; peatlands in the British Isles, Sweden, and Finland; grasslands in Venezuela, Colombia, and Argentina; brush lands in Brazil; former sea bottom in The Netherlands; and spoil banks resulting from mining operations in many countries around the world.

Plantation establishment following clear-cutting of the same species is standard practice in many areas where intensive silviculture is practiced. Examples of such reforestation efforts are found in the Douglas fir forests and southern pine forests of the United States, Australian *Eucalyptus* forests, and the *Chamaecyparis* forests of Japan and China.

Plantation establishment on sites still supporting considerable forest cover often involves conversion from native to exotic species. Examples include the pine forests of Australia, replacement of marquis vegetation with pines or eucalypts in the Mediterranean, conversion of scrub hardwood and coppice in central Europe, and improvement of degraded tropical forests in several equatorial countries (Evans, 1992). Often, natural forests are converted to plantations of widely different species, and from highly complex associations to exotic monocultures. These include plantations of *Pinus caribaea, Gmelina arborea, and Eucalyptus* in Brazil; *Pinus radiata* in Australia, Chile, and New Zealand; *Pinus elliottii* x *taeda* hybrids in Australia; *Pinus caribaea* in Fiji; *Tectona grandis, Eucalyptus tereticornus,* and *Cryptonieria japonica* in India; *Triplochiton scleroxylon, Eucalyptus saligna*, and *Pinus* species in Africa. Whether regeneration is accomplished with native or exotic species, plantation forestry offers a unique opportunity to locate species, selections within species, and specific clones according to their soil-site requirements. It is not sufficient to plant hardwoods on "hardwood" sites and conifers on "conifer" sites because of the tremendous differences in growth potential on different soils for different genera, species, selections, and clones.

Some reasons for the trend toward plantation forestry include:

- greater demands for forest products;
- introduction of genetically improved stock;
- development of clonal material;
- easy manipulation of the stand for the most desired assortment of products;
- higher yields;
- full use of land by complete stocking;
- improved conditions for the operation of machinery;
- reduced cost of logging;
- prompt restocking after harvesting;
- more uniformity of tree size;
- improved log quality.

Plantation forestry generally involves an integrated system of cultural practices, including slash disposal, site preparation, and careful planting of cultivars. It may include such other silvicultural practices as soil drainage, weed and pest control, pruning, thinning, fertilization, and the use of genetically superior or clonal planting stock. It usually involves even-age management and clear-cut harvesting of monocultures.

Plantations are literally man-made forests in the sense that they are established and maintained as the result of site manipulation. Such efforts to improve the site and increase tree survival and growth may have profound influences on certain soil physical, chemical, and biological properties, particularly properties of the O horizon. Whenever heavy machinery is used in harvesting or site preparation, there is an opportunity for undesirable soil disturbance. Soil disturbance is lower for tracked than for wheeled vehicles, lower on flat than on steep terrain, lower on sandy than on clayey soils, and lower on dry than on wet sites. However, long-term effects on soil properties and long-term site productivity are in the realm of conjecture, because few well-controlled studies have been conducted over sufficient time periods to give a definitive answer.

PRIOR LAND USES CAN LEAVE LEGACIES IN SOILS

Many plantations and some natural forests occur on land that was not previously forested. Natural forests become established and plantations are created on abandoned agricultural and pastoral lands, on mining spoils, on stabilized sand dunes, on artificially drained wetlands, and even on abandoned industrial lands. When managing forest soils, it is important to understand the prior land use on the site (Oheimb et al., 2008). Erosion may have removed topsoil from one area and deposited it in another (Lafon et al., 2000). If the land was converted from forest to agriculture, stump removal may have created a soil with discontinuous and convoluted horizon structure. Farming or grazing practices may have depleted the soil of nutrients, compacted the surface soil, altered the soil's acidity, or increased soil nutrient content, particularly phosphorus content (Standish et al., 2008; Boussougou et al., 2010). Mine spoils may be high in heavy metals or in sulfides. Sulfides lead to soils with very low pHs that are difficult to ameliorate. Stabilized sand dunes need to be managed more carefully than other sandy soils, because they are more susceptible to further wind erosion. Artificially drained wetlands are often subject to periodic flooding, and frequently have elevated soil organic matter levels that will decline rapidly after draining. Old industrial lands may have high levels of polychlorinated biphenyls (PCBs) and heavy metals, including arsenic, cadmium, lead, and mercury (Morin and Calas, 2006). It is always advisable to try to discover the prior land use of any soil you wish to manage.

HARVESTING IMPACTS LONG-TERM SITE FERTILITY

Forest harvesting systems remove varying amounts of biomass from the site. A variety of long-term site productivity experiments have attempted to quantify the effects of biomass removal and soil compaction during harvesting on site productivity. The results have been quite variable. Biomass removal appears to have little or no effect on site productivity. Soil compaction has increased productivity on some sites and decreased it on others. Many soil properties have been shown to be altered, at least temporarily, in these studies, but no systematic decrease in site productivity has been found

(Powers et al., 2005). This appears to be because the removal of essential nutrients during harvesting is small and the organic matter removed has minimal effects on soil organic matter content. The Henderson long-term site productivity study in North Carolina has shown that cultural practices applied after harvesting, during stand establishment and stand development actually increase site productivity (Allen et al., 1995; Forest Nutrition Cooperative, 2011).

LAND CLEARING CAN LOWER SOIL PRODUCTIVITY

Considerable quantities of vegetation or slash must often be removed from a site so that a plantation can be established. Development of detailed plans prior to logging can be invaluable in solving slash disposal problems. Such plans should take into consideration the soils, terrain, climate, and cover type. The type, amount, and distribution of logging waste or unwanted vegetation and terrain conditions will influence the choice of disposal method and machinery. Proper location of roads and selection of techniques and tools suited to each vegetation–terrain combination may avoid costly damage to the soil and site.

Machines are widely used for disposal of slash and excess vegetation in preparation for plantation establishment. Machines are used extensively, along with fire, in the conversion of eucalypt stands to pine in Australia, in the disposal of low-quality hardwoods in preparation for pine plantations in the southeastern United States, and in site clearing for Douglas fir plantations in the Pacific Northwest.

Crawler tractors (continuous-tracked tractors) with V-shaped shearing blades are often used to sever stems of residual vegetation. Root rakes and Young's teeth attached to crawler tractor blades may be used to uproot trees and windrow debris. The windrows are often burned after a few months, but they still occupy a considerable portion of the land area that would otherwise be available for planting. Unfortunately, these windrows often contain considerable topsoil as well (Figure 12.1).

Figure 12.1 Excessive soil disturbance and surface soil movement from a shearing and piling site preparation operation.

Haines *et al.* (1975) reported that approximately 5 cm of the surface soil was deposited in windrows during shearing and piling operations in the Carolina Piedmont. After 19 years, loblolly pine in an area that was not sheared and piled contained 85% more volume than trees in the sheared and piled area.

Considerable concern has arisen over the precipitous decline in soil organic matter that often accompanies stand conversion efforts, particularly in the tropics (Lundgren, 1978; Chijicke, 1980). Morris *et al.* (1983) found that an extremely large amount of topsoil ended up in windrows of a sheared and piled slash pine flatwoods site even when the clearing had been done with care. The effects of land-clearing methods on soil erosion rates and subsequent site productivity can be quite significant (Dissmeyer and Greis, 1983). A major problem with the use of heavy equipment for uprooting and windrowing brush and logging debris is that the quality of the operation is heavily dependent on the skill and dedication of the heavy equipment operator. No matter how well planned land-clearing operations may be, the potential for site degradation can be controlled only through careful supervision of skilled and dedicated equipment operators.

Successful conversion of wiregrass (*Aristida stricta*) and scrub oak (*Quercus laevis*) sandhills in Florida and the Carolinas to southern pines has been accomplished by chopping. This operation generally consists of passing a heavy (up to

20 Mg) drum roller over the soil surface one or more times. Sharp parallel blades attached along the length of the drum cut debris and vegetation, including small trees, and press it to lie conformally with the soil surface. Because only a small part of the chopped vegetation and debris is incorporated into the mineral soil, this organic material acts as mulch. It retards erosion, moderates soil temperature fluctuations, and reduces soil moisture losses. Chopping appears to be a practical means of conserving soil nutrients and is particularly useful on dry, sandy soils (Burns and Hebb, 1972).

SITE PREPARATION PRECEDES STAND ESTABLISHMENT

Harvesting and the elimination of logging slash and unwanted vegetation may be considered the first steps in preparing a site for regeneration. However, the term "site preparation" is commonly used for those soil-manipulation techniques designed to improve conditions for seeding or planting that result in increased germination or seedling survival and tree growth. While this may be accomplished by manual operations in some remote localities and on relatively clean sites, soil preparation is increasingly achieved by a variety of machines such as scarifiers, disks, trenchers, bedding plows, subsoil rippers, and so on.

SCARIFICATION IS A MEANS OF EXPOSING MINERAL SOIL

In cool to cold regions, exposure of mineral soil, elimination of competing vegetation, and improvement of water relations are primary goals of soil preparation operations. Because of the slow rate of litter decomposition, reactivation of the soil and release of nutrients for the young seedlings are also important. Scalping is one method of scarification. It is accomplished with a trailing plow or V-blade that produces a furrow about 5 cm deep and 0.5 to 1 m wide, in which the seedlings are planted. Scalping displaces much of the organic-enriched topsoil from the immediate reach of the young plants and may result in prolonged slow growth on sandy soils.

Spot scarification has been accomplished with a number of devices ranging in complexity from anchor chains to the Bracke mounder. The objective of such treatments is to expose or mound mineral soil. Mounds of mixed mineral soil and organic matter are generally superior for seedling survival and growth to mounds of mineral soil, which are superior to mineral soil exposed by scraping, which is superior to planting through the undisturbed O horizon (Sutton and Weldon, 1993, 1995). Unless scarification is carried out parallel with the slope, it may initiate erosion. This type of soil manipulation has not been shown to lead to site degradation.

Figure 12.2 A site where a dense B horizon has been ripped prior to planting.

CULTIVATION INCREASES SURVIVAL AND EARLY GROWTH OF SEEDLINGS

Harrowing or disking differs from plowing mainly in the depth and degree of disturbance of the top soil. Harrowing consists of drawing multiple disks through the surface soil and displacing, but not necessarily inverting, the disk slice. Plowing, which inverts a slice of soil, can be done by either a disk or a moldboard plow. The latter is most often used on sites with a thick sod. Both methods are effective in mixing organic and mineral materials and reducing compaction. Harrowing and plowing may result in a temporary immobilization of soil nutrients due to the incorporation of large amounts of organic material. Soil preparation for planting is sometimes accomplished by double moldboard plows. Trees are then planted in the furrow. In wet areas the furrow may serve as a water trap and reduce survival and early growth.

Deep loosening of subsoil layers, often called ripping, is a relatively recent development in forest soil management. It is accomplished with chisel-like bars drawn up to 0.5 to 1 m into the soil by powerful tractors and by three-in-one plows that both rip and bed the site (Figure 12.2). Subsoil loosening is aided by a winged foot at the base of the chisel arm that vibrates as it is drawn through the soil. Soil fracturing is maximized if ripping is done when soils are relatively dry. Tree roots commonly find old root channels and proliferate in them. Ripping provides artificial channels, and trees have been shown to proliferate in such channels, particularly in compact soils (Nambiar and Sands, 1992). The fractures or artificial channels gradually disappear (Carter *et al.*, 1996), but roots often penetrate along the fracture lines long before closure occurs.

There have been attempts to improve the productivity of degraded forest Spodosols with ortstein layers. Ripping to rupture the hardpan, plus the addition of lime, organic materials, and mineral fertilizers, improves the growth of pines and mixed stands. Attempts to improve the growth of pine by deep placement of lime and phosphorus fertilizers during site preparation of an Aquod soil met with only limited success (Robertson *et al.*, 1975).

Bedding, the use of opposing disks to create a mound, is a standard soil preparation practice in many areas. Bedding concentrates surface soil, litter, and logging debris into a ridge 20 to 30 cm high and 1 to 2 m wide at the base. Planting on beds usually improves the survival and early growth of both conifers and hardwoods on a variety of sites. The beneficial effects of bedding have been attributed to improved drainage, improved microsite environment for tree roots in regard to nutrients, aeration, temperature, and moisture, and reduced competition from weedy species.

On wet sites, survival and growth responses from bedding derive primarily from removal of excess

Table 12.1 Effects of site preparation and fertilization on height and volume of slash pine after nine years on a Typic Albaquult.

N (kg/ha)	P	Nonprepared Height (m)	Nonprepared Volume (m³/ha)	Harrowed Height (m)	Harrowed Volume (m³/ha)	Bedded Height (m)	Bedded Volume (m³/ha)
0	0	2.17	1.94	2.87	3.57	4.30	8.85
90	0	2.07	1.60	3.26	4.63	4.88	12.82
0	90	5.95	14.39	5.89	20.33	7.66	49.80
90	90	6.74	21.43	7.02	32.68	7.29	37.28

surface water. Surface drainage results in accelerated decomposition of organic matter and mineralization of nutrients. Where the A horizon is thin and the B horizon is either impervious to roots or nutrient-poor, the concentration of topsoil, humus, and litter provides more favorable conditions for root development and, subsequently, top growth. Haines et al. (1975) reported that the incorporation of 73 Mg ha⁻¹ into 15-cm-high mounds of screened surface soil significantly increased slash pine growth over that obtained where no debris was added. However, incorporation of debris in excess of 73 Mg ha⁻¹, the approximate amount left after logging, resulted in no additional improvement after four years.

Pritchett and Smith (1974) combined three types of site preparation with varying amounts of mineral fertilizers (Table 12.1).

In this test, harrowing reduced ground vegetation, but it resulted in only slightly better tree growth than at the control site. Competition for moisture was not a critical factor on this wet site. Bedding, on the other hand, resulted in a 98% increase in tree heights and a 74% increase in total volume after nine years of growth on unfertilized plots. The response to the fertilizer applications indicated that, while part of the growth improvement from bedding derived from improved nutrition, equally important benefits came from reduced competition and improved surface drainage. Wilde and Voigt (1967) reported that at age 14, white pine planted on beds were significantly taller (5.2 m) and had better survival than trees planted in furrows (1.4 m). They found that planting on scalped soil resulted in improved survival, but average tree

heights were only 3.2 m. There have been a number of other reports of improved tree growth from harrowing or bedding in the southern pine region (Haines and Pritchett, 1964; Worst, 1964; Derr and Mann, 1970; Klawitter, 1970a; Mann and McGilvray, 1974; Harrington and Edwards, 1996), for maritime pine in southwest France (Sallenave, 1969), and for radiata pine in Australia (Woods, 1976; Hoopmans et al., 1993; Costantini et al., 1995) (Figure 12.3).

Bedding often increases survival and early growth, but growth gains dissipate with time. This phenomenon is often associated with density-dependent mortality within the stand. Because of high survival and rapid growth, the canopy in the bedded areas closes faster than in unbedded control areas. Soon after canopy closure, density-dependent mortality or, in some southern pine species, loss of crown volume begins. Either of these processes soon leads to growth losses that allow the control plots to catch up to the bedded plots. Obviously, density management is important if one is to take advantage of growth gains from cultivation.

Improvements in tree growth in response to different types of soil management practices depend greatly on soil properties, site conditions, and the tree species involved. Soils information can be a valuable aid in deciding which intensive management practices to employ. Fisher (1980) presented a guide for the use of soils information in silviculture in the southeastern coastal plain. He found that a system of soil groups based on soil drainage class, depth to and nature of the B horizon, and the character of the A horizon was a valuable aid in

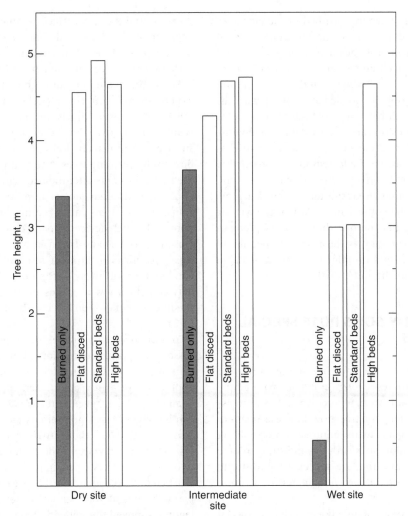

Figure 12.3 Height of 8-year-old slash pine as influenced by seedbed treatment and soil moisture conditions (Mann and McGilvray, 1974). Used with permission.

determining soil management practices for intensive silviculture. This system allows the use of existing soils information in deciding which species to plant, what harvest system to use, what water management techniques to employ, how to prepare the site, and what fertilization regime to follow.

By applying this model, many proprietary decision support systems have been developed for use in plantation forestry. The development of detailed geographic information systems, the use of the global positioning system, the creation of detailed soil maps, our increased knowledge of site

productive capacity, and our understanding of site-specific responses to cultural practices have greatly increased the use and value of these systems. Intensive forestry has moved from its equipment-intensive phase to a knowledge-intensive phase.

EXTREME SITES MAY REQUIRE INTENSIVE SOIL MANAGEMENT

The vast majority of forest soils require no special management techniques for continued high productivity. But when disturbed or adverse sites are planted with trees in an effort to ameliorate soil conditions and improve aesthetic values, special soil management techniques are usually required. Afforestation of spoil banks resulting from mining operations, seasonally flooded areas and peatlands, moving sand dunes, and old cultivated fields degraded by erosion cannot be accomplished without special effort. Some of these soils are nutrient-deficient but can be made reasonably productive through the addition of fertilizers. Soils of some adverse sites contain toxic substances or lack mycorrhizal fungi, while others are too dry or too wet for good tree growth. The one thing that all problem sites have in common is the need for special soil management efforts in order to obtain acceptable plant establishment and growth.

DRY, SANDY SOILS POSE SPECIAL PROBLEMS

Rainfall exerts a major influence on the worldwide distribution of forests. Even in regions with ample rainfall, there are areas where available soil moisture is too low for good tree growth. These areas are mostly deep sands with minimal water-holding capacity or shallow soils with low moisture-storage capacity. Dry sands that are also exposed to strong winds offer a special problem in terms of dune stabilization.

The stabilization of moving sands is a special problem in certain coastal and inland areas when marine or glaciofluvial deposits of sand are disturbed. Blow sands that have required special afforestation efforts are found in France, the Netherlands, New Zealand, North Africa, North America, and the Middle East (Mailly *et al.*, 1994), as well as in some coastal and dry, sandy areas on most continents. Dune fixation by tree planting was underway in Europe by the middle of the eighteenth century and in America by the early nineteenth century (Lehotsky, 1972). One early attempt at dune stabilization in America was on Cape Cod (Baker, 1906), and other projects have been undertaken in the Great Lakes region of North America and on Cape Hatteras. The sandy

areas around the Great Lakes became unstable and began to move following timber removal and farming in the late 1800s, and extensive dune stabilization programs began in the early 1900s.

Sand dune fixation requires the establishment of a permanent plant cover (Kellman and Kading, 1992). The cover directs or lifts the prevailing wind currents away from the easily eroded sandy surface. This usually cannot be accomplished directly by planting trees. The young trees simply are not sufficient to alter the wind currents, and the abrasive sands often damage unprotected seedlings. Consequently, blow sands are first stabilized temporarily by such devices as parallel rows of fencing, closely spaced stakes driven into the sand, surface trash, plantings of drought-resistant grasses, or a combination of these techniques. The distance between the rows of stakes or grass depends on the intensity of erosion but may be no more than 1 to 2 m.

Xerophytic clump grasses have been used successfully in stabilization efforts, but normally they cannot be established in infertile sands without the application of a mixed fertilizer. In New Zealand (Gadgil, 1971), nitrogen fertilizer was used to establish bunch grass as a preliminary step in stabilizing blow sands. After about two years, lupine was broadcast seeded without additional fertilizer. After two or three years, the lupine was mechanically crushed or disked in order to reduce competition for planted pines. It was estimated that the reseeding legume supplied the soil with 200 to 300 kg ha^{-1} of nitrogen during the five to six years before the pine overtopped it.

As soon as the sand movement has been halted, the dunes are normally planted with a hardy tree species. Conifers have generally been superior to hardwoods on adverse sites, although both types have been used. Radiata pine is planted on dune sands in New Zealand, and maritime pine is often used in southwest France. Jack pine, red pine, Scots pine, pitch pine, Virginia pine, and sand pine have been used on less favorable sandy sites in other areas, while white pine, Norway spruce, locust, and sometimes alder have been planted on the moister sands between dunes. Planting distances for pine seem to vary from 1.5×1.5 m on eroding sites to 2.0×2.0 m on other sites. Hand application of a fertilizer equivalent to about 200 kg ha^{-1} of diammonium phosphate, or spot application of the same material applied in an opening 20 to 30 cm on either side of the seedling and 15 cm deep, is often done soon after planting. The amount of fertilizer at each spot should not normally exceed about 50 g in order to reduce the danger of salt damage to the seedling root system. The nutrient ratio should be based on soil fertility conditions, as determined by experience or by a soil test.

There are vast areas of deep, dry sands in many countries that require special management techniques for the establishment of desirable tree species. These marginal sites often support woody scrub or grass vegetation and are therefore well stabilized. Droughty soils of the southeastern United States sandhills are examples of these adverse sites. These sites, mostly Psamments, support remnants of longleaf pine stands harvested around the turn of the twentieth century, plus scrub hardwoods and wiregrass (*Aristida stricta*). Early harvesting methods made no provision for regeneration of the pine, and subsequent efforts at reforestation, especially with slash pine, have not been highly successful. The understory scrub oaks and wiregrass prevent natural establishment of most pines (Burns and Hebb, 1972; Outcalt, 1994).

IRRIGATION ALONG WITH FERTILIZATION ON DRY SITES IS INCREASING

Irrigation has long been practiced in forests, but only on a limited scale. There are nearly 250 000 ha of irrigated plantations in Pakistan (Sheikh, 1986) and probably a somewhat larger area of irrigated plantations in India. Some of these plantations were established as early as the 1860s. Production on these sites has been excellent, but it often declines due to increasing salinity (Ahmed *et al.*, 1985). Irrigated fuel wood plantations are common in the arid regions of Africa and the Middle East (Wood *et al.*, 1975; Jackson, 1977). Some investigations on the use of irrigation or fertigation, irrigation with a nutrient solution, have been carried out in

the United States (Leaf *et al.*, 1970; Baker, 1973; Harrington and DeBell, 1995; Dougherty *et al.*, 1998). In general, these practices result in improved growth, but they are costly and run the risk of nutrient loss into surface water or groundwater. Horticulturists seem to have perfected fertigation regimes for orchard crops, and forestry could probably apply these regimes to forest plantations if growth increases warranted the expense incurred.

SEASONALLY FLOODED SOILS PRESENT UNIQUE PROBLEMS

These periodically wet soils occupy a rather large percentage of the tropical and monsoon rainforests, but because of the high evapotranspiration rate of these ecosystems, excess water is seldom a limiting factor for tree growth. Other extensive areas of seasonally wet forests are found in coastal flats, basins, and soils with underlying impervious layers. Examples of wetland forests are the nearly 8 million ha along the coastal plain of the southeastern United States (Klawitter, 1970b). These complex wetlands are very broad and nearly level interstream areas of coastal flats. They are interspersed with bays and pocosins, which are characterized by high centers with natural drainage impeded by elevated rims, sluggish outlets, and impermeable subsoils, as well as swamps and ponds. These wetlands support unique vegetation, act as filters for water moving from uplands to streams, and have gained protection through environmental legislation in many countries. Such legislation often allows "normal silvicultural operations" to be carried out, but controversy as to what constitutes "normal silvicultural operations" continues (Messina and Conner, 1998).

The bays and swamps are generally wet year round and are often covered with one to several feet of peat. However, an excessively high water table during only a few months of the year appears to be the primary limiting factor to intensive woodland management of most wet plains. During seasonally dry periods, the plains can be exceedingly dry because their primary source of water is precipitation. Tree cover may consist of scattered suppressed stems of one of the southern pines or locally dense stands of pines and hardwoods, with an understory of grasses and shrub clumps. Occasional rises of a few centimeters in the land surface may retard surface water flow over large areas, except in the immediate vicinity of shallow streams. Water moves slowly though the soils because of slowly permeable subsoil layers and small hydraulic heads between the soil surface and the stream.

Some forms of water control, mostly through ditch drainage, have been applied to thousands of hectares of forests on wet flats and in the Southeast and Gulf Coastal Plains of the United States since 1960. The objectives have been to (1) improve the trafficability of woods roads so as to provide ready access for harvesting, fire protection, and tending; (2) reduce soil compaction and puddling; (3) facilitate regeneration of pine on prepared sites and ensure early survival of planted seedlings; and (4) increase growth or reduce rotation age (Hewlett, 1972). Whether or not such practices qualify as normal silvicultural operations is still hotly debated.

Surface drainage ditches may not remove water from soils of the wet flats very effectively because the drawdown of the groundwater table along the ditch may extend only a few meters inland from the ditch. The effective distance from the ditch is a function of the ease with which water moves laterally through the soil. The higher the percentage of large pores in the soil, the more effective ditch drainage will be.

Young and Brendemuehl (1973) reported that a drainage system consisting of main drains 1.2 m deep and 800 m apart with lateral collecting ditches 0.5 m deep spaced about 200 m apart improved the site productivity of a Rains loamy sand (Typic Paleaquult), but the system was not intensive enough to provide the pole-sized slash pine with continuous protection from excess water. Furthermore, natural aging of the system tended to return the area to its original state following several wet years. In drainage tests in northeastern France on pseudogley soils with impermeable layers at 25 to 50 cm, drainage ditches at 20-m intervals improved the growth of Norway spruce, Douglas fir, and larch almost as much as spacings of 10 m and significantly more than spacings of 40 m (Levy, 1972).

In some cases, the use of water by a semi-mature stand of trees will eventually make the ditch system unnecessary or even detrimental to stand water relations. Since tree growth is best when water table depth is constant (White and Pritchett, 1970), it is wise to have some control on the level to which a drainage system lowers the water table (Fisher, 1980).

The benefits from draining wetlands generally derive from increased rooting depth, which in turn reduces wind-throw, improves soil aeration, and increases the nutrient supply. The last result may come about through accelerated oxidation of soil organic matter, as well as through increases in the volume of root-exploitable soil. However, excessive drainage of permeable soils may actually reduce growth and inhibit regeneration of certain types of wet-site trees, such as tupelo and cypress. In a drainage experiment on Leon fine sand (Aerie Haplaquod), seven-year slash pine growth was not improved by 0.5- or 1.5-m-deep ditches. Poor growth near ditches apparently resulted from excessive removal of groundwater, particularly during dry periods (Kaufman et al., 1977).

Bedding, or row mounding, may be an alternative to lateral ditching of wet flats and is certainly a normal silvicultural operation. Klawitter (1970b) found that four-year-old slash pines planted on beds were 13% taller than trees on unbedded plots in a Typic Paleaquult, even though the experimental area was adjacent to a main drainage ditch. In another part of the coastal plain on an undrained soil with similar properties, slash pine averaged 67% taller on bedded plots than on unbedded plots after three years (Pritchett and Smith, 1974). Early advantages of bedding may not persist into the later stages of stand development on some sites; however, on wet flats, where trees stagnate without some relief from excess water, the response to bedding persists throughout the rotation.

An improvement in accessibility and an increase in the growth rate of stagnant stands of pine often found on the seasonally flooded flatlands appear to be desirable objectives. However, it is not at all clear whether these objectives can or should be obtained through major drainage projects. A combination of water control, through perimeter ditching for the removal of excess surface water, and bedding and fertilization, where needed for improved tree survival and growth, appears to be a more appropriate management scheme. The forest manager must strive for multiple-use land productivity while maintaining a relatively stable ecosystem on these delicate sites.

PEATLANDS ARE OFTEN MANAGED FOR FOREST PRODUCTION

Peatlands occupy vast areas of permanently wet lands, especially in cool climate regions such as those found in Scandinavia, Siberia, Canada, and parts of the British Isles. These organic soils occupy as much as 15–30% of the total land area of Sweden and Finland, where they often support poor stands of spruce and pine. In Finland, swamp drainage for forestry purposes has been carried out on over 2.5 million ha. Another 3 to 4 million ha of peatlands are worthy of drainage in regions in which forest production is favorable, while the remaining 3 million ha will be left to nature conservancy, according to Huikari (1973). Holmen (1971) reported that about 7 million ha, 17% of the land area of Sweden, may be regarded as peatlands with little or no forest production because of adverse water and nutrient conditions. Another 1 million ha have been drained. Water is generally controlled at 20 to 40 cm below the soil surface with a series of interconnecting ditch drains and check dams.

Canada has 130 million ha of peatlands, much of which is forested (MacFarlane, 1969; Kuhry et al., 1993), and large areas of forested peatlands occur in Minnesota, Michigan, and parts of New England. These forested "muskegs" illustrate nicely the two major interests of foresters and ecologists in these unique soils. The first interest is in their stability in the face of perturbation (Sayn-Wittgeostein, 1969); the second is in the productive capacity of the muskeg forests (Vincent, 1965). Many of these organic soils are shallow accumulations over bedrock. These shallow peatlands are often quite productive of timber, wildlife habitat, and water; but when they are disturbed, they break down quickly and are reformed at an astonishingly slow rate. On

deep peat, there is often such a high water table rise after logging that reforestation of the site is very difficult. The management of these complex lands is difficult, and our understanding of the proper techniques for the protection and utilization of peatlands remains incomplete.

Holmen (1969) pointed out that afforestation of drained peatlands is not generally successful without the use of fertilizers. Phosphorus and potassium are the elements most often deficient, and applications of 45 kg phosphorus and 120 kg potassium per hectare often suffice for 15 to 20 years. On poorly decomposed peat, it is also advisable to apply about 50 kg nitrogen per hectare. Micronutrients are not generally deficient.

Water control in peatlands can do more than increase forest production. It can improve the habitat and supply of food for wildlife and enhance recreation possibilities and other types of multiple uses. However, drainage and site preparation may increase fire danger, subsidence, and nutrient loss and can lead to irreversible drying (Charman et al., 1994; Trettin et al., 1997). There has been concern that drainage and site preparation might lead to increases in CO_2 and CH_4 emissions from peatlands (Roulet and Moore, 1995; Minkkinen et al., 1997). Although the emission of these gases is increased by forest practices, the additional CO_2 fixation by rapidly growing forest plantations more than offsets the increased emissions and makes drained and site-prepared peatland sites net carbon sinks (Laine et al., 1997). Many peatlands are now threatened by global warming (Trettin et al., 2006). Lower water tables and higher temperatures, associated with global warming, can cause significantly increased decomposition, disturbing the balance between accrual and depletion necessary to sustain peatlands.

GELISOLS PRESENT UNIQUE CHALLENGES

In regions with a long period of winter cold and a short period of summer warmth, a layer of frozen ground develops that does not completely thaw during summer. This perennially frozen ground is termed permafrost. Soils with permafrost are placed in a separate order, Gelisols, in the USDA classification system (Bockheim *et al.*, 1997). Gelisols, largely a Northern Hemisphere phenomenon, occur in a circumpolar zone in North America, Greenland, Europe, and Asia (Margesin, 2009). Permafrost may also occur in the periglacial environments of high mountains in either hemisphere. Nearly half of the area of Canada and the former Soviet Union is affected by permafrost (Baranov, 1959; Brown, 1967). All of Greenland and 80% of the land surface of Alaska are underlain by some type of permafrost (Ferrians, 1965).

Permafrost can occur in either mineral or organic soils. It is sometimes divided into three types: (1) seasonally frozen ground that is not actually permafrost since these areas may thaw during a small portion of the year; (2) discontinuous permafrost, where the frozen layer is discontinuous in horizontal extent; and (3) continuous permafrost, where the frozen layer is continuous in horizontal extent. In Alaska, the tree line and the division between continuous and discontinuous permafrost generally coincide; however, in Asia, the zone of continuous permafrost extends far south of the tree line (Figure 12.4). Climate change has greatly increased our awareness of the importance of permafrost, since deterioration of established permafrost can contribute large amounts of CO_2 to the environment (Koven *et al.*, 2011).

The management problems associated with permafrost generally arise from the important role that vegetation plays in insulating the soil from temperature change and thus perpetuating the permafrost layer. When forests are harvested or burned or when tundra vegetation is destroyed by vehicular traffic during the summer season, the permafrost layer may thaw. This leads to soil subsidence and either flooding or cryoturbation of the soil (Nicholas and Hinkel, 1996; Swanson, 1996; Arseneault and Payette, 1997; Bridgham *et al.*, 2001). Either of these changes in soil physical properties usually precludes new plant growth for decades while the soil structure and permafrost layer redevelop.

The placement of structures such as roads, buildings, and pipelines on permafrost soils also presents

Discontinuous permafrost

Continuous permafrost

Sea bottom permafrost

——— Treeline

Figure 12.4 Distribution of permafrost and location of the tree line in the northern hemisphere (French, 1976). Used with permission.

great difficulties if the soil is to be prevented from thawing (Brown, 1970). It is imperative that management practices that preserve the permafrost layer be developed. This means only partial removal of the tree canopy, the construction of insulating pads beneath buildings, and the restriction of vehicular travel over the ground.

TURNING MINE SPOIL INTO SOIL IS A LONG-TERM CHALLENGE

Revegetation of land that has been drastically disturbed by open pit mining for coal, iron ore, sand, phosphates, and other resources is a slow process at best. These wastelands must often be graded, partially resurfaced with fertile soil, fertilized and limed, and stabilized before afforestation is attempted. The kind and degree of amelioration depend on the physical, chemical, and biological properties of the spoils. These properties vary, depending on lithology, spoilage, degree of weathering, and erosion processes. Bare spoils cause serious environmental problems in many places by drastically affecting adjacent land, water, forests, or aesthetic values. Because of the widespread concern about their degrading effect on the environment, reclamation of spoil banks has received considerable research attention in recent years (Burger and Torbert, 1992). There are grass, shrub, and herbaceous species adapted to most classes of spoils, but there are few tree species adapted to the most severe spoil sites.

It is common practice to establish grasses or mixtures of grasses and legumes on severe spoil sites. After these species are seeded, slopes are generally mulched with straw (4 to 5 Mg ha^{-1}) and the mulch is anchored with asphalt emulsion. After a vegetative cover becomes well established, trees are often planted. Care must be taken to avoid cover crops that offer severe competition with trees (Fisher and Adrian, 1980).

Some spoils support excellent tree growth without any site preparation or soil amendments, while others require grading, terracing, harrowing, or ripping (Dunker et al., 1995). Grading of spoils high in clay content may cause compaction, thereby reducing air and water infiltration, root penetration, and tree growth. By contrast, Czapowskyj (1973) reported on studies in Pennsylvania indicating that coarse-textured anthracite spoils were improved by compaction resulting from grading operations. He also found evidence that calcium, magnesium, and potassium concentrations were higher in certain graded spoil types than in their ungraded counterparts.

Removal of impeding soil layers to extract material from beneath them often changes soil drainage class. Replacing these soil layers often does little to re-create sites that will support the tree species that occupied the area before mining. Darfus and Fisher (1984) found that when Spodosol soils that

supported slash pine were dredge mined, the spodic horizon was disturbed and post-mining soil moisture regimes did not support slash pine.

All essential elements are found in most overburden, but concentrations vary from very low to toxic. Most mine spoils are deficient in nitrogen, and vegetative growth on these spoils results from nitrogen added by rainfall, biological fixation, or fertilizer addition. However, Cornwell and Stone (1968) found that certain anthracite and black shale spoils supplied sufficient nitrogen for good tree growth as a result of weathering of these materials. Spoils of other mining operations are generally devoid of organic material and nitrogen. May et al. (1969) reported that while kaolin mining spoils were extremely variable, most of them were deficient in nitrogen, as well as in phosphorus, potassium, and calcium. Marx (1977) obtained improved survival and growth of pine seedlings planted on spoil banks when the seedlings were inoculated in the nursery with select strains of ectomycorrhizal fungi.

Levels of available iron, manganese, zinc, and other metals may be sufficiently high in very acid soil materials to pose a serious problem in the establishment of vegetation. Fortunately, toxic strip mine spoils are not widespread, and most toxic conditions can be eliminated with applications of limestone. Although lime is a relatively inexpensive material, the extremely high application rates often required on acid mine waste can be quite costly. From 10 to 90 Mg ha^{-1} may be required to neutralize the acids contained in spoil materials before leguminous plants and grasses can be grown successfully (Czapowskyj, 1973).

Vegetation usually grows better on reclaimed spoils if a variety of amendments are used to ameliorate the site (Schoenholtz et al., 1992). A wide variety of organic amendments have been tested including papermill sludge, municipal sewage sludge, mushroom compost, pulverized fuel ash, wood chips and sawdust, and shredded municipal garbage (Sabey et al., 1990; Feagley et al., 1994; Pichtel et al., 1994; Stark et al., 1994; Sort and Alcaniz, 1996; Perkins and Vann, 1997). All of these help promote vegetation establishment to some degree.

Inoculation of the sites with mycorrhizal fungi appears to be as important as the addition of organic matter (Shetty *et al.*, 1994). Ectomycorrhizae reinvade the site rapidly if there are undisturbed areas nearby to serve as sources of spores (Gould *et al.*, 1996). The soil-borne spores of arbuscular mycorrhizae (endomycorrhizae) do not reinvade a site readily. In addition, the spores of these organisms have usually been eaten or killed in stockpiled topsoil, so it is usually necessary to plant mycorrhizal seedlings or to inoculate the site artificially (Allen, 1991; Mehrotra, 1998).

NURSERY SOILS ARE MANAGED AS INTENSIVELY AS AGRICULTURAL SOILS

The successful operation of a forest nursery involves large investments in labor, buildings, and equipment, and only an operation sufficiently large to make effective use of the specialized machinery can normally be justified. Except in special situations where part of the equipment may be rented or made to serve double duty, a nursery of 5 to 10 ha might be considered a minimum size for an economic operation (Figure 12.5).

A nursery of this size can produce 8 to 16 million 1-0 pine seedlings per year, providing seedlings

for 4000 to 8000 ha of plantations. A substantially larger area is needed if a rotation with a green manure or cover crop is followed, or in cool climates where two years or more are needed to produce seedlings of satisfactory size for field planting. Armson and Sadreika (1974) have presented excellent instructions for nursery soil management for northern coniferous species, and Duryea and Landis (1984) have produced a valuable general manual of bare-root forest nursery practices.

Although seedling production in the United States has been steadily increasing, especially in the private sector, the production of both hardwood and coniferous seedlings must increase to meet the need for expanded areas of plantation forestry. The production of containerized nursery stock has also been steadily increasing, but in many parts of the temperate zone, "bare root" production of nursery stock still predominates.

NURSERY SITE SELECTION IS IMPORTANT

Because nurseries are not easily moved from one site to another, it is especially important that much thought be given to soil and site factors that make for efficient operations in the production of quality seedlings (Morby, 1984). Soil depth and texture are of prime importance for nurseries. A soil depth of 1.0 to 1.5 m is normally desired, without a radical textural change between horizons, to ensure adequate drainage and aeration and space for root development.

Areas exposed to extreme temperatures and strong winds should be avoided. In cold climates, location near large bodies of water will moderate temperature extremes. On the other hand, terrain should be selected that permits free drainage of cold air, thus reducing the risk of frost injury. The nursery site should be located near a reliable transportation network and as close as possible to major areas of reforestation, both for convenience and to minimize differences in growing season length and other climatic conditions between the two areas.

Figure 12.5 A southern pine nursery on a deep, sandy loam soil shortly after seed germination.

SOIL TEXTURE IS OFTEN THE KEY TO SUCCESS OR FAILURE

Soil texture influences water regime, cation exchange capacity and nutrient retention, susceptibility to sand splash, surface washing, compaction, and the ease of lifting. Loamy sands to sandy loams are generally preferred textures for nursery soils. Coarse-textured sands are acceptable for nurseries, but coarse textures place special demands on management of water, fertility, and organic matter. Soils at the other end of the textural range, those containing more than 30–40% silt plus clay, are definitely to be avoided. Soils containing a high percentage of silt often become very hard when dry, erode easily, and demand a meticulous watering regime. Soils high in clay are often difficult to cultivate, make lifting difficult, and require considerable effort to adjust moisture, acidity, and nutrient levels. They are also slow to warm in the spring, and in cold climates, seedlings may suffer from frost heaving. Furthermore, seedling root systems are likely to be injured during lifting from fine-textured soils.

SITE LEVELING MAY BE NECESSARY BUT IS PROBLEMATIC

Level terrain is preferred for nurseries on sandy soils, but a slight slope may be necessary to provide surface drainage for finer-textured soils. Sufficient grading and leveling should be done to minimize the erosion hazard to beds, translocation of fertilizer salts, and accumulation of water in depressions. However, exposing large areas of subsoil should be avoided. The movement of large amounts of topsoil may result in nutrient-poor or even calcareous subsoil being near the surface. Where extensive areas of topsoil must be removed during leveling operations, care should be taken to stockpile the topsoil and then redistribute it over areas that were exposed during grading. If such redistribution is not possible, a special effort must be made to adjust soil acidity and to increase the organic matter content of the disturbed soil before any attempt at seedling production is made.

Ideally, a soil map will have been used in siting the nursery. A more detailed soil map of the nursery area, including any reserve area for later expansion, should be made as soon as possible after site selection. The map should indicate the soil texture in both surface and subsurface horizons, depth to restricted drainage, changes in soil acidity and nutrient status, and other properties critical to seedling growth. For example, hardwood seedlings generally require more fertile and less acid soils than conifer seedlings, so that if both hardwoods and conifers are grown in the same nursery, it may be possible to reserve the more fertile areas for hardwoods. Furthermore, a detailed map makes a convenient base for planning and record keeping – two essential steps in any nursery operation.

SOIL FERTILITY AND ACIDITY MANAGEMENT ARE CRUCIAL TO NURSERY SUCCESS

A reasonable concentration of nutrients in the original soil is desired, but since they can be added as fertilizers, the soil reserve of nutrients is not of overriding importance. It is more important that the soil not contain high concentrations of soluble salts, free carbonates, or toxic materials, and that it contain a reasonable level of organic matter.

Optimum soil acidity for most tree species lies between pH 5.2 and 6.2. The lower half of this range appears to be best for conifers, while many deciduous trees grow best in the pH range from 5.8 to 6.2 (Figure 12.6). Although many species thrive when soil acidity is high, the efficiency of fertilizer use is decreased in strongly acid soils. The amounts of native calcium, magnesium, and potassium are usually low in very acid soils; furthermore, added potassium is more easily leached and the availability of added phosphorus is reduced under these conditions. Aluminum, iron, and manganese tend to be much more soluble in acid soil. Moderately high concentrations of these elements probably have little direct effect on conifer growth, but they may well affect some deciduous seedlings and the cover or green manure crops grown in rotation with the seedlings.

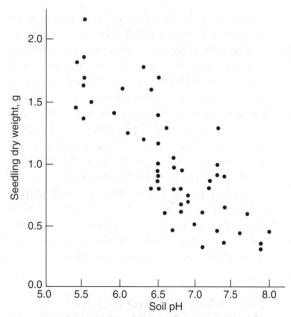

Figure 12.6 Relationship of soil acidity to mean total dry weight of 2-0 red pine seedlings (Armson and Sadreika, 1974). Courtesy of the Ontario Ministry of Natural Resources.

At the other end of the range, near neutrality, some species have difficulty obtaining sufficient iron, and sometimes manganese, for normal growth. Pine seedlings are particularly subject to iron deficiencies in soils above approximately pH 6.5. Most tree species, including pines, can be grown in soils above pH 7.0 if their nutritional requirements for iron, manganese, and phosphorus can be met with chelates or foliar sprays. In fact, they can be grown successfully in some finer-textured soils near neutrality without micronutrient additions if the soils are well supplied with micronutrients and organic matter. Some root diseases are more prevalent in near-neutral and alkaline soils. Obviously, soils selected for nursery sites must permit control of acidity. If irrigation is anticipated, the water should be checked for soluble salts. For example, 25 cm water containing 500 ppm soluble salts adds 1250 kg ha^{-1} of these materials to the soil. This is more than enough salt to significantly affect the pH value of weakly buffered sandy soils.

Simple indicator test kits and pH meters are useful for checking soil acidity, but it is well to keep in mind that changes in soil acidity of as much as 0.5 to 1.0 pH unit can take place seasonally. Furthermore, microsite differences caused by leveling or organic matter applications often result in wide variations in pH value among soil samples. Where adjustments in soil acidity are needed, agricultural limestone or sulfur is usually added. However, small changes in soil acidity can be made by using fertilizer sources that have acidic or basic residues. The application of 200 kg ha^{-1} of nitrogen as ammonium sulfate, for example, can increase the acidity of sandy soil as much as 1.5 pH units during a season.

Care should be exercised in using limestone or sulfur to alter the acidity of sandy soils to prevent excessive change in acidity. The kinds and amounts of clay and organic matter control the soil's cation exchange capacity, which in turn determines the amount of lime or sulfur required to change soil acidity by a given amount. Generalized curves useful in determining the amount of agricultural limestone required to lower the acidity of some soil textural groups are presented in Figure 12.7. The same curves may be used to predict the amount of sulfur required to raise soil acidity from the present level to a desired level. About one-third as much sulfur as agricultural

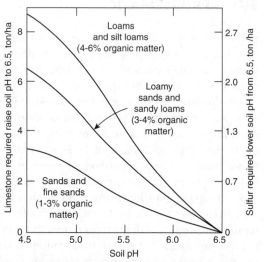

Figure 12.7 Tons of agricultural limestone required to lower soil acidity or tons of sulfur required to increase soil acidity. Differences in requirements correspond to differences between present and desired pH values.

limestone is required to lower the soil acidity by a given amount.

SOIL ORGANIC MATTER IS ALSO A KEY TO PRODUCTIVE NURSERIES

The maintenance of a reasonable level of organic matter is particularly important in sandy nursery soils to retain fertilizer elements against leaching and to buffer the soil against rapid changes in acidity. Because these soils contain little clay, their exchange capacity resides almost entirely with the organic fraction. Organic matter additions also improve soil structure, friability, and water intake and retention and reduce soil crusting and erosion. The organic matter of a soil is the seat of microbial activity, and it contains the reserve of most nutrients in sandy soils.

The practical level of organic matter that can be maintained in sandy soils is largely dictated by climatic conditions. For example, maintaining 4–5% organic matter in Oregon nursery soils may be less difficult than maintaining 1–2% in southeastern coastal plain soils. Soil organic matter can be increased briefly by the use of green manure crops. Significantly longer increases can be achieved by the addition of compost, sawdust, peat, or other organic material. However, such increases are temporary because of the decomposition of these materials by soil organisms, and they must be replaced on a regular schedule (Davey, 1984; Rose *et al.*, 1995).

COVER CROPS SERVE SEVERAL PURPOSES

Nursery seedlings are often grown in rotation with green manure and cover crops in attempts to replace organic matter lost by the high biological activity associated with well-aerated and fertilized nurseries. Each 1% of organic matter in the top 20 cm of a sandy soil is equivalent to 24 to 28 Mg of dry matter per hectare. Therefore, a nursery soil may normally contain 50 or 60 mg of organic matter. While a good green manure crop may add 20 to 30 Mg of fresh material to the soil, only 5 to 6 Mg of this is dry

matter. Moreover, when mixed into the soil, 50–75% of this material will decompose within a few months, leaving only 1 or 2 Mg of relatively stable humus to add to the soil's organic matter reserve. This is little more than that required to replace normal decomposition losses during a cropping season. While the benefits from green manure organic materials are rather short-lived, the use of green manure is, nonetheless, the most practical means of maintaining the productivity of many nursery soils.

Rotations with green manure crops can also be an effective means of reducing infestations of nematodes and soil-borne diseases that often infest nursery soils, provided that the crops selected are resistant to the growth and development of such organisms. A cover crop may also be selected that will increase the population of certain arbuscular mycorrhizal fungi useful to the hardwood seedlings that follow. The selection of a suitable green manure crop should also be related to soil and climatic conditions. For example, nursery soils are often too acid for good growth of many leguminous crops. Some species of vetch, lupine, soybeans, and field peas may be used, but to obtain rapid plant growth and abundant production of organic material, one must usually resort to crops such as oats, rye, millet, corn, sudangrass, sorghum, or a sorghum–sudan hybrid.

Cover crops also protect the soil from wind and water erosion and reduce nutrient leaching during any extended period when an area is not used for the production of nursery stock. A cover crop should be plowed under before it reaches maturity to avoid the nuisance of a volunteer crop in later nursery beds. However, the crop should be allowed to grow as long as feasible because of the increased dry matter production and because mature and lignified materials do not decompose as rapidly after being turned into the soil.

OTHER ORGANIC ADDITIONS ARE GENERALLY NECESSARY

Composted and raw organic materials such as sawdust, straw, peat, and ground bark have long been important sources of materials for improving the physical conditions of soil and increasing organic matter in nurseries. Since these are often waste materials, they are relatively inexpensive to obtain but costly to handle. More recently, the supply of these materials has diminished to the point where they are used principally on nursery soils with critically low levels of organic matter or to correct problem areas, such as those resulting from leveling operations. At the same time, digested sewage sludge has become a common amendment for increasing soil organic matter content.

Composted materials consisting of sawdust, straw, and other materials with a wide carbon to nitrogen ratio are mixed with topsoil and/or mineral fertilizers or animal manure and allowed to partially decompose or ferment. These materials range in C:N ratio from over 1000:1 for raw sawdust, to 500:1 for fresh bark, 60:1 for peat, and 10:1 for digested sewage sludge. During the fermentation process, carbon dioxide develops, the C:N ratio is decreased, and the mineral constituents are made available. Composted materials can be applied at fairly heavy rates of 40 to 60 Mg ha^{-1} and incorporated into the seedbed without danger of significant nitrogen immobilization. There is a danger of introducing root pathogens with some composted materials. Fumigation of the materials prior to application, or of the nursery soils after application, will minimize this problem. Moreover, composted materials are quite expensive to use, so raw organic materials are the more common amendments for nursery soils. In using raw organic materials, care must be exercised to prevent excessive nutrient immobilization during critical periods.

Relatively inert materials such as acid peat, sawdust, and digested sludge are slow to decompose and thus persist in the soil for longer periods. While they are less expensive to use than composted materials, they can result in nitrogen shortages at critical periods in seedling development, and more caution is required in their use. Allison (1965) reported that the nitrogen requirements of microorganisms that decomposed sawdust ranged from 0.3% of the dry weight of Douglas fir wood to 1.4% for red oak wood. Digested sludge often has a high content of heavy metals that may damage tree seedlings or cause other environmental problems (Rose *et al.*,

1995). Great care must be taken in choosing the source of digested sludge for use in nursery soils.

When raw sawdust or other highly carbonaceous materials are to be used directly in nursery soils, they should be applied sufficiently in advance of the seeding operation so that partial decomposition takes place prior to seed germination. These applications may be made prior to planting of a cover crop or green manure crop or more than a month prior to seeding the nursery beds. In any event, 10 to 20 kg of extra nitrogen is needed per milligram of dry organic materials added in order to hasten decomposition and minimize the danger of a nitrogen shortage to the nursery crop. Since this extra nitrogen will become available to nursery stock after the added organic material has largely decomposed, care must be taken to prevent excessive levels of nitrogen in the planting bed at later stages of seedling development. This can be accomplished by monitoring the concentration of nitrogen in seedling tissue, or the available nitrogen in the soil, and applying supplemental nitrogen only when the results indicate a need exists.

A FERTILIZATION PROGRAM IS ESSENTIAL TO EXCELLENT NURSERY STOCK PRODUCTION

While added composted materials, manure, and organic residues are acceptable means of maintaining soil organic matter levels that contribute to good physical properties, they generally do not contain sufficient nutrients to replenish those lost during cropping. Consequently, the maintenance of nursery soil fertility depends largely on the use of commercial fertilizers, applied alone or in combination with organic materials and green manure crops (van den Driessche, 1984).

Relatively large amounts of fertilizers are often used in nurseries to replace nutrients removed in the harvested crop, as well as those lost by leaching or fixed in the soil. Leaching losses can be rather high in irrigated sandy soils even where the organic matter content is maintained at a reasonable level. Crop removal may also be substantial because both seedling roots and tops are removed. Slash pine

Table 12.2 Nutrient concentrations and contents in 8-month-old slash pine seedlings.

Component	N	P	K	Ca	Mg	Al	Cu	Fe	Mn	Zn
					Nutrient Concentration (mg/kg)					
Tops	1.20	0.15	0.60	0.43	0.09	408	4.8	118	161	48
Roots	0.86	0.23	0.59	0.36	0.08	1947	10.8	617	66	40
					Nutrient Content (kg/ha)					
Tops	44.6	5.6	22.3	16.0	3.3	1.5	0.18	0.44	0.60	0.18
Roots	8.7	2.3	6.0	3.6	0.8	2.0	0.01	0.62	0.06	0.04
Total	53.3	7.9	28.3	19.6	4.1	3.5	0.19	1.06	0.66	0.22

seedlings from a Florida nursery, grown at the rate of 2.2 million plants per hectare, weighed 16 500 kg (green weight) when harvested at eight months of age. Oven-dry weights averaged 3715 and 1012 kg ha^{-1} for seedling tops and roots, respectively. This results in the removal of large amounts of nutrients (Table 12.2). Armson and Sadreika (1974) reported even larger removals of nutrients in two-year-old red pine and spruce (Table 12.3).

SOIL TESTING IS A MUST

Addition of relatively large amounts of organic materials and mineral fertilizers, along with removal of nutrients in the crop and by leaching, results in considerable fluctuation in nutrient levels in nursery soils during a cropping period. A soil test provides information on the fertility status of soils that is particularly useful prior to seedbed preparation. Soil acidity and nutrient status are

Table 12.3 Amounts of nitrogen, phosphorus, and potassium (kg ha^{-1}) in 2-year-old red pine and white spruce seedlings planted at four seedbed densities.

Nutrient	Seedbed Density (Plants/m^2)			
	108	215	430	861
	Red Pine			
N	113	119	164	220
P	16	22	27	35
K	56	80	109	133
	Write Spruce			
N	88	107	150	167
P	19	24	31	41
K	44	60	71	114

normally adjusted prior to seeding the tree crop. Since plants are not available for tissue tests at this time, soil tests are the principal diagnostic tool. Soil samples should be collected from seedbeds four to six weeks prior to seeding to allow adequate time for sample analyses and the application of any necessary fertilizers.

A composite soil sample should be taken from each nursery management unit or from areas within a management unit that differ in soil type, past management, or productivity. The composite sample should consist of 12 to 15 subsamples (cores) taken at random over a uniform area. Cores are normally taken to a 15- to 20-cm depth with a 2.5-cm-diameter soil tube and collected in a clean container. After air-drying and mixing, about 250 ml is packaged, labeled, and submitted to a laboratory for testing. The location from which the sample was collected should be identified on a map of the nursery on which each management unit is designated. A detailed soils map can serve as a basis for record keeping as well as management planning. Chronological records of soil test results, fertilizer treatments, fumigation, seeding date and rate, irrigation, and other cropping practices are useful to the manager in diagnosing problem areas and planning for future crops. Soil fertility tests should be carried out annually.

FERTILIZATION IS A TWO-STEP PROCESS

Limestone or sulfur, when used to adjust soil acidity, and base fertilizers are normally applied broadcast before fumigation and seeding or transplanting. In this way, the amendments can be well mixed throughout the rooting zone. If limestone or sulfur is needed, it should be mixed with the soil two or three months in advance of seeding, to allow time for them to react with soil components. By contrast, most fertilizer materials are applied near the time of seeding or transplanting to minimize the danger of leaching losses.

Phosphorus deficiencies are generally corrected by the application of 40 to 100 kg ha^{-1} of phosphorus as ordinary or concentrated superphosphate. The actual amount needed depends on soil-fixing or retention properties and the degree of deficiency, as indicated by a soil test. In acid sandy soils, it is often more economical to apply larger amounts of less soluble forms of phosphorus such as ground rock phosphate. Phosphorus in this form is less soluble and less likely to leach beyond the seedlings' shallow root system than phosphorus in soluble phosphates. Rates of 500 to 2000 kg ha^{-1} are normally applied at one- to three-year intervals, or as indicated by soil test results.

Nitrogen and potassium needs may be met by mixing part of the material into the seedbed prior to planting and applying the remainder in two or more "top-dress" applications during the growing season. Split applications are particularly effective for nitrogen because of the ease with which it can be lost from the system and the high salt index of most nitrogen sources. Soluble fertilizer salts may result in drainage to transplants when present in high concentrations. Usually, no more than 30 to 50 kg of mineral nitrogen is applied to sandy soils prior to seeding. Higher rates may be used on finer-textured soils or when less soluble sources, such as urea-formaldehyde or sulfur-coated urea, are used.

Most sources of nitrogen and potassium are completely water-soluble, and their solutions can be injected into the irrigation system for direct application. Solid materials can be applied as a top-dress to the seedlings. The number of top-dress applications needed depends on crop performance, soil conditions, rainfall, and irrigation patterns. However, three or four applications per year are not uncommon for sandy soils. Potassium is sometimes top-dressed on seedlings a few months prior to lifting from the nursery bed in an effort to harden

and condition them to withstand cold temperatures and transplanting shock. Because the application of nitrogen can induce other nutrient deficiencies (Teng and Timmer, 1995; van den Driessche and Ponsford, 1995), assays of seedling foliar nutrient content during the growing season are useful for maintaining balance.

TREE CONDITIONING FOR TRANSPLANTING IMPROVES SURVIVAL AND GROWTH

Undercutting, wrenching, and root pruning are widely used methods of conditioning seedlings of certain species. When properly scheduled and conducted, these cultivation treatments can be used to control top–root ratio, carbohydrate reserve accumulation, and the onset of dormancy. A reduction in top–root ratio may improve the success of plantings on droughty soils. Increased food reserves and early dormancy generally reduce cold damage and ensure better survival on adverse sites but, most importantly, an efficient root system is the key to success where transplants are stressed by low soil moisture.

Undercutting consists of passing a thin, flat, sharp blade beneath seedlings at depths of 15 to 20 cm. Undercutting is usually performed only once in a bed of seedlings several months before lifting. The purpose of the operation is to sever seedling taproots, reduce height growth, and promote development of fibrous root systems. It has been used successfully in nurseries for pines, oaks, and other species that tend to form taproots.

Wrenching has been described in detail by Dorsser and Rook (1972) as a method of conditioning and improving forest seedlings for outplanting. Their research on wrenching was largely confined to radiata pine, but the technique has also been used on other species, including hardwood seedlings. Wrenching can be performed with a sharp spade inserted beneath the seedlings at an angle that severs their taproots and then slightly lifting to aerate the root zone. However, it is usually accomplished by passing a sharp, thin, tilted blade beneath seedbeds at certain intervals following undercutting to prevent renewal of deep rooting or height growth.

Timing of the conditioning sequence is important, and it varies with tree species and soil and climate factors. The sequence generally begins when seedlings are near the height desired for outplanting, but with at least two months of growing season remaining. With radiata pine in New Zealand, the operation begins when seedlings are about 20 cm in height. A thin, flat blade is passed beneath seedlings at a depth of about 8 to 10 cm with minimum disturbance to lateral roots. After undercutting, wrenching is performed at one- to four-week intervals for the remainder of the seedlings' growing season, using a thicker, broader blade, tilted 20° from horizontal, with the front edge lower than the rear. The type of seedling desired dictates the interval between wrenchings.

The term root pruning is sometimes used to describe the undercutting operation outlined above. It has also been used to describe the trimming operation performed after seedlings have been removed from the ground. However, in New Zealand and other countries where wrenching is commonly practiced, root pruning is accomplished by vertical cutting with a coulter, or rolling circular blade, between rows of seedlings at four- to six-week intervals. The vigorous lateral root growth initiated after the taproot is severed is partially controlled by cutting the roots 8 to 9 cm from the trees. Wrenching and vertical root pruning are often used with slow-growing conifer species, such as spruce and northern pines, to produce 2 + 0 seedlings with a root area index equal to that of 1 + 1 or 2 + 1 transplants (Armson and Sadreika, 1974; Duryea and Landis, 1984).

BIOCIDES PROTECT SEEDLINGS BUT ALTER NURSERY SOIL PROCESSES

Nursery crops are often grown as a single species of densely populated, rapidly growing succulent plants. Consequently, they are particularly vulnerable to the ravages of diseases, insects, and competing vegetation. Many nursery pests can be controlled, or their damage minimized, by manipulating the soil environment to ensure sturdy, vigorous seedlings. For example, the soil acidity should be kept below

pH 6.0 to discourage the development of pathogenic fungi. Damping-off injury of very young seedlings, commonly associated with *Pythium, Rhizoctonia, Phylophthora, Fusarium,* and *Sclerotium* fungi, is rarely serious if the nursery soil is maintained at pH 5.5 or lower. High soil nitrogen concentrations at the time of germination should also be avoided. Low to moderate plant density and adequate moisture combined with good drainage of the surface soil generally result in optimum seedling growth without favoring disease and pests. Rotating tree seedlings with green manure and cover crops that are not hosts of disease organisms, insects, and nematodes is an accepted method of reducing the pest problem. However, in older nurseries, pests often cannot be controlled by soil management techniques alone, and biocides must be employed.

Soil fumigation to control weeds, soil-borne diseases, nematodes, and soil-inhabiting insects is a common and economically feasible practice in tree nurseries in many parts of the world. In addition to controlling damping-off organisms, fumigation with such compounds as methyl bromide, ethylene dibromide bichloropropene, and chloropicrin controls most root rots, probably the most destructive of all nursery soil pathogens. Fumigants are applied as gases under plastic covers or injected into the soil as a solution. In either event, the best results are obtained if the chemicals are applied when surface soil temperatures are between 100 and 30 °C, when soil moisture is adequate for good seed germination, and when the soil has been recently cultivated to ensure good penetration. Penetration is not as good in fine-textured or compacted soils as in sands and sandy loams, and organic matter reduces the effectiveness of fumigants by sorption and degradation. Therefore, higher than normal amounts of fumigants may be needed on fine-textured soils with high levels of organic matter.

Methyl bromide use is being phased out because of its potential to destroy the ozone layer; however, many other broad-spectrum biocides are under development to take its place. All biocides alter the soil flora and fauna to a greater or lesser extent. Many microorganisms reestablish themselves in the soil within a few weeks after fumigation. The biological diversity of the soil microbial population is reduced, but the functional diversity is apparently not altered and decomposition processes are not greatly affected (Degens, 1998). However, mycorrhizal fungi may not recolonize nursery soils in some regions for several months or even years. This is particularly true of arbuscular mycorrhizal fungi that produce soil-borne, rather than air-borne, spores.

Long time lags for the reestablishment of ecto-trophic mycorrhizae are more likely to be found in cold climates or where exotic species have been recently introduced. The time-honored methods of inoculating seedbeds by mixing a small amount of forest soil into the beds or using forest litter as protective mulch can hasten reestablishment. Because these methods may introduce pests into the nursery, commercial inoculum, which is now available in some areas, may be a better source of mycorrhizal fungi (Bowen, 1965; Marx, 1977). Unfortunately, no source of endotrophic mycorr-hizal inoculum other than soil or litter exists. Nur-series that produce arbuscular mycorrhizal tree species – most hardwoods – must either avoid the use of broad-spectrum biocides or systematically inoculate their nursery beds.

CHAPTER 13
Fire Influences

OVERVIEW

Fires are major features of the life, death, and rebirth of most forests around the world. Repeated low-intensity fires may shape the forest by killing small trees, rejuvenating fire-tolerant grasses, and accelerating the cycling of nutrients. High-intensity fires kill the majority of trees, oxidize large quantities of nutrients such as nitrogen, and alter plant–soil interactions for decades. Fire oxidizes organic matter to form carbon dioxide and water, releasing tremendous amounts of energy as heat. Nitrogen in organic matter is oxidized (not volatilized) to N_2 and various N oxides and is lost from the system. Calcium in organic matter is converted to calcium oxides and bicarbonates, which are lost in wind-blown ash or retained on site. Phosphorus in organic matter may be lost as a gas or released as barely soluble phosphate salts. Fires drastically alter the composition and activity of the soil biota, though little information is available on these changes and their implications for nutrient cycling. Despite the loss of nutrients in fires, nutrient availability to plants typically increases after fire as a result of heat-induced release of nutrients, reduced competition among plants, and perhaps through sustained changes in soil conditions (such as temperature and water content).

MOST FORESTS BURN

Most forests of the world experience fires, with climate-driven differences in fire frequency and intensity. Seasonally dry regions such as portions of the coastal plain of the Southeastern U.S. often experience low-intensity fires every few years, and longleaf pine forests are well adapted to frequent fires. Most temperate conifer forests experience fire regimes with moderate-to-high severity fires every 50 to 200 years. Montane and boreal conifer forests often burn at intervals of several hundred years. Deciduous temperate forests have variable fire regimes, with fire return intervals of a few decades (or less) to several centuries (or longer). In tropical regions, fires are common in seasonally dry environments, and rare in perennially moist areas. Even the wettest tropical rainforests may burn at a timescale of thousands of years (Sanford *et al.*,1985), and in response to direct and indirect forest clearing by people (Cochrane *et al.*,1999). In some years, the extent of fire may be greater in tropical and subtropical regions than in the rest of the globe (Goldammer, 1993).

Frequent, low-intensity fires (Figures 13.1, 13.2) tend to cover small areas, up to a few hundred hectares. Less common, severe fires burn much larger areas (Figure 13.3). Given that fires have been a natural part of the long-term evolution of many tree species and forest types, it's not surprising that soil impacts, even when severe, do recover over time (Certini, 2005).

In recent times, direct human use of fire may have had a greater effect on forests and forest soils than natural fires. Humans have historically used fire to clear land and manage wildlife; with the development of major agriculture in forested regions, fire became the tool of choice for removing trees and wood from soils in Europe, North America, South America, and around the tropics. Fire is also a tool for forest management, where it is used to reduce the risk of wildfire, to alter species

Ecology and Management of Forest Soils, Fourth Edition. Dan Binkley and Richard F. Fisher.
© 2013 John Wiley & Sons, Ltd. Published 2013 by John Wiley & Sons, Ltd.

Figure 13.1 Control burning in a loblolly pine/hardwood stand using a back fire with a light fuel load.

Figure 13.2 Periodic burning under controlled conditions results in park-like stands of pine, such as this never-harvested, frequently burned stand of longleaf pine in northern Florida, USA (Wade Tract Preserve, Tall Timbers Research Station).

composition, and to reduce the woody debris left after logging (Figure 13.4).

Fire effects on soils have intrigued scientists since the beginning of forest soil science. In Sweden, Hesselman (1917) showed that fire oxidized and lost a large portion of a forest's N capital, while also increasing N availability. The production of ammonium seemed highest between pH 4.5 and 4.9, whereas nitrate production was favored in the range of pH 5.5 to 6.9. In the western United States, Isaac and Hopkins (1937) reported that about 500 kg N ha^{-1} were lost from a slash fire in a Douglas fir clear-cut. Heyward and Barnette (1934) examined the effects of repeated surface fires in mature longleaf

pine forests, and concluded that the total N content of the soil was not substantially reduced.

FIRE PHYSICS LARGELY DETERMINE FIRE IMPACTS ON FOREST SOILS

The oxidation of organic matter releases large amounts of energy as heat. The amount of energy released in a fire depends on fuel consumption and

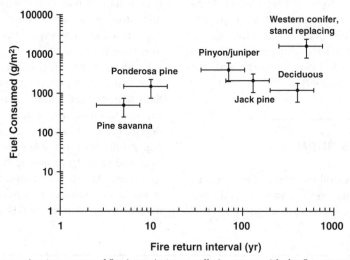

Figure 13.3 Fuel consumption (a measure of fire intensity) generally increases with the fire return interval in temperate forests of North America (based on Olson, 1981; Christensen, 1987).

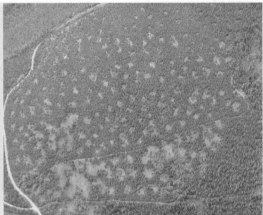

Figure 13.4 Fire has been used to reduce fuel loadings after harvesting, but piled debris burns with such high intensity that nutrient removal is maximized and soils are typically damaged. Left, pile burning in the Manitou Experimental Forest in Colorado (Courtesy of Lance Asherin). Fifteen months after the pile was burned, microbial activity beneath the burned pile was about one-fourth of that of soil from outside the pile (Jiménez Esquilín *et al.*, 2007). The aerial photo on the right shows the scars of pile burning 50 years later – a combination of effects of intense fire on soils, seeds, and plants (Courtesy of C. Rhoades). This practice is declining in many parts of the world, because of evidence of site degradation, poor revegetation by native species, and increased establishment of invasive weeds (e.g. Haskins and Gehring, 2004; Korb *et al.*, 2004).

the amount of moisture in the fuel (substantial energy can be absorbed in evaporating water from fuels). Plant biomass typically contains about 18 MJ kg^{-1} of dry material (Agee, 1993). How much energy is released during a wildfire? A severe wildfire in an old-growth Douglas fir forest may consume about 17% of the total biomass of 1000 Mg ha^{-1}, for a fuel consumption of about 170 Mg ha^{-1}. The energy release would be on the order of 3 million MJ ha^{-1}, or 300 MJ m^{-2}. Gasoline has an energy content of about 30 MJ L^{-1}, so the forest fire's energy release would equal that of burning about 10 L of gasoline across every square meter of ground area.

This tremendous release of energy as heat raises the temperature of the soil. The pattern of temperature increase depends on the rate of burn, the amount of fuel consumed, soil moisture, and conductivity properties of the soil. Rapidly advancing fires that consume little fuel have little effect on soil temperatures. Slower fires that consume more fuel may have temperatures exceeding 700 °C at the soil surface (particularly under slash piles), declining to 200 °C a few cm into the mineral soil, to normal levels below 15 or 30 cm depth. The heating of soils depends

heavily on the water content of the soil (Figure 13.5); water absorbs large amounts of heat without large increases in temperature, and vaporization of water consumes even more energy (2.5 MJ L^{-1}).

HEAT ALTERS THE ORGANIC MATTER REMAINING IN SOILS

Fire of course reduces the organic matter content of forests and soils, but the heat of fire also transforms some of the organic matter left behind. The size of organic matter particles and compounds may be reduced, and the overall chemical characteristics (oxygen-containing functional groups decrease) and reactivity change (Figures 13.6, 13.7). The functional implications of these changes may be large, but generalizations are not yet available. How much longer will altered compounds remain in the soil? Charcoal is typically viewed as lasting a very long time (see Chapter 4), and fire-altered humic materials do seem to be more resistant to microbial breakdown, but a great deal more work will be needed to provide strong evidence of typical changes, and deviations from typical responses.

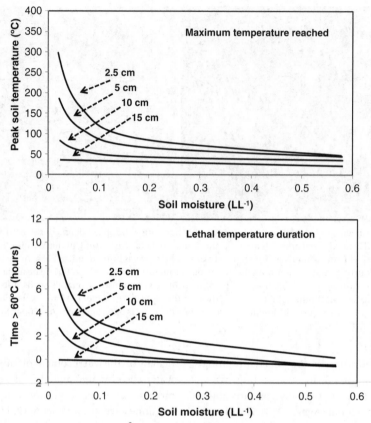

Figure 13.5 Burning of $13.5 \, kg$ of dry fuel m^{-2} led to heat penetration into a variety of soils. Higher soil moisture moderated the heating of the soil, as water vaporization consumed energy (upper). The duration of root-lethal temperatures also depended strongly on soil moisture (lower; from data of Busse *et al.*, 2010).

FIRE REMOVES NUTRIENTS BY FIVE PROCESSES

Losses of nutrients in fires result from the combined effects of these processes:

1. Oxidation of compounds to a gaseous form (gasification).
2. Vaporization (volatilization) of compounds which are solid at normal temperatures.
3. Convection of ash particles in fire-generated winds.
4. Leaching of ions in solution out of the soil following fire.
5. Accelerated erosion following fire.

The relative importance of these processes varies for each nutrient, and is modified by differences in fire intensity, soil characteristics, topography, and climatic patterns.

NITROGEN LOSSES ARE PRIMARILY FROM OXIDATION (NOT VOLATILIZATION)

Organic compounds contain nutrients, and some of these nutrients are in a reduced state ($R-NH_2$, $R-SH_2$). These reduced forms of N and S are oxidized at the temperatures reached in fires, releasing energy as gaseous, oxidized compounds are created.

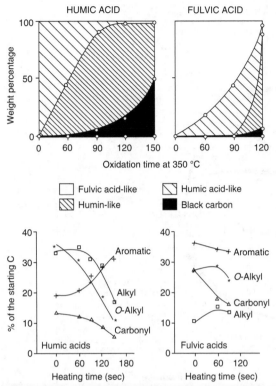

Figure 13.6 Heating soil organic matter (upper) shifts the humic acid compounds (alkali-soluble) toward increasing humin material (insoluble), and with sufficient time up to half becomes black carbon (charcoal). The same treatment shifts fulvic acid compounds (acid-soluble) to humic acids and humin. Within these broad classes of organic compounds, the proportions of the major functional groups are very sensitive to heating time (lower). Reprinted from Environment International, 30, José A. González-Pérez, Francisco J. González-Vila, Gonzalo Almendros, Heike Knicker, The effect of fire on soil organic matter—a review. 855–870, 2004, with permission from Elsevier.

For example, the combustion of an amino acid (cysteine) releases N_2 and SO_2 gases:

$$4CHCH_2NH_2COOHSH + 15O_2 \rightarrow 8CO_2 + 14H_2O + 2N_2 + 4SO_2 + energy$$

For some reason, these oxidation losses of nitrogen came to be mistakenly referred to as "volatilization" losses in the literature, as if the N simply evaporated when heated. Wood doesn't evaporate in fires, and neither does the N contained in organic matter.

Three approaches have been used to examine nutrient losses from fires. The simplest method involves heating samples in a furnace and measuring the change in nutrient content. Knight (1966) used this method to examine N losses from small samples of forest floor materials heated for 20 minutes at various temperatures (Figure 13.8). No N oxidized to gas at 200 °C, 25% was lost at 300 °C, and about 55% disappeared at 700 °C. Tiedemann et al. (1979) examined S losses in a furnace from the foliage of various species and found from 25–90% of the S was lost as temperature increased from 375 to 575 °C.

One problem with using furnaces is that a sample is heated from all directions, whereas fires generate gradients of temperatures within the forest floor. Mroz et al. (1980) placed forest floor samples in clay pots and then placed the pots in a furnace preheated to 500 °C for 30 minutes. This allowed rapid heating of the surface while lower levels remained cool. They found a substantial loss of N from the upper portion of the samples, but most of this N was recovered in the lower portion. On average, the net loss of N was only 3–10% of the total. In contrast, Knight's pattern in Figure 13.8 would indicate that about 45% of the N should have been lost. Losses of N in actual fires are probably much lower than would be suggested by furnace experiments on small samples which do not allow for the importance of temperature gradients.

The second approach attempts to quantify the change in nutrient content of fuels burned in the field. The high variability in fuel loading and in fire intensity poses a stiff challenge for accurate measurements. In general, N loss under field conditions appears related to the amount of organic matter consumed. Table 13.1 summarizes nutrient losses from fires, and Figure 13.9 illustrates this relationship for N from a variety of fires reported in the literature. The variation in organic matter loss accounts for about 85% of the variation in N loss, which equaled about 0.5% of organic matter loss (for more discussion, see Raison et al., 1985). Most studies of nutrient losses from fires are from relatively low-intensity prescribed fires and high-intensity slash fires (good examples include Harwood and Jackson, 1975; Covington and Sackett, 1984; Raison et al., 1985; and Little and Ohmann, 1988).

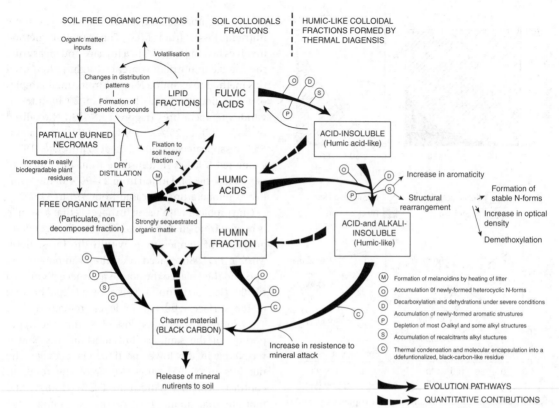

Figure 13.7 Diagram of how fire alters the organic matter chemistry of a soil. Free soil organic matter is changed in size, chemical composition, and likelihood of binding onto mineral particles. Mineral-bound (colloidal) compounds show changes, and newly formed soluble and insoluble compounds may be converted to charcoal. Reprinted from Environment International, 30, José A. González-Pérez, Francisco J. González-Vila, Gonzalo Almendros, Heike Knicker, The effect of fire on soil organic matter—a review. 855–870, 2004, with permission from Elsevier.

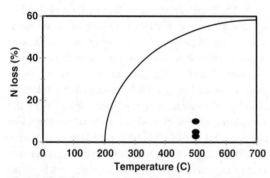

Figure 13.8 Nitrogen oxidation and loss rise rapidly as temperatures increase above 200 °C (data from Knight, 1966) when small samples are heated from all sides in a furnace. Losses are much smaller when samples are heated from above for three different intervals of time at 500 °C. (from data of Mroz *et al.*, 1980)

The third approach is simply comparing the nutrient content of burned areas with unburned areas. Several studies have documented the long-term effects of repeated, low-intensity fires on N availability. In a 30-year study in a loblolly pine–longleaf pine ecosystem, fires at intervals of one, three or four years reduced the N content of the forest floor by as much as 85% (Figure 13.10). But even 30 years of annual fires reduced the total N content of the forest floor and 0–20 cm mineral soil by only $300\,kg\,N\,ha^{-1}$ (or $10\,kg\,N\,ha^{-1}$ for each fire). A second 30-year study found no net loss of N.

Annual losses of carbon from ecosystems due to fire are a function of fire intensity and fire return interval. Based on rough estimates of these parameters, losses of organic matter to fire vary from less than $0.01\,Mg\,ha^{-1}\,yr^{-1}$ (averaged across a whole

Table 13.1 Nutrient losses from fires.

Fire type/vegetation	Nutrient Loss (kg ha^{-1})					
	N	P	Ca	Mg	K	S
Prescribed/loblolly pine (Richter *et al.*, 1982)	10–40	—	—	—	—	2–8
Slash fire/Hemlock, Douglas fir (Feller, 1983)						
Hot	980	16	154	29	37	—
Moderate	490	9	87	7	17	—
Slash fire/*Eucalyptus* (Raison *et al.*, 1985)	75–100	2–3	19–30	5–10	12–21	—
Slash fire/*Eucalyptus* (Harwood and Jackson, 1975)	—	10	100	37	51	—
Wildfire/Douglas fir (Grier, 1975)	855	—	75	33	282	—

fire return interval) in deserts and mesic deciduous forests to over 1 Mg ha^{-1} yr^{-1} in ecosystems where intense fires with comparatively short return times are the norm. Assuming that a loss of 6 kg N Mg^{-1} fuel is a reasonable estimate for most canopy fires, annual nitrogen losses due to fire can be estimated. Thus,

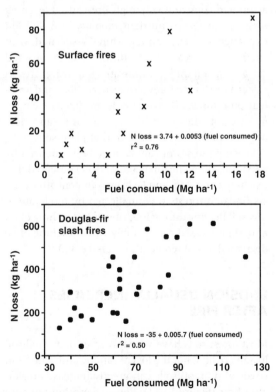

Figure 13.9 Nitrogen loss from fires is generally proportional to fuel consumption, with about 5 to 6 kg of N lost per Mg of fuel burned (after Binkley, 1999; data for the lower graph from Little and Ohmann, 1988).

average nitrogen losses due to fire may vary from approximately 0.12 to 12 kg ha^{-1} yr^{-1}. Soil leaching commonly removes about 1–3 kg N ha^{-1} yr^{-1} from forests. Depending on the fire return interval and the quantity of fuel consumed, fires remove about as much N as leaching (Johnson *et al.*, 1998).

Gasification losses of P compounds are more complex. All P is in the phosphate (PO_4^{3-}) form in ecosystems, but high temperatures in fires generate a variety of oxide (and organic oxide) compounds. The most important oxide formed during fires is probably P_4O_{10}, which can volatilize at 360 °C (Cotton and Wilkinson, 1988).

Vaporization (or volatilization) losses occur when a compound evaporates with no chemical change. For example, nitrate (NO_3^-) present in soils vaporizes at temperatures as low as 80 °C (Greenwood and Earnshaw, 1984), and amino acids may vaporize below 200 °C (Weast, 1982), well below their combustion temperatures (though concentrations of these acids are too low to have significance for the fire effects on N budgets). Metal cations such as potassium, calcium, and magnesium remain stable at higher temperatures. Potassium hydroxide (KOH) vaporizes above 350 °C (Cotton and Wilkinson, 1988), whereas oxides of calcium and magnesium are stable even at 2500 °C (Greenwood and Earnshaw, 1984). The actual vaporization temperature may be lower for these metals when they are part of organic molecules (Raison *et al.*, 1985).

Convection losses of nutrients in ash can be very large owing to the high concentrations of nutrients in ash, and the large quantities of ash that may be lost from a site. For example, the calcium

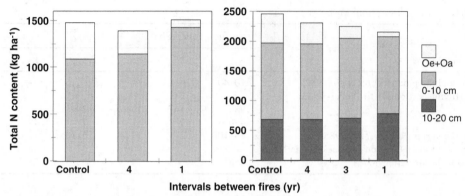

Figure 13.10 Thirty years of repeated prescribed fires in loblolly pine/longleaf pine forests reduced the N content of the forest floor, and increased the N content of the mineral soil, for no net loss in one study (left, from data in Wells, 1971), or a net loss of up to $300\,kg\,N\,ha^{-1}$ over 30 years (right, from data in Binkley *et al.*, 1992a).

concentration in leaf litter of *Eucalyptus pauciflora* was $470\,mg\,kg^{-1}$, compared with $5740\,mg\,kg^{-1}$ in black ash, and $14\,750\,mg\,kg^{-1}$ in gray ash (Raison *et al.*, 1985). The higher concentration in ash indicates large gaseous losses of carbon compounds, and little or no gaseous loss of calcium.

NUTRIENT LEACHING RATES INCREASE AFTER FIRE

The leaching of nutrients from soil after fire is influenced by the increased quantity of ions available, changes in uptake and retention by plants, the adsorptive properties of the forest floor and soil (both microbial and mineral), and patterns of precipitation and evapotranspiration. However, even in the most extreme cases, the extent of losses by this process is generally small relative to other loss pathways and the total nutrient capital. Leaching losses following wildfire in a western Washington old-growth Douglas fir forest appeared to be small; the majority of ions released from ash were retained in the upper 20 cm of soil (Grier, 1975). Leaching of NO_3-N from an intensely burned Douglas fir-pine-grass ecosystem in eastern Washington accounted for a loss of approximately 0.5% of the total soil N capital (Tiedemann *et al.*, 1979). These leaching losses appear to have been due to fire-caused acceleration of rates of net nitrification and the high mobility of the nitrate ion.

Knight *et al.* (1985) found large increases in soil solution concentrations of some chemical species in recently burned stands of lodgepole pine. They attributed comparatively high fire-caused leaching losses to increased nutrient mobility, low soil ion exchange capacity, and flushing associated with snowmelt. Increased concentrations of anions and cations in soil and groundwater following fires in eastern pitch pine forests (Boerner, 1982) and longleaf pine forests (Lewis, 1974) are likely a consequence of the very coarse texture (and low adsorption capacity) of the coastal plain soils.

Leaching losses of cations may be influenced by variations in soil heating. Stark (1977) observed that leaching of Ca, Mg, and Fe from soils beneath Douglas fir-larch forests was unaffected by moderate to low soil heating during fire. However, leaching of Ca and Mg increased considerably while leaching of Fe decreased at soil temperatures above 300 °C.

EROSION USUALLY INCREASES AFTER FIRE

Most forest soils have very low rates of annual soil loss in erosion, but the loss of canopy and O horizon cover of mineral soils can greatly increase erosion after fires (Figure 13.11). Elevated erosion rates are usually small enough to have limited ecological impact, but in some cases the losses may be severe. High losses tend to be associated with severe fires

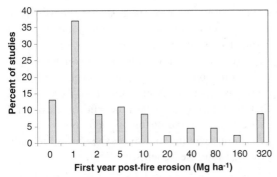

Figure 13.11 Most unburned soils show near-zero rates of erosion, and two-thirds of burned forests in the United States show less than 5 Mg ha^{-1} loss in the first year after fire. In extreme situations, especially where slope failures occur, first-year losses can top 100 Mg ha^{-1} (from data of Robichaud *et al.*, 2000).

that remove all cover from the mineral soil surface (Figure 13.12), intense storms, and slope failures.

Accelerated rates of erosion following fire can cause significant nutrient losses (Wright and Bailey, 1982), because of changes in vegetation, soil properties, hydrology, and geomorphic processes (Swanson, 1981). The actual amount and duration of any increase in erosion varies widely among sites as a consequence of fire intensity, soil infiltration

Figure 13.12 Post-fire erosion on this moderate slope averaged about 10 Mg ha^{-1} yr^{-1} for this granite-derived soil in central Colorado following a severe wildfire that consumed the entire O horizon (as well as forest canopy; MacDonald and Larsen, 2009). The first erosion-catching apron filled with sediment necessitating an additional apron to catch the overflow; the sediment pile beside the aprons shows the magnitude of material transported downslope.

capacity, topography, climate, and patterns of vegetation recovery. Loss of plant cover and forest floor exposes soil to increased kinetic energy of raindrops, which may increase sediment movement. Sediment and nutrient losses may be ameliorated by surface litter and post-fire needlecast. Fires may increase the erodibility of soils if mineral soil is exposed, or if the capacity for water infiltration is substantially reduced. Porosity and infiltration rates may decrease, and soil aggregates may be dispersed by beating rains. Pores may become clogged by tiny particles. The amount of erosion following a fire shows great variability across landscape positions, among soil types, and fire intensities. An intense rainstorm could generate large amounts of erosion from a burned watershed that would show little erosion in the absence of a large storm.

Where surface organic layers are not completely consumed, changes in the pore space and infiltration rates may be too small to be detected, yet even relatively mild fires can expose mineral soil and reduce infiltration rates. For example, Ursic (1970) found that annual burning in watersheds supporting hardwoods on hilly terrain increased streamflow and sediment yields by 50–100%. DeBano *et al.* (1998) tabulated results of post-fire studies of erosion, and the four available studies showed large increases; burned sites yielded between 3 and 200 Mg ha^{-1} more sediment than unburned sites. More studies would be needed to flesh out the distribution of frequency of small increases versus large increases.

Sykes (1971) reported that there was little information or agreement on the hydrological effects of fire in northern forests. Burning off the heavy moss layer in these forests may alter the distribution of summer runoff, with erosion accompanying flash floods. However, there are indications of increases in water infiltration rates in burned sites compared to unburned soils, which may result in surprisingly little erosion in boreal forests (Sykes, 1971). He pointed out that these infiltration data are in contrast to those reported by workers in temperate zones, where infiltration rates on burned areas have been slower than on unburned areas.

What is the significance of soil lost to erosion after fires? Few studies have compared the productivity of soils in eroded and uneroded sites after forest fires. An intriguing study by Amaranthus and Trappe (1993) provides some clues about possible effects. They used a post-logging site in Oregon that was swept by a severe wildfire which completely consumed the forest floor across 85–95% of the area. Erosion estimates found that about $100\,\text{Mg}\,\text{ha}^{-1}$ of mineral soil moved downslope in the year following the fire. The scientists planted seedlings of incense cedar and Douglas fir in soils from which the top had eroded. In the soil that accumulated in sediment traps (i.e., the former top-most mineral soil) and the trapped soil that had been sterilized to kill mycorrhizal inoculum, the Douglas fir seedlings formed only a few ectomycorrhizal root tips in any of the soils, but they grew well in all three soils (height increments of about $28\,\text{mm}\,\text{yr}^{-1}$). The incense cedar seedlings showed a slight (but significant) effect of soil treatment on height growth ($14\,\text{mm}\,\text{yr}^{-1}$ for topsoil, whether sterilized or not, compared with $12\,\text{mm}\,\text{yr}^{-1}$ for eroded soil). Colonization of incense cedar root tips was highest in the topsoil (37% of root tips), intermediate in the sterilized topsoil (21%), and lowest in the eroded soil. Survival of incense cedars followed the pattern of mycorrhizal infection, with twice as many surviving in the topsoil than in the eroded soil. This story may indicate that recolonization of the site by Douglas fir may be relatively unconstrained by erosion and mycorrhizal inoculum, but that the establishment of incense cedar may depend strongly on competition with faster-growing Douglas fir seedlings as well as the effects of erosion on mycorrhizal inoculum.

EROSION PREVENTION TREATMENTS OFTEN SHOW LITTLE VALUE

Fires generate great concern among forest managers and the public, and motivation may be strong to minimize the "damage" after the fire. Emergency treatments often include seeding with annual grasses (in hopes they will not dominate future understories), native seed mixtures, and exotic species, mulching with straw or other materials, and even using large, soil-disturbing machines to maneuver fallen logs into horizontal positions on slopes! These treatments cost on the order of US$600 to $10\,000\,\text{ha}^{-1}$, or $30 to $1000\,\text{Mg}^{-1}$ of avoided erosion. A thorough review of over 300 such treatments in the United States concluded that any protection offered by these expensive treatments was small (Robichaud et al., 2000), and that monitoring to determine effectiveness (and ineffectiveness) would be a good investment to determine whether massive expenditure actually provided tangible benefits.

EROSION LOSSES MAY BE MOST SEVERE WHEN POST-FIRE STORMS GENERATE DEBRIS FLOWS

The erosion of surface soil material happens after every fire, with losses ranging from negligible to large. However, the mass loss from erosion at the very surface of the soil can be dwarfed by the truly huge mobilization of soil (and subsoil) materials when heavy rains lead to slope failures and debris flows. Post-fire debris flows occur about every 350 years in Yellowstone National Park in the United States, shaping stream morphology and riparian ecosystems (Meyer and Pierce, 2003). A great example comes from Idaho, where Kirchner et al. (2001) compared rates of erosion measured in the past century with those from the past $10\,000$ years. Recent erosion rates averaged about $0.1\,\text{Mg}\,\text{ha}^{-1}$ yr^{-1} (regardless of watershed size), but the long-term average was $1.4\,\text{Mg}\,\text{ha}^{-1}\,\text{yr}^{-1}$. If the longer-term record indicated mass wasting events after fires, the post-fire losses would average something like a meter of material lost ($400\,\text{Mg}\,\text{ha}^{-1}\,\text{yr}^{-1}$) if severe fires and erosion episodes occurred every 300 years, or $1400\,\text{Mg}\,\text{ha}^{-1}\,\text{yr}^{-1}$ if the big events happened once every 1000 years.

MAJOR EROSION EVENTS MAY BE VITAL TO HEALTHY STREAMS AND RIPARIAN ECOSYSTEMS

Erosion is typically considered to be undesirable, as losing fertile topsoil and clogging streams and riparian ecosystems with debris seem to be self-evidently bad. As with many value judgments in ecology and soils, a closer look may change perspectives. In the absence of erosion, especially periodic mass wasting events, stream channels would tend to be V-shaped without riparian terraces, riffles, sinuosity, and other traits that typically characterize healthy streams and riparian ecosystems (Benda et al., 2003).

VERY SEVERE HEATING CAN ALTER SOIL MINERALOGY

Very hot fires can fracture rocks as a result of the energy released from heating water within the rocks (sedimentary rocks) or within cracks (such as in igneous rocks). Where temperatures are severe enough (> 500 °C), kaolinite can be decomposed (into aluminum oxides and silica). Ulrey et al. (1996) examined four burned forests in California, and found that heavy fuel loadings covered about 1–2% of the areas, and these locations burned intensely enough to decompose kaolinite and halloysite.

Soils often have a reddish hue beneath areas where large amounts of fuels led to intense heating. At high temperatures (and limited oxygen), Fe-OOH compounds can be altered to Fe_2O_3; significant generation of Fe_2O_3 may cover 1–15% of a burned area (Goforth et al., 2005).

FIRES MAY OR MAY NOT DECREASE WATER INFILTRATION INTO SOILS

Water infiltration rates are often diminished following fire owing to the plugging of surface pores and to increased fire-induced water repellency (Krammes and DeBano, 1965). Soils can also show hydrophobic tendencies in response to simple drying, so fires are not the only factor in determining soil hydrophobicity. Indeed, a severe fire in Spain removed hydrophobicity that had developed under normal (no-fire) forest development (Cerdà and Doerr, 2005). Doerr et al. (2006) reported a similar pattern for eucalyptus forests near Sydney, Australia; prefire hydrophobicity was substantially reduced by wildfire.

Increased sediment loss as a consequence of the formation of hydrophobic soil layers has been documented in ponderosa pine forests (Campbell et al., 1977; White and Wells, 1981). Hydrophobic properties appear to develop when organic molecules volatilize (evaporate) as a result of heating; as the vapor comes in contact with cooler soil surfaces at depth, it condenses and forms non-wettable surfaces.

The nature of such hydrophobicity has been worked out in laboratory experiments (DeBano et al., 1998). For example, heating soils to less than 175 °C has little effect on rates of water infiltration into soils. Temperatures between 175 and 200 °C may substantially reduce infiltration rates, while temperatures over 290 °C lead to combustion of the organic molecules that would create hydrophobic conditions at lower temperatures.

The spatial pattern of hydrophobicity may be very important to any influence on post-fire erosion rates. Patches of hydrophobic soils surrounded by high-infiltration soils may have little chance for increased erosion, as water does not travel far (or in large quantities) before infiltrating soil. Woods et al. (2007) studied the pattern of hydrophobicity at small (1 m²) and moderate (225 m²) scales after fires in Colorado and Montana, USA, and found high variability at both scales (Figures 13.13, 13.14). The high erosion losses

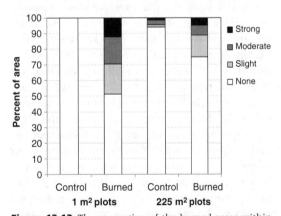

Figure 13.13 The proportion of the burned areas within the Hayman fire, Colorado, USA (near Figure 13.12) that showed hydrophobicity depended on the scale of interest. Small plots showed about half of the sampling points had some degree of hydrophobicity (with none evident in unburned plots), compared with about 25% of larger plots (with some evidence of hydrophobicity even in unburned plots; after data from Woods et al., 2007).

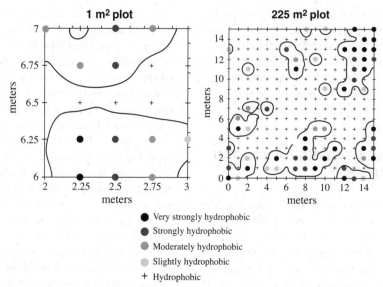

Figure 13.14 The spatial distribution of hydrophobic soils after the Hayman fire was very patchy at the scale of 1 m² (left) and 225 m² (right). Patchiness can reduce the influence of hydrophobicity on erosion by allowing water to infiltrate before moving far enough to gather larger volumes of runoff. Reprinted from Geomorphology, 86, Scott W. Woods, Anna Birkas, Robert Ahl, Spatial variability of soil hydrophobicity after wildfires in Montana and Colorado, 465–479, 2007, with permission from Elsevier.

that followed the Hayman fire in Colorado (similar to Figure 13.12) probably related more to the loss of soil cover (canopy and O horizon) than to hydrophobicity (MacDonald and Larsen, 2009).

Fire characteristics that promote hydrophobicity include high (but not extreme) fire severity, coarse texture, and low soil water content. Despite these insights from controlled experiments, the contribution of hydrophobicity to overall post-fire erosion of forest soils remains largely unknown. The generation of hydrophobic patches in burned landscapes is relatively straightforward, and water infiltration rates can be measured in hydrophobic patches and compared with those for unburned or burned-but-not-hydrophobic soils. Water infiltration rates may be lower in hydrophobic patches for a period of months to several years under field conditions (Dyrness, 1976; McNabb et al., 1989). Saturated conductivity in hydrophobic areas may be reduced by 10–40%, but the effects last for only a few storm events (Robichaud, 2000).

How much erosion would result from a burned watershed if hydrophobic patches did not develop? The present state of knowledge has too much uncertainty to quantify the contribution of each hydrologic factor in generating post-fire erosion. Larsen et al. (2009) concluded that the extent of cover on the mineral soil was the major determinant of erosion, with any contribution of hydrophobicity being difficult to verify. This underscores the importance of re-establishing vegetation to diminish both erosion and nutrient loss (Knight et al., 1985).

ASH MAY CONTAIN LARGE QUANTITIES OF NUTRIENTS

Fires leave behind large amounts of ash that can be subject to wind erosion (see Ewel et al., 1981); typical masses of post-fire ash range from 2 to 15 Mg ha^{-1} (Raison et al., 1985). The concentrations of nutrients are unusually high in ash, giving nutrient contents of post-fire ash layers of 20 to 100 kg N ha^{-1}, 3 to 50 kg P ha^{-1}, and 40 to 1600 kg Ca ha^{-1}. Most of the nutrients in ash are water soluble or readily released by microbial activity, and this should provide a large supply of available nutrients for recovery of vegetation.

DIRECT NUTRIENT RELEASE FROM SOIL HEATING MAY BE IMPORTANT

The obvious layers of ash following fire have led many researchers to focus on the nutrient content of ash as the major pool of nutrients released by fire. However, the direct effects of heating may release comparable quantities of some nutrients that were formerly bound in soil organic matter. Heat may release ammonium and phosphorus from organic matter, even if the organic matter is not consumed in the fire.

DECOMPOSITION AND MICROBIAL ACTIVITY CHANGE AFTER FIRE

The environmental conditions of the residual forest floor and soil may be very different after the fire. Charred and darkened organic materials may absorb radiation better than do unburned materials, resulting in warmer conditions. For example, Neal *et al.* (1965) reported that soil temperatures (at 5 cm depth) in a burned portion of a Douglas fir clear-cut averaged 6 °C higher than in unburned portions. Fires may also increase soil moisture by decreasing water use by vegetation.

A pulse of increased nutrient availability typically follows fires as a result of reduced competition among plants, the release of elements from organic matter, and altered activity of soil biota. Changes in microbial properties of soils after fire have been documented, but few generalizations seem supportable. Decomposition is generally expected to increase after fire because of increased temperatures at the soil surface and increased soil moisture. Such an increase was demonstrated by van Cleve and Dyrness (1985), who placed cellulose strips in mesh bags in the upper soil horizons of unburned and adjacent burned white spruce stands. After two months, most of the cellulose had decomposed in the burned plots, whereas the cellulose in the forest remained practically undecomposed. In a more detailed study, Bissett and Parkinson (1980) showed that higher rates of cellulose decomposition in burned sites occurred only in the field; laboratory incubations actually had slower

decomposition rates for burned soils. They concluded that the microenvironmental effects of fire were more important than the chemical effects. Decomposition studies using leaf materials generally have not shown increased rates of decomposition in burned areas. For example, Grigal and McColl (1977) found that aspen leaves decayed more slowly in burned than unburned plots in Minnesota, and Weber (1987) found no differences in the rate of decomposition of jack pine needles on burned and unburned plots in Ontario. Covington and Sackett (1984) reported that eight months of decomposition after a fire in a ponderosa pine stand released 108 kg N ha^{-1} more than that found for an unburned stand. Schoch and Binkley (1986) reported a similar increase in N release of 60 kg ha^{-1} during six months after a fire in a stand of loblolly pine.

The impact of high-intensity fires on soil microbial activity has received little attention. Bissett and Parkinson (1980) found no difference in microbial biomass between burned and unburned plots in a spruce–fir forest in Alberta after six years, but the ratio of bacteria to fungi was higher in burned plots. Modern genetic and physiological-profiling techniques allow investigators to go beyond lumping all bacteria, or all fungi, into a single pool. For example, Hart *et al.* (2005) found that the diversity of the bacterial community dropped one month after a severe wildfire in a ponderosa pine forest, yet the fungal community diversity increased dramatically (Figure 13.15). Any implications of these changes in community composition remain unknown.

Plant nutrition after fire depends on changes in both nutrient availability and in the distribution and activity of plant roots and mycorrhizal fungi. Few studies have examined root responses to the nutritional effects of fire. Chapin and Bloom (1976) and Chapin and van Cleve (Chapin and Van Cleve,1981) found that excised roots from plants in young post-fire ecosystems displayed higher rates of phosphate adsorption than did the same species in older ecosystems, suggesting greater P limitation in the post-fire soil. Increased nutrient availability may result in higher concentrations of nutrients in plant tissues following fire, and these higher

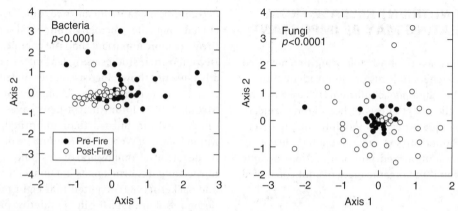

Figure 13.15 The diversity of the bacterial community contracted one month after a severe wildfire in pine forest in New Mexico, USA (left). In contrast, the diversity of the fungal community increased greatly (right). The functional implications of major changes in the diversity of microbial communities remain to be explored. Reprinted from Forest Ecology and Management, 220, Stephen C. Hart, Thomas H. DeLuca, Gregory S. Newman, M. Derek MacKenzie, Sarah I. Boyle, Post-fire vegetative dynamics as drivers of microbial community structure and function in forest soils, 166–184, Copyright 2005, with permission from Elsevier.

concentrations may (or may not!) increase rates of decomposition and nutrient turnover (Chapin and van Cleve, Chapin and Van Cleve,1981).

SOIL ACIDITY DECLINES AND PH RISES AFTER FIRE

Soil pH typically increases immediately after fire, then declines back to pre-fire levels over a period of months, years, or decades. Soil pH influences the availability of some nutrients, both through direct geochemical effects and indirect effects on microbial activity. For example, Montes and Christensen (1979) raised the pH of incubated soils from beneath several different vegetation types by 0.5–1.7 pH units, and found that nitrate production increased by several-fold.

Soil pH depends upon the equilibrium between the exchange complex and the soil solution (see Chapter 8). An exchange complex dominated by H^+ and Al^{3+} maintains a low (strongly acidic) pH in the soil solution, whereas domination of the exchange complex by so-called base cations (K^+, Ca^{2+}, and Mg^{2+}) maintains a higher pH (less acidic) soil solution.

Fires increase soil pH by two processes. The combustion of undissociated organic acids (such as

acetic acid) in litter and soil removes the organic acids from the ecosystem:

$$CH_3COOH + 2O_2 \rightarrow 2CO_2 + 2H_2O$$

In this case, no H^+ ions were removed, but the removal of the organic acid component of the exchange complex may increase the pH of the soil solution.

The second fire process actually consumes H^+ from the soil, essentially titrating the soil. The extent of this process is commonly calculated as the release of base cations by the fire (Chandler *et al.*, 1983). Although these cations are not bases in any chemical sense, their release is associated with the consumption of H^+. For example, the combustion of an organic compound (acetate) containing K^+ consumes one H^+ for every K^+ released:

$$CH_3COOK + 2O_2 + H^+ \rightarrow 2CO_2 + 2H_2O + K^+$$

The H^+ is consumed in the production of water, not by any reaction with the K^+. Through a series of reactions with water and carbon dioxide, the released cations form dissolved bicarbonate salts that may readily leach into the mineral soil, where the cations may be exchanged for other cations (such as aluminum) on the exchange complex.

The fire-induced change in pH depends on the consumption of organic acids (with H^+ associated with the anion), the consumption of organic anions (with other cations associated with the anions), the original pH of the soil, and the buffer capacity of the soil. Forest floor fuels generally contain about 1–3 moles of negative charge per kilogram. Depending upon pH, perhaps half of the charge will be balanced by H^+ (representing undissociated acids), and about half will be balanced by cations such as K^+, Ca^{2+}, and Mg^{2+} (representing dissociated acids). Combustion of 20 000 kg ha^{-1} of such fuel would consume about 20 kmol$_c$ ha^{-1} of undissociated acid, and also consume 20 kmol$_c$ of free H^+ ha^{-1} associated with the release of the nutrient cations. Binkley (1986) estimated that a series of 12 surface fires during 24 years in a loblolly pine stand lowered the H^+ content of the soil by about 120 kmol$_c$ ha^{-1}, resulting in an increase in pH from 3.8 to 4.1.

The increase in pH depends upon the original soil pH for two reasons. Soils with pH values about 6.5 tend to be strongly buffered by the presence of carbonate minerals, diminishing the effect of the fire. Soils with a very low pH (< 4.5) are strongly buffered by the presence of either organic acids or aluminum. The largest changes in pH after canopy fires would probably occur in the range of pH 4.5 to 6.5.

BASE CATIONS MAY INCREASE AFTER FIRE

The supply of nutrient cations (calcium, magnesium, and potassium) generally increases following fire. This increase results from direct release from burning organic matter, and from any increase in subsequent rates of organic matter decomposition. Heating may also directly alter soil exchange properties, causing cation release (Khanna and Raison, 1986). In fact, the nutrient cation content of the forest floor and ash layer may be higher after fires, due to the addition of material from the combustion of the vegetation and organic debris. For example, van Cleve and Dyrness (1985) examined the nutrient content of forest floors in an unburned white spruce forest, and in an area where the spruce forest

was consumed by a canopy fire. Both unburned and lightly burned areas had about 1300 kg Ca ha^{-1} in the forest floor, whereas the heavily burned areas had over 6700 kg Ca ha^{-1}. As a side note, it is difficult to explain such a large increase in forest-floor Ca in the heavily burned areas, as the vegetation likely contained < 1000 kg Ca ha^{-1}; this might be a case where a mass-balance approach would raise skepticism about the adequacy of the experiment (see Chapter 8). Dyrness *et al.* (1989) found that cations were increased in the forest floor and mineral soil in Alaskan white spruce and black spruce stands only in intense fires. After a canopy fire in an old-growth Douglas fir forest, Grier (1975) found no increase in total forest floor calcium.

A portion (typically 10–30%) of the cations contained in ash dissolves readily in water, and typically leaches into the mineral soil (Walker *et al.*, 1986). Grier (1975) estimated that 670 mm of snowmelt moved 75 kg Ca ha^{-1} out of the ash layer into the upper mineral soil, and also moved 15 kg ha^{-1} deeper than 20 cm into the mineral soil. Portions of the site with a heavier ash layer leached about four times more calcium. A similar study of the effects of a severe canopy fire in a jack pine stand showed about 12 kg ha^{-1} of Ca leaching from the forest floor into the mineral soil (Smith, 1970).

Leaching losses of nutrient cations also increase after canopy fires. For example, the loss of Ca^{2+} and K^+ to streamwater rose by 26% and 265% after the Little Sioux fire in northern Minnesota (Wright, 1976). However, even the large relative increase in K^+ loss amounted to only 1.5 kg ha^{-1} of extra loss, which is unlikely to affect site fertility.

Changes in cation nutrition of plants after fires have received little attention, but some insights are available from more extensive research on slash-burned areas. Vihnanek and Ballard (Vihnaneck and Ballard, 1988) found that 5 to 15 years after slashburning, the concentrations of calcium and potassium in Douglas fir foliage on burned sites exceeded the concentrations on unburned sites in 80% of the sites examined. The availability of calcium, magnesium, and potassium does not limit forest growth in most temperate ecosystems (Chapter 14), so the effects of fires on losses or cycling rates of these cations are likely unimportant to ecosystem recovery and productivity in most cases. However, the nutrient cation supply in many Ultisols and Oxisols is very low, and cation losses from fires on these highly weathered soils may be critical. We know of no data on the effects of fires in tropical forests on supply rates of nutrient cations, but expect that impacts of fires could be critical to ecosystem recovery and future production.

PHOSPHORUS AVAILABILITY INCREASES AFTER FIRE, AT LEAST TEMPORARILY

The availability of phosphorus limits production in many ecosystems around the world, yet the effects of fire on P availability have been poorly characterized. In laboratory studies, Humphreys and Lambert (1965) found that available phosphorus increased when soils were heated in a furnace to between 200 and 600 °C. Humphreys and Craig (1981) heated soils in furnaces, and showed that much of the reduction of organic P was matched by increases in aluminum and iron phosphates; very little P was lost from the samples at temperatures up to 600 °C. In a field study of a slash-and-burn treatment in a wet tropical rainforest, the concentration of readily available P in the upper soil declined following the first heavy rains after burning, probably resulting from conversion of P to less available forms rather than to actual losses of P (Ewel *et al.*, 1981). In a study of harvesting and burning in *Eucalyptus*, Ellis and Graley (1983) found that elevated availability of P relative to uncut forests lasted only about one year. In addition to the direct losses of P in fire, subsequent effects of fire on P cycling include changes in pH, which may alter the solubilities of inorganic phosphates as well as microbial P transformations.

Increases in pH should generally increase P availability in situations where P is bound with iron and aluminum, and decrease the availability of P bound with calcium (Lindsay, 1979). Increased microbial activity in burned soils may increase the release of P from organic matter, and competition between microbes and plants for the increased supply of P could determine the degree of enhancement of plant nutrition.

In soils high in iron and aluminum oxides, fires may increase P sorption and decrease P availability. For example, Sibanda and Young (1989) found that heating soils to just 200 °C doubled the P sorption capacity of a soil from Zimbabwe, and heating to 400 °C drastically increased P sorption. Increased sorption resulted from removal (oxidation) of organic matter that was bound with soil iron and aluminum, allowing these metals to bind P.

NITROGEN AVAILABILITY ALSO INCREASES AFTER FIRE

The concentration of soil ammonium generally increases greatly after fire, sometimes by an order of magnitude or more. Walker *et al.* (1986) demonstrated that soil ammonium may increase as a direct result of soil heating. Ammonium may also be added to soil in ash. Concentrations of ammonium may remain elevated, resulting from both the increase in ammonium production and any decrease in ammonium consumption by plants and microbes (Knight *et al.*, 1985; Dyrness *et al.*, 1989). A severe canopy fire in a *Eucalyptus* forest in Australia increased soil ammonium concentrations by four-fold, but concentrations declined to pre-fire levels within six months.

Nitrate concentrations sometimes rise over a period of weeks or months following fire. Immediate increases are generally slight due to the low vaporization temperature for nitric acid (as low as 80 °C) and the very low concentration of NO_3-N in ash. Net nitrification increases following fire in a wide array of ecosystems as a consequence of increased pH and availability of ammonium.

Pulses of ammonium and nitrate availability are ephemeral, often lasting a few growing seasons or less (Adams and Attiwill, 1986; Christensen, 1987). Subsequent changes in N availability following fire depend upon the quantity of N lost in the fire, changes in rates of microbial mineralization after the fire, and competition between microbes and plants for mineralized N. The eucalyptus fire examined by Adams and Attiwill (1986) increased mineralization of N during in-field incubations by four-fold for a period of six months, after which no effect of the fire was apparent. Grady and Hart (2006) found that a wildfire almost doubled net N mineralization in the O horizon and upper mineral soil, though thinning followed by broadcast burning had little effect.

NITROGEN FIXATION INCREASES AFTER FIRE ONLY IF SYMBIOTIC N-FIXING SPECIES INCREASE

Some authors have speculated (and cited hopeful papers) that non-symbiotic nitrogen fixation increases after fire (Woodmansee and Wallach, 1981; Boerner, 1982; Christensen, 1987), but available evidence does not support such speculation. Most studies of fire effects on non-symbiotic nitrogen fixation have found that the rates are so low as to be important only on a timescale of decades or centuries, and none has shown ecologically meaningful increases in rates following fire. For example, Wei and Kimmins (1998) estimated that free-living N fixation would average about $0.3\,kg\,N\,ha^{-1}\,yr^{-1}$ in a lodgepole pine forest without fire, or $0.6\,kg\,N\,ha^{-1}\,yr^{-1}$ if the forest burned in a wildfire.

On the other hand, burned areas are often colonized by plants capable of symbiotic nitrogen fixation. Longleaf pine forests in the southeastern U.S. may develop large populations of understory legumes following regular prescribed fires. Relatively low densities of herbaceous legumes probably contribute $< 1\,kg\,N\,ha^{-1}\,yr^{-1}$, but where densities are very high ($> 2\,plants\,m^{-2}$), rates of $5\text{–}10\,kg\,N\,ha^{-1}\,yr^{-1}$ are likely (Hendricks and Boring, 1999). Even relatively low rates of N fixation may accelerate cycling of N. Nitrogen-fixing plants often require greater quantities of phosphorus than other plants, and the potential effects of fire-related changes in P availability to N cycling should be explored.

LONG-TERM EFFECTS OF FIRE ON SOIL PRODUCTIVITY REMAIN UNCERTAIN

Fires remove nutrients from forests, but typically increase nutrient turnover rates at least in the short term. But what are the overall, longer-term effects?

Surprisingly little information has been collected on this fundamental question, and the available information suggests that no single generalization will be appropriate across forests and soils.

Studies with prescribed, surface fires in pine stands have generally shown little, if any, change in foliar chemistry (Landsberg and Cochran, 1980; Binkley *et al.*, 1992a). Few studies have documented the changes in growth in pine stands following surface fires, and no broad generalizations are supported. One study in longleaf pine stands in Alabama found that burning (five fires in ten years) reduced volume increments by one-quarter to one-third (Boyer, 1987), whereas fire in an oak forest in New Jersey increased growth by one-third to one-half (Boerner *et al.*, 1988).

Slash fires are typically hotter and consume more fuel than surface fires, so the impacts of slash fires might be larger than those of prescribed, surface fires. W. G. Morris initiated a study of effects of slash fires in Douglas fir forests in 1947, where he recorded and mapped fire intensities in 62 pairs of burned and unburned plots. Resampling showed that severely burned plots had 7–50% less total N in 0–10 cm depth mineral soil relative to lightly burned or unburned soils (Kraemer and Hermann, 1979). Miller and Bigley (1990) examined growth on 44 of these plots, comparing tree characteristics. The mean annual increment of Douglas fir was significantly higher on burned plots, where growth exceeded that on unburned plots by $0.85\,m^3\,ha^{-1}\,yr^{-1}$ (a 27% difference). Was the greater growth a result of fire-induced increases in nutrient supplies, or to the effect of fire in reducing competition from non-conifer vegetation? Burned plots did, in fact, have less non-conifer vegetation, but no experiment was done to separate the contribution of different aspects of fire's impact on conifer growth.

In another study, Vihnanek and Ballard (Vihnaneck and Ballard, 1988) examined Douglas fir growth and nutrition on 20 sites (age 5–15 years after harvest) on burned and unburned plots. Nutrient contents in foliage were generally greater on burned plots and were not lower than those on unburned plots at any location. This study also could not separate any effects of fire on soils and

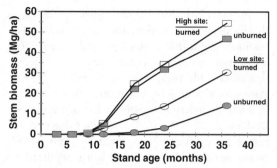

Figure 13.16 Productivity of clonal *Eucalyptus grandis x urophylla* was greater on sites prepared with slash fires than on sites without burning; photo is from the low-productivity site (means of four replicate plots/site; J. L. Stape, personal communication).

nutrients from the effect of fire in reducing tree competition with non-tree vegetation.

A good example of the growth effects of post-harvest slash fires comes from *Eucalyptus* plantations in Brazil (Figures 13.16, 13.17). Post-harvest fires of low-to-moderate intensity increased growth by about 15% on a high-productivity site and by more than 100% on a low-productivity site. The greater growth on burned plots resulted from better development of leaf area and from less allocation of carbohydrates to belowground production. In this case, competing vegetation was thoroughly controlled by herbicides, so greater

Figure 13.17 *Eucalyptus grandis x urophylla* clones three and a half years after planting, with and without use of fire for site preparation for a low-productivity site (see growth information in Figure 13.13). Better growth on burned plots resulted from increased P supply from both the resin-extractable and NaOH-extractable pools. Higher supply of P allowed greater canopy development (top panels, looking up through canopies). Fertilization also increased both labile pools of P and growth. Treatments did not affect bicarbonate-extractable P, HCl-extractable P, or residual P. (data from J. L. Stape and J. Spears).

growth on the burned plots probably resulted from better nutrition of the trees. On the lower-quality site, the P concentration in leaves was almost twice as high on burned plots as on unburned plots, so the major nutritional benefit of fire was probably increased P supply in the soil, particularly in the pools extractable by resins and by NaOH (N concentrations in foliage did not differ). Many more studies of this sort (including competition control) will be needed to provide a clear picture of the situations where fire increases or decreases soil productivity.

CHAPTER 14
Nutrition Management

OVERVIEW

The productivity of most forests is limited by the supply of one or more nutrients, with nitrogen and phosphorus limitations spread across the largest area. Severe nutrient limitations can be diagnosed by visual symptoms of leaves, by very low stand leaf area, or by very low nutrient concentrations in chemical analyses of leaves. Most forests that are nutrient limited have no overt visual symptoms, and fertilization experiments are needed to identify limiting nutrients. Responses to fertilizer applications may correlate with foliar chemistry, soil classes, productivity classes, or soil analyses. No method has proven useful in all situations, but in most cases, one or more approaches have been shown to correlate usefully with stand nutrient limitations. Effective research programs need to evaluate nutrient limitation and fertilization responses at a landscape scale.

Most forests increase in growth rate following fertilization, and many cases provide substantial profits. The increased stem growth following fertilization derives from increases in leaf area and net primary production, increases in foliage photosynthesis, and shifts in allocation of photosynthate from root production to stem production. Fertilization also changes stand characteristics, accelerating growth, dominance, and self-thinning of stands. Nitrogen fertilizers are synthesized from air, using natural gas as an energy source. Other types of fertilizers come from mining, with varying degrees of purification and treatment before use. Most of the fertilizer applied to forests remains in the soil, with about 20–25% entering trees. Experiments in forest nutrition and fertilization include uncertainties about the responsiveness of stands and of groups of stands. Formal methods of approaching management decisions are available to guide the application of imperfect information from experiments to real-world decisions about stand management.

Soil chemistry and forest biogeochemistry combine to influence the nutrition of trees, and tree nutrition has a large influence on productivity, partitioning of productivity to stem growth, and even resistance to some diseases and insects. Forest managers may passively alter tree nutrition as an unintended consequence of harvesting, burning, or other operations. In intensively managed plantations, direct application of mineral fertilizers improves tree nutrition and boosts stem growth. This chapter focuses on the active end of the spectrum, including approaches to diagnosing tree nutritional status and operational aspects of forest fertilization.

NITROGEN AND PHOSPHORUS ARE THE MOST COMMON LIMITING NUTRIENTS

Across the globe, most forests show strong N limitations; native forests in both temperate and tropical regions increase growth by about 20% when N fertilizer is added (LeBauer and Treseder, 2008). Phosphorus deficiency is also common in some temperate regions and is widespread in fast-growing plantations in the tropics (Gonçalves et al., 1997). Many tropical soils are low in P supply (as are many temperate soils), so P limitations may be expected to be very common in native tropical forests. Too few

Ecology and Management of Forest Soils, Fourth Edition. Dan Binkley and Richard F. Fisher.
© 2013 John Wiley & Sons, Ltd. Published 2013 by John Wiley & Sons, Ltd.

fertilization trials have looked for P limitations in native tropical forests to support strong generalizations, and P limitations do not appear to be pervasive (Grubb, 1995; Adams, 1996, 2007; Tanner *et al.*, 1998; Davidson *et al.*, 2004). Perhaps the prevalence of P limitation in fast-growing tropical plantations becomes apparent after silvicultural operations reduce other limitations by controlling weeds and adding other nutrients.

ASSESSMENTS OF NUTRIENT LIMITATIONS REQUIRES A CHOICE OF RESPONSE VARIABLE

The core idea of nutrient limitation is that some environmental factor constrains some aspect of tree (or stand) growth. The growth of most trees is constrained by the light available for canopies to photosynthesize, and the growth of most stands of trees is limited by the amount of water available for transpiration. Limitations can be defined in ways other than growth, such as reproductive success. Nutrient limitations may differ between the scales of individual trees and stands, and among individual species in mixed-species stands. If a tree species were planted in monoculture on a low-N soil, fertilization with N might increase the stand-level growth. However, increased rates of growth typically benefit dominant trees and increase death rates of suppressed trees. In such a case, the suppressed trees might not be considered N limited, even if the overall stand responded strongly. Similarly, increasing N supply in a mixed-species forest may benefit some species at the expense of others (for example, see Siddique *et al.*, 2010).

The growth or reproductive success of some trees and stands might be constrained strongly by the supply of a single resource. For example, the growth of slash pine trees on some sandy soils in Florida, USA is severely constrained by the availability of P (Figure 14.1); with such an extreme limitation,

Figure 14.1 Phosphorus deficiency in 20-year-old slash pine (left side) severely limits the development of canopy leaf area and growth, compared with P-fertilized trees (right side).

alleviation of the P deficiency is necessary before trees can respond to increases in supplies of other nutrients.

In many cases (perhaps half of all ecosystems? See Harpole *et al.*, 2011), a synergistic response follows the simultaneous alleviation of two limiting resources, with the response exceeding the additive response to alleviation of each resource separately. Operational fertilization programs sometimes focus on widespread application of single-element fertilizers, but most intensive forest management focuses on application of a suite of elements that yield the most profitable growth responses.

SEVERE NUTRIENT LIMITATIONS CAN BE DIAGNOSED VISUALLY

When nutrient limitations are severe, visible symptoms such as very low leaf area (Figure 14.1) or leaf yellowing become apparent. In the southeastern United States, forest managers may prescribe fertilization treatment based on visual estimates of leaf area index. Color pictures of nutrient-deficient tree leaves can be found in Benzian (1965), Bengtson (1968), Baule and Fricker (1970), Will (1985), and Dell (1996). Other symptoms may include stem deformities and loss of leaves, such as boron deficiency in N-fertilized Scots pine stands in Sweden (Figure 14.2).

Figure 14.2 Boron deficiency in Scots pine in northern Sweden is evident from the deformity and death of leaders, followed by dominance of side shoots that become leaders, with the overall result of deformed stems (from the experimental site described by Högberg *et al.* 2006).

Table 14.1 Visible symptoms of nutrient deficiencies in some eucalypt species (based on Dell, 1996).

Leaf Age	Symptoms	Likely Deficient Nutrient
Old leaves:	Leaf coloration even, green to yellow, small reddish spots may develop secondarily	Nitrogen
	Leaf coloration even, with reddish blotches; or leaves uniformly purple to red	Phosphorus
	Leaf coloration patterned, interveinal chlorosis	Magnesium
	Leaf coloration patterned, scorched margins or interveinal necrosis	Potassium
Newly expanding leaves	Dieback at shoot apex, nodes enlarged, leaves with corky abaxial veins, apical chlorosis, malformed with incomplete margins	Boron
	Dieback at shoot apex, nodes enlarged, leaves with irregular or undulate margins, some interveinal chlorosis	Copper
	Dieback at shoot apex, nodes normal, leaves buckled from impaired marginal growth	Calcium
	No dieback, leaves normal size:	
	Leaves pale green to yellow	Sulfur
	Leaves yellow with green veins	Iron
	Leaves with marginal or mottled chlorosis, small necrotic spots	Manganese
	No dieback, leaves small and crowded	Zinc

Some symptoms may relate to a specific nutrient limitation, but in many cases chemical analysis of foliage is needed. For example, twisted, deformed leaders in Douglas fir have been found to relate to copper deficiency (Will, 1972) and boron deficiency (Carter *et al.*, 1983). A general scheme has been developed for the visual symptoms of nutrient deficiencies in *Eucalyptus* (Table 14.1), depending on whether the symptoms appear primarily on old leaves or newly expanding leaves. If such visible symptoms of nutrient deficiency are present, growth is probably severely impaired.

MAJOR DEFICIENCIES OFTEN SHOW UP IN FOLIAR CONCENTRATIONS OF NUTRIENTS

Simple measures of the nutrient concentration in foliage may be sufficient for identification of nutrient limitation in some situations. "Critical" concentrations have been proposed for most major commercial forest species (Table 14.2). For conifers, nitrogen concentrations less than 10 to 12 $mg\,g^{-1}$ of leaf mass may indicate nitrogen deficiency, and critical phosphorus concentrations equal about 10% of the nitrogen concentrations.

Foliar analysis can be complex, as nutrient concentrations in leaves vary with leaf age and development, location within the canopy, age and competitive status of the tree, and even year-to-year variations in precipitation. A decrease in the percentage of phosphorus throughout a growing season may result from removal of phosphorus or an increase in carbohydrate content (a dilution effect). This effect can be removed by accounting for changes in leaf weight. Some methods include this factor indirectly by expressing nutrient content per leaf, per needle fascicle, or per 100 leaves. Direct measurements of leaf area have allowed nutrient content to be expressed per unit of leaf area. This approach allows direct assessment of the change in leaf nitrogen content by removing the effects of varying leaf weight (density).

Careful attention to sources of variability can allow fairly precise characterization of foliar nutrient levels, but studies have not verified the ability of even precise characterization to predict growth responses to fertilization. Part of the problem may derive from the importance of two factors: varying leaf nutrient concentrations and varying biomass of the total canopy.

Given the variations in nutrient concentrations with the crowns of individual trees, some

Table 14.2 Typical recommended critical concentrations of foliar nutrients (mg g^{-1}); sites with trees falling below these concentrations often respond well to fertilization (as might some with concentrations marginally above these values).

Species	N	P	K	Ca	Mg
Loblolly pine	11	1.0			
Slash pine	8–12	0.9	2.5–3.0	0.8–1.2	0.4–0.6
True firs	11.5	1.5			
Scots pine	12–14	1.4–1.8	3.5–4.5		0.8
Douglas fir	10–12	0.8–1.0	3.5–4.5	1.5–2.0	0.6–0.9
Lodgepole pine	10–12	0.9–1.2	3.5–4.0	0.6–0.8	0.6–0.8
Western hemlock	10–12	1.1–1.5	4.0–4.5	0.6–0.8	0.6–0.8
White spruce	10.5–12.5	1.0–1.4	2.5–3.0	1.0–1.5	0.5–0.8
Western red cedar	11–13	1.0–1.3	3.5–4.0	1.0–2.0	0.5–0.9
Norway spruce	12–15	1.2–1.5	3.5–5.0	0.4–0.6	0.4–0.6
Eucalyptus grandis, South Africa	12.5	1.0	3.6	5.6	3.5
Eucalyptus maculata, Australia	12–16	0.4–0.5	4	1.5–2.0	0.3
Radiata pine	12–15	1.2–1.4	3.0–3.5	1.0	0.6–0.8

standardization choices are typically made to allow comparisons among stands, treatments, and sites. Samples are often taken of upper crown foliage of the current or 1-year-old age classes (for conifers retaining more than one year of foliage) during a period of relatively stable concentrations, with the idea that these leaves are contributing much of the photosynthesis of the whole crown. Sampling during stable periods allows greater repeatability among sites and years but does not necessarily give the best indication of nutrient limitations.

Critical concentrations for micronutrients are not well established. Possible deficiency in conifer needles might be indicated by levels less than 25 mg Mn kg^{-1}, 50 mg Fe kg^{-1}, 15 mg Zn kg^{-1}, 3 mg Cu kg^{-1}, 12 mg B kg^{-1}, and 0.1 mg Mo kg^{-1} (Carter, 1992).

Once critical concentrations are known, foliage sampling across landscapes can identify which stands would be expected to respond to fertilization (Figure 14.3).

RATIOS OF NUTRIENT CONCENTRATIONS MAY BE MORE INFORMATIVE THAN CONCENTRATIONS OF INDIVIDUAL ELEMENTS

It may seem logical that the degree of N limitation may depend in part on whether a tree (or stand) has a sufficient supply of N, and in part on whether the current supply of another element (such as P) could

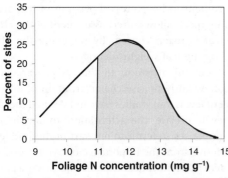
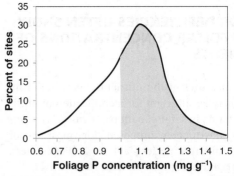

Figure 14.3 Distributions of concentrations of N (left) and P (right) in loblolly pine needles across more than 100 studies in the southeastern US (after Albaugh *et al.* 2008). Shaded areas indicate proportion of all sites that would considered deficient in each nutrient, based on critical levels in Table 14.2.

support greater growth. Multiple nutrients are always present in leaves, so the ratios among them (sometimes called stoichiometry, see Sterner and Elser, 2002) may provide more information on potential limitations.

The idea of nutrient ratios as indicators of limitations was popularized in forestry by work on seedlings by Tørsten Ingestad. He developed a hydroponic facility where seedling roots were suspended in a chamber and misted with high volumes of very dilute nutrient solutions. By supplying nutrients directly to the roots at high rates and low concentrations, he could evaluate optimum supply rates and ratios of all nutrients. Two general conclusions emerged from a large number of these studies. The potential growth rates of seedlings receiving optimum nutrient supplies greatly exceed those found when seedlings are grown even in very fertile soils. Ingestad concluded that the ratio of nutrients supplied to the root must be balanced precisely to match the relative levels required by the seedling. Further, these relative ratios are remarkably similar for a wide range of species (Table 14.3; Ingestad, 1982).

These "Ingestad" ratios cannot be extrapolated directly to larger trees growing in soils, as the hydroponic levels appear to represent "luxury" levels. Later work in Sweden, for example, used ratios of 10 for phosphorus, 35 for potassium, 2.5 for calcium, and 4 for magnesium (Linder, 1995) rather than the notably higher values listed in Table 14.3 from hydroponic experiments.

Table 14.3 Ratio of nutrients to nitrogen, with the ratio of nitrogen normalized to 100. A ratio of 10 would mean the nutrient concentration would be 0.1 of the value for N (after Ingestad, 1982).

Species	P	K	Ca	Mg
Scots pine	14	45	6	6
Norway spruce	16	50	5	5
Sitka spruce	16	55	4	4
Japanese larch	20	60	5	9
Western hemlock	16	70	8	5
Douglas fir	30	50	4	4

A MORE ACTIVE APPROACH: ASK A TREE WHICH NUTRIENTS LIMIT GROWTH

The critical values listed in Table 14.2 developed from field trials where large numbers of trees (and plots of trees) were fertilized, with growth responses compared to concentrations in unfertilized leaves. A smaller version of this general approach is to plant seedlings in representative soil samples, and then test for growth responses to added nutrients. This bioassay approach is a fertilization field trial in miniature, without needing to wait for several years to measure the growth response. Seedlings substitute for trees, and potted soil samples (often well sieved) substitute for on-site soils. For example, a five-month bioassay of soils from plots with eucalyptus or nitrogen-fixing albizia showed a higher nitrogen supply in albizia soils but a higher phosphorus supply in eucalyptus soils (Binkley, 1996).

Bioassays are valuable for identifying nutrients that limit tree growth, but they often fail to provide quantitative predictions of the response to fertilization. Potential problems with bioassays include the artificial environment of the greenhouse and patterns inherent in seedling development. Mead and Pritchett (1971) found that bioassays with slash pine seedlings correlated only moderately with the response to fertilization in the field; a maximum of 45% of the variability in the field response to phosphorus fertilization could be predicted from the weight of 8-month-old seedlings.

An intriguing twist on the bioassay theme uses root ingrowth bags (mesh bags filled with artificial soil) that are fertilized with various nutrients. If more roots grow into bags containing nitrogen than into control bags, then nitrogen may limit forest growth. This technique has not been used extensively, but Cuevas and Medina (1983) found that root growth responded to bags fertilized with calcium and phosphorus in a lateritic soil, and to nitrogen in a sandy soil, which matched their expectations based on foliar chemistry, soil type, and nutrient content of litterfall. Root response in a study in Hawaii also matched predictions based on the tree growth response to fertilization (Raich *et al.*, 1994). This approach could also be tried with decomposition litter bags; if the disappearance rate from standard paper strips in mesh bags placed in the soil was greater if nitrogen (or other nutrient) was added to the bags, this might indicate a nitrogen deficiency.

VECTOR ANALYSIS MAY IDENTIFY LIMITING NUTRIENTS IN CONIFERS WITHIN A SINGLE GROWING SEASON

Vic Timmer of the University of Toronto developed a simple, rapid approach for identifying nutrient limitations under field conditions (Timmer and Stone, 1978; Timmer and Morrow, 1984). The method takes advantage of the determinate growth habit of many conifers. In these species, the number of needles that can be produced in the current year is predetermined in the bud set in the previous year. Trees that are fertilized at the beginning of the current season can have larger needles, but not more needles, than unfertilized trees. The increase in needle weight correlates well with later increases in growth (Timmer and Morrow, 1984); trees that respond to fertilization with large increases in first-season needle size also respond with large stem growth increases over the next several years. Timmer expanded on this idea by including the nutrient content of foliage.

To use this approach to identify nutrient limitations, treatment plots are fertilized in the spring with a combination of potentially limiting nutrients, and control plots are left untreated. In the late summer, needles are collected, dried, weighed, and analyzed for nutrient concentrations. Nutrient concentration and needle mass are used as axes on a graph, and the nutrient content (concentration times mass) isolines are plotted (Figure 14.4). The trajectory of the vector extending from the control to the fertilized value is used to diagnose the limiting status of each nutrient.

How successful is this approach? Early tests with conifers in Canada found good correspondence between changes in foliar nutrients and wood growth responses (Weetman and Fournier, 1982; Timmer and Morrow, 1984). Valentine and Allen (1990) found that an application of nitrogen + phosphorus led to different interpretations of limitations than the foliar response to single-element fertilization. They also compared predictions of nutrient limitation with actual stem growth responses, contrasting the graphical approach with the simpler critical concentration predictions (see Table 14.2). Five of nine stands responded best to nitrogen + phosphorus; three of these responses were predicted correctly by the foliar response to fertilization, and four were predicted correctly by simple critical nutrient concentrations in foliage.

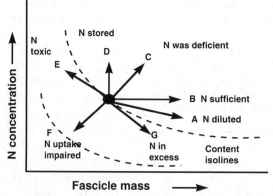

Figure 14.4 The graphical approach to gauging nutrient limitation based on foliar response. An increase in needle mass and nutrient concentration indicates that nutrients may limit growth (vector C). An increase in concentration without an increase in needle mass indicates storage of extra nitrogen (vector D). Other vectors are uncommon and indicate negative effects (after Valentine and Allen's (1990) version of Timmer's graphical analysis method).

Four of the stands responded best to nitrogen alone, and these were all identified correctly by the foliar response to fertilization, while the critical concentration level predicted that two of these stands were also deficient in phosphorus. Binkley *et al.* (1995) tried the foliar response approach in a replicated age sequence of lodgepole pine in Wyoming. They found that needle mass increased with complete fertilization in older stands and that concentrations of nitrogen, phosphorus, and potassium all increased. However, factorial fertilization with nitrogen, phosphorus, and potassium showed responses only to nitrogen, which would have been consistent with simpler predictions based on critical concentrations in foliage from unfertilized trees. Overall, it's not clear that foliar responses to fertilization identify nutrient limitations much better than critical concentrations.

DRIS COMBINES FOLIAR CONCENTRATION, NUTRIENT RATIOS, AND TREE GROWTH

The Diagnosis and Recommendation Integrated System (DRIS) is a multifactorial approach to identifying optimal tree nutrition (Beaufils, 1973). The "norms" are identified by the foliar nutrient concentrations and ratios of foliar nutrients for high-productivity stands. Stands with lower than normal concentrations or ratios are expected to respond well to fertilization. Hockman and Allen (1990) tested the DRIS approach using 62 fertilization trials spread across 52 soil series. The foliar characteristics from the upper 25th percentile in growth were used to define normal values. These values were then used to make predictions about the growth response to nitrogen and phosphorus fertilization. About 80% of the stands predicted to respond to nitrogen fertilization actually did, and no site that was evaluated as unresponsive actually responded. The DRIS norms correctly identified 90% of the phosphorus-responsive stands, but the analysis mistakenly predicted that 25% of the non-responding stands would respond to phosphorus fertilization.

FIELD TRIALS ARE NECESSARY TO MOVE FROM DIAGNOSIS TO RECOMMENDED TREATMENTS

Once nutrient limitations have been identified, it's important to know which sites would respond, and how large the response to a treatment might be. In some cases, the diagnosis of nutrient limitations, such as critical values, would be undertaken in a set of research plots. The insights would then be applied across a landscape based on some feature, such as foliar concentrations determined for stands across the landscape. For example, about 80% of the forest fertilization decisions in Australia are informed by the concentrations of nutrients in foliage (May *et al.*, 2009). Soil analyses are relied upon heavily for deciding fertilizer applications at the time of planting for eucalypts, but not as often for pines. Fertilization decisions are sometimes based on soil types.

Historically, researchers often focused on very few (or single) research sites, with three or four replicate treatment plots at one location. This sort of replication nests plots within single stands and has limited statistical power. A strong inference may be made about treatment response for that single site, but provide little guidance on the response that would develop across a variety of sites. Minimizing variation in the suite of ecological factors that vary among sites can give a strong answer for a very small area, and a very weak answer for a large area. Nesting all the research in a single location would be equivalent to assessing an entire forest's growth and yield by studying one stand. The ecological factors that vary across landscapes influence stand nutrition just like rates of growth and yield; nutrition needs to be assessed at the scale at which the insights will be applied (Binkley, 2008). An example of an efficient "twin-plot" design is presented in Chapter 16.

The current state of knowledge allows useful predictions to be made about nutrient limitations for most of the major commercial forest areas of the world. For example, most (but not all) mid-rotation stands of loblolly pine in the southeastern United States will respond to combined additions of nitrogen and phosphorus, and intensive cropping of *Eucalyptus* plantations requires copious additions of nitrogen, phosphorus, and, commonly, other nutrients. Within this general framework, many case-specific exceptions are important. Some sites respond well to nitrogen alone, whereas others require phosphorus fertilization before a nitrogen response can develop (Figure 14.5).

Where local experts cannot yet provide information on likely nutrient limitations, the major approaches to relating tree nutrition to site-specific conditions are concentrations of nutrients in foliage, concentrations of labile nutrient pools in soils, and site classification. The value of each approach depends on site-specific characteristics (how well does the approach capture the variance in forest response?) and costs. Growth responses to fertilization may be correlated with site characteristics, or a simpler approach may involve defining a critical response level for classifying stands into responding and non-responding groups. For example, Comerford and Fisher (1982) used discriminate analysis to classify slash pine stands by their response to nitrogen and phosphorus fertilization. They found that extractable soil phosphorus or the nitrogen:phosphorus ratio of foliage could classify about 80% of stands correctly into responsive and non-responsive groups. In this case, the precise magnitude of response was not critical, only whether stands responded above a defined level. The acceptability of this level of precision depends on the relative cost of fertilization and the value of

response. If fertilization is cheap and response is very valuable, then the cost of fertilizing unresponsive stands will be more than offset by the value of the response that occurs in responsive stands that may not have been fertilized due to imperfect predictions.

OPERATIONAL FERTILIZATION AROUND THE WORLD

Fertilization is used to supplement inherent soil nutrient supplies, particularly (almost exclusively) in plantation forests where managers anticipate profitable increases in tree growth. Fast-growing pine plantations can use more than $100 \, kg \, N \, ha^{-1} \, yr^{-1}$ and $10 \, kg \, P \, ha^{-1} \, yr^{-1}$, and some eucalyptus plantations can use even more (Fox et al., 2011). Fertilization near the time of planting was a focus in pine forests of the southeastern U.S. through the early 1990s, and then fertilization practices expanded to adding both N and P near the middle of a rotation (Figure 14.6). The shift to emphasis on N + P at mid-rotation developed following recognition of substantial nutrient limitations in this phase of stand development. Interestingly, fewer than half the stands respond to N or P alone, but 85% respond strongly when both are added (Fox et al., 2011). Across the region, a one-time fertilization with $225 \, kg \, N \, ha^{-1}$ plus $30 \, kg \, P \, ha^{-1}$ led to an average increase of $3.5 \, m^3 \, ha^{-1} \, yr^{-1}$ ($28 \, m^3 \, ha^{-1}$ over eight years). A key to the profitability of fertilization is the increased value of the products that can be derived from larger trees. Intensive fertilization of loblolly pine in Georgia, USA increased the amount of sawnwood from 9% (control treatment) to 25% (Jokela et al., 2010), which combines with the increase in total harvest volume for a very profitable response (see the example presented in Chapter 16).

Douglas fir forests of the Pacific Northwestern United States and western Canada are commonly limited by the soil nitrogen supply. Alleviation of this limitation by fertilization increases growth rates by $2 \, to \, 4 \, m^3 \, ha^{-1} \, yr^{-1}$ for 8 to 15 years (Chappell et al., 1991). In some situations, other nutrients, such as sulfur or boron, may become limiting following nitrogen fertilization (Blake et al., 1990; Mika et al., 1992).

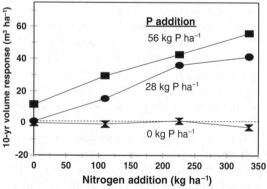

Figure 14.5 Mid-rotation stands of loblolly pine on well-drained sites did not respond to N alone only marginally to P alone; the response to N depended heavily on the addition of P (North Carolina State Forest Nutrition Cooperative, 1997).

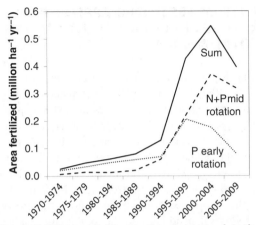

Figure 14.6 Operational fertilization of pine stands in the southeastern United States (from data presented by Albaugh *et al.* 2012).

In the same region, stands of western hemlock may or may not be nitrogen limited; nitrogen fertilization decreases the growth of hemlock in as many stands as it increases this growth (Chappell *et al.*, 1991). Nutrient limitations in hardwood forests in this region have not been characterized, with the exception of intensively managed plantations of hybrid poplars, which are heavily fertilized with nitrogen and zinc. Nitrogen also commonly limits forest growth in the Rocky Mountains and in the southeastern United States, often in combination with phosphorus, potassium, or sulfur (Binkley *et al.*, 1995).

The productivity of conifer forests in Europe is typically limited by N supply, though operational fertilization is uncommon because of environmental concerns about too much enrichment of ecosystems by acid deposition. The productivity of conifer plantations in the United Kingdom is typically limited by phosphorus or phosphorus + potassium; nitrogen limitations are uncommon, perhaps as a result of moderate levels of nitrogen deposition in rainfall (Taylor, 1991). Nitrogen limitation is the most common problem in forests of Norway and Sweden, where nitrogen fertilization may increase volume growth by 1.5 to 3 m^3 ha^{-1} yr^{-1} for 5 to 10 years. Very large, chronic additions of nitrogen may cause limitations of phosphorus and potassium, and some stands respond very well to low doses of boron (especially if nitrogen fertilized). Conifer forests in Denmark are less

commonly limited by nitrogen, perhaps as a result of high rates of nitrogen deposition.

A great deal of interest in Germany in the 1980s focused on magnesium deficiencies in conifers, especially Norway spruce (Hüttl and Schaaf, 1997). This interest derived from occasional yellowing of crowns, sometimes in association with needle abscission and reduced canopy leaf area. Much of the supposed "decline" of spruce in Germany in the 1980s was thought to be derived from magnesium deficiencies, exacerbated either by excessive leaching from the soil or by nitrogen-induced interference with magnesium nutrition (see Oren *et al.*, 1988; Buchmann *et al.*, 1995). Fertilization with magnesium often reduced the yellowing of older age classes of spruce needles (Kaupenjohann, 1997). Interestingly, studies focused on determining whether alleviation of needle yellowing led to improved canopy appearance. In the few cases where actual growth was examined, no significant growth increase was found after magnesium fertilization (Makkonen-Spiecker and Spiecker, 1997). Magnesium concentrations apparently have to be very low before tree growth is reduced. The actual influence of the magnesium supply on tree growth in Europe may need more direct evaluation.

Maritime pine stands in France appear to be most strongly limited by the supply of soil phosphorus (Trichet *et al.*, 2009). Five decades of research have shown that application of 15 to 35 kg P ha^{-1} provides 20–40% increases in growth that last for 10 to 20 years. Responses are stronger on wetter sites than drier sites.

Nutrient limitations are pronounced enough in conifer plantations in New Zealand and Australia that most stands are fertilized at least once in a rotation. Over 100 000 ha are fertilized annually in Australian plantations of eucalypts and pine, with N (50–70 kg ha^{-1}), P (25–40 kg ha^{-1}), and sometimes, K (25–35 kg ha^{-1}), S (20–25 kg ha^{-1}), Cu, and Zn (May *et al.*, 2009). Timing of fertilization is about equally split among planting, early-rotation, and mid-rotation, with typically about a 20% increase in volume growth (3 to 10 m^3 ha^{-1} yr^{-1}).

High productivity of conifer forests in Japan depends on alleviation of the nitrogen limitation; nitrogen fertilization typically raises productivity to the point where fertilization with phosphorus and potassium allows further increases in growth (Kawana and Haibara, 1983).

Almost all tropical forest plantations are fertilized with phosphorus, and most are also fertilized with other elements including nitrogen, calcium, magnesium, and potassium (Figure 14.7). The actual rates applied for any particular stand depend on local site factors, such as soil type, texture, topographic position, and the economic assessments of likely response. Typical growth responses to fertilizer additions are 4 to $8\,m^3\,ha^{-1}\,yr^{-1}$ for five years or longer (Barros *et al.*, 1990; Gonçalves *et al.*, 1997). Growth responses after fertilization with nitrogen, phosphorus, and potassium in South Africa are commonly 6 to $8\,m^3\,ha^{-1}\,yr^{-1}$ (Herbert and Schönau, 1989).

RATE OF FERTILIZER APPLICATION MAY BE MORE IMPORTANT THAN THE TOTAL AMOUNT

A large portion of the cost of fertilization in forests deals with the logistical challenge of spreading tons of materials uniformly across landscapes. The application cost per kilogram of added fertilizer goes down with increasing rates of application, so it might seem logical to add as much as possible in one (or a few) applications. This would be an optimal decision if the nutritional response of the trees was a linear function of the amount of fertilizer added; however, growth responses depend both on the total quantity added and the rate of addition. A three-decade fertilization experiment in a Scots pine stand in northern Sweden showed that the growth response to addition of

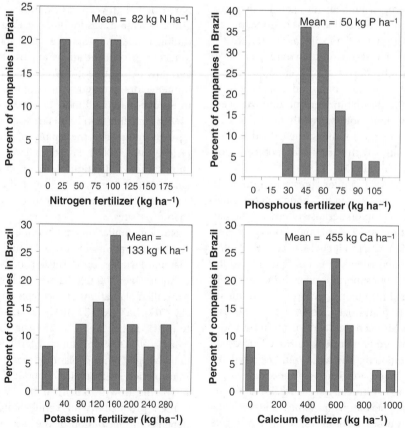

Figure 14.7 Rates of fertilization for eucalypt plantations in Brazil, based on current practices of 20 companies in 2010 (accounting for about 60% of Brazil's total eucalypt area; data provided by J.L. Stape).

900 kg N ha^{-1} could range from a 15% increase if applied at a heavy dose over just six years to 105% if applied at a low dose rate over 25 years (Figure 14.8). Clearly, the response of the forest depended on the rate of application more than the cumulative amount. This finding has important implications for examining the effects of chronic, low levels of N addition from atmospheric deposition: the effects of low rates over many years cannot be mimicked accurately by high doses over a shorter period.

FERTILIZATION INCREASES NET PRIMARY PRODUCTION AND PARTITIONING TO WOOD

Improved nutrition could increase wood growth in two ways: greater net primary production or higher partitioning of photosynthate to wood growth (and less below ground). Net primary production might increase as a result of greater capture of resources (such as higher light absorption by larger canopies) or greater efficiency of using resources. Improved nutrition often leads to either a constant or declining allocation of photosynthate to the production of roots and mycorrhizae, so the relative partitioning to wood growth may increase.

The nutrient (and enzyme) concentrations in foliage typically increase with fertilization, contributing to the increase in net primary production. Heavily fertilized *Eucalyptus* trees averaged about 20% higher nitrogen concentration than control trees and had, on average, a 15% greater capacity for photosynthesis (Figure 14.9). Whether the trees were fertilized or not, their photosynthetic capacity related moderately well to leaf nitrogen content.

The increase in foliar nutrient concentrations and leaf area may or may not be associated with substantial increases in total production. Not all of the nitrogen in leaves is present in photosynthetically active enzymes, not all of the canopy is well illuminated, and overall canopy transpiration may be restricted by the water supply.

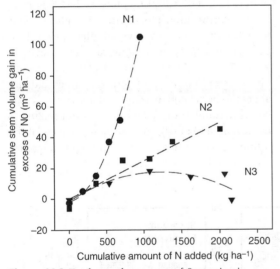

Figure 14.8 Total growth response of Scots pine in northern Sweden to nitrogen fertilization at different rates. N0 was the control (unfertilized) treatment; growth responses to very heavy doses of N (N3, 120 kg N ha^{-1} yr^{-1} or more over a few years) led a weak initial increase in growth, then a decline back to the level of the control treatment. Intermediate rates of addition (N2, about 60 kg N ha^{-1} yr^{-1}) provided 40% greater volume than controls by the time 2000 kg N ha^{-1} had been added. The low rate (N1, about 30 kg N ha^{-1} yr^{-1}) and a total addition of just 900 kg N ha^{-1} more than doubled volume. (Reproduced with permission from Global Change Biology, Tree growth and soil acidification in response to 30 years of experimental nitrogen loading on boreal forest, Peter Hogberg, Houbao Fan, Maud Quist, Dan Binkley, Carl Olof Tamm, 2006, 489–499, John Wiley and Sons.)

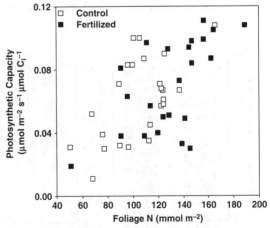

Figure 14.9 The photosynthetic capacity of 29-month-old *Eucalyptus saligna* trees in Hawaii related strongly to foliar nitrogen concentrations (80 mmol m^{-2} of leaves ∼1.2 %N, 160 mmol m^{-2} ∼ 2.5 percent nitrogen); fertilized plots had about 20% more N in leaves and 15 percent greater photosynthetic capacity (data from M. Ryan, personal communication).

An intensive fertilization regime in a stand of lob-lolly pine increased wood production by 4.1 Mg ha^{-1} yr^{-1}; this extra growth resulted in part from greater net primary production, but also to a greater partitioning of total growth into wood (wood growth = 30% of NPP for control treatment versus 37% for fertilized treatment; Albaugh *et al.*, 1998). Intensive fertilization of a young stand of *Eucalyptus saligna* also increased wood production by increasing net primary production and reducing allocation to fine root production. Fertilization of a snowgum (*Eucalyptus pauciflora*) stand with 500 kg N ha^{-1} in a single dose led to an increase in wood production of 1.5 Mg ha^{-1} yr^{-1}, even though net primary production increased by only about 0.9 Mg ha^{-1} yr^{-1} (Figure 14.10). Fine root production dropped in the fertilized stand, providing a large part of the carbo-hydrate supply for wood growth.

ARE FERTILIZED STANDS MORE SUSCEPTIBLE TO DROUGHT THAN UNFERTILIZED STANDS?

One can imagine that higher leaf areas might be associated with greater water use and greater sensitivity of large-canopied trees to cavitation of stemwood during drought periods. If this risk were of major importance, a variety of studies would have reported drought-related mortality from fertilized stands. Few studies have reported any interaction between drought tolerance and fertilization. Some evidence indicates that water-use efficiency increases after fertilization and that fertilized trees generally are not at greater risk of dying during droughts than trees with poorer nutritional status. Some studies have reported evidence of higher mortality in fertilized plots. During the most severe drought on record, a fer-tilized plot of radiata pine in Australia showed 10% mortality (concentrated in smaller-diameter classes of trees, not dominants) compared with no mortality in a control plot (Snowdon and Benson, 1992). This study was not replicated, so confidence in the cause of the difference between plots is hard to establish. In Brazil, a moderately severe drought killed 24% of *Eucalyptus* trees in control plots, 31% of trees in low-fertilization plots, and 42% of trees in high-fertilization plots; with six replicate plots for each treatment, the overall effect of fertiliza-tion on mortality was significant at a level of $p = 0.07$ (J. L. Stape, personal communication). The greater mortality of fertilized trees did not appear to be related directly to increases in canopy size (the number of dying trees was not related to the leaf area of plots), so a mechanistic view of the influence of fertilization on drought tolerance remains unclear.

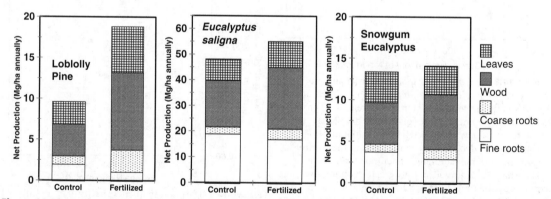

Figure 14.10 Fertilization of loblolly pine in North Carolina (470 kg/ha of nitrogen, 135 kg/ha of phosphorus, 275 kg/ha of potassium) doubled net primary production, and wood production was 2.4 times the control value (Albaugh *et al.*, 1998). Fertilization of *Eucalyptus saligna* in Hawaii increased wood growth by increasing total production and decreasing root production (Ryan *et al.* 2004). Fertilization of snowgum increased net primary production by just 6 percent and wood production by 30 percent (Keith *et al.*, 1997).

ELEMENTAL CONCENTRATIONS DIFFER AMONG FORMULATIONS OF FERTILIZERS

The formulation of fertilizer is important for calculating the amount of an element to be applied to an area. For example, 100 kg of ammonium nitrate (NH_4NO_3) contains 35 kg of nitrogen, 60 kg of oxygen, and 5 kg of hydrogen. An application of 100 kg ha^{-1} of nitrogen therefore requires 285 kg of ammonium nitrate. Most discussions of fertilization mention the quantity of the nutrient applied rather than the weight of the fertilizer. For example, a rate may be reported as "100 kg of nitrogen as urea"' or simply as "100 kg urea nitrogen."

Agricultural fertilizers in the United States are sold with the percentages of "nitrogen–phosphorus–potassium" listed on the bag. A 10-kg bag of "10–10–10" would appear to have 1 kilogram of each nutrient. However, the phosphorus and potassium values are expressed as the antiquated "oxide" forms: P_2O_5 and K_2O (potash). To obtain the true percentage of phosphorus, the listed value is multiplied by 0.44; for potassium the multiplier is 0.83. A bag containing 10 kg of 10–10–10 fertilizer actually contains 1 kg nitrogen, 0.44 kg phosphorus, and 0.83 kg potassium.

NITROGEN FERTILIZERS ARE SYNTHESIZED USING NATURAL GAS

Nitrogen fertilizers are commonly synthesized by the Haber–Bosch process, in which N_2 and CH_4 are passed over a metal catalyst at high temperature and pressure to produce NH_3. The ammonia may then be transformed into a variety of fertilizers, such as urea (($NH_2)_2CO$), nitrate, or ammonium. Fertilizer form affects chemical reactions in the ecosystem, and availability of the fertilizer to trees. For example, ammonium nitrate dissolves quickly in water; ammonium may be retained on cation exchange sites in the O horizon or mineral soil, and nitrate may be taken up quickly or leached into the mineral soil. The ammonium may gradually become available to plants and nitrifying bacteria, and the nitrate may quickly be absorbed by plants or leached from

the soil profile. Urea must first be hydrolyzed to form ammonium:

$$(NH_2)_2CO + 2H_2O + 2H^+ \rightarrow 2NH_4^+ + H_2CO_3$$

Note that one H^+ is consumed from the soil solution for each ammonium formed; this consumption is offset in part by dissociation of carbonic acid (H_2CO_3) to form H^+ and bicarbonate (HCO_3^-). Plant uptake of ammonium also will release one H^+ back to the soil solution. Urea fertilization usually raises soil pH for a few months but has little effect more than one year later. The H^+ effects of ammonium and nitrate fertilizers are not balanced, like those resulting from urea transformation and use. Uptake of ammonium generates H^+, which is not balanced by previous consumption from the soil solution. Nitrate uptake consumes H^+ from the soil solution without balancing previous production (as is the case for on-site nitrification). Annual fertilization of agricultural soils can alter H^+ budgets substantially, but fertilization of forest soils is typically infrequent enough that any change is negligible. Seedling nurseries present an exception to this generalization. Annual fertilization with certain forms of fertilizer can have a significant acidifying effect.

How important are these differences in chemistry? Urea contains 45% nitrogen and 55% hydrogen, carbon, and oxygen, and ammonium nitrate is about 34% nitrogen. Fertilizer weight is important in aerial applications, and in this case, urea would appear to be the better choice. Any savings in application costs, however, could be offset by differences in growth responses. For example, the Scots pine growth response on some soils may be about 45% greater with ammonium nitrate than with an equivalent amount of urea nitrogen (Malm and Moller, 1975). In North America, up to a 20% greater response to ammonium nitrate was reported in some Douglas fir studies from British Columbia (Dangerfield and Brix, 1979; Barclay and Brix, 1985), but other experiments showed no difference (Brix, 1981). Jack pine stands in eastern Canada have shown no significant difference between nitrogen forms in growth response (Weetman and Fournier, 1982). Loblolly and slash pines in the southeastern United States have shown little difference in response between the two forms (Allen and Ballard, 1982; Fisher and Pritchett, 1982).

Fertilizer formulation may affect the loss of fertilizer from the ecosystem. Nitrate may be subject to rapid leaching if heavy rainfall precedes uptake by plants. Gaseous losses from urea fertilization may occur if urea is added during a fairly dry period. Hydrolysis to ammonium can produce a rapid, very localized increase in soil pH around the fertilizer granule that will cause some ammonium to deprotonate, forming ammonia gas. Rainfall after fertilization helps prevent volatilization losses by dissolving the urea granules and by preventing localized increases in pH. The season of fertilization may be an important component of the tree response if rainfall patterns affect retention of the fertilizer. In any case, volatilization losses from urea fertilization are usually less than 10% of the applied nitrogen.

Some special formulations of nitrogen fertilizers (such as urea-formaldehyde or iso-butylidene diurea) have been developed to promote slow release of nitrogen, maximizing the time course of availability to trees and minimizing the losses from high application rates. These formulations are expensive, and their use in forestry has been limited to high-value nursery and Christmas tree soils.

The size of fertilizer granules is another important consideration in forest fertilization. The precision of aerial fertilization is increased with the use of relatively large (0.5 to 1.0 cm) granules. Large granules are affected less by wind patterns and may penetrate canopies better (reducing foliage burning).

PHOSPHATE FERTILIZERS ARE MINED

Phosphorus fertilizers come in three general forms: ground rock phosphate, acid-treated phosphates, and mixtures with other nutrients. The original source of most phosphorus fertilizer materials is mined rock phosphate. In general, phosphorus fertilizers have little effect on soil acidity, but weathering of ground rock phosphate may slightly increase soil pH.

Ground rock phosphates are mined from deposits of various types of calcium apatite minerals:

$$Ca_{10}(PO_4CO_3)_6(F, Cl, or OH)_2$$

The solubility of apatite varies with the anion (F^-, Cl^-, or OH^-). Fluoride gives the lowest solubility, so a high content of fluoride in rock phosphate fertilizer reduces the rate of release of phosphorus for plant uptake. The rate at which various types of rock phosphate dissolve and become available for plant uptake can be determined from solubility with citric acid (see Bengtson, 1973). The rate at which ground rock phosphorus becomes available to plants is also affected by soil pH and the particle size of the rock; low pH and small particle size favor rapid dissolution. Large grains dissolve very slowly, especially in neutral and high-pH soils.

Phosphate minerals may be treated with acids to produce fertilizers with a range of phosphorus contents. Superphosphate is generated by mixing ground rock phosphate with sulfuric acid, and contains about 8% phosphorus, 20% calcium, and 12% sulfur:

$$3Ca(H_2PO_4)_2 \cdot H_2O + 7CaSO_4 \cdot 2H_2O$$

Rock phosphate can also be treated with phosphoric acid (H_3PO_4) rather than sulfuric acid to produce concentrated (or triple) superphosphate with 2% phosphorus, 13% calcium, and no sulfur:

$$Ca(H_2PO_4)_2 \cdot H_2O$$

These forms dissolve readily and are quickly available to plants.

Under agricultural conditions, rapid fertilizer availability is an asset. In forestry, prolonged availability at low levels may be preferred to rapid availability. Therefore, ground rock phosphate (with high citrate solubility) may have a bit of an edge over superphosphate, especially on soils with high phosphorus sorption capacity. Differences in response to various forms of phosphorus are usually slight, and relative fertilizer costs usually should guide decisions in forest nutrition management (Allen and Ballard, 1982; Hunter and Graham, 1983; but see also Bengtson, 1976; Torbert and Burger, 1984).

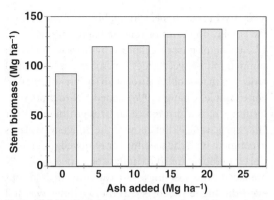

Figure 14.11 Stem biomass response (age 6.5 years) to addition of forest biomass ash in Brazil (data from Moro and Gonçalves, 1995, and Guerrini *et al.*, 1999).

Some forms of fertilizer can be bulk blended for single applications of multiple nutrients. One of the blends used most commonly in forestry, diammonium phosphate (DAP), contains about 24% phosphorus and 21% nitrogen. In most cases, more nitrogen than phosphorus is needed, so DAP is often mixed with urea or ammonium nitrate to achieve the desired nitrogen:phosphorus ratio in the fertilizer.

FERTILIZATION WITH "LIME" CAN REDUCE SOIL ACIDITY AND IMPROVE CALCIUM AND MAGNESIUM NUTRITION

Various forms of fertilizer contain calcium and magnesium. The most commonly used forms are types of "lime," which include a basic anion such as carbonate or sometimes hydroxide. Other forms of calcium and magnesium fertilizers work well, but most applications come from mined deposits (or industrial wastes) that are basic. For example, application of 200 kg of gypsum (calcium sulfate) increased the volume of *Eucalyptus grandis* by 50% on an Oxisol in Brazil (Barros and Novais, 1996).

A wide variety of industrial wastes can be applied to forest soils to provide cation nutrients and increase growth (Gonçalves *et al.*, 1997). These include "slags" that result from blast-furnace production of iron, "dregs" and "grits" from Kraft processing of wood for pulp, "extinct lime" (sodium hydroxide plus

calcium carbonate) from pulp processing, and ash from burning of forest biomass to generate electricity (used either directly as ash or as "burnt ash" following greater oxidation). For example, Moro and Gonçalves (1995) added a range of wood ash to a plantation of *E. grandis* in Brazil (Figure 14.11). Stem growth increased up to application rates of about 15 Mg ash ha^{-1} (which should contain about 270 kg ha^{-1} of calcium and 25 kg ha^{-1} of magnesium). Higher rates of application did not increase growth.

RAISING SOIL pH REQUIRES VERY LARGE AMOUNTS OF LIME

In some cases, forest managers may seek to raise soil pH in the expectation of improving tree growth. The growth response to gypsum by Barros and Novais (1996) did not result from an increase in soil pH, as the application rate was far too low to change the pH of an Oxisol. The response resulted from improved calcium and sulfur supplies.

Massive amounts of lime are required to substantially alter soil pH (Figure 12.7). Addition of 1 to 2 Mg ha^{-1} of limestone is commonly required to change soil pH by 1 unit, and this effect is restricted to the O horizon and the uppermost mineral soil. Liming does not generally increase tree growth. In Finland, experimental liming of forest soils showed that about 65–75% of the added calcium remained in the O horizon and upper mineral soil even 25 years after treatment, and reduced extractable aluminum by about 20% (Derome *et al.*, 1986). The liming treatments did not benefit the trees; in fact, growth declined by an average of 3% for Scots pine and 10% for Norway spruce. The effects of liming in Finland were matched by those in Sweden, leading to the conclusion that liming experiments indicate an overall reduction in growth of about 10% for 15 to 20 years (Popovic and Andersson, 1984; Andreason, 1988). In a review of experience from Europe and Sweden, Nihlgård and Popovic (1984) concluded that soil acidification could be prevented with applications of 5000 kg of lime per hectare but that stand growth would probably decline. Some German studies have reported better growth responses to liming. For example, Spiecker (1995) found that growth of Norway spruce was increased about six years after a liming treatment, but the treatment also included addition of nitrogen, so the response may have resulted from either the liming or the nitrogen.

Liming has strong influences on forest biogeochemistry, especially within the O horizon. Liming often (but not always) increases decomposition rates in the O horizon and may double rates of nitrogen mineralization (Persson *et al.*, 1989). In Bavaria, Germany, Kreutzer (1995) found little effect of 4000 kg of dolomitic lime per hectare on the pH of the mineral soil, but the rate of decomposition on the O horizon increased dramatically. The nitrogen content of the O horizon dropped by $170 \, kg \, ha^{-1}$, and soil leachate contained extremely high concentrations of nitrate nitrogen ($> 10 \, mg \, L^{-1}$). Liming increased earthworm biomass by eight-fold and also increased root biomass in the O horizon (and decreased root biomass in the mineral soil).

If liming does not increase tree growth, why would foresters invest in the expense of liming? Beese and Meiwes (1995) summarized German perspectives on liming, emphasizing ideas about improving soils through liming, perhaps reversing expected negative impacts of acid deposition even if tree growth does not improve.

TYPICALLY ABOUT 10 TO 20% OF FERTILIZER ENTERS TREES

Most of the nutrients applied in fertilization never enter the trees. For a variety of pine species, recovery of nitrogen and phosphorus fertilizers in trees has been reported to range between 5% (at very high application rates) and 25% (Ballard, 1984). The average across 30 studies of fertilizer retention in a range of conifer ecosystems was about 50%, with about half of the retained nitrogen entering the trees (Johnson, 1992). A summary of acid-rain related studies with labeled nitrogen showed that about 25% of the added nitrogen was taken up by trees (Nadelhoffer *et al.*, 1999). As with many features in forest soils, the variability around these averages is large enough that individual situations may not relate at all to the average. The highest rates of

recovery in trees appear in applications with either low application rates or repeated applications.

The rate of nitrogen capture by trees depends on competition between trees, microbes, and chemical processes that stabilize nitrogen in soil organic matter. For example, Chang (1996) added ^{15}N to plots with seedlings of western red cedar, western hemlock, or Sitka spruce and either allowed the understory plants to compete with the seedlings or removed the understory. Removal of the understory increased ^{15}N uptake by the seedlings by two- to eight-fold, but even the highest recovery in the seedlings was still just 15% of the added nitrogen, compared with 50% or greater recovery of the ^{15}N in the soil.

The retention of nitrogen fertilizer with increasing doses was examined in an experiment in Billingsjön, Sweden. Nitrogen was applied at rates of 0 to $600 \, kg \, ha^{-1}$ on three occasions to an old (80- to 100-year-old) Scots pine stand over a period of 20 years (Nohrstedt, 1990). The soil (O horizon down to the B2 horizon) retained about 50% of the added nitrogen up to a cumulative dose of $1500 \, kg \, ha^{-1}$; addition of $1800 \, kg \, ha^{-1}$ decreased nitrogen recovery to about 35% (Figure 14.12). The absolute quantity of nitrogen recovered in the soils appeared to reach a maximum of about $700 \, kg \, ha^{-1}$ over 20 years. Nitrogen leaching (in soil water at 50 cm depth) was not increased by nitrogen rates of up to $1500 \, kg \, ha^{-1}$, but it increased from background levels of about $1 \, kg \, ha^{-1}$ annually to about $4 \, kg \, ha^{-1} \, yr^{-1}$ in the plots that received $1800 \, kg \, N \, ha^{-1}$. Clear-cutting these stands generated no increase in nitrogen leaching from plots that had received less than $800 \, kg \, ha^{-1}$, and plots that had received higher nitrogen additions lost up to $17 \, kg \, ha^{-1} \, yr^{-1}$ for three years after the harvest. Losses of nitrogen through denitrification were negligible ($< 0.1 \, kg \, ha^{-1} \, yr^{-1}$) for all treatments before and after harvesting (Nohrstedt *et al.*, 1994).

Where did the missing fertilizer go?

Two approaches are commonly used to follow the fate of applied nutrients: mass balance of the nutrients held in the biomass and soils, and the use of isotopes. Using the biomass and nutrient pool method, a

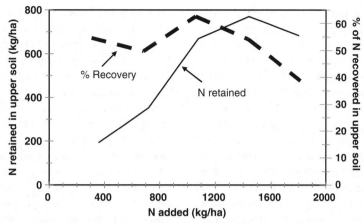

Figure 14.12 Cumulative nitrogen fertilization (over 20 years) of an 80-year-old Scots pine stand in Billingsjön, Sweden, showed high relative retention of fertilizer from the O horizon to the B2 horizon (data from Nohrstedt, 1990).

researcher attempts to measure changes in pool sizes and identify any increases resulting from fertilization. The "pool change" method requires accurate sampling and low within-pool variability. For example, fertilization of loblolly pine trees at age 3 or 4 years showed that about 10% of the added nitrogen was found in the pines two years later, with about 5–15% in the herbaceous plants. This approach works for small vegetation with fairly low nitrogen content, but the nitrogen content of large trees may be difficult to estimate with enough precision to determine the presence of any extra nitrogen. Some very intensive studies have met with success in accounting for fertilizer recovery. For example, addition of $400 \, kg \, ha^{-1}$ of nitrogen to a radiata pine forest increased plant nitrogen uptake by $170 \, kg \, ha^{-1}$ (this 40% recovery in tree is unusually high), another $145 \, kg \, ha^{-1}$ remained in the soil, and $60 \, kg \, ha^{-1}$ was lost in soil leachate (Raison *et al.*, 1990). The remaining $35 \, kg \, ha^{-1}$ (9% of the applied nitrogen) remained unaccounted for.

Nitrogen fertilizer can be labeled with the stable isotope ^{15}N and the ratio of the isotopes ^{15}N and ^{14}N can be determined later for each pool and the contribution of fertilizer nitrogen calculated. For example, Melin *et al.* (1983) applied $100 \, kg \, ha^{-1}$ of ammonium nitrate to 130-year-old Scots pine, with either the ammonium or the nitrate labeled with ^{15}N. The total recovery in their single-tree plots was about 80% of the applied nitrogen for both forms, but about 10% more of the nitrate fertilizer

was found in the vegetation after two years. The missing 20% of the fertilizer was either volatilized or leached below the 30-cm depth of the soils. Nadelhoffer *et al.* (1999) synthesized results from nine forests in North America and Europe that received additions of labeled ^{15}N to investigate ideas about excessive levels of nitrogen inputs. Less than 10% of the added nitrogen was lost in soil leachates at seven of the nine sites, and the other two sites (which experienced large rates of nitrogen deposition from the atmosphere) leached about 15–50% of the added nitrogen. Across all sites, the authors estimated that about 20–25% of the added nitrogen entered trees, 70% was retained in the soil, and 10% was lost to leaching or as gases.

Overall, it appears that, following fertilization, less than a quarter of the fertilizer is taken up by trees in the first few years; another quarter-to-half is immobilized in microbial biomass and soil organic matter, and a variable and difficult-to-measure amount is lost from the forest ecosystem through leaching and volatilization.

Why doesn't more fertilizer get into the trees?

This is a difficult question, but a classic set of calculations can be used to illustrate some of the factors (after Miller, 1981). Consider a forest soil with $5000\,kg\,ha^{-1}$ of nitrogen, 1% of which ($50\,kg$ ha^{-1}) mineralizes annually and is taken up by trees. About one-quarter of the $50\,kg\,ha^{-1}$ goes into biomass increment, and the rest returns to the soil in litter and dead roots. In the first growing season following fertilization with $200\,kg\,ha^{-1}$ of nitrogen, availability could be $250\,kg\,ha^{-1}$. This rate would exceed the uptake capacity of the trees. If uptake increased by 50%, then less than 15% of the added nitrogen would move into the trees. Of the remaining fertilizer, most is immobilized by the microbes; only a small fraction is lost. The soil nitrogen capital is now about $5175\,kg\,ha^{-1}$, and if mineralization patterns have not been altered, the nitrogen available to trees in the second growing season after fertilization would be 1% of 5150, or $51.5\,kg\,ha^{-1}$. In general, tree recovery of added nitrogen can be explained by increased uptake in the first season after treatment (Miller, 1981). Fertilization may have more of a residual effect on nitrogen mineralization than this sample calculation would indicate, but the general trend probably holds unless additions are large relative to total soil nitrogen.

Given such low recovery of nitrogen fertilizers by trees, why are such large amounts of fertilizer added in operational fertilization programs? Smaller quantities may be largely immobilized by soil microbes, leaving little for tree uptake. Johnson and Todd (1985) applied urea nitrogen to recently planted stands of loblolly pine and yellow poplar at a rate of $100\,kg\,ha^{-1}\,yr^{-1}$ once a year for three years, or $25\,kg\,ha^{-1}$ four times a year for three years. Both species responded better to the once-a-year applications than to the quarterly applications, probably because microbial immobilization removed most of the $25\,kg\,ha^{-1}$ dose but a lower proportion of the $100\,kg\,ha^{-1}$ dose.

FERTILIZER RESPONSE DIFFERS AMONG GENOTYPES AND SPECIES

Genetic differences among individuals and strains within species play a major role in tree growth and nutrition. Most genetic selection programs focus on overall growth and stem form, without direct attention to interactions between genotype and nutrients. The importance of nutrition to genetic selection is illustrated by comparing two clones of *Eucalyptus grandis* from Brazil (Barros and Novais, 1996). The better-performing clone obtained twice as much phosphorus from the soil (and fertilizer) than the poorer clone and produced twice as much stem biomass. The poorer clone obtained only 30% of the potassium used by the better clone.

Superior genotypes may not be superior in all environments, so tree breeders are often concerned with Genotype x Environment interactions (Zobel and Talbert, 1984). Nutrient availability is part of this environmental component. A number of experiments have examined the Genotype x Nutrient interaction by measuring the growth response to fertilization of genetically selected trees. Some species, such as loblolly and slash pines, have shown no significant interactions (Matziris and Zobel, 1976; Rockwood *et al.*, 1985). For these species, superior trees respond to fertilization with the same volume increment exhibited by other genotypes. Other species, such as radiata pine in Australia, show marked differences in fertilization response among genotypes (Waring and Snowdon, 1977; Nambiar, 1985). For example, the poorest-responding family of radiata pine in one study increased its growth by only 9% in response to nitrogen fertilization, whereas the best-responding family doubled its growth (Fife and Nambiar, 1995). Fertilization improved water use efficiencies of all families, with no interaction between water use and nitrogen response among families.

What mechanisms account for Genotype $\times$ Nutrient interactions? It is possible that some genotypes are more efficient at utilizing nutrients. In agricultural plants, this commonly involves both greater growth per kilogram of nutrient obtained from the soil and greater uptake from the soil (Saric and Loughman, 1983). However, variations in nutrition among tree genotypes have so far been related simply to differences in uptake rather than to different nutrient-use efficiencies (Nambiar, 1985; Barros and Novais, 1996). The primary mechanism appears to be genetic variations in root systems that result in variations in nutrient uptake. Increased rooting density (centimeters of roots and mycorrhizae per volume of soil) should enhance the uptake of nutrients, especially those of intermediate or low mobility, such as phosphate or ammonium (Bowen, 1984).

FERTILIZATION AFFECTS NON-TARGET VEGETATION

Understory production usually increases after fertilization unless the overstory is too dense to allow a response. In the "Garden of Eden" study in California, Powers and Ferrell (1996) provided an abundant nutrient supply to young ponderosa pine trees, with and without herbicide control of competing vegetation. At their Whitmore site, fertilization (with 1000 kg ha^{-1} of nitrogen plus a full suite of other elements) had little effect on pine volume, but the biomass of understory shrubs increased from 6.2 kg ha^{-1} in control plots to 16.4 kg ha^{-1} in the fertilized plots. Removal of competing vegetation increased pine volume by 3.5-fold by increasing the water and nutrient supply to the pines. Competition control combined with fertilization further increased the pine volume to 4.5 times the volume of the control plots.

Browse for deer and other animals is usually increased by fertilization, especially when applied in combination with thinning (Rochelle, 1979). Forage production for cattle may also increase; studies in the southeastern United States have found that fertilization at the time of plantation establishment increases forage yield by 350 to 5500 kg ha^{-1} yr^{-1} annually for five years. Fertilization at the time of thinning may increase forage yields by 650 to 2000 kg ha^{-1} yr^{-1} for a couple of years (Shoulders and Tiarks, 1984).

A very interesting effect of fertilization in Douglas fir stands was reported by Prescott *et al.* (1993); they found that repeated fertilization greatly reduced understory salal shrubs, a major competitor with the trees.

FOREST FERTILIZATION USUALLY HAS MINOR EFFECTS ON WATER QUALITY

Streamwater concentrations of nutrients may increase following fertilization as a result of direct input into streams or leaching through the soil. Most increases remain well below drinking water standards (10 mg N L^{-1} as nitrate in the United States and Canada, 11.3 mg N L^{-1} as nitrate in Europe), but some studies have found peak concentrations of nitrate nitrogen as high as 10 to 25 mg N L^{-1} for short periods (Figures 14.13 and 14.14) and for short distances downstream.

Average concentrations increase much less than peak concentrations, and forest fertilization studies have found that average concentrations have not risen to levels that would threaten drinking water quality (Binkley *et al.*, 1999). Nutrient input to streams varies with the formulation of the fertilizer; ammonium nitrate typically leads to higher nitrate nitrogen concentrations than does urea (Figure 14.13). Repeated fertilization may lead to sustained, high concentrations in streamwater, but this possibility has not been explored. Phosphate fertilization poses no drinking water threat, but increased concentrations may allow for transient increases in the productivity of aquatic ecosystems. All of the studies summarized in Figure 14.14 examined temperate forests; given massive rates of fertilization in tropical regions such as Brazil, it will be important to determine if these patterns hold for warmer, wetter situations.

FERTILIZATION MAY OR MAY NOT INCREASE SOIL ORGANIC MATTER

It may seem logical that if addition of nutrients increases tree growth, then more organic matter should be added to the soil, increasing soil organic matter pools. Chapter 4 discussed the factors that influence the processing and stabilization of C in soils, and the rate of input is only one factor. Changing the chemistry of organic matter added to soils might alter the short-term and long-term decomposition rates, and it might affect the stabilization of organic matter into long-term pools. Many studies of N-fixing species have shown higher C accumulation than under non-N-fixing trees (see Chapter 11), but the available studies on forest fertilization seem to indicate a wider variety of responses. For example, McFarlane *et al.* (2010) found that fertilization of ponderosa pine forests

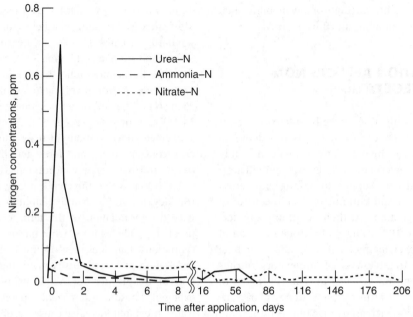

Figure 14.13 Average concentration of nitrogen forms following fertilization of two watersheds with 225 kg/ha in Washington (ppm = mg L^{-1}; from Moore, 1975).

across a productivity gradient in California increased O horizons by 50–100%, and also increased the pool of mineral soil C at some sites. The most productive site did not show an increase in soil C in response to fertilization. Hyvönen *et al.* (2008) examined the long-term (14–30 year) accumulation of soil C in response to repeated, heavy doses of fertilizer, and found that soil C increased by about 0.5 Mg ha^{-1} yr^{-1} in fertilized plots across the decades (with no direct relationship with fertilizer dose). In contrast, heavy fertilization of a eucalyptus plantation on a fertile soil in Hawaii showed no net change in soil C (despite high precision in resampling of pool sizes; see Figure 15.6).

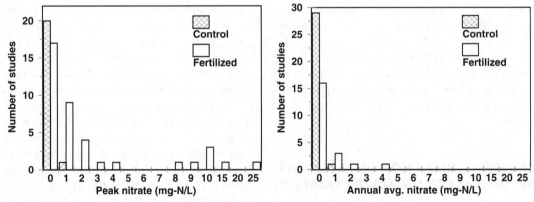

Figure 14.14 Nitrogen fertilization usually increases the peak concentration of nitrate with maximum values < 2 mg N L^{-1}. Several studies have documented peak increases of more than 10 mg/L. Fertilization has much less effect on the annual average concentrations of nitrate (from Binkley *et al.*, 1999).

FERTILIZATION IS AN INVESTMENT OF ENERGY AS WELL AS MONEY

Decisions about fertilization are typically based on costs and expectations of the return on investment. The response to fertilization can also be examined in terms of energy invested and energy returned in the form of wood biomass. From the simplest standpoint, the energy costs can be tabulated and compared with the energy content of the extra wood obtained from fertilization. The internationally accepted unit for energy is the Joule (J); 1 million J (MJ) equals 950 BTU, or 240 kcal. The synthesis of nitrogen fertilizers involves the use of natural gas energy to reduce N_2 to NH_3, which is then processed into urea or ammonium nitrate. The total energy cost of synthesis is about 50 MJ per kilogram of nitrogen in urea. Mining phosphate minerals has an energy cost of about 1.6 MJ kg^{-1} of phosphorus, and transformation into superphosphate (8% phosphorus) or triple superphosphate (20% phosphorus) adds 2.8 MJ kg^{-1} and 8.0 MJ kg^{-1} of phosphorus, respectively (Pimentel and Pimentel, 1979).

As an example, consider an application of 100 kg N ha^{-1} and 50 kg P ha^{-1} (as triple superphosphate). The energy content of this application would be 5000 MJ ha^{-1} for nitrogen and 480 MJ ha^{-1} for phosphorus. The next step is transport to the forest. Switzer (1979) estimated the cost at about 0.0064 MJ kg^{-1} of nitrogen in urea for each kilometer; assuming that the factory is 800 miles away, the transportation energy for 100 kg of nitrogen sums to 50 MJ. Including the phosphorus fertilizer would raise transportation energy costs to 75 MJ. Application from a helicopter would cost another 180 MJ (Switzer, 1979). If fertilized by tractor rather than helicopter, the application cost is close to 50 MJ ha^{-1}. Fertilization increased growth by 17 m^3 ha^{-1}, and harvesting this amount (assuming that harvest energy is proportional to biomass removed) would require about 9000 MJ. The energy content of softwood is about 15 000 MJ m^{-3}, or 250 000 MJ for 17 m^3 ha^{-1}. Summing the costs of synthesis, transportation, application, and harvest, the energy cost of fertilization + harvest is about 14 600 MJ ha^{-1}. The ratio of wood energy to the energy cost of the operation is 250 000:14 600, or about 17:1. In contrast, the return in energy in wheat or corn in response to fertilization typically falls between 5:1 and 10:1 (calculated from Pimentel and Pimentel, 1979).

CHAPTER 15

Managing Forest Soils for Carbon Sequestration

OVERVIEW

Carbon sequestration in forest soils depends simply on the balance between rates of C addition to soils and rates of C losses. This simple balance, however, depends on many interacting processes that vary over space and time, preventing simple insights from being very informative. The results of hundreds of studies show that soil C may or may not increase when forest vegetation replaces grass pastures or row-crop agricultural crops. This uncertainty depends not so much on how poorly scientists have studied these questions, but on real-world differences that vary among situations. Even in situations where planted forests do not increase soil C, the forests actually contribute large amounts of new soil C; these large additions may simply offset losses of C that accumulated under previous vegetation and land use. Rising temperatures may increase decomposition rates, but any changes would likely depend in part on broad ecosystem changes including rates of litter addition that also change in response to climate. Perhaps the best indicators of future soil C pools come from current geographic patterns that already span ranges in soil factors and climate; in general, warmer soils accumulate more C than cooler soils. The complete long-term picture of C sequestration in forest soils may depend on how frequent and intense periodic disturbances become; a fire may quickly override any pattern of C sequestration that took a century to develop.

_____

Tree canopies absorb short-wave sunlight, and convert about 1% to 3% of the absorbed energy into carbohydrates through photosynthesis. How much of this total gross primary production enters the soil? The answer depends very much on partitioning of the carbohydrate among tree tissues, and the timescale under consideration. Trees tend to partition about 15–30% of their total production into the growth and maintenance of leaves; about one-third to one-half of the material may accumulate as leaf biomass, with the rest oxidized to CO_2 as part of the leaf metabolism for growth and maintenance (Litton _et al._, 2007). Some of the carbohydrates may arrive in the roots within a few days, supporting the production of root biomass and the maintenance of roots, mycorrhizae and the rhizosphere community (Chapter 6). Tissues such as woody stems may not enter the soil for decades to millennia (Chapter 4). The overall accumulation (sequestration) of carbon is a very complex result of all these avenues and rates of organic matter inputs, and the equally complex processes and rates involved in decomposition and stabilization of organic matter.

EDDY FLUX SYSTEMS MEASURE NET CHANGES IN THE C CONTENT OF FORESTS WITH GREAT PRECISION

The net balance between all the processes that add and remove C from forests can be determined with precision at near-instantaneous timescales with eddy flux systems. Sonic anemometers measure air velocity in three dimensions, and CO_2 samplers

Ecology and Management of Forest Soils, Fourth Edition. Dan Binkley and Richard F. Fisher.
© 2013 John Wiley & Sons, Ltd. Published 2013 by John Wiley & Sons, Ltd.

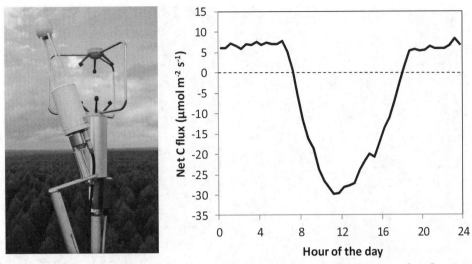

Figure 15.1 An eddy flux system depends on rapid, precise measurements of three-dimensional air flow (sonic anemometer on the right) coupled with rapid measurements of CO_2 (sensor on the left). Negative fluxes mean accumulation of C in the ecosystem; through the course of the day and night, this eucalyptus plantation gained about $5\,g\,C\,m^{-2}$. (Courtesy of Yann Nouvellon).

match the air movement with precise concentrations of CO_2 (Figure 15.1). The net movement of air can be coupled with the CO_2 concentration of the air to determine the net flow of C up from the sensors (out of the forest) or down from sensors (into the forest).

Eddy flux systems have two major limitations for understanding C flux in forests. The first is methodological; the system requires that CO_2 exchange between the ecosystem and atmosphere happens by convection (blowing wind), as diffusion would not be detected. This requirement is often met, but can be problematic during nights with no wind. The second deals with resolution; a very good number that integrates a very large number of interacting, countervailing processes may not give much insight on relative importance of processes. If a forest is gaining more C than it is losing, is the C accumulating in fast-turnover pools, or long-term pools? Is a change in C accumulation resulting from short-term effects of drought on current plant uptake or on decomposition rates?

Some of the carbohydrate production by forest canopies reaches soils quickly and is rapidly utilized to support the growth of fine roots,

mycorrhizae, and the rhizosphere community. A girdling experiment in Sweden demonstrated this rapid flow. Girdling removes the bark and the cambium from stems, cutting off the flow of carbohydrates from the canopy to the roots (Figure 15.2). The stemwood xylem remains intact, and canopies remain well supplied with water until the roots eventually die from starvation. Girdling the Scots pine trees in Sweden led to a one-third reduction in the CO_2 rising from the soil in just five days, and the CO_2 efflux dropped by about half in one to two months (Högberg *et al.*, 2001). As with many cases in forest soils, the results from one situation may, in some cases, be consistent with another case (such as an oak forest example in New York, USA, Levy-Varon *et al.*, 2011), but not in others. Some forests, such as eucalyptus and other hardwood forests, may store such large quantities of carbohydrates in the root systems that: (1) the flow of recently produced carbohydrates from the canopy is "diluted" into a large storage pool (delaying their use), and (2) the release of CO_2 from roots may not decline for months after girdling (Edwards and Ross-Todd, 1979; Binkley *et al.*, 2006).

Figure 15.2 Girdling of Scots pine trees in Sweden (left) led to rapid declines in CO_2 efflux from the soil, indicating that carbohydrates flow from the canopy and are used in the soil within a few days to weeks. In contrast, girdling in a *Eucalyptus* plantation in Brazil showed only slight decreases in CO_2 efflux after girdling, indicating large stores of carbohydrates that "buffer" the current supply of carbohydrates from the canopy.

TOTAL BELOWGROUND CARBON ALLOCATION (TBCA) IS WHAT TREES SEND TO ROOTS

Just as carbon-focused scientists use eddy flux approaches to estimate the net flux of C for forests, a carbon-balance approach can be used to estimate the total amount of C that is funneled below ground from tree canopies. The total belowground carbon allocation (TBCA) approach is a simple mass balance approach (Raich and Nadelhoffer, 1989; Giardina and Ryan, 2002). A mass balance approach means that because matter isn't created or destroyed, we can be sure that:

$$\text{Inputs} - \text{Outputs} \pm \text{Change in storage} = 0$$

If only one piece of this equation is unknown, it can be calculated easily. The outputs of C from a forest soil are almost exclusively in the form of CO_2 efflux from the soil surface, and this can be measured with reasonable (and quantifiable) precision. The inputs to the soil come from aboveground

litterfall (again, easily measured with quantifiable precision), plus all the C that trees send below ground to support root maintenance, grow new roots, and support mycorrhizal fungi. This is difficult (or impossible) to measure directly. The change in soil C can be measured, at least in some cases, with sufficient (and quantifiable) precision. Therefore, the C sent below ground by trees (TBCA, Figure 15.3) equals the efflux of soil CO_2 minus litterfall plus or minus any change in C storage in the soil (in both the pool of live roots and dead organic matter).

The TBCA approach works very well in situations where the rate of change in soil C pools can be measured with high precision because of low spatial variance. For example, Binkley *et al.* (2004) measured the C content of 0–45 cm depth soil four times through a rotation of *Eucalyptus* in Hawaii; the pool averaged 138 Mg ha^{-1}, with no significant change over time. If there was a real change in soil C that was not detected by the sampling precision of the experiment, there was a 95% probability that the change was within plus or

Total belowground C input = CO$_2$ efflux - Litterfall + Change in storage

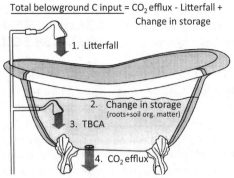

Figure 15.3 The flow of C from tree canopies into the belowground portion of a forest can be estimated by knowing the rates of litterfall input, the change in storage of C in live roots and dead organic matter, and the rate of CO$_2$ efflux from the soil (based on a diagram from Giardina and Ryan 2002).

minus $0.6\,Mg\,C\,ha^{-1}\,yr^{-1}$. The other components of the belowground C story were much larger, including soil CO$_2$ efflux of about $20\,Mg\,C\,ha^{-1}\,yr^{-1}$, litterfall input of $4.2\,Mg\,C\,ha^{-1}\,yr^{-1}$, and accumulation of live root biomass of $2.1\,Mg\,C\,ha^{-1}\,yr^{-1}$ (Giardina and Ryan, 2002). Even in forests where changes in soil C content cannot be measured with precision, reasonable rates of soil change (from the literature or other sources) would usually be small enough relative to the other fluxes that the TBCA approach can be used to test ideas about the effects of tree species, management, and other factors. The key limitation of the TBCA approach is that it cannot identify what happens to the C once it has gone below ground, such as use in respiration to maintain roots, or in growth of new roots, or in allocation to mycorrhizal fungi.

LONGER-TERM CHANGES CAN BE ESTIMATED BY MEASURING SIZES OF POOLS RATHER THAN FLUXES

The change in soil C storage might be estimated by measuring the rate of CO$_2$ efflux from the soil, versus the rate of C added to the soil through root production and aboveground litter input. For example, a set of intensively managed eucalyptus plantations in Brazil sent about $1000\,g\,C\,m^{-2}\,yr^{-1}$ into the atmosphere, and aboveground litterfall added about $300\,g\,C\,m^{-2}\,yr^{-1}$ into the soil (Ryan et al., 2010). If we knew the rate of root production we could do the arithmetic and determine if the soils were gaining or losing C. Unfortunately, root production could not be measured with enough precision to come close to providing a useful value for soil C change. Fortunately, a well-designed sampling of soils can provide direct estimates of rates of C change (at least for soils with low spatial variability) without a need to know the rates of C inputs or outputs. In this case, the release of $1000\,g\,C\,m^{-2}\,yr^{-1}$ into the atmosphere was associated with a net increase in soil C of about $17\,g\,C\,m^{-2}\,yr^{-1}$. Most of the focus of this chapter will be on net rates of change in soil C, with details about the processes found in other chapters.

SOIL CARBON STORES MAY CHANGE WHEN TREES INVADE GRASSLANDS

Afforestation of pastures leads to dramatic changes in the aboveground vegetation. A first expectation might be that this much accumulation of forest biomass would likely be associated with large increases in soil organic matter and carbon. However, the annual productivity of forests, agricultural fields, and pastures may not differ much, and we know remarkably little about the comparative additions of aboveground and belowground carbon, or about how differences in chemical characteristics of freshly dead organic matter lead to different rates of C turnover and accumulation. Forests typically accumulate O horizons, which are thin or non-existent in pastures. Empirical studies sometimes have shown higher C accretion under forests, but not always.

A classic tree invasion story comes from grassland–aspen forest ecotones in Alberta, Canada. Hans Jenny (1980) reported a study that examined a vegetation/age sequence of grassland, invading aspen (45 years old), balsam poplar (85 years old) and Douglas fir (150 years old). The soils differed substantially among vegetation communities. The grasslands had black chernozem (Mollisol) soils, and the old Douglas fir site was a degraded (sic) brown-wooded soil (Alfisol). Across this sequence, mineral soil C (0–80 cm depth) dropped by more than half. If the soils in the sequence had all been equal prior to tree establishment, and if the trees currently found on a site were the first generation, then the rate of C loss would have been on the order of 90 to 180 g $C m^{-2} yr^{-1}$. The C storage in the O horizon was overlooked, but would likely make up for less than half of the apparent C lost from the mineral soil.

The assumptions of an age sequence can be checked for reasonableness by calculating the implications for mass balance. For example, exchangeable bases declined by two-thirds along the sequence, and total nitrogen dropped by almost half. The sum of exchangeable bases differed by 450 $kmol_c ha^{-1}$, and soil nitrogen (0–80 cm) by 13 500 $kg ha^{-1}$. If such changes happened over 150 years, the rate of 3 $kmol_c ha^{-1}$ annual loss of base cations would be five times the rate that would be expected to accumulate in trees and in the forest floor. The loss rate of nitrogen of 90 $kg ha^{-1} yr^{-1}$ is also several times the rate of accumulation in biomass. If these rates were true, then massive losses in soil leachates would need to be accounted for, which would be very difficult in such a dry climate. The most likely explanation of these rapid soil changes is that they didn't occur so rapidly or that the selection of sites spanned major differences in vegetation that pre-dated the establishment of the trees. (See Chapter 16 for some discussion of the challenges of chronosequence studies.)

REFORESTATION AFTER PASTURE ABANDONMENT MAY OR MAY NOT INCREASE SOIL C STORAGE

When pastures are created in forested areas, it is common for the trees to return if grazing and other land management activities do not prevent tree reestablishment. Several chronosequence studies have looked at soil changes following reforestation of pastures, and most investigators expected to see substantial gains in soil carbon pools. Thorough documentation of changes has not always supported these expectations. For example, planting *Eucalyptus urophylla* into a plowed pasture in Brazil led to wood accumulation of 350 $Mg ha^{-1}$ over three decades, even with two thinnings (Figure 15.4). The change in soil C (0–45 cm) was not significant. Was the lack of change a result of a very static pool of soil C, or did high rates of inputs and outputs cancel out? Reforestation of agricultural species sometimes follows cropping of species with the C4 photosynthetic pathway, such as corn (maize), sugarcane, and tropical pasture grasses. These C4 species differ in the natural abundance of ^{13}C (relative to ^{12}C; both are stable isotopes) from C3 trees. The prior crops essentially "label" the soil organic matter with a ^{13}C signature, which allows researchers to follow the loss of old soil carbon and the gain of new soil carbon under the influence of trees. Analysis of ^{13}C patterns revealed that the lack of change did not result from a lack of interesting dynamics. Over 30 years, the 0–15 cm soil lost an average of 0.21 $Mg C ha^{-1}$ yr^{-1} from the pool of C that had accumulated under the former pasture, balanced precisely by a gain of 0.21 $Mg C ha^{-1} yr^{-1}$ from tree-derived organic matter (data from Cook, 2012). The development of an O horizon contributed about 0.17 $Mg C ha^{-1} yr^{-1}$; any comparison of pasture and forest soils that forgot to include the O horizon (as noted in Chapter 4) would have missed the major change.

In general, tree establishment in pastures shows only slow rates of C accumulation. A pair of chronosequences of forests developing after pasture abandonment in Puerto Rico showed no apparent carbon accretion across 60 years of forest development (Garcia-Montiel, 1996). Another three

Figure 15.4 A pasture in São Paulo, Brazil was plowed and planted with *Eucalyptus urophylla*. After 30 years and two thinnings, the stem biomass of the forest was about 350 Mg ha^{-1} (white arrow points to the same native tree). Although this large of a change in vegetation might be expected to drive large changes in soil, the total soil C did not change significantly (based on data of Cook 2012; photos courtesy of Jose Luiz Stape).

chronosequences from Puerto Rico found very high variation in two series, and some apparent carbon accretion in the third. The overall accretion was about 0.0 to 0.1 Mg C ha^{-1} yr^{-1} (Lugo *et al.*, 1986).

In Costa Rica, a replicated chronosequence found that active pastures had 16 Mg C ha^{-1} in soil depth of 0 to 10 cm, 5- to 10-year-old regenerating forests had 15 Mg C ha^{-1}, and 10- to 15-year-old regenerating forests had $21\,\mathrm{Mg\,C\,ha}^{-1}$ (Reiners *et al.*, 1994). The rate of carbon accretion for that five-year period would average $1.5\,\mathrm{Mg\,C\,ha}^{-1}\,\mathrm{yr}^{-1}$; this anomalously high rate does not warrant confidence because the overall chronosequence did not show a significant time trend (so any differences among individual plots may well result from soil factors other than time).

REFORESTATION AFTER PLOWED AGRICULTURE MAY ALSO SHOW LITTLE CHANGE IN SOIL C STORAGE

Foresters and some soil scientists commonly believe that forests are the best type of vegetation for maintaining or improving soil conditions. Forest land uses are expected to protect soils better than agricultural land uses, and trees appear to protect soils better than pasture grasses. Such generalizations may not be supported by evidence for a variety of reasons. The first is that few generalizations apply to all forest soils, and even fewer apply to all agricultural situations. When forests replace row crops, land management practices such as annual fertilization stop; the change in vegetation and land management may also include a change in practices that favored sustained soil fertility at a high level.

In some cases, formal "meta-analyses" of the literature can provide clear conclusions about the typical effect of some factor (such as afforestation), including the typical size of the effect, the variability around the typical size, and some factors that influence whether a case may likely fall above or below the typical effect size. For example, Berthrong *et al.* (2009) aimed to determine the effect of afforestation on soil properties. They compiled patterns from 153 independent examples, and concluded that afforestation on average reduces soil C by 7%. This pattern only applies to the 0–30 cm depth mineral soil, because only 5% of the studies they tallied had included sampling of the O horizon. As noted in Chapter 4, this is a common oversight and a major reason why it's important to conceive of the O horizon as part of the soil rather than something sitting on top of "the soil." Laganière *et al.* (2010) also did a similar meta-analysis (with more sites), and

reached several conclusions: afforestation after plowed agriculture leads to larger soil C increases than post-pasture afforestation, and inclusion of O horizons shows that afforestation does *not* lower total soil C.

How reliable are the estimates of soil C change that go into meta-analyses? Answering long-term questions with short-term experiments is a challenge (as in the Alberta example above), and a thorough case study from the northeastern USA illustrates this clearly. Hamburg (1984) sampled a chronosequence of sites that had been abandoned after agriculture. Linking the points on the chronosequence showed an impressive C accretion of about 250 to $500\,\mathrm{g\,m}^{-2}\,\mathrm{yr}^{-1}$. However, a chronosequence assumes that sites would be very similar except for whatever changes are attributable to time. An extra century of agricultural land use might have depleted soil C relative to the sites where forest recovery started sooner, and other potential confounding factors might be important. With this in mind, Fuller (2007) resampled three of the sites used in the chronosequence. The slopes of the 25-year change should match the chronosequence slope if the chronosequence reliably captured the effect of time (Figure 15.5). The flat slopes indicated that no confidence would be warranted in a high rate of C accumulation with forest age, and that confounding problems prevented the chronosequence from providing reliable insights.

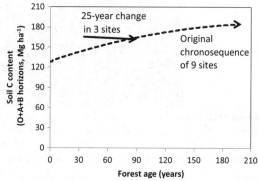

Figure 15.5 A chronosequence of forest sites in New England, USA showed an increase of 0.25 to $0.5\,\mathrm{Mg\,ha}^{-1}\,\mathrm{yr}^{-1}$ in soil C, but a resampling of three of the sites after 25 years showed no change (or a slight decline) in soil C (from data in Hamburg 1984, Fuller 2007).

A strong case study from South Carolina, USA documented rates of soil change by thorough sampling soils within a single site over time (Richter *et al.*, 1999; Richter and Markewitz, 2001). Carol Wells designed a very thorough soil-sampling scheme for a 5-year-old plantation of loblolly pine on a former agricultural field, and then repeated the sampling at five-year intervals and archived the soils. After 30 years, the O horizon had accumulated just 35 Mg C ha^{-1} (1.2 Mg C ha^{-1} yr^{-1}), and the mineral soil showed almost no accumulation (2.1 Mg C, or 0.07 Mg C ha^{-1} yr^{-1}; Richter *et al.*, 1999). Intriguingly, the C added to the mineral soil had a very short residence time (based on patterns of ^{14}C following atmospheric testing of nuclear bombs); the O horizon had longer residence time C than did the mineral soil.

In Hawaii, USA, reforestation with *Eucalyptus* trees showed no net increase in soil carbon after 10 to 15 years in both chronosequence and repeated-sampling studies (Bashkin and Binkley, 1998; Binkley *et al.*, 2004c). However, the lack of net change resulted from large offsetting losses of old cane-derived carbon (about 1.5 kg m^{-2}) and the gain of new *Eucalyptus-derived* carbon (Figure 15.6). Heavy annual fertilization in this experiment increased the canopy leaf area by 25% and increased gross primary production by about 1 kg C m^{-2} yr^{-1} (Ryan *et al.*, 2004); however, allocation of

C to belowground production was not affected by fertilization, and both the gain of C3-derived C (from trees) and retention of C4-derived C (from previous sugarcane agriculture) showed no response to fertilization. This lack of response to fertilization contrasts strongly with effects of N-fixing trees at nearby sites on the same soil series (Kaye *et al.*, 2000; Figures 11.17 and 11.19).

The patterns of ^{13}C allowed some detailed investigations of both C pools and fluxes (Giardina *et al.*, 2004). The trees added 0.4 kg C m^{-2} yr^{-1} to the soil in the form of falling litter from the canopy, but aboveground litterfall accounted for less than 20% of the C added to the belowground portion of the ecosystem. The trees added another 2.0 kg C m^{-2} yr^{-1} through their root systems, supporting the growth and respiration of root cells, the mycorrhizal community, and feeding the soil detrital system with dead roots. Of the 2.4 kg C m^{-2} yr^{-1} added to the soil, 2.2 kg C m^{-2} yr^{-1} returned to the atmosphere, so there was a net accumulation within the belowground portion of the ecosystem of 0.2 kg C m^{-2} yr^{-1}. About 90% of the accumulating C was located in live (coarse-diameter) roots and 10% in soil organic matter.

In this *Eucalyptus* plantation, about 80% of the C sent below ground by trees (in litterfall and via roots) returned to the atmosphere within one year. This high "flow through" rate for the overall

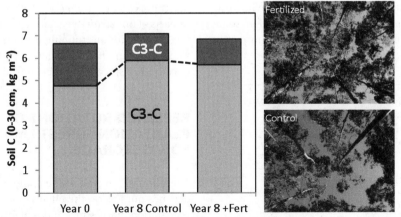

Figure 15.6 Low variability in soil properties provided high precision for estimating changes in soil C in a eucalyptus plantation in Hawaii. The 95% confidence interval on total change in C was +/− 55 g C m^{-2} yr^{-1}. This lack of changes resulted from offsetting losses of 140 g C m^{-2} yr^{-1} of older, cane-derive C and an equal gain of new C from trees. Heavy fertilization had no effect on either loss or gain of soil C. (from data of Binkley *et al.* 2004).

C budget showed little net change in soil organic matter. However, a pattern of little net change can result from either very little change, or from large offsetting gains and losses. The O horizon increased by 0.05 kg C m^{-2} yr^{-1}, but this small change resulted from very large offsetting inputs (0.41 kg C m^{-2} yr^{-1}) and losses (0.36 kg C m^{-2} yr^{-1}). The net change in mineral soil organic matter (to 30 cm depth) was a (non-significant) loss of 0.02 kg C m^{-2} yr^{-1}, which might seem to indicate a static system. On the contrary, stable isotope methods showed that the soil lost 0.24 kg C m^{-2} yr^{-1} of old soil C (C accumulated in the soil prior to the plantation's establishment), at the same time it gained 0.22 kg C m^{-2} yr^{-1} of the eucalyptus-derived C.

HARVESTING TYPICALLY LOWERS THE MASS OF O HORIZONS

Major events such as intense fire, severe winds, and harvesting dramatically alter the structure and function of forests, with legacies that can last for centuries. Soil organic matter can decline as a result of removal of existing material, reductions in the inputs of material, and any increase in rates of decomposition. Nave *et al.* (2010) summarized the findings of hundreds of studies, and concluded that forest harvesting typically leads to about a 30% drop in the mass of the O horizon (Figure 15.7). Given the activities of large machines during the harvesting operation, and perhaps in site preparation, some of the O horizon may be expected to be transported

into the mineral soil rather than lost from a site. However, upper mineral horizons generally show little change (gain or loss) in C. Combining all horizons, forest harvesting seems to lower soil C by about 7–10%. The losses should not be considered as unavoidable or permanent; harvesting techniques can be modified to reduce impacts (see Chapter 12), and subsequent forest recovery may rebuild the soil C capital.

FOREST PLANTATION MANAGEMENT OFTEN INVOLVES REDUCING COMPETING VEGETATION

The growth of trees is commonly reduced when other competing vegetation is present, and various silvicultural operations (site preparation, planting, herbicide application) aim to reduce this competition. Reduced understory production and biomass typically lead to greater growth of the desired tree species; does this shift affect the accumulation of soil organic matter? Powers *et al.* (2012) summarized the results from more than two dozen long-term studies, and concluded the effects were typically minor, with any significant change occurring only in the uppermost (0–10 cm) mineral soil (Figure 15.8). The studies with the best designs (including repeated sampling within plots) tended to show no effect of vegetation control treatments. In a broader context, Powers *et al.* (2012) noted that actual impacts of understory vegetation on C sequestration might depend on the implications for fires, as understory fine fuels can have a controlling influence on fire spread and fuel consumption.

REPEATED ROTATIONS OF PLANTATION FORESTS MAY CHANGE SOIL C STORAGE

The ultimate, overall sequestration of C in forests depends not just on the effects of afforestation, site preparation, harvesting, and replanting, but on the summed and interactive effects of the full suite of factors that go along with forest development and silviculture. Two factors may have opposing

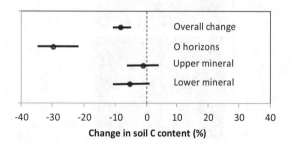

Figure 15.7 Meta-analysis of the impact of forest harvesting shows large losses of O horizon C, but little change overall (from data of Nave *et al.* 2010).

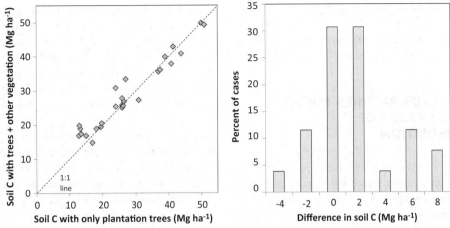

Figure 15.8 A synthesis of long-term studies that examined the influence of vegetation control treatments on soil C found a small tendency for higher soil C in the presence of trees + competing vegetation (averaging about 1 Mg C ha^{-1}), with no apparent difference in other portions of the soil profile (from data in Powers *et al.* 2012).

influences, such as C loss in site preparation and C gain through tree growth, or may both push overall sequestration in the same direction. Only a few studies have taken on the challenge of characterizing the overall rate of change in soil C as forests develop and are managed over time.

One example is the nationwide forest soil-sampling program in Sweden (Olsson *et al.*, 2009). With almost 1500 sites sampled across the forest region of the country, future sampling will have an opportunity to characterize the response of forest soils to the full suite of factors that change soils. This approach essentially uses the full geography of the country to develop the "map" of soil influences, rather than investigating the contribution of one or several factors at one or several locations and then making guesses about regional patterns.

Perhaps the most thorough study yet available comes from a 30-year study with over 300 sites in Brazil. Cook (2012) provided a 30-year assessment of changes in soil C for these sites. This time span covered three to four full rotations of *Eucalyptus* silviculture across a very wide range of geography (sites across a 1000 km gradient), soils, and management details. The grand average of soil C content (0–30 cm) was 31.7 Mg ha^{-1} in the original sampling, and 27.7 Mg ha^{-1} in the most recent sampling, representing a net loss of 0.2 Mg C ha^{-1} yr^{-1}. Of course, some sites had gained C and some lost C,

but the 95% confidence interval for the full diversity of sites and operations ranged from a net loss of 0.1 to 0.3 Mg C ha^{-1} yr^{-1}.

As time goes by, we should have more well-designed studies that shed light on the overall rates of C sequestration in forest soils, and which factors tend to accentuate or reverse gains in soil C.

RISING TEMPERATURES MAY INCREASE TREE GROWTH IN BOREAL AND TEMPERATE FORESTS

The growth of trees is, of course, influenced by many environmental factors, including the supply of resources (water, soil nutrients, and atmospheric CO_2) and temperature. If the future climate warms as expected, how will tree growth respond? Way and Oren (2010) reviewed over 60 studies and reached three conclusions: trees will likely grow faster, change allometry (taller, thinner, less root development), and changes will be more pronounced for trees growing below their optimum temperatures, and slight for trees in the warm tropics. Increased tree growth would likely mean increased total productivity of forests, as photosynthesis was found to increase more with rising temperature than plant respiration. What would these changes mean for soil C accumulation? Two factors would be important: any shift in below-ground C flux, and any change in the suite of processes that determine the net balance between decomposition and production of stable soil C.

RISING TEMPERATURES MIGHT INCREASE RATES OF DECOMPOSITION

The release of CO_2 from soils tends to be high in warm, summer months and low in cool and cold months. This would seem to indicate a strong temperature sensitivity of decomposition. Seasonal patterns confound two very different processes, however: decomposition and root growth (and maintenance). These two processes were separated in an experiment in a Norway spruce stand in Austria by using trenched plots (with no live roots) in comparison with untrenched plots. The release of CO_2 in decomposition did indeed rise with temperature, but the CO_2-generating activity of roots showed a much stronger effect, as root growth showed strong seasonal patterns as part of overall plant phenology (Figure 15.9).

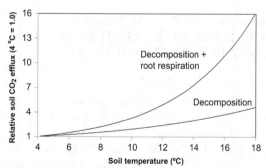

Figure 15.9 As soils warm through the seasons in a Norway spruce in Austria the rate of CO_2 release rises steeply, increasing by about 7 fold as temperature climbs from 4 to 14 °C (upper line). The pattern results from the seasonal pattern of root growth and respiration (including associated mycorrhizae and rhizosphere microbes feeding on new C released by roots), and the decomposition of soil organic matter. Removing the contributions of roots (and newly released tree carbon) leads to a smaller temperature response, with decomposition increasing about 3-fold as temperature climbs from 4 to 14 °C (lower line; from data of Schindlbacher et al. 2009).

The influence of temperature on decomposition is not straightforward, even after the confounding fluxes of root generation of CO_2 are accounted for. Conant et al. (2011) emphasized that decomposition entails the breakdown of large molecules into short molecules through the action of enzymes, and that the effects of temperature may differ for each of these features. The accessibility of large molecules may be limiting if they are bound to soil surfaces or occluded within stable aggregates, so any influence of temperature on rates of turnover of soil–mineral connections could be important. The rate-limiting step for decomposition may vary with temperature and over seasons (and years?); complex interactions between substrate supplies and enzyme activity may obscure any simple temperature effect (Brzostek and Finzi, 2011).

Fissore et al. (2009) used a geographic gradient in soil temperatures to examine turnover rates of soil organic matter under pine and hardwood forests in the eastern USA. About 2% of the total carbon in the soil was defined as "active," with a turnover time of about a month; the active pool slowed down in turnover rate with increasing temperature. The "slow" pool of soil organic matter showed turnover rates of decades to thousands of years, with no clear pattern with temperature. This study does not mean that within a single site there would be no responsiveness of decomposition to temperature; rather, it indicates that when temperature effects also encompass the broad ecosystem suite of factors, any simple influence is small compared to the influence of confounding factors.

WHAT WILL BE THE EFFECT OF RISING TEMPERATURES ON SOIL C STORAGE?

This heading is an example of a poorly phrased question that might lead to unproductive thinking. Temperatures do not have a single, consistent effect on any aspect of the C budget of forests, and we should not expect any single, robust "effect" to be found in any research project. Complex interactions often provide average overall effects, along with variability around those overall averages. The two features (average and variance) may be determined by investigating the response of individual pieces of the C budget to warming temperatures, such as the response of tree growth or root growth or litter decay. These responses need to be integrated across long enough periods of time that contributions to the accumulation of C in soils may develop. The scientific literature probably has hundreds (or thousands) of papers dealing with aspects of these topics in recent years.

One approach to evaluating the combined effects of interacting factors is to warm a forest soil artificially in the field and evaluate process responses such as soil CO_2 efflux and pools of soil C (Rustad et al., 2001). Soil temperatures can be raised by shining infrared lamps from above, and by burying heating wires within the soil. These approaches have unavoidable methodological challenges, including potential drying of the O horizon by infrared lamps, and soil disturbance as heating wires are buried.

A second type of problem arises in the interpretation of effects: if warming increases CO_2 efflux from the soil, did this come from increased decomposition or increased root growth (and maintenance respiration)? In some cases these two potential sources are separated by treatments (such as trenching) that remove living roots as CO_2 sources. For example, Schindlbacher et al. (2009) found that warming soils beneath Norway spruce in Austria increased soil CO_2 efflux by about $2.8\,\mathrm{Mg\,C\,ha^{-1}}$ $\mathrm{yr^{-1}}$ in untrenched plots, and $1.8\,\mathrm{Mg\,C\,ha^{-1}\,yr^{-1}}$ in trenched plots, indicating that living roots contributed 35–40% of the soil CO_2 efflux.

A third type of problem is identifying whether the measured responses to soil warming are transient adjustments to new soil conditions, or representative of long-term changes that would be expected to represent patterns generated by warmer climates. A soil-warming experiment in a deciduous forest in the northeastern USA (Melillo et al. 2002; Butler et al., 2012) found that CO_2 release from decomposition increased quickly with heating, but declined substantially after five years as the supply of heat-sensitive organic matter substrates was exhausted. In contrast, soil N continued to remain higher in warmed plots throughout the study.

A final limitation for soil-warming experiments relates back to the initial question for this section; there is no reason to expect a general, robust effect of soils to warming, and warming experiments are very expensive and can be done on too few sites to give us good insights about the actual effects that climate warming will have across any sizable forest landscape.

GEOGRAPHY MAY GIVE THE BEST INSIGHTS FOR POTENTIAL CLIMATE CHANGE

A simpler approach might be to ask how current patterns of soil C storage relate to recent patterns of temperature, perhaps with other factors (such as precipitation) included. The scale of evidence is often most powerful when it fits the scale of a question. Broad geographic patterns that result from the same driving factors that will determine future soil conditions may give us much better insights than short-term studies on the temperature response of decomposition of fresh litter. A good example comes from Scandinavia, where Callesen et al. (2003) evaluated patterns of forest soil C for relationships with temperature and other factors. For well-drained soils, the storage of soil C (O horizon to 1 m) increased substantially with increasing temperature (Figure 15.10; see also Figure 4.9). Sites with higher precipitation, and with coarser soils, tended also to have higher soil C. The analysis provides strong evidence that warmer temperatures should be expected to lead to accumulation of C in Scandinavian forests on well-drained soils. The patterns may be different on wet sites, where any confounding effects of changes in soil aeration could override the sorts of patterns found currently across geographic gradients.

SIMPLE ANNUAL FLUXES MAY BE OVERWHELMED BY EPISODIC FLUXES

Annual changes might be the focus of typical concepts and measurements, but the real responses of soil C to changing climate might

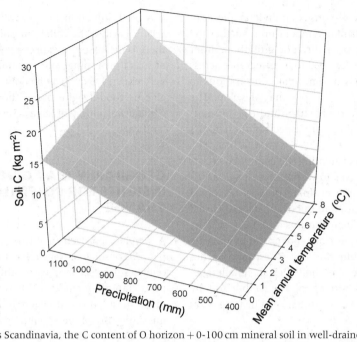

Figure 15.10 Across Scandinavia, the C content of O horizon + 0-100 cm mineral soil in well-drained locations increases with increasing temperature and increasing precipitation (from relationships in Callesen *et al.* 2003).

depend more on losses that happen through episodic events. A warmer climate would lead to permafrost melting, exposing centuries-old organic materials to decomposition. The material may be relatively resistant to decomposition, but the change in aeration and temperature could be important (for a good review, see Schuur *et al.*, 2008). Perhaps more important would be changes in fire regimes. Fires reduce the store of C in surface wood and O horizons (and perhaps mineral horizons), but also add large amounts of coarse material (including char). Forest recovery after fires may include species with different litter chemistries that foster different soil communities, leading to different rates of stabilization of soil C. When potential interactions may be so important and so difficult to determine, the best prediction of future soil conditions may depend on waiting for the future to arrive and then measuring the changes.

PART VI

Thinking Productively about Forest Soils

CHAPTER 16

Evidence-Based Approaches

OVERVIEW

The power of information depends in large part on how well the information is used. Insights from research may be applied effectively, ignored, or used inappropriately. Some common ways to use information poorly (to be wrong) include basing decisions on incomplete accounting, misunderstanding the limitations of patterns for understanding processes, and misunderstanding how variation in responses can be as important as the average size of responses. Formal approaches have been developed in other fields for evaluating wise decisions based on uncertain (and incomplete) information, and for expressly harnessing evidence (rather than theories and hopes) in management decisions. In many cases, managers have great opportunities to apply the techniques of "pocket science" to gain insights, including realizing when business-as-usual leads to undesired outcomes.

For every complex and difficult problem, there is a solution that is simple, straightforward and wrong. (H. L. Mencken)

Gaining reliable knowledge is the basis of all good science and effective management. When we begin to think about a problem in science or a management decision, we would be wise to ponder the essence of this story:

A man was riding on the train from Reading, a university town in England, to London. His compartment mate, he discovered, was a scientist. He made many attempts to engage this gentleman in conversation, but he only received nods or one-word answers. As they rode along, they passed a verdant green pasture with a beautiful white horse standing in it. The man said, "look at that beautiful white horse." The scientist replied, "well it's white on this side anyway."

As scientists or managers we must always be skeptical. We must be curious about why and how, but at some point we must move forward based on the knowledge that we have at hand. That is why we need to be assured that the knowledge we have at hand is reliable. And if we have to guess the color of the other side of the horse, we need to remember this is a guess and not based on the strongest evidence.

We commonly use two of the three scientific methods when we think about scientific problems or management decisions – *induction* and *retroduction*. Induction is useful for finding laws of association between classes of facts. For example, if we observe that bald cypress (*Taxodium disticum* (L) Rich) commonly occurs on very poorly drained soil, we might conclude that bald cypress is restricted to such soils and establish a law of association. However, bald cypress grows very well on well-drained soils if there is adequate precipitation.

We often use retroduction to develop explanations or reasons for facts when it is actually best used to develop hypotheses. For example, if we observe that pines commonly occur on soils with spodic horizons, our best guess (our hypothesis) might be that pines cause spodic horizons. Retroduction is the method of circumstantial evidence used in courts of law. It may be reliable enough for the courts, but, by itself, it is not sufficient for

Ecology and Management of Forest Soils, Fourth Edition. Dan Binkley and Richard F. Fisher.
© 2013 John Wiley & Sons, Ltd. Published 2013 by John Wiley & Sons, Ltd.

obtaining reliable knowledge. Retroduction is a reliable method for generating hypotheses; however, these hypotheses must be tested to gain reliable knowledge, and that brings us to the third scientific method.

The third method is the hypothetical-deductive (H-D) method. First we must define terms that are necessary to understanding the H-D method: theory, research hypothesis, statistical hypothesis, and test consequence. A theory is a broad, general conjecture about a process. A research hypothesis is a theory that is intended to be tested experimentally. A research hypothesis is more specific than a theory, since, for example, the location and species of interest must be specified. It must be tested indirectly because it embodies a process, and experiments can only give facts entailed by a process. The process itself is abstract and non-factual. The indirect test is conducted by logically deducing one or more test consequence(s); that is, predicted facts, such that if the research hypothesis is true, then the test consequence(s) must be true, and the test consequence(s) must correspond to a feasible experiment, e.g., one that is not technologically impossible or too costly to perform.

INCOMPLETE ACCOUNTING MAY GIVE WEAK INSIGHTS

Does afforestation of agricultural lands lead to increased storage of organic matter and carbon in the soil? Only a complete accounting could answer this question, and too often the sampling designs used to address this question are incomplete. For example, Wang *et al.* (2011) conducted a very intensive study to characterize how larch plantations changed soil organic matter relative to agriculture. The repeated-sampling and chronosequence sites indicated that C in the mineral soil increased at a rate of about $0.1 \, \text{kg} \, \text{C} \, \text{m}^{-2} \, \text{yr}^{-1}$ under larch, compared with an annual loss of about $0.3 \, \text{kg} \, \text{C} \, \text{m}^{-2} \, \text{yr}^{-1}$ in agricultural land uses. How much did the accumulation of an O horizon contribute to C sequestration? We can only guess, as the O horizons weren't sampled. If the O horizon had a C content of $2 \, \text{kg} \, \text{m}^{-2}$ ($20 \, \text{Mg} \, \text{ha}^{-1}$), and it took 40 years to accumulate, this would be a rate of C accumulation of $0.05 \, \text{kg} \, \text{C} \, \text{m}^{-2}$

yr^{-1}; this rough estimate suggests the authors may have missed about one-third of the actual C accumulation.

Just as accounting may be incomplete when the upper part of the soil is omitted, neglecting lower horizons can lead to the same problem. Many forest soils are quite deep, with soil development and root penetration extending far below ground. A *Eucalyptus* plantation in Brazil on a highly weathered Oxisol had more than half of its total soil C located deeper than 2 m (Figure 16.1). The trees sent fine roots at least 10 m into the soil; about 80% of the fine roots occurred in the upper meter of soil, but more than 10% developed deeper than two meters.

Another example of poor accounting would be following the drop in elevation of the soil surface in a peatland after draining, and jumping to the conclusion that drainage led to massive decomposition of peat as soil aeration improves. Indeed, Minkkinen and Laine (1998) examined over 250 drained sites in Finland, and after about 60 years the soil surface had subsided by about 22 cm (about $0.3 \, \text{cm} \, \text{yr}^{-1}$). This would appear to be a very high rate of organic matter loss following draining, but in fact the subsidence resulted from an increase in bulk density of the drained soil horizons, and the combination of increased bulk density and the greater productivity of forests led to an apparent increase in soil C (perhaps $0.1 \, \text{kg} \, \text{C} \, \text{m}^{-2} \, \text{yr}^{-1}$) rather than a loss (Minkkinen and Laine, 1998).

THERE ARE THREE EASY WAYS TO BE WRONG

Forest soil studies, even if they are only aimed at improving management decisions, are best begun with a hypothesis. There are two major things that can go wrong in a study or experiment. One is to say that A is better than B when it is not; in statistical terms, this is a Type I error. The second way to be wrong is to say that A is not better than B when it actually is, a Type II statistical error. Using a classical hypothesis-testing approach, the correct definition of the population to which you wish to make inferences, an adequate sampling approach, and statistically sound analytics will save you from these common errors. The third way to be wrong is to

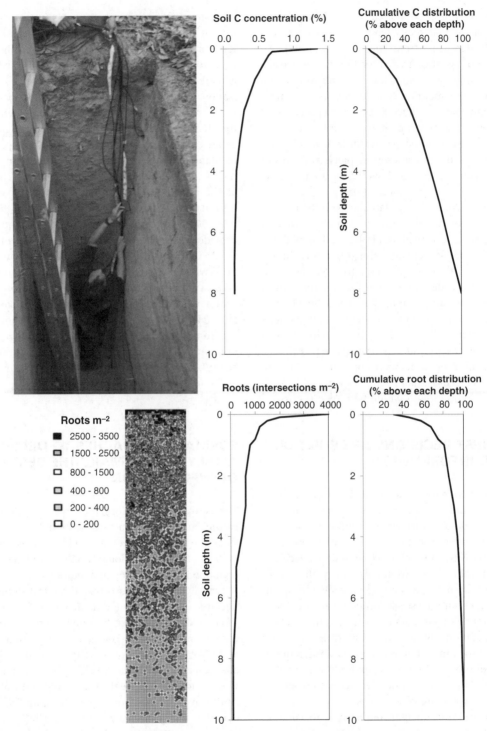

Figure 16.1 Some forest soils are very deep, and sampling only the upper 30 or 50 cm of this Oxisol in Brazil would miss two-thirds of the soil carbon (upper graphs) and a quarter of the fine roots (1-3 mm diameter, lower graphs). (Courtesy of Jean-Paul Laclau.)

have a hypothesis supported, but to interpret the wrong factor that really led to the outcome (this Type III idea came from Powers *et al.*, 1994). An example of "getting it right for the wrong reason" might be that soil removal into windrows has no effect on subsequent tree growth because soil nutrient supply was not affected. The prediction may be supported in an experiment, but the inferred mechanism may not have been true; the lack of reduction in tree growth on windrowed plots could result from a combination of lowered nutrient supply but improved competition control.

From statistics, we are familiar with Type I and Type II errors. A Type I error is committed when a hypothesis is accepted (and the null hypothesis rejected) that is actually false, or, more simply, believing in things that aren't true. As noted in Chapter 14, a decision to fertilize a stand with a nutrient that is not limiting would be a Type I error. A Type II error involves rejecting a hypothesis (and accepting the null hypothesis) when in fact it was true, or simply failing to believe in true things. With fertilization, a decision not to invest $100 to fertilize a hectare that would respond with $250 worth of extra growth would be a Type II error.

THE BEST DECISIONS ARE BUILT ON GOOD INFORMATION

Any question in forest soils could be examined with a wide variety of experimental designs. Experiments are designed to test the confidence and size of a treatment, and these effects are most easily detected if other potentially confounding effects are held constant. In the twentieth century, this sort of thinking often led to nesting all replicates of an experiment within a single location, because differences among sites might modify the treatment effect and make it difficult to detect. Unfortunately, these studies also limited their populations of inference to single locations! When geographically varying features are important, knowledge needs to be developed across varying geographies.

A particularly efficient approach is to conduct research experiments as a part of ongoing operational forestry. Stape *et al.* (2006) developed a "twin-

plot" approach where a company's inventory plot system was used to assess nutritional status and fertilization responsiveness. A random subset of inventory plots is chosen, and each selected plot is paired with an adjacent plot that receives fertilization. Within a single inventory plot location, there are no degrees of freedom to evaluate whether the fertilized plot grew better than might be expected at random (pre-fertilization comparison of stand characteristics between the twin plots might be useful). Replication in this case occurs at the landscape scale: the same scale at which information will be applied. A twin-plot design with 50 or 100 pairs of plots across a landscape provides the information that decision makers need to decide if fertilization is profitable anywhere, everywhere, or only for certain types of sites (Figure 16.2). Additional tests for the interaction of fertilization with weed control can use a "triplet" design, and additional factors could be applied with increasing numbers of plots at each site (without replication within sites). The approach also allows sites to be ranked by likely size of growth response, if funds are not available to fertilize all the responsive stands.

FORMAL APPROACHES TO DECISION ANALYSIS CAN MAKE THE BEST USE OF INFORMATION

The use of uncertain information in making decisions can be illustrated by a case study: should loblolly pine stands be fertilized? The case study involves a population of 30-year-old loblolly pine stands across a region, with a site index of about 20 m at 25 years. A typical stand structure would be 750 stems per hectare, with an average diameter and height of 25 cm and 28 m. The fertilization prescription under consideration includes 100 kg ha^{-1} nitrogen as urea and 50 kg ha^{-1} phosphorus as triple superphosphate, at a total cost of $120 ha^{-1}. Fertilization, on average, would increase the yield (in this hypothetical case study) by 17 m^3 ha^{-1} over five years, and average diameter would increase to 30 cm. The following analysis assumes that the management of the unfertilized stand is profitable by itself and that the

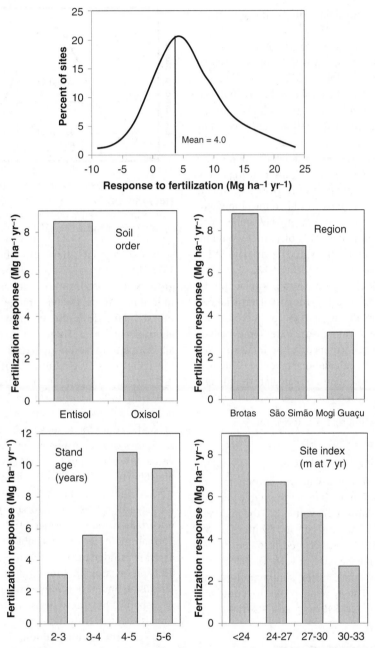

Figure 16.2 A twin plot design was used to estimate fertilization responsiveness of 35,000 ha area in Brazil. Growth response in the first 2 years averaged 4 Mg ha^{-1} yr^{-1} (top); this design with 131 pairs of plots allows managers to double the average growth response by choosing to fertilize only the most responsive soils types, regions, ages, and site productivities (bar graphs; from data of Ferreira and Stape 2009).

added value of fertilization can be judged on an incremental basis.

If all biomass were harvested at age 35 years and used as pulpwood, the extra wood would be worth $106 (at $6.25 m^{-3}, or about $15 per cord) at the time of harvest. This increased value would not cover the cost of fertilization, so complex calculations are not needed to show that the investment would lose money.

Fertilization becomes more attractive if some of the biomass can be used as chip 'n' saw (cutting a few small boards and chipping the rest) or sawtimber. About 25% of the biomass would go for pulpwood, 45% to chip 'n' saw, and 30% to sawtimber. The added value of pulpwood (25% of 17 m^3) would be $27. The chip 'n' saw value would be an extra $111, and sawtimber would add another $122. The total increased stumpage value from fertilization would be $260 ha^{-1}, or $140 more than the cost of fertilization.

Since money in the present is considered more valuable than money at a future date, some form of interest or discount rate needs to be included. Net present value (NPV) calculations take the value of a resource at some future date and calculate the current value based on a chosen interest (or discount) rate:

$$\text{NPV} = \text{Future value} \, x \, (1 + \text{interest rate})$$
$$- \text{year} - \text{current costs}$$

In the loblolly pine example, $260 ha^{-1} would be discounted over five years at the chosen rate of interest. Fertilization would be profitable only if the present (discounted) value of the harvest exceeded the present cost of fertilization. If an annual return of 7% were desired, $260 ha^{-1} after five years would have a current value of $185 ha^{-1}, for an NPV of $65 ha^{-1} after subtraction of the fertilization cost. The NPV at a 5% discount rate would be $83 ha^{-1}, or $49 ha^{-1} at a 9% rate.

GROWTH RESPONSES ARE UNCERTAIN

In the loblolly pine case study, the gain from fertilization was given as 17 m^3 ha^{-1}. This value is the average of a series of 104 installations of the North Carolina State Forest Nutrition Cooperative. Some

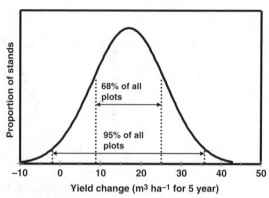

Figure 16.3 Distribution of the 5-year-growth response to fertilization, with an average response of 17 m^3 ha^{-1} across stands and a standard deviation of 14 m^3 ha^{-1}.

stands increased their growth by more than 17 m^3 ha^{-1}, and a few even showed reductions in growth after fertilization (due largely to increased mortality). Across the wide range of sites examined, the standard deviation of growth responses was about 14 m^3 ha^{-1}. Therefore, about two-thirds of all installations responded within ±14 m^3 ha^{-1} of the mean of all sites, and about 95% fell within ± 28 m^3 ha^{-1} (Figure 16.3).

About 11% of the stands responded with a decrease in growth. These trials covered a wide range of stand conditions, so stratification of sites into categories of responsiveness might reduce response variability. Some stands are not limited by nitrogen or phosphorus supplies, and others are so densely stocked that fertilization accelerates mortality. For these reasons, the fertilization of each stand involves uncertainty about the magnitude of the response. Formal methods are available to aid in making decisions under uncertainty and are well suited to decision making in forest nutrition management. The discussion here is limited to uncertainty in level of response, but these decision-making approaches can also be used to evaluate the importance of uncertainty about costs and values.

BREAKEVEN ANALYSIS IS THE SIMPLEST APPROACH

If the magnitude of the growth response to fertilization is uncertain, a manager may simply ask how large the response must be to cover the cost of fertilization (including the discount rate for the investment). In the loblolly pine case study, fertilization cost $120 ha^{-1}, and the value of extra growth was about $15.30 m^{-3} (without considering taxes). To account for the five-year time period of the investment, a 7% discount rate would mean that the NPV of 1 m^3 five years in the future would be $10.90. Therefore, the breakeven point would occur where the NPV of the extra growth equaled $120, or about 11 m^3 ha^{-1}. Since this value is about 35% less than the regionwide average response to nitrogen + phosphorus fertilization, a decision to fertilize would probably be wise. In fact, if a very large number of stands is to be fertilized, enough stands should respond to cover the cost of fertilizing those that do not. If only a small area is to be fertilized, the odds of hitting very responsive or very unresponsive stands are high. As more areas are fertilized, the overall response should approach an average level. Further, even if a large area would ensure overall profitability, it might be desirable to identify unresponsive stands and increase profits by efficient allocation of fertilizer. These issues illustrate the economic incentive behind developing a framework for making the best decisions under uncertainty.

DECISION TREES IDENTIFY CHOICES, PROBABILITIES, AND OUTCOMES

If a manager faced a decision on fertilizing all stands or none, the breakeven analysis presented above would indicate that she should fertilize if she expected an average response of 11 m^3 ha^{-1} or more. In reality, she would be happiest to fertilize only stands that respond with more than 11 m^3 ha^{-1}, saving the investment on stands that do not reach the breakeven response level. If the true average of 17 m^3 ha^{-1} were distributed normally across all stands with a standard deviation of 14 m^3 ha^{-1}, then about 67% of the stands would exceed the breakeven point (Figure 16.4). Integrating the area under the curve above 11 m^3 ha^{-1} shows that the NPV of fertilizing

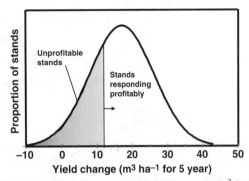

Figure 16.4 If the breakeven response were 11 m^3/ha after 5 years, then about 67 percent of the stands would respond profitably.

only responding stands is about $145 ha^{-1}. The average response would now be about 24 m^3. What would be the cost savings if the unresponsive stands were not fertilized? At first glance, it might appear that the savings would match the cost of fertilization ($120 ha^{-1}). However, the true savings would be somewhat lower because many stands that did not achieve the breakeven point would still show some increased growth. Integrating the curve below 11 m^3 ha^{-1} shows an average saving of $98 for each unresponsive hectare not fertilized.

These numbers can be put into a decision-tree framework to illustrate the choices faced by the manager (Figure 16.5). If she decides to fertilize all stands (the upper fork of the tree), 67% of the stands will respond above the breakeven point and 33% will not. Multiplying these proportions by the value of each type of stand gives an overall value of this decision:

$$0.67\,(\$145\,ha^{-1}) + 0.33\,(-\$98\,ha^{-1}) = \$65\,ha^{-1}$$

Note that the combined value of $65 ha^{-1} is the same value that was calculated originally for this case study. If the manager had decided not to fertilize (the lower fork of the tree), no cost would be incurred and no response obtained. As the upper fork of the decision tree yielded a value $65 ha^{-1} greater than the lower fork, deciding not to fertilize would have an opportunity cost of $65 ha^{-1}.

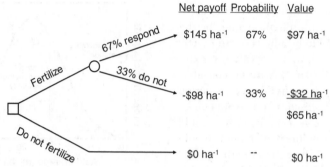

Figure 16.5 Decision tree representing the choice between fertilizing all stands or none if the average response is $17\,\text{m}^3\,\text{ha}^{-1}$.

FERTILIZATION ENTAILS TWO KINDS OF RISK

This example shows that there are two kinds of wrong decisions that managers might make (Figure 16.6). The first type (I) would be to fertilize a stand that would not respond with at least a breakeven increase in yield. Equally real is the second type (II), where a potentially responsive stand is not fertilized. Fertilizing a non-responsive stand in this case costs $98\,\text{ha}^{-1}$ on average; accounting for the fact that only 33% of the area falls into this

category gives an average cost (Type I error) of $32\,\text{ha}^{-1}$ across the entire area. Failure to fertilize a responsive stand (with a value of $145\,\text{ha}^{-1}$ on 67% of the area) is a Type II error with an average cost of $97\,\text{ha}^{-1}$. In this case, the Type II error is three times as costly as the Type I error, and an optimal manager would rather err on the side of fertilizing unresponsive areas to obtain the increased value from fertilizing every responsive area.

In reality, managers often prefer to incur Type II errors rather than Type I errors, as only Type I errors lose money out of pocket. Managers are typically

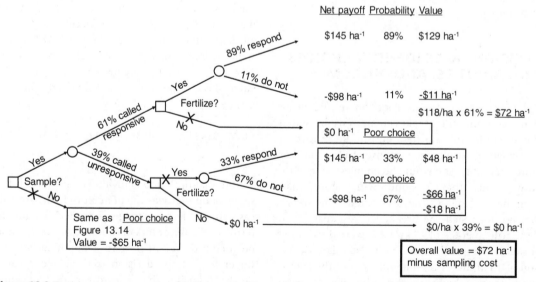

Figure 16.6 Decision tree representing the choice to fertilize with 80 percent accuracy in a response prediction, assuming that an average response of $17\,\text{m}^3\,\text{ha}^{-1}$ and 67 percent of stands are truly responsive. An "X" marks the poorer choice of each fork.

less accountable for Type II errors, in which potential profits are missed but no current funds are lost. Nonetheless, this basic decision framework provides the best available estimates of costs for both types of wrong decisions, allowing managers to apply their own criteria for the acceptability of each type of error.

DETAILS ARE VERY IMPORTANT IN DETERMINING THE OVERALL PROFITABILITY OF OPERATIONAL NUTRITION MANAGEMENT

A twin-plot fertilization experiment (as in Figure 16.2) was developed in Venezuela with pairs of control and fertilized plots of *Eucalyptus* (Carrero, 2012). The growth response averaged 6 Mg ha^{-1} (about 12 m^3 ha^{-1}) in the first two years after fertilization, though some sites showed no fertilization response and others responded with increases up to 20 Mg ha^{-1}. The cost of fertilization (and weed control) was very high, about \$650 ha^{-1}. The value of the extra wood was high, \$470 ha^{-1}, but not high enough to cover the cost of the fertilization and weed control. About 30% of the stands responded strongly enough (> 9 Mg ha^{-1} over two years) to more than cover the costs, so a decision to fertilize only the most responsive 30% of sites would have an average growth response of 14 Mg ha^{-1} over two years, and a net present value of \$330 ha^{-1}. In fact, the distribution of responses for sites that responded with > 9 Mg ha^{-1} was skewed, with a large portion of sites responding very strongly, so the actual net present value (integrating under the curve) was \$360 ha^{-1}. Achieving this level of return on investment depends on having accurate information on distribution of responses, and the ability to distinguish which sites are most responsive. Further, any continued growth response (beyond two years) would add to the value of fertilization without increasing costs, raising both the proportion of responding sites and the value generated within each site. All these details are fundamental in operational programs of nutrition management, and well-designed twin-plot experiments can provide the

necessary information on both average response and the important aspects of variation.

CHRONOSEQUENCES TRACE PATTERNS IN SPACE TO MAKE INFERENCES ABOUT TIME

Many fascinating questions about the development of forest soils cannot be addressed experimentally because they address spans of time that are too long for direct observations. Patterns of soil development at the scale of centuries are largely outside the realm of active experimentation, so scientists may rely on "space for time" substitutions, or chronosequences. Riverside terraces are an attractive location for chronosequence studies, because recently deposited sediments may provide an example of soils that have just begun to develop, while a few meters away (and somewhat higher in elevation), older terraces show characteristics that take time to develop. The ages of the terraces may be determined by the ages of trees growing in the soil, and sampling soils over a distance of 50 to 100 m may capture changes in soils that occur over centuries (Figure 16.7).

Two major questions arise for any chronosequence study. The first is whether the observed

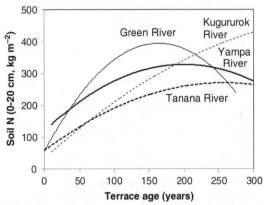

Figure 16.7 The pattern of soil N in relation to the age of floodplain terraces for two cottonwood chronosequences in Colorado (Green and Yampa Rivers), and two poplar/white spruce chronosequences in Alaska (from data of Kaye *et al.* 2003, Adair *et al.* 2004, Rhoades *et al.* 2008).

pattern truly resulted from the accumulated influence of soil-forming factors over time, or whether other factors differed substantially among the different-aged sites. For example, age sequences on riverside terraces always have high elevations for older soils, and higher elevations may be associated with changes in depth to water tables. Less obviously, the texture of materials deposited by floods tends to change with elevation, with higher terraces formed by slower-moving water that deposit finer-textured sediments. These covarying factors may or may not obscure a pattern that would result from increasing spans of time.

The second question is what processes led to the observed patterns with soil age? The accumulation of nitrogen in the Alaska chronosequences might result from nitrogen fixation, because early development along these rivers had alder (Tanana) and *Shepherdia* (Kugururok). The cottonwood chronosequences in Colorado had no major nitrogen-fixing plants, so the accumulation of N probably results from the flood deposits of sediments (which contain N). Nitrogen entering the soil in the litter of N-fixing plants may have different opportunities for cycling and incorporation into long-term stabilized N pools than N that arrives with flood sediments.

The assumptions underlying chronosequence studies can be difficult to test within the timeframe of a single experiment, but sometimes later re-measurements of the same locations can test whether the apparent time trends result from time alone or whether confounding factors were problematic. For example, Zou and Bashkin (1998) sampled a chronosequence of *Eucalyptus* plantations in Hawaii, and concluded that earthworm populations increased strongly with stand age, along with a significant increase in soil carbon and a decrease in active fungal biomass (Figure 16.8). Repeated sampling of one of the stands, however, showed no trend in carbon, indicating the age sequence was confounded with other soil differences unrelated to time. This sort of resampling of sites used in earlier studies has also shown that many classic stories about ecosystem succession are not supported (see Johnson and Miyanishi, 2008). Longer-term chronosequence studies cannot be cross-checked by following changes within a single stand over a period of just a few years or decades, so these examples of "erroneous" chronosequences should be kept in mind when looking for time trends over long periods.

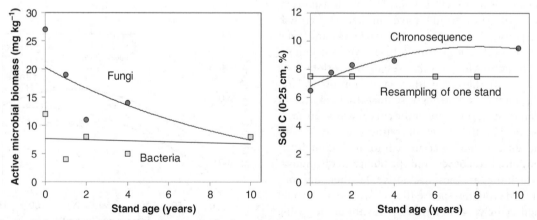

Figure 16.8 A chronosequence of adjacent stands of *Eucalyptus* in Hawaii showed decreasing active fungal biomass, and constant bacterial biomass, as soil carbon concentration increased significantly. However, a repeated sampling within a single stand showed no increase in soil carbon, raising the question of whether the pattern of fungi and bacteria would be a trend to expect over time, or an artifact of confounding differences in soils among the stands used for the chronosequence (from data of Zou and Bashkin 1998, Binkley *et al.* 2004). For other repeat-sampling tests of a chronosequence, see Figures 12.14, 12.15, and 15.5.

SPATIAL PATTERNS CAN BE USED TO GAIN INSIGHTS ON OTHER SOIL-FORMING FACTORS

Just as nearby soils that differ in age can be used to infer changes that likely happen in individual soils over time, the influence of other soil-forming factors may be examined by comparing nearby soils that are the same age but differ in other factors. One of the best examples of this sort of gradient study

comes from the work of Oliver Chadwick and colleagues (Chadwick and Chorover, 2001) on an ancient lava flow in Hawaii. These scientists used an elevational gradient on a single-age (~170 000 year) mountainside to infer the changes that would happen in an individual soil as a result of a 20-fold gradient in precipitation (Figure 16.9). Base saturation remains high when annual precipitation is less than evapotranspiration. Higher precipitation brings the opportunity for leaching of base cations,

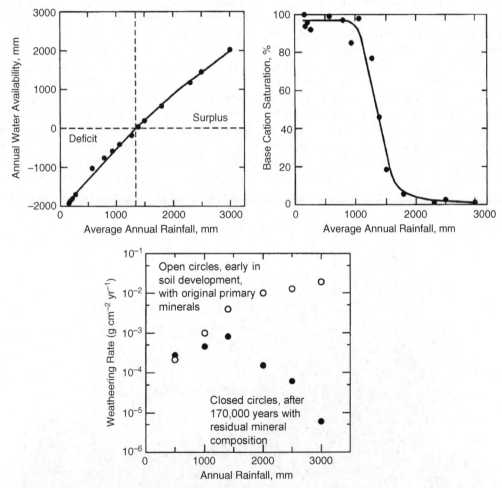

Figure 16.9 Along a 170,000 year old lava flow, an elevational gradient shows rising precipitation; above about 1300 mm yr^{-1}, enough water is available to exceed evapotranspiration losses and base cation leaching becomes important. Weathering of the original minerals in the lava would increase with increasing precipitation, but the alteration of minerals over 170,000 now shows the highest rates of weathering where soil water is high enough to foster weathering, but low enough that more weatherable minerals remain (Reprinted from Geoderma, 100, Oliver A. Chadwick, Jon Chorover, The chemistry of pedogenic thresholds, 321–353, 2001, with permission from Elsevier).

depleting base saturation. The current rate of mineral weathering depends on the current moisture conditions, and the legacy of previous weathering that depleted the original stocks of weatherable minerals.

EXTRAPOLATING FROM ONE SCALE TO ANOTHER CAN BE MISLEADING

The longest record of change in a forest soil comes from the Rothamsted Broadbalk "wilderness" plots, where, in 1880, an agricultural field was allowed to revegetate with trees and shrubs. Soil carbon accumulated at a rate of about $3.5 \, Mg \, ha^{-1}$ annually (Jenkinson *et al.*, 1994), and this classic experiment has influenced the way many soil scientists expect soil C to increase with afforestation. Unfortunately, the size of the wilderness is too small to separate the normal changes in forest soils from edge effects of neighboring fields (Figure 16.10), and the observed rates of change in the soils are probably much larger than would be found in a stand that was large enough to avoid edge effects.

SYSTEMATIC REVIEWS ARE FORMAL SOURCES FOR THE BEST AVAILABLE INFORMATION

In the late twentieth century, the healthcare community developed a suite of formalized approaches under the theme of "evidence-based medicine" (Straus *et al.*, 2010). Physicians began to acknowledge that the amount of information available in the literature, and in the clinical practices of thousands of colleagues, greatly exceeded their own individual base of knowledge and experience. Before the development of the Internet, physicians needed a system for generating the best possible insights from available research, including some way to account for the varying quality of different research projects. The Cochrane Collaboration (now an international association with tens of thousands of contributing scientists) developed formal protocols for effective evaluation of the state of knowledge in various medical disciplines. These "systematic reviews" developed clear definitions of the problem to be examined, with quantifiable questions. Criteria were developed in

Figure 16.10 The Broadbalk "wilderness" is too small to separate actual changes over time under trees from the edge effects of adjacent fields(photo by S. Hamburg).

advance for which sorts of experimental designs would be strong enough to warrant confidence, and then independent groups of scientists would collect all the available studies and evaluate the evidence (often with a formal, quantitative meta-analysis); all steps in the reviews are transparent. In some ways, the proliferation of access to knowledge on the Internet has made it easier for anyone to access original information from research projects using search engines; however, the quality-assurance aspects of systematic reviews ensure this approach to evidence-based practices will remain important.

Systematic reviews have been developed for a wide range of conservation-related issues, such as the effects of liming on stream biota (Mant et al., 2011). The effects of changing hydrology on greenhouse gas generation in peat soils (Bussell et al., 2010) corroborated the pattern reported near the beginning of this chapter (from Minkkinen and Laine, 1998): no significant increase in CO_2 efflux was found for drained peatlands. The Collaboration for Environmental Evidence (along with the Center for Evidence-Based Conservation at Bangor University in the United Kingdom) provides guidance and resources for the development of these carefully crafted reviews.

SCIENCE IS NOT JUST FOR SCIENTISTS

Unfortunately, most of the important challenges in forest soils have not (yet) been addressed by systematic reviews. Some of the most useful information may be found currently in textbooks, in journal articles that report the results of individual experiments, and a variety of review papers that may or may not have the clarity and rigor of a formal systematic review. Hundreds of forest soil scientists are also a source of valuable insights, and an Internet search followed by an email may lead to helpful advice.

A fundamental challenge in managing forest soils is simply that no two soils are alike. At just about every measurement scale, from microns to kilometers, the structure and key processes of soils show high variation. Most forest

management approaches deal with units on the order of 1 ha to thousands of hectares, and managers want to make the best decisions about how to prescribe and implement the best management practices. Knowledge generated for somewhat similar soils and forests may have some general value for application to a new, specific case, but how good is good enough? People entrusted with managing soils and forests need to tap into the best available information from general sources, and figure out how to apply this knowledge to specific cases.

POCKET SCIENCE WONT GET YOU TO THE MOON, BUT ITS AN EVEN BETTER INVESTMENT

It takes billions of dollars (and euros) to develop science that can launch vehicles into orbit, to the moon, and across the solar system. Rocket science has given us great insights about the universe around us, but rocket-scale science is not the only sort of powerful science.

"Pocket science" recognizes that land management decisions always have to be case-specific, and that each operational treatment of a site provides an opportunity for learning. For example, forest practices in a region may commonly include post-logging slash fires. A curious manager may have read that, in some cases, slash fires degrade soil fertility, but would that be true in this specific case? A pocket science approach would entail applying the typical treatment to the majority of a 200 ha unit, but using a fireline to prevent burning on half of one hectare. A few photographs before and after burning might be useful, and with good GPS documentation of boundaries, the manager could return in a few years to see if the unburned areas showed higher or lower tree growth. This approach would not be very strong from a statistical point of view, but it would provide the manager with the opportunity to learn if any major surprises developed. An investment of two work-days each year to revisit pocket science trials (in the company of colleagues) may pay large dividends.

The pocket science approach builds in learning opportunities to routine management operations. The insight from any single, unreplicated pocket science experiment might grow substantially when combined with other similar endeavors across a landscape, and throughout a professional's career. Perhaps the greatest value of pocket science is how it sharpens a manager's thinking about the effectiveness of current practices and the value of always considering how current practices can be improved.

References

Adair, E. C., Binkley, D. and Anderson, D. C. (2004) Patterns of nitrogen accumulation and cycling in riparian floodplain ecosystems along the Green and Yampa rivers. *Oecologia*, **139**, 108–116.

Adams, M. A. (1996) Distribution of Eucalypts in Australian landscapes: landforms, soils, fire and nutrition. In: P. M. Attiwill and M. A. Adams (eds) *Nutrition of Eucalypts*, CSIRO, Collingwood, pp. 61–76.

Adams, M. A. (2007) Nutrient cycling in forests and heathlands: an ecosystem perspective from the water-limited South. *Soil Biology*, **10**, 333–360.

Adams, M. A. and Attiwill, P. M. (1986) Nutrient cycling and nitrogen mineralization in eucalypt forests of southeastern Australia. II. Indices of nitrogen mineralization. *Plant and Soil*, **92**, 341–362.

Agee, J. (1993) *Fire Ecology of Pacific Northwest Forests*. Island Press, Washington, D.C., 493 pp.

Ahmed, R., Khan, D. and Ismail, S. (1985) Growth of *Azadiracata india* and *Melia azederach* on coastal sand using highly saline water for irrigation. *Pakistan Journal of Botany*, **17**, 229–233.

Albaugh, T. J., Allen, H. L., Dougherty, P. M., Kress, L. and King, J. S. (1998) Leaf area and above- and belowground growth responses of loblolly pine to nutrient and water additions. *Forest Science*, **44**, 1–12.

Albaugh, J. M., Blevins, L., Allen, H. L., Albaugh, T. J., Fox, T. R., Stape, J. L. and Rubilar, R. A. (2010) Characterization of foliar macro- and micronutrient concentrations and ratios in loblolly pine plantations in the southeastern United States. *Southern Journal of Applied Forestry*, **34**, 53–64.

Albaugh, T. J., Vance, E., Gaudreault, W., Fox, T., Allen, H. L., Stape, J. L. and Rubilar, R. (2012) Carbon emissions and sequestration from fertilization of pine in the southeastern United States. *Forest Science*, in press.

Alexander, L. T. and Cady, J. G. (1962) *Genesis and hardening of laterite in soils*. USDA Technical Bulletin 1282.

Alig, R. J. and Butler, B. J. (2004) *Area changes for forest cover types in the United States, 1952 to 1997, with projections to 2050*. USDA Forest Service General Technical Report PNW-GTR-613

Allen, H. L. and Ballard, R. (1982) Fertilization of loblolly pine. In: Symposium on the loblolly pine ecosystem (east region), School of Forest Resources, North Carolina State University, Raleigh, pp. 163–181.

Allen, H. L., Blevins, D. P. and Morris, L. A. (1995) *Effects of Harvest Utilization, Site Preparation and Vegetation Control on Growth of Loblolly Pine on a Piedmont Site*. NCSFNC Research Note No. 11. College of Forest Resources, North Carolina State University, Raleigh.

Allen, M. F. (1991) *The Ecology of Mycorrhizae*. Cambridge University Press, Cambridge, UK.

Allison, F. F. (1965) *Decomposition of wood and bark sawdusts in soil, nitrogen requirements and effects on plants*. USDA Technical Bulletin 1332 related practices.

Amaranthus, M. P. and Trappe, J. M. (1993) Effects of erosion on ecto- and VA-mycorrhizal inoculum potential of soil following forest fire in southwest Oregon. *Plant and Soil*, **150**, 41–49.

Amer, F. D. D., Bouldin, D., Black, C. and Duke, F. (1955) Characterization of soil phosphorus by anion exchange resin adsorption and ^{32}P equilibration. *Plant and Soil*, **6**, 391–408.

Andreason, O. (1988) Suiting forest management to a changed environment. In: *Forest Health and Productivity*, The Marcus Wallenberg Foundation Symposia Proceedings #5, Falun, Sweden, pp. 67–75.

Andrews, M., James, E. K., Sprent, J. I., Boddey, R. M., Grosse, E. and Bueno dos Reis, F. Jr. (2011) Nitrogen fixation in legumes and actinorhizal plants in natural ecosystems: values obtained using ^{15}N natural abundance. *Plant Ecology and Diversity*, **4**, 131–140.

Armson, K. A. and Sadreika, V. (1974) *Forest Nursery Soil Management and Related Practices*. Ontario, Canada.

Arseneault, D. and Payette, S. (1997) Landscape change following deforestation at the Arctic tree line in Quebec, Canada. *Ecology*, **78**, 693–706.

Ecology and Management of Forest Soils, Fourth Edition. Dan Binkley and Richard F. Fisher.
© 2012 John Wiley & Sons, Ltd. Published 2012 by John Wiley & Sons, Ltd.

Aubert, G. and Tavernier, R. (1972) Soil Survey. In: *Soils of the Humid Tropics*. U.S. National Academy of Sciences, Washington, D.C, pp. 17–44.

Augusto, L., Turpault, M.-P. and Ranger, J. (2000) Impact of forest tree species on feldspar weathering rates. *Geoderma*, **96**, 215–237.

Augusto, L., Ranger, J., Binkley, D. and Rothe, A. (2002) Impact of several common tree species of European temperate forests on soil fertility. *Annals of Forest Science*, **59**, 233–254.

Ayers, A. S., Takahashi, M. and Kanehiro, Y. (1947) Conversion of non-exchangeable potassium to exchangeable forms in a Hawaiian soil. *Soil Science Society of America Proceedings*, **11**, 175–181.

Baeten, L., Verstraeten, G., de Frenne, P., Vanhellemont, M., Wuyts, K., Hermy, M. and Verheyen, K. (2011) Former land use affects the nitrogen and phosphorus concentrations and biomass of forest herbs. *Plant Ecology*, **212**, 901–909.

Baker, H. P. (1906) The holding and reclamation of sand dunes and sand wastes. *Forests Quarterly*, **4**, 282–288.

Baker, J. B. (1973) Intensive cultural practices increase growth of juvenile slash pine in Florida sandhills. *Forest Science*, **19**, 197–202.

Baldwin, M., Kellogg, C. E. and Thorp, J. (1938) *Soil Classification*. Soils and Men: Yearbook of Agriculture 1938. U.S. Government Printing Office, Washington, D.C.

Ballard, R. (1984) Fertilization of plantations. In: G. D. Bowen and E. K. S. Nambiar (eds) *Nutrition of Plantation Forests*, Academic Press, London, pp. 327–360.

Baranov, I. Y. (1959) Geographical distribution of seasonally frozen ground and permafrost. In: V. A. Obruchev (ed.) *Institute of Permafrost Studies, Academy of Sciences*, Moscow. Natyral Resources Council of Canada Technical Translation No. 1121 (1964), pp. 193–219.

Barber, S. A. (1995) *Soil Nutrient Bioavailability: A mechanistic approach*. John Wiley & Sons, Inc., New York, NY.

Barclay, H. J. and Brix, H. (1985) Effects of urea and ammonium nitrate fertilizer on growth of a young thinned and unthinned Douglas fir stand. *Canadian Journal of Forest Research*, **14**, 952–955.

Bardgett, R. D. and McAlister, E. (1999) The measurement of soil fungal:bacterial biomass ratios as an indicator of ecosystem self-regulation in temperate meadow grasslands. *Biology and Fertility of Soils*, **29**, 282–290.

Barnes, B. V., Pregitzer, K. S., Spies, T. A. and Spooner, V. H. (1982) Ecological forest site classification. *Journal of Forestry*, **80**, 493–498.

Barnes, B. V., Zak, D. R., Denton, S. R. and Spurr, S. H. (1998) *Forest Ecology*, 4th edition. John Wiley & Sons, Inc., New York. NY.

Barnes, R. (1980) *An allocation and optimization approach to tree growth modeling: concepts and application to nitrogen economy*. Unpublished manuscript.

Barron, A. R., Wurzburger, N., Bellenger, J. P., Wright, S. J., Kraepiek, A. M. L. and Hedin, L. O. (2008) Molybdenum limitation of asymbiotic nitrogen fixation in tropical forest soils. *Nature Geoscience*, **2**, 42–45.

Barros, N. F. and Novais, R. F. (1996) Eucalypt nutrition and fertilizer regimes in Brazil. In: P. M. Attiwill and M. A. Adams (eds) *Nutrition of Eucalyptus*, CSIRO, Melbourne, pp. 335–355.

Barros, N. F., Novais, R. F. and Neves, L. C. L. (1990) Fertilição e correção do solo para plantio de Eucalipto. In: N. F. Barros and R. F. Novais (eds) *Relação solo-Eucalipto*, Department of Forest Soils, Federal University of Viçosa, M.G., Brazil, pp. 127–186.

Bashkin, M. A. and Binkley, D. (1998) Changes in soil carbon following afforestation in Hawaii. *Ecology*, **79**, 828–833.

Battigelli, J. and Berch, S. (2002) *Soil Fauna in the Sub-Boreal Spruce (SBS) Installations of the Long-Term Soil Productivity (LTSP) Study of Central British Columbia: One-Year Results for Soil Mesofauna and Macrofauna*. Note LTSPS-05. British Columbia Ministry of Forests, Research Branch, Victoria, BC Canada.

Baule, H. and Fricker, C. (1970) *The Fertilizer Treatment of Forest Trees*. BLV Verlagsgesellshcaft, Munich. 259 pp.

Beaufils, E. R. (1973) *Diagnosis and recommendation integrated system (DRIS)*. Soil Science Bulletin #1, University of Natal, Pietermaritzburg, South Africa. 132 pp.

Beck, D. F. (1971) *Polymorphic site index curves for white pine in the southern Appalachians*. USDA Forest Service Research Paper SE-80.

Beck, D. F. and Trousdell, K. B. (1973) *Site index: accuracy and prediction*. USDA Forest Service Research Paper SE-108.

Beck, T., Joergensen, R. G., Kandeler, E., Makeschin, F., Oberholzer, H. R., Nuss, E. and Scheu, S. (1997) An inter-laboratory comparison of ten different ways of measuring soil microbial biomass. *Soil Biology and Biochemistry*, **129**, 1023–1032.

Beese, J. O. and Meiwes, K. J. (1995) 10 Jahre Waldkalkung – Stand und Perspektiven. *Allgemeine Forstzeitschrift*, **50**, 946–949.

Bélanger, N. and VanRees, K. J. C. (2008) Sampling Forest Soils. In: M. R. Carter and E. G. Gregorich (eds) *Soil*

Sampling and Methods of Analysis, 2nd edition. CRC Press, Boca Raton, FL, pp. 15–24.

Benckiser, G. (1997) *Fauna in Soil Ecosystems*. Marcel Dekker, Inc., New York.

Benda, L., Miller, D., Bigelow, P. and Andras, K. (2003) Effects of post-wildfire erosion on channel environments, Boise River, Idaho. *Forest Ecology and Management*, **178**, 105–119.

Bengtson, G. W. (1968) *Forest Fertilization: Theory and practice*. Tennessee Valley Authority, Muscle Shoals.

Bengtson, G. (1973) Fertilizer use in forestry: materials and methods of application. In: *Proceedings of the International Symposium on Forest Fertilization*, FAO-IUFRO, Paris, December, pp. 97–153.

Bengtson, G. (1976) Fertilizers in use and under evaluation in silviculture: a status report. In: *Proceedings XVI IUFRO World Congress*, Working Group on Forest Fertilization, Oslo, Norway.

Benzian, B. (1965) *Experiments on nutrition problems in forest nurseries*. Forestry Commission Bulletin #37. Her Majesty's Stationery Office, London.

Berg, B. and Laskowski, R. (2006) *Litter Decomposition: A guide to carbon and nutrient turnover*, Academic Press, Burlington.

Berg, B. and Matzner, E. (1997) Effect of N deposition on decomposition of plant litter and soil organic matter in forest systems. *Environmental Reviews*, **5**, 1–25.

Berger, T. W., Neubauer, C. and Glatzel, G. (2002) Factors controlling soil carbon and nitrogen stores in pure stands of Norway spruce (*Picea abies*) and mixed species stands in Austria. *Forest Ecology and Management*, **159**, 3–14.

Berggren Kleja, D., Olsson, M., Syensson, M. and Jansson, P.-E. (2010). Soil C dynamics in Swedish forest soils – gradients from south to north. In: R. Jandl and M. Olsson (eds) *Cost Action 639, Greenhouse gas budget of soils under changing climate and land use (BurnOut)*, Federal Research and Training Centre for Forests, Natural Hazards and Landscape (BFW), Vienna, pp. 81–84.

Bergholm, J., Braekke, F. Frank, J., Hallbäcken, L., Ingerslev, M. and Mälkönen, E. (1997) Soil chemistry and weathering. In: *Imbalanced forest nutrition – vitality measures*. Section of Ecology, Swedish University of Agricultural Sciences, SNS Project 1993–1996, Uppsala, pp. 74–89.

Bergkvist, B. and Folkeson, L. (1995) The influence of tree species on acid deposition, proton budgets and element fluxes in south Swedish forest ecosystems. *Ecology Bulletin*, **44**, 90–99.

Berry, E. C. (1994) Earthworms and other fauna in the soil. In: J. L. Hatfield and B. A. Stewart (eds) *Soil Biology: Effects on soil quality*. Lewis Publ. Boca Raton, FL, pp. 61–90.

Berthrong, S. T., Jobbágy, E. G. and Jackson, R. B. (2009) A global meta-analysis of soil exchangeable cations, pH, carbon, and nitrogen with afforestation. *Ecological Applications*, **19**, 2228–2241.

Beskow, G. (1935) Soil freezing and frost heaving. *Sveriges Geologiska Undersokning*, **26**, 1–242.

Bestelmeyer, B. T., Tugel, A. J., Peacock, G. L., Jr., Robinett, D. G., Shaver, P. L., Brown, J. R., Herrick, J. E., Sanchez, H. and Havstad, K. M. (2008) State-and-Transition Models for Heterogeneous Landscapes: A Strategy for Development and Application. *Rangeland Ecology and Management*, **62**, 1–15.

Bezkorovaynaya, I. N. (2005) The formation of soil invertebrate communities in the Siberian afforestation experiment. In: D. Binkley and O. Menyailo (eds) *Tree Species Effects on Soils: Implications for Global Change*. NATO Science Series, Springer, Dordrecht, pp. 307–316.

Bigger, C. M. and Cole, D. W. (1983) Effects of harvesting intensity on nutrient losses and future productivity in high and low productivity red alder and Douglas fir stands. In: R. Ballard and S. P. Gessel (eds) *IUFRO Symposium on forest site and continuous productivity*. USDA Forest Service General Technical Report PNW-163, Portland, pp. 167–178.

Binkley, D. (1983) Interaction of site fertility and red alder on ecosystem production in Douglas fir plantations. *Forest Ecology and Management*, **5**, 215–227.

Binkley, D. (1984) Douglas-fir stem growth per unit of leaf area increased by interplanted Sitka alder and red alder. *Forest Science*, **30**, 259–263.

Binkley, D. (1986) Soil acidity in loblolly pine stands with interval burning. *Soil Science Society of America Journal*, **50**, 1590–1594.

Binkley, D. (1992) H+ budgets. In: D. W. Johnson and S. E. Lindberg (eds) *Atmospheric Deposition and Nutrient Cycling*, Springer-Verlag, New York, pp. 450–466.

Binkley, D. (1995) The influence of tree species on forest soils: processes and patterns. In: D. J. Mead and I. S. Cornforth (eds) *Proceedings of the Trees and Soils Workshop*, Lincoln University. Agronomy Society of New Zealand Special Publication #10, Lincoln University Press, Canterbury, pp. 1–34.

Binkley, D. (1996) Bioassays of the influence of *Eucalyptus saligna* and *Albizia falcataria* on soil nutrient supply and limitation. *Forest Ecology and Management*, **91**, 229–234.

Binkley, D. (1999) Disturbance in temperate forests of the northern hemisphere. In: L. Walker (ed.), *Ecosystems of Disturbed Ground*, Elsevier Science, Amsterdam, pp. 469–482.

Binkley, D. (2002) Ten-year decomposition in a loblolly pine forest. *Canadian Journal of Forest Research*, **32**, 2231–2235.

Binkley, D. (2005) How nitrogen-fixing trees change soil carbon. In: D. Binkley and O. Menyailo (eds) *Tree Species Effects on Soils: Implications for Global Change*. NATO Science Series, Springer, Dordrecht, Chapter 8.

Binkley, D. (2006) Soils in ecology and ecology in soils. In: B. Warkentin (ed.), *Footprints in the Soil: People and Ideas in Soil History*, Elsevier, Amsterdam, pp. 259–278.

Binkley, D. (2008) Three key points in the design of forest experiments. *Forest Ecology and Management*, **255**, 2022–2023.

Binkley, D. and Giardina, C. (1997) Biological nitrogen fixation in plantations. In: E. K. S. Nambiar and A. G. Brown (eds) *Management of soil, water, and nutrients in tropical plantation forests*. ACIAR Monograph 43, Canberra, pp. 297–337.

Binkley, D. and Giardina, C. (1998) Why do tree species affect soils? The warp and woof of tree-soil interactions. *Biogeochemistry*, **42**, 89–106.

Binkley, D. and Hart, S. C. (1989) The components of nitrogen availability assessments in forest soils. *Advances in Soil Science*, **10**, 57–116.

Binkley, D. and Matson, P. (1983) Ion exchange resin bag method for assessing forest soil N availability. *Soil Science Society of America Journal*, **47**, 1050–1052.

Binkley, D. and Menyailo, O. (2005) Gaining insights on the effects of tree species on soils. In: D. Binkley and O. Menyailo (eds) *Tree Species Effects on Soils: Implications for Global Change*. NATO Science Series, Springer, Dordrecht, Chapter 1.

Binkley, D. and Richter, D. (1987) Nutrient cycles and H$^+$ budgets. *Advances in Ecological Research*, **16**, 1–51.

Binkley, D. and Sollins, P. (1990) Acidification of soils in mixtures of conifers and red alder. *Soil Science Society of America Journal*, **54**, 1427–1433.

Binkley, D. and Valentine, D. W. (1991) Fifty-year biogeo-chemical effects of Norway spruce, white pine, and green ash in a replicated experiment. *Forest Ecology and Management*, **40**, 13–25.

Binkley, D., Burnham, C. and Allen, H. L. (1999) Water quality impacts of forest fertilization with nitrogen and phosphorus. *Forest Ecology and Management*, **121**, 191–213.

Binkley, D., Cromack, K., Jr. and Baker, D. (1994) Nitrogen fixation by red alder: biology, rates, and controls. In: D. Hibbs (ed.) *The Biology and Management of Red Alder*. Oregon State University Press, Corvallis, pp. 57–72.

Binkley, D., O'Connell, A. M. and Sankaran, K. V. (1997) Stand development and productivity. In: E. K. S. Nambiar and A. G. Brown (eds) *Management of soil, nutrients and water in tropical plantation forests*. Australian Centre for International Agricultural Research Mono-graph #43, pp. 419–442.

Binkley, D., Smith, F. W. and Son, Y. (1995) Nutrient supply and limitation in an age-sequence of lodgepole pine in southeastern Wyoming. *Canadian Journal of Forest Research*, **25**, 621–628.

Binkley, D., Sollins, P. and McGill, W. G. (1985) Natural abundance of N-15 as a tracer of alder-fixed nitrogen. *Soil Science Society of America Journal*, **49**, 444–447.

Binkley, D., Stape, J. L. and Ryan, M. (2004a) Thinking about resource use efficiency in forests. *Forest Ecology and Management*, **193**, 5–16.

Binkley, D., Ice, G. G., Kaye, J. and Williams, C. A. (2004b) Nitrogen and phosphorus concentrations in forest streams of the United States. *Journal of the Water Resources Association*, **2004**, 1277–1291.

Binkley, D., Kaye, J., Barry, M. and Ryan, M. G. (2004c) First rotation changes in soil carbon and nitrogen in a Eucalyptus plantation in Hawaii. *Soil Science Society of America Journal*, **68**, 1713–1719.

Binkley, D., Richter, D., David, M. B. and Caldwell, B. (1992a) Soil chemistry in a loblolly/longleaf pine forest with interval burning. *Ecological Applications*, **2**, 157–164.

Binkley, D., Stape, J. L., Takahashi, E. N. and Ryan, M. G. (2006) Tree-girdling to separate root and heterotrophic respiration in two *Eucalyptus* stands in Brazil. *Oecologia*, **148**, 447–454.

Binkley, D., Valentine, D. W., Wells, C. and Valentine, U. (1989) An empirical analysis of the factors contributing to 20-year decrease in soil pH in an old-field plantation of loblolly pine. *Biogeochemistry*, **7**, 39–54.

Binkley, D., Sollins, P. Bell, R., Sachs, D. and Myrold, D. (1992b) Biogeochemistry of adjacent conifer and alder/conifer ecosystems. *Ecology*, **73**, 2022–2034.

Birk, E. and Vitousek, P. (1984) Patterns of N retranslocation in loblolly pine stands: response to N availability. *Bulletin of the Ecological Society of America*, **65**, 100.

Bissett, J. and Parkinson, D. (1980) Long-term effects of fire on the composition and activity of the soil microflora of a subalpine, coniferous forest. *Canadian Journal of Botany*, **58**, 1704–1721.

Blake, J., Chappell, H. N., Bennett, W. S., Webster, S. R. and Gessel, S. P. (1990) Douglas fir growth and foliar nutrient responses to nitrogen and sulfur fertilization. *Soil Science Society of America Journal*, **54**, 257–262.

Bloomfield, C. (1954) A study of podzolization. III. The mobilization of iron and aluminum by Rimu (*Dacrydium cupressium*). *Journal of Soil Science*, **5**, 9–45.

Bockheim, J. G., Tarnocai, C. Kimble, J. M. and Smith, C. A. S. (1997) The concept of gelic materials in the new Gelisol order for permafrost affected soils. *Soil Science*, **162**, 927–993.

Boerner, R. (1982) Fire and nutrient cycling in temperate ecosystems. *BioScience*, **32**, 187–192.

Boerner, R., Lord, T. and Peterson, J. (1988) Prescribed burning in the oak–pine forest of the New Jersey Pine Barrens: effects on growth and nutrient dynamics of two *Quercus* species. *American Midland Naturalist*, **120**, 108–119.

Boggie, R. (1972) Effects of water-table height on root development of *Pinus contorta* on deep peat in Scotland. *Oikos*, **23**, 304–312.

Bohn, H. L., McNeal, B. L. and O'Conner, G. A. (2001) *Soil Chemistry*, 3rd edition. John Wiley & Sons, Inc., New York.

Bonner, F. T. (1968) *Responses to soil moisture deficiency by seedlings of three hardwood species*. USDA Forest Service Research Note 50–70.

Bouillet, J. P., Laclau, J. P., Gonçalves, J. L. M., Moreira, M. Z., Trivelin, P.C.O., Jourdan, C., Silva, E. V., Piccolo, M. C., Tsai, S. M. and Galiana, A. (2008) Mixed-species plantations of *Acacia mangium* and *Eucalyptus grandis* in Brazil 2: Nitrogen accumulation in the stands and biological N$_2$ fixation. *Forest Ecology and Management*, **255**, 3918–3930.

Boussougou, I. N. M., Brais, S., Tremblay, F. and Gaussiran, S. (2010) Soil quality and tree growth in plantations of forest and agricultural origin. *Soil Science Society of America Journal*, **74**, 993–1000.

Bowen, G. D. (1965) Mycorrhiza inoculation in nursery practice. *Australian Forestry*, **29**, 231–237.

Bowen, G. D. (1984) Tree roots and the use of soil nutrients. In: G. D. Bowen and E. K. S. Nambiar (eds) *Nutrition of Plantation Forests*, Academic Press, London.

Bowen, G. D. and Rovira, A. D. (1969) The influence of microorganisms on growth and metabolism of plant roots. In: W. J. Whittington (ed.) *Root Growth*, Butterworths, London, pp. 170–199.

Boyer, W. D. (1987) Volume growth loss: a hidden cost of periodic prescribed burning in longleaf pine? *Southern Journal of Applied Forestry*, **11**, 154–157.

Boyle, J. R. and Voigt, G. K. (1973) Biological weathering in silicate minerals. *Plant and Soil*, **38**, 191–201.

Boyle, J. R., Voigt, G. K. and Saweney, B. L. (1974) Chemical weathering of biotite by organic acids. *Soil Science*, **117**, 42–45.

Brady, N. and Weil, R. (2002) *The Nature and Properties of Soils*. Pearson, Upper Saddle River, NJ.

Brandhorst-Hubbard, J. L, Flanders, K. L., Mankin, R. W., Guertal, E. A. and Crocker, R. L. (2001) Mapping of Soil Insect Infestations Sampled by Excavation and Acoustic Methods. *Journal of Economic Entomology*, **94**, 1452–1458.

Braun-Blanquet, J. (1932) *Plant Sociology: the Study of Plant Communities* (English translation). McGraw Hill, New York, NY.

Bray, R. H. and Kurtz, L. T. (1945) Determination of total, organic, and available forms of phosphorus in soils. *Soil Science*, **59**, 39–45.

Bredemeier, M., Blanck, K., Lamersdorf, N. and Wiedey, G. A. (1995) Response of soil water chemistry to experimental "clean rain" in the NITREX roof experiment at Solling, Germany. *Forest Ecology and Management*, **71**, 31–44.

Bridgham, S. D., Ping, C.-L., Richardson, J. L. and Updegraff, K. (2001) Soils of Northern Peatlands: Histosols and Gelisols. In: J. L. Richardson and M. J. Vepraskas (eds) *Wetland Soils*. CRC Press, Boca Raton, FL, pp. 343–370.

Brix, H. (1981) Effects of nitrogen fertilizer source and application rates on foliar nitrogen concentration, photosynthesis and growth of Douglas fir. *Canadian Journal of Forest Research*, **11**, 775–780.

Broadfoot, W. M. (1969) Problems in relating soil to site index for southern hardwoods. *Forest Science*, **15**, 354–364.

Brockett, B. F. T., Prescott, C. E. and Grayston, S. J. (2012) Soil moisture is the major factor influencing microbial community structure and enzyme activities across seven biogeoclimatic zones. *Soil Biology and Biochemistry*, **44**, 9–20.

Brown, A. G., Nabiar, E. K. S. and Cossalter, C. (1997) In: E. K. S. Nambiar and A. G. Brown (eds) *Management of Soil, Nutrients and Water in Tropical Plantation Forests*. Australian Centre for International Agricultural Research Monograph 43, Canberra, pp. 1–24.

Brown, B. A. (1995) Toward a theory of podzolization. In: W. W. McFee and M. J. Kelly (eds) *Carbon Forms and Functions in Forest Soils*. Soil Science Society of America, Madison, WI, pp. 253–274.

Brown, R. J. E. (1967) Permafrost in Canada. Map 1246A. Geological Survey Canada. National Research Council of Canada, Ottawa, Canada.

Brown, R. J. E. (1970) *Permafrost in Canada: Its influence on northern development*. University of Toronto Press, Toronto, Canada.

Bruckner, A., Kandeler, E. and Kampichler, C. (1999) Plot-scale spatial patterns of soil water content, pH, substrate-induced respiration and N mineralization in a temperate coniferous forest. *Geoderma*, **93**, 207–223.

Bruijnzeel, L. A. (1982) *Hydrological and biogeochemical aspects of man-made forests in South-Central Java, Indonesia*. PhD thesis, Free University of Amsterdam, The Netherlands. 256 pp.

Bruijnzeel, L. A. (1991) Nutrient input–output budgets of tropical forest ecosystems: a review. *Journal of Tropical Ecology*, **7**, 1–24.

Brundrett, M., Bougher, N., Dell, B., Grove, T. and Malajczuk, N. (eds) (1996) *Working with mycorrhizas in forestry and agriculture*. ACIAR Monograph 32 Canberra.

Bryant, R. B. and Arnold, R. W. (1994) *Quantitative modeling of soil forming processes*. Soil Science Society of America, Madison, WI.

Brzostek, E. R. and Finzi, A. C. (2011) Substrate supply, fine roots, and temperature control proteolytic enzyme activity in temperate forest soils. *Ecology*, **92**, 892–902.

Buchmann, N., Oren, R. and Zimmermann, R. (1995) Response of magnesium-deficient saplings in a young, open stand of *Picea abies* (L.) Karst. to elevated soil magnesium, nitrogen, and carbon. *Environmental Pollution*, **87**, 31–43.

Bunnell, F. L., Tait, D. E. N., Flanagan, P. W. and vanCleve, K. (1977) Microbial respiration and substrate weight loss. I. A general model of the influences of abiotic variables. *Soil Biology and Biochemistry*, **9**, 33–40.

Buol, S. W., Southard, R. J., Graham, R. C. and McDaniel, P. A. (2003) *Soil Genesis and Classification*, 5th edition. Iowa State University Press, Ames.

Burger, J. A. and Torbert, J. L. (1992) *Restoring forests on surface mined land*. Technical Extension Publication. Virginia Polytechnic Institute, Blacksburg, VA.

Burns, R. M. and Hebb, E. A. (1972) *Site preparation and reforestation of droughty, acid sands*. United States Department of Agriculture, Agricultural Handbook 426. Washington, DC.

Burrough, P. A. and McDonnell, R. A. (1998) *Principles of Geographic Information Systems*. Oxford University Press, London.

Busse, M. D., Shestak, C. J., Hubbert, K. R. and Knapp, E. E. (2010) Soil physical properties regulate lethal heating during burning of woody residues. *Soil Science Society of America Journal*, **74**, 947–955.

Bussell, J., Jones, D. L., Healey, J. R. and Pullin, A. S. (2010) *How do draining and re-wetting affect carbon stores and greenhouse gas fluxes in peatland soils? A systematic review*. Collaboration for Environmental Evidence Review #08-012, Center for Evidence-based Conservation, Bangor University. available online at: http://www.environmentalevidence.org/SR49.html.

Butler, S. M., Melillo, J. M., Johnson, J. E., Mohan, J., Steudler, P. A., Lux, H., Burrows, E., Smith, R. M., Vario, C. L., Scott, L., Hill, T. D., Aponte, N. and Bowles, F. (2012) Soil warming alters nitrogen cycling in a New England forest: implications for ecosystem function and structure. *Oecologia*, **168**, 819–828.

Butterbach-Bahl, K. and Kiese, R. (2005) Significance of forests as sources for N_2O and NO. In: D. Binkley and O. Menyailo (eds) *Tree Species Effects on Soils: Implications for Global Change*. NATO Science Series, Kluwer Academic Publishers, Dordrecht, pp. 173–191.

Cain, M. L., Subler, S., Evans, J. P. and Fortin, M.-J. (1999) Sampling spatial and temporal variation in soil nitrogen availability. *Oecologia*, **18**, 397–404.

Cairns, M. A., Brown, S., Helmer, E. H. and Baumgardner, G. A. (1997) Root biomass allocation in the world's upland forests. *Oecologia*, **111**, 1–11.

Callesen, I., Liski, J., Raulund-Rasmussen, K., Olsson, M. T., Tau-Strand, L., Vesterdal, L. and Westman, C. J. (2003) Soil carbon stores in Nordic well-drained forest soils – relationships with climate and texture class. *Global Change Biology*, **9**, 358–370.

Campbell, R. E., Baker, M. B., Jr. and Ffolliott, P. F. (1977) *Wildfire effects on a ponderosa pine ecosystem: an Arizona case study*. U.S.D.A. Forest Service Research Paper RM-191. 12 pp.

Carmean, W. H. (1961) Soil survey refinements needed for accurate classification of black oak site quality in southeastern Ohio. *Soil Science Society of America Proceedings*, **25**, 394–397.

Carmean, W. H. (1975) Forest site quality evaluation in the United States. *Advances in Agronomy*, **27**, 209–269.

Carrero, O. (2012) *Environmental and financial risk analysis in eucalyptus forest production*. PhD dissertation, North Carolina State University, Raleigh.

Carter, M. R. and Gregorich, E. G. (2008) *Soil Sampling and Methods of Analysis*, 2nd edition. CRC Press, Boca Raton, FL.

Carter, M. R., Holmstrom, D. A., Cochrane, L. M., Brenton, P. C., VanRoestel, J. A., Langille, D. R. and Thomas, W. G. (1996) Persistence of deep loosening of naturally compacted subsoils in Nova Scotia. *Canadian Journal of Soil Science*, **76**, 541–547.

Carter, R. E. (1992) Diagnosis and interpretation of forest stand nutrient status. In: H.N. Chappell, G. F. Weetman and R. E. Miller (eds.), *Forest fertilization: sustaining and improving nutrition and growth of western forests*. College of Forest Resources Contribution #73, University of Washington, Seattle, pp. 90–97.

Carter, R. E., Otchere-Boateng, J. and Klinka, K. (1983) Dieback of a 30-year-old Douglas-fir plabntation in the Britain River Valley, British Columbia: symptoms and diagnosis. *Forest Ecology and Management*, **7**, 249–263.

Cerdà, A. and Doerr, S. H. (2005) The influence of vegetation recovery on soil hydrology and erodibility following fire: an eleven-year investigation. *International Journal of Wildland Fire*, **14**, 449–455.

Certini, C. (2005) Effects of fire on properties of forest soils: a review. *Oecologia*, **143**, 1–10.

Chadwick, O. A. and Chorover, J. (2001) The chemistry of pedogenic thresholds. *Geoderma*, **100**, 321–353.

Chamblin, M. A., Paasch, R. K., Lytle, D. A., Moldenke, A. R., Shapiro, L. G. and Dietterich, T. G. (2011) Design of an Automated System for Imaging and Sorting Soil Mesofauna. *Biological Engineering Transactions*, **4**, 17–41.

Chandler, C., Cheney, P., Thomas, P., Trabaud, L. and Williams, D. (1983) *Fire in Forestry, Volume 1: forest fire behavior and effects*. John Wiley & Sons, Inc., New York.

Chang, S. X. (1996) *Fertilizer N efficiency and incorporation and soil N dynamics in forest ecosystems of northern Vancouver Island*. PhD thesis, Faculty of Forestry, University of British Columbia, Vancouver. 190 pp.

Chapin, F. S. H. and Bloom, A. (1976) Phosphate adsorption: adaptation of tundra graminoids to a low temperature, low phosphorus environment. *Oikos*, **26**, 111–121.

Chapin, F. S. H., III and Van Cleve, K. (1981) Plant nutrient absorption and retention under differing fire regimes. In: H. A. Mooney, T. M. Bonnicksen, N. L. Christensen, J. E. Lotan and W. A. Reiners (eds) *Fire regimes and ecosystem properties*. U.S.D.A. Forest Service General Technical Report WO-26, pp. 301–321.

Chapin, F. S. III, Bloom, A. J., Field, C. B. and Waring, R. H. (1987) Plant responses to multiple environmental factors. *Bioscience*, **37**, 49–57.

Chappell, H. N., Cole, D. W., Gessel, S. P. and Walker, R. B. (1991) Forest fertilization research and practice in the Pacific Northwest. *Fertilizer Research*, **27**, 129–140.

Charman, D. J., Aravema, R. and Warner, B. G. (1994) Carbon dynamics in a forested peatland in north-eastern Ontario, Canada. *Journal of Ecology*, **82**, 55–62.

Chen, Q. (1987) *Nitrogen transformations in adjacent cypress and loblolly pine ecosystems*. MS thesis, Duke University, Durham. 71 pp.

Chesworth, W. (ed.) (2008) *Encyclopedia of Soil Science*. Springer, The Netherlands.

Chijicke, E. O. (1980) *Impact on soils of fast-growing species in lowland humid tropics*. Food and Agriculture Organization Forestry Paper No. 21, Rome.

Christensen, N. L. (1987) The biogeochemical consequences of fire and their effects on the vegetation of the coastal plain of the southeastern United States. In: L. Trabaud (ed.) *The Role of Fire in Ecological Systems*. SPB Academic Publishing, The Hague, The Netherlands, pp. 1–21.

Christina, M., Laclau, J.-P., Gonçalves, J. L. M., Jourdan, C., Nouvellon, Y. and Bouillet, J.-P. (2011) Almost symmetrical vertical growth rates above and below ground in one of the world's most productive forests. *Ecosphere*, **2**, 1–10.

Clayton, J. L. (1979) Nutrient supply to soil by rock weathering. In: *Impact of Intensive Harvesting on Forest Nutrition*. State University of New York, Syracuse, pp. 75–96.

Clayton, J. L. (1984) A rational basis for estimating element supply rate from weathering. In: E. L. Stone (ed.) *Sixth North American Forest Soils Conference*, University of Tennessee Press, Knoxville, pp. 405–419.

Cleaves, E. T., Fisher, D. W. and Bricker, O. P. (1974) Chemical weathering of serpentinite in the eastern Piedmont of Maryland. *Geological Society of America Bulletin*, **85**, 437–444.

Cleaves, E. T., Godfrey, A. E. and Bricker, O. P. (1970) Geochemical balance of a small watershed and its geomorphic implications. *Geological Society of America Bulletin*, **81**, 3015–3032.

Clements, F. E. (1916) *Plant Succession*. Carnegie Institute Publication 242. Washington, D.C.

Cline, M. G. (1944) Principles of soil sampling. *Soil Science*, **58**, 275–288.

Cochran, P. H. (1969) *Thermal properties and surface temperatures of seedbeds*. USDA Forest Service PNW Forest and Range Experiment Station, Portland. 19 pp.

Cochrane, M. A., Alencar, A., Schulze, M. D., Souza, C. M., Jr., Nepstad, D. C., Lefebvre, P. and Davidson, E. A. (1999) Positive feedbacks in the fire dynamic of closed canopy tropical forests. *Science*, **284**, 1832–1835.

Coile, T. S. (1952) Soil and the growth of trees. *Advances in Agronomy*, **4**, 329–398.

Coile, T. S. and Schumaher, F. X. (1964) *Soil-site relations, stand structure, and yields of slash and loblolly pine in the southern United States*. T. S. Coile, Inc., Durham, NC.

Cole, D. W. and Rapp, M. (1981) Elemental cycling in forest ecosystems. In: D. E. Reichle (ed.) *Dynamic Properties of Forest Ecosystems*, Cambridge University Press, England, pp. 341–409.

Cole, L., Dromph, K. M., Boaglio, V. and Bardgett, R. D. (2004) Effect of density and species richness of soil mesofauna on nutrient mineralisation and plant growth. *Biology and Fertility of Soils*, **39**, 337–343.

Coleman, D. C. (2001) Soil biota, soil systems, and processes. *Encyclopedia of Biodiversity, Volume 5*. Academic Press, San Diego, CA.

Coleman, D. C. and Crossley, D. A., Jr. (1996) *Essentials of Soil Ecology*. Academic Press Inc., San Diego, CA, 205 pp.

Coleman, S. M. and Dethier, D. P. (eds) (1986) *Rates of Chemical Weathering of Rocks and Minerals*. Academic Press, Orlando. 603 pp.

Coleman, D. C., Crossley, D. A., Jr. and Hendrix, P. F. (eds.) (2004) *Fundamentals of Soil Ecology*, 2nd edition, Elsevier Academic Press, New York, NY.

Comerford, N. B. and Fisher, R. F. (1982) Use of discriminant analysis for classification of fertilizer-responsive sites. *Soil Science Society of America Journal*, **46**, 1093–1096.

Compton, J. E. and Cole, D. W. (1998) Phosphorus cycling and soil P fractions in Douglas fir and red alder stands. *Forest Ecology and Management*, **110**, 101–112.

Conant, R. T., Ryan, M. G., Agren, G., Birge, H. E., Davidson, E. A., Eliasson, P. E., Evans, S. E., Frey, S.,

Giardina, C. P., Hopkins, F., Hyvönen, R., Kirschbaum, M. U. F., Lavallee, J. M., Leifield, J., Parton, W. J., Steinweg, J. M., Wallenstein, M. D., Wetterstedt, J. A. M. and Bradford, M. A. (2011) Temperature and soil organic matter decomposition rates – synthesis of current knowledge and a way forward. *Global Change Biology*, **17**, 3392–3404.

Cook, R. L. (2012) *Soil organic carbon in tropical plantation forestry: Long-term effects of Eucalyptus silviculture and vegetation selection*. PhD dissertation, Department of Forestry and Environmental Resources, North Carolina State University, Raleigh.

Cornelis, W. M., Khlosi, M., Hartmann, R., VanMeirvenne, M. and De Vos, B. (2005) Comparison of Unimodal Analytical Expressions for the Soil-Water Retention Curve. *Soil Science Society of America Journal*, **69**, 1902–1936.

Cornwell, S. M. and Stone, E. L. (1968) Availability of nitrogen to plants in acid coal mine spoils. *Nature*, **217**, 768–769.

Costantini, A., Nester, M. R. and Podberscek, M. (1995) Site preparation for *Pinus* establishment in southeastern Queensland. 1. Temporal changes in bulk density. *Australian Journal of Experimental Agriculture*, **35**, 1151–1158.

Cote, B. and Camire, C. (1984) Growth, nitrogen accumulation, and symbiotic dinitrogen fixation in pure and mixed plantings of hybrid poplar and black alder. *Plant and Soil*, **78**, 209–220.

Cotta, B. (1852) *Praktische Geohnoise für Land- und Forstwirte und Techniker*. Dresden.

Cotta, H. (1809) *Systematische Anleitung zur Taxoation der Waldungen*. Berlin. **202**, 281–308, 361–391.

Cotton, F. A. and Wilkinson, G. (1988) *Advanced Inorganic Chemistry*, 5th edition. John Wiley & Sons, Inc., New York.

Courchesne, F. and Gobran, G. R. (1997) Mineralogic variations of bulk and rhizosphere soils from a Norway spruce stand. *Soil Science Society of America Journal*, **61**, 1245–1249.

Courty, P.-E., Buée, M., Diedhiou, A. G., Frey-Klett, P., LeTacon, F., Rineau, F., Turpault, M.-P., Uroz, S. and Garbaye, J. (2010) The role of ectomycorrhizal communities in forest ecosystem processes: new perspectives and emerging concepts. *Soil Biology and Biochemistry*, **42**, 679–698.

Covington, W. W. and Sackett, S. (1984) The effect of a prescribed fire in Southwestern ponderosa pine on organic matter and nutrients in woody debris and forest floor. *Forest Science*, **30**, 183–192.

Cowels, H. C. (1899) The ecological relations of the vegetation. *Botanical Gazette*, **27**, 95–129.

Creasey, J., Edwards, A. C., Reid, J. M., MacLeod, D. A. and Cresser, M. S. (1986) The use of catchment studies for assessing chemical weathering rates in two contrasting upland areas in northeast Scotland. In: S. M. Colman and D. P. Dethier (eds) *Rates of Chemical Weathering of Rocks and Minerals*. Academic Press, Orlando, pp. 468–502.

Cressie, N. (1993) *Statistics for Spatial Data*. Wiley Interscience, New York, NY.

Cross, A. F. and Schlesinger, W. H. (1995) A literature review and evaluation of the Hedley fractionation: applications to the biogeochemical cycle of soil phosphorus in natural ecosystems. *Geoderma*, **64**, 197–214.

Crossley, D. A., Mueller, B. R. and Perdue, J. C. (1992) Biodiversity of micrarthropods in agricultural soils: relations and processes. *Agricultural Ecosystems and Environment*, **40**, 37–46.

Crow, S. E., Lajtha, K., Filley, T. R., Swanston, C. W., Bowden, R. D. and Caldwell, B. A. (2009) Sources of plant-derived carbon and stability of organic matter in soil: implications for global change. *Global Change Biology*, **15**, 2003–2019.

Cstantini, A. M., Nestor, R. and Podberscek, M. (1995) Site preparation for *Pinus* establishment in southeast Queensland. *Australian Journal of Experimental Agriculture*, **35**, 1159–1164.

Cuevas, E. and Medina, E. (1983) Root production and organic matter decomposition in a terra firme forest of the upper Rio Negro basin. In: *International Symposium on Root Ecology and Its Applications*, Gumpenstein, Irdning, Austria, pp. 653–666.

Currie, W. S., Harmon, M. E., Burke, I. C., Hart, S. C., Parton, W. J. and Silver, W. (2010) Cross-biome transplants of plant litter show decomposition models extend to a broader climatic range but lose predictability at the decadal time scale. *Global Change Biology*, **16**, 1744–1761.

Czapowskyj, M. M. (1973) Establishing forest on surface-mined lands as related to fertility and fertilization. In: *Symposium on Forest Fertility*. USDA Forest Service General Technical Report NE-3, pp. 132–333.

Dalla-Tea, F. and Marco, M. A. (1996) Fertilizers and Eucalypt plantations in Argentina. In: P. M. Attiwill and M. A. Adams (eds) *Nutrition of Eucalypts*, CSIRO, Collingwood, Australia, pp. 327–333.

Dangerfield, J. and Brix, H. (1979) Comparative effects of ammonium nitrate and urea fertilizers on tree growth and soil processes. In: *Forest fertilization conference*, Contribution #40, Institute of Forest Resources, University of Washington, Seattle, pp. 133–139.

Daniels, R. B. and Hammer, R. B. (1992) *Soil Geomorphology*. John Wiley & Sons, Inc., New York.

Danjon, F. and Reubens, B. (2008) Assessing and analyzing 3D architecture of woody root systems, a review of methods and applications in tree and soil stability, resource acquisition and allocation. *Plant and Soil*, **303**, 1–34.

Darfus, G. H. and Fisher, R. F. (1984) Site relations of slash pine on dredge mine spoils. *Journal of Environmental Quality*, **13**, 457–492.

Darwin, C. W. (1883) *The formation of vegetable mould, through the action of worms with observations on their habit*. John Murray, London. Available at: http://darwin-online.org.uk/EditorialIntroductions/Freeman_VegetableMouldandWorms.html.

Davey, C. B. (1984) Nursery soil organic matter: management and importance. In: M. L. Duryea and T. D. Landis (eds) *Forest Nursery Manual*, Martinus Nijhoff/Dr W. Junk Publishers, Boston, MA, pp. 81–86.

Davidson, E., Myrold, D. D. and Groffman, P. M. (1990) Denitrification in temperate forest ecosystems. In: S. P. Gessel, D. S. Lacate, G. F. Weetman and R. F. Powers (eds) *Sustained Productivity of Forest Soils*, University of British Columbia Faculty of Forestry Publication, Vancouver, pp. 196–220.

Davidson, E. A., Reis de Carvalho, C. J., Guimarães Vieira, I. C., deO. Figueiredo, R., Moutinho, P., Ishida, F. O. Y. and Primo dos Santos, M. T. (2004) Nitrogen and phosphorus limitation of biomass growth in a tropical secondary forest. *Ecological Applications*, **14**, S150–S163.

DeAngelis, D. L. (1992) *Dynamics of Nutrient Cycling and Food Webs*. Chapman & Hall, London.

DeBano, L. F., Neary, D. G. and Ffolliott, P. F. (1998) *Fire's Effects on Ecosystems*. John Wiley & Sons, Inc., New York. 333 pp.

DeBell, D. S. and Radwan, M. A. (1979) Growth and nitrogen relations of coppiced black cottonwood and red alder in pure and mixed plantings. *Botanical Gazette Supplement*, **40**, S97–S101.

Degens, B. P. (1998) Decreases in microbial functional diversity do not result in corresponding changes in decomposition under different moisture conditions. *Soil Biology and Biochemistry*, **20**, 1989–2000.

Dell, B. (1996) Diagnosis of nutrient deficiencies in euca-lypts. In: P. M. Attiwill and M. A. Adams (eds) *Nutrition of Eucalypts*, CSIRO, Collingwood, Australia, pp. 417–440.

DeMontigny, L. E., Preston, C. M., Hatcher, P. G. and Kögel-Knaber, I. (1993) Comparison of humus horizons from two ecosystem phases on northern Vancouver Island using ^{13}C CPAMS NMR spectroscopy and CuO oxidation. *Canadian Journal of Soil Science*, **73**, 9–25.

Derome, J., Kukkola, M. and Malknen, E. (1986) *Forest liming on mineral soils. Results of Finnish experiments.* National Swedish Environmental Protection Board Report 3084, Solna, Sweden.

Derr, H. J. and Mann, W. F. (1970) *Site preparation improves growth of planted pines.* USDA Forest Service Research Note SO-106.

Dijkstra, F. A. (2003) Calcium mineralization in the forest floor and surface soil beneath different tree species in the northeastern US. *Forest Ecology and Management*, **175**, 185–194.

Dijkstra, F. A., vanBreemen, N., Jongmans, A. G., Davies, G. R. and Likens, G. E. (2003) Calcium weathering in forested soils and the effect of different tree species. *Biogeochemistry*, **62**, 252–275.

Dindal, D. L. (1990) *Soil Biology Guide.* John Wiley & Sons, Inc., New York.

Dinkelaker, B., Hehgler, C., Neumann, G., Eltrop, L. and Marschner, H. (1997) Root exudates and mobilization of nutrients. In: *Trees: Contributions to modern tree physiology.* Backhuys Publishers. Leiden, The Netherlands, pp. 441–452.

Dissmeyer, G. E. and Greis, J. G. (1983) Sound soil arid water management good economics. In: E. L. Stone (ed.) *The Managed Slash Pine Ecosystem.* School of Forest Resources and Conservation, University of Florida, Gainesville, FL, pp. 194–202.

Dixon, J. B. and Schulze, D. G. (2002) *Soil Mineralogy with Environmental Applications.* Soil Science Society of America, Madison, WI.

Doerr, S. H., Shakesby, R. A. and MacDonald, L. H. (2009) Soil Water Repellency: A Key Factor in Post-Fire Erosion. In: A. Cerda and P. R. Robichaud (eds) *Fire Effects on Soils and Restoration Strategies.* Science Publishers, Enfield, New Hampshire, pp. 197–224.

Doerr, S. H., Shakesby, R. A., Blake, W. H., Chafer, C. J., Humphreys, G. S. and Wallbrink, P. J. (2006) Effects of differing wildfire severities on soil wettability and implications for hydrological response. *Journal of Hydrology*, **319**, 295–311.

Doran, J. W. and Jones, A. (eds) (1997) *Methods of Assessing Soil Quality.* Science Society of America, Madison, WI.

Dorsser, J. C. vanand Rook, D. A. (1972) Conditioning of radiate pine seedlings by undercutting and wrenching: description of methods, equipment and seedling response. *New Zealand Journal of Forestry*, **17**, 61–73.

Dougherty, P. M., Allen, H. L., Kress, K. W., Murthy, R., Maier, C. A., Albaugh, T. L. and Sampson, D. A. (1998) An investigation of the impacts of elevated carbon dioxide, irrigation and fertilization on the physiology and growth of loblolly pine. In: *The productivity and sustainability of southern forest ecosystems in a changing environment. Ecological studies*, Vol. 128. Springer-Verlag, New York, NY, pp. 149–168.

Driscoll, C. T., vanBreeman, N. and Mulder, J. (1984) Aluminum chemistry in a forested spodosol. *Soil Science Society of America Journal*, **49**, 437–444.

Dunker, R. E., Hooks, C. L., Vance, S. L. and Darmody, R. G. (1995) Deep tillage effects on compacted surface-mine land. *Soil Science of America Journal*, **59**, 192–199.

Duryea, M. L. and Landis, T. D. (1984) *Forest Nursery Manual: Production of bareroot seedlings.* Martinus Nijhoff/Dr W. Junk Publishers. Boston, MA.

Duxbury, J. M., Smith, M. S. and Doran, J. W. (1989) Soil organic matter as a source and a sink of plant nutrients. In: D. Coleman, J. M. Oades and G. Uehara (eds) *Dynamics of Soil Organic Matter in Tropical Ecosystems*, University of Hawaii, Honolulu, pp. 3–67.

Dyrness, C. T. (1969) *Hydrologic properties of soils on three small watersheds in the western Cascades.* USDA Forest Service Research Note PNW-111.

Dyrness, C. T. (1976) *Effect of wildfire on soil wettability in the high Cascades of Oregon.* USDA Forest Service Research Paper PNW-202, Portland.

Dyrness, C. T., VanCleve, K. and Levison, J. D. (1989) The effect of wildfire on soil chemistry in four forest types in interior Alaska. *Canadian Journal of Forest Research*, **19**, 1389–1396.

Dyson, F. (2010) *The Best American Science and Nature Writing. 2010.* Houghton Mifflin, Boston.

Ebermayer, E. (1876) *Die gesammte Lehre der Waldstreu mit Rueksicht auf die chemische Statik des Waldbaues.* J. Springer, Berlin.

Edwards, C. A. (1991) The assessment of populations of soil-inhabiting invertebrates. *Agriculture, Ecosystem and Environment*, **34**, 145–176.

Edwards, C. A. (1998) *Earthworm Ecology.* St. Lucie Press, New York. 389 pp.

Edwards, N. T. and Ross-Todd, B. M. (1979) The effects of stem girdling on biogeochemical cycles within a mixed deciduous forest in eastern Tennessee. I. Soil solution chemistry, soil respiration, litterfall and root biomass studies. *Oecologia*, **40**, 247–257.

Ekblad, A. and Huss-Danell, K. (1995) Nitrogen fixation by *Alnus incana* and nitrogen transfer from *A. incana* to *Pinus sylvestris* influenced by macronutrients and ectomycorrhiza. *New Phytologist*, **131**, 453–459.

Ekschmitt, K., Liu, M., Vetter, S., Fox, O. and Wolters, V. (2005) Strategies used by soil biota to overcome soil organic matter stability – why is dead organic matter left over in the soil? *Geoderma*, **128**, 167–176.

Ellis, R. C. and Graley, A. M. (1983) Gains and losses in soil nutrients associated with harvesting and burning eucalyptus rainforest. *Plant and Soil*, **74**, 437–450.

Emsley, J. (1984) The phosphorus cycle. In: O. Hutzinger (ed.) *The Natural Environment and the Biogeochemical Cycles*, Volume 1, Part A. Springer Verlag, Berlin, pp. 147–162.

Enloe, H. A., Graham, R. C. and Sillett, S. C. (2006) Arboreal histosols in old-growth redwood forest canopies. *Soil Science Society of America Journal*, **70**, 408–418.

Enoki, T. and Kawaguchi, H. (1999) Nitrogen resorption from needles of *Pinus thunbergii* Parl. growing along a topographic gradient of soil nutrient availability. *Ecological Research*, **14**, 1–8.

Eriksson, H. M. (1996) *Effects of tree species and nutrient application on distribution and budgets of base cations in Swedish forest ecosystems*. PhD thesis, Swedish University of Agricultural Sciences, Uppsala.

Eswaran, H., Rice, T., Ahrens, R. and Stewart, B. A. (eds) (2003) *Soil Classification. A Desk Reference*. CRC Press. Boca Raton, Florida.

Evans, J. (1992) *Plantation Forestry in the Tropics*, 2nd edition. Clarendon Press, Cambridge, UK.

Ewel, J., Berish, C., Brown, B., Price, N. and Raich, J. (1981) Slash and burn impacts on a Costa Rican wet forest site. *Ecology*, **62**, 816–829.

Eyk, J. J. Van der (1957) *Reconnaissance soil survey in northern Surinam*. PhD dissertation, Wageningen.

Eyre, S. R. (1963) *Vegetation and Soils*. Aldine, Chicago.

Fahey, T. J., Hughes, J. W., Pu, M. and Arthur, M. A. (1988) Root decomposition and nutrient flux following whole-tree harvest of northern hardwood forest. *Forest Science*, **34**, 744–768.

Farrelly, N., Reamonn, M. F. and Radford, T. (2009) The use of site factors and site classification methods for the assessment of site quality and forest productivity in Ireland. *Irish Forestry*, **66**, 21–38.

Feagley, S. E., Valdez, M. S. and Hudnall, W. H. (1994) Papermill sludge, phosphorus, potassium and lime effect on clover grown on a mine soil. *Journal of Environmental Quality*, **23**, 759–765.

Feller, M. C. (1981) Catchment nutrient budgets and geological weathering in *Eucalyptus regnans* ecosystems in Victoria. *Australian Forestry*, **44**, 502–510.

Feller, M. C. (1983) Impacts of prescribed fire (slashburning) on forest productivity, soil erosion, and water quality on the coast. In: Prescribed *Fire-Forest Soils Symposium Proceedings*. Land Management Report #16, Ministry of Forests, British Columbia, Victoria, pp. 57–91.

Fellin, D. G. and Kennedy, P. C. (1972) *Abundance of arthropods inhabiting duff and soil after prescribed burning on forest clearcuts in northern Idaho*. USDA Forest Service Research Note INT-162. 8 pp.

Fernow, B. (1907) *History of Forestry*. University Press, Toronto.

Ferreira, J. M. de A. and Stape, J. L. (2009) Productivity gains by fertilisation in *Eucalyptus urophylla* clonal plantations across gradients in site and stand conditions. *Southern Forests*, **71**, 253–258.

Ferrians, O. (1965) Permafrost map of Alaska. U.S. Geological Survey, Miscellaneous Map 1-445.

Fierer, N. and Jackson, R. B. (2006) The diversity and biogeography of soil bacterial communities. *Proceedings of the National Academy of Science*, **103**, 626–631.

Fierer, N., Breitbart, M., Nulton, J., Salamon, P., Lozupone, C., Jones, R., Robeson, M., Edwards, R. A., Felts, B., Rayhawk, S., Knight, R., Ohwer, F. and Jackson, R. B. (2007) Metagenomic and small-subunit rRNA analyses reveal the genetic diversity of bacteria, archaea, fungi, and viruses in soil. *Applied and Environmental Microbiology*, **73**, 7059–7066.

Fife, D. N. and Nambiar, E. K. S. (1995) Effect of nitrogen on growth and water relations of radiata pine families. *Plant and Soil*, **168**, 279–285.

Fink, S. (1992) Physiologische und strukturelle Veranderungen an Baumen unter Magnesiummangel. In: G. Glatzel, R. Jandel, M. Sieghardt and H. Hager (eds) *Magnesiummangel in Mitteleuropischen Wald'kosystemen*, Forstliche Schriftenreihe Band 5, Universitat fuhr Bodenkultur, Vienna, pp. 16–26.

Fisher, R. F. (1972) Spodosol development and nutrient distribution under Hydnaceae fungal mats. *Soil Science Society of America Proceedings*, **30**, 492–495.

Fisher. R. F. (1980) Soils: interpretations for silviculture in the southern coastal plain. In: J. P. Barnett (ed.) *Proceedings 1st biennial southern silvicultural research conference*. USDA Forest Service Technical Report SO-34, pp. 223–300.

Fisher, R. F. (1984) Predicting tree and stand response to cultural practices. In: E. L. Stone (ed.) *Forest Soils and Treatment Impacts*, University of Tennessee Press, Knoxville, TN, pp. 53–66.

Fisher, R. F. (1995a) Amelioration of degraded rainforest soils by plantations of native trees. *Soil Science Society of America Journal*, **59**, 544–549.

Fisher, R. F. (1995b) Soil organic matter: clue or conundrum? In: W. W. McFee and J. M. Kelly (eds) *Carbon Forms and Functions in Forest Soils*. Soil Science Society of America. Madison, WI, pp. 1–12.

Fisher, R. F. and Adrian, F. W. (1980) Bahiagrass reduces slash pine seedling survival and growth. *Tree Planters Notes*, **32**, 19–21.

Fisher, R. F. and Eastburn, R. B. (1974) Afforestation alters prairie soil nitrogen status. *Soil Science Society of America Proceedings*, **38**, 366–368.

Fisher, R. F. and Juo, A. S. R. (1995) Mechanisms of tree growth in acid soils. In: D. O. Evans and L. T. Szott (eds) *Nitrogen Fixing Trees for Acid Soils*. Nitrogen Fixing Tree Research Reports (Special Issue). Winrock International and NFTA, Morrilton, AR, pp. 1–18.

Fisher, R. F. and Miller, R. J. (1980) Description of soils having disturbed or discontinuous horizons. *Soil Science Society of America Journal*, **44**, 1279–1281.

Fisher, R. F. and Pritchett, W. L. (1982) Slash pine response to different nitrogen fertilizers. *Soil Science Society of America Journal*, **46**, 113–136.

Fisher, R. F. and Stone, E. L. (1968) Soil and plant moisture relations of red pine growing on a shallow soil. *Soil Science Society of America Proceedings*, **32**, 725–728.

Fisher, R. F. and Stone, E. L. (1969) Increased availability of nitrogen and phosphorus in the root zone of conifers. *Soil Science Society of America Proceedings*, **33**, 955–961.

Fisk, M. G., Fahey, T. J. and Groffman, P. M. (2004) Earthworm invasion, fine-root distributions, and soil respiration in north temperate forests. *Ecosystems*, **7**, 55–62.

Fissore, C., Giardina, C. P., Swanston, C. W., King, G. M. and Kolka, R. K. (2009) Variable temperature sensitivity of soil organic carbon in North American forests. *Global Change Biology*, **15**, 2295–2310.

Fölster, H. and Khanna, P. K. (1997) Dynamics of nutrient supply in plantation soils. In: E. K. S. Nambiar and A. G. Brown (eds) *Management of Soil, Nutrients and Water in Tropical Plantation Forests*. Australian Centre for International Agricultural Research Monograph 43, Canberra, pp. 339–378.

Forest Nutrition Cooperative (2011) http://www.forestnutrition.org/henderson.htm.

Forge, T. A., Hogue, E., Neilsen, G. and Neilsen, D. (2003) Effects of organic mulches on soil microfauna in the root zone of apple: implications for nutrient fluxes and functional diversity of the soil food web. *Applied Soil Ecology*, **22**, 39–54.

Forrester, D. I., Schortemeyer, M., Stock, W. D., Bauhus, J., Khanna, P. K. and Cowie, A. L. (2007) Assessing nitrogen fixation in mixed- and single-species plantations of *Eucalyptus globulus* and *Acacia mearnsii*. *Tree Physiology*, **27**, 1319–1328.

Fox, T. (1995) Forest soils and low-molecular-weight organic acids. In: W. W. McFee and J. M. Kelly (eds) *Carbon Forms and Functions in Forest Soils*, Soil Science Society of America, Madison, WI, pp. 53–62.

Fox, T. and Comerford, N. (1990) Low-molecular-weight organic acids in selected forest soils of the southeastern USA. *Soil Science Society of America Journal*, **54**, 1139–1144.

Fox, T. R., Allen, H. L., Albaugh, T. J., Rubilar, R. A. and Carlson, C. A. (2007) Tree nutrition and forest fertilization of pine plantations in the southern United States. *Southern Journal of Applied Forestry*, **31**, 5–11.

Fox, T. R., Miller, B. W., Rubilar, R., Stape, J. L. and Albaugh, T. J. (2011) Phosphorus nutrition of forest plantations: the role of inorganic and organic phosphorus. In: E. K. Bünemann *et al.* (eds) *Phosphorus in Action*, Springer-Verlag, Berlin, Chapter 13.

Franklin, J. F., Dyrness, C. T., Moore, D. G. and Tarrant, R. F. (1968) Chemical soil properties under coastal Oregon stands of alder and conifers. In: J. M. Trappe, J. F. Franklin, R. F. Tarrant and G. H. Hansen (eds) *Biology of Alder*. USDA Forest Service, Portland, pp. 157–172.

Franklin, O., Högberg, P., Ekblad, A. and Ågren, G. (2003) Pine forest floor carbon accumulation in response to N and PK additions: bomb ^{14}C modeling and respiration studies. *Ecosystems*, **6**, 6–658.

Frelich, L. E., Calcote, R. R., Davis, M. B. and Pastor, J. (1993) Patch formation and maintenance in an old-growth hemlock-hardwood forest. *Ecology*, **74**, 51–527.

French, H. M. (1976) *The Periglacial Environment*. Longman, London.

Frostegård, A. and Bååth, E. (1996) The use of phospholipid fatty acid analysis to estimate bacterial and fungal biomass in soil. *Biology and Fertility of Soils*, **22**, 50–65.

Frostegård, Å., Bååth, E. and Tunlio, A. (1993) Shifts in the structure of soil microbial communities in limed forests as revealed by phospholipid fatty acid analysis. *Soil Biology and Biochemistry*, **25**, 723–730.

Fuller, C. (2007) *The legacy of subsistence agriculture on New England's successional forest soils: A 25-year direct measurement perspective*. BS thesis, Environmental Science, Brown University, Providence.

Gadgil, R. L. (1971) The nutritional role of *Lupinus arboreus* in coastal sand dunes forestry: III. Nitrogen distribution in the ecosystem before tree planting. *Plant and Soil*, **35**, 113–126.

Gahagan, A., Giardina, C. P., King, J. S., Pregitzer, K. S., Binkley, D. and Burton, A. J. (2012) Carbon fluxes and storage after 60 years of stand development in red pine (*Pinus resinosa*) plantations and mixed hardwood stands in northern Michigan old fields. Submitted to *Global Change Biology*.

Gahoonia, T. S. and Nielsen, N. E. (1992) The effects of root-induced changes on the depletion of inorganic and organic phosphorus in the rhizosphere. *Plant and Soil*, **143**, 185–191.

Garcia-Montiel, D. (1996) *Changes in nutrient cycling during tropical reforestation*. PhD thesis, Colorado State University, Ft. Collins. 92 pp.

Garcia-Montiel, D. C. and Binkley, D. (1998) Effect of *Eucalyptus saligna* and *Albizia falcataria* on soil processes and nitrogen supply in Hawaii. *Oecologia*, **113**, 547–556.

Gardner, C. M. K., Robinson, D., Blyth, K. and Cooper, J. D. (2000) Soil water content. In: K. A. Smith and C. E. Mullins (eds) *Soil and Environmental Analysis*, Marcel Dekker, New York, NY, pp. 1–64.

Gedroiz, K. K. (1912) Colloidal chemistry in relation to the problems of soil science. *Zhur. Opit. Agron.*, **13**

Gehl, R. J. and Rice, C. W. (2007) Emerging technologies for *in situ* measurement of soil carbon. *Climatic Change*, **80**, 43–54.

Gerdemann, J. W. and Trappe, J. M. (1975) Taxonomy of Endogonaceae. In: F. F. Sanders, B. Mosse and P. B. Tinker (eds) *Endomycorrhizas*, Academic Press, New York, pp. 35–51.

Gessel, S. P. and Balci, A. H. (1965) Amount and composition of forest floors under Washington coniferous forests. In: C. T. Youngberg (ed.) *Forest–Soil Relationships in North America*. Oregon State University Press, Corvallis, OR, pp. 11–23.

Gholz, H. L., Fisher, R. F. and Pritchett, W. L. (1985) Nutrient dynamics in slash pine plantation ecosystems. *Ecology*, **63**, 1827–1839.

Giardina, C. P. and Ryan, M. G. (2002) Total Belowground Carbon Allocation in a fast-growing Eucalyptus plantation estimated using a carbon balance approach. *Ecosystems*, **5**, 487–499.

Giardina, C., Ryan, M. G. and Hubbard, R. M. (1999) Do carbon turnover rates in mineral soil vary with temperature and texture? Manuscript submitted to *Nature*.

Giardina, C., Huffman, S., Binkley, D. and Caldwell, B. (1995) Alders increase phosphorus supply in a Douglas fir plantation. *Canadian Journal of Forest Research*, **25**, 1652–1657.

Giardina, C. P., Ryan, M. G., Hubbard, R. M. and Binkley, D. (2001) Tree species and soil textural controls on carbon and nitrogen mineralization rates. *Soil Science Society of America Journal*, **65**, 1271–1279.

Giardina, C. P., Binkley, D., Ryan, M. G., Fownes, J. H. and Senock, R. S. (2004) Belowground carbon cycling in a humid tropical forest decreases with fertilization. *Oecologia*, **139**, 545–550.

Giesler, R., Högberg, M. and Högberg, P. (1998) Soil chemistry and plants in fennoscandian boeal forest as exemplified by a local gradient. *Ecology*, **79**, 119–137.

Gill, R. A. and Jackson, R. B. (2000) Global patterns of root turnover for terrestrial ecosystems. *New Phytologist*, **147**, 13–31.

Glatzel, G. (1991) The impact of historic land use and modern forestry on nutrient relations of Central European forest ecosystems. *Fertilizer Research*, **27**, 1–8.

Glinka, K. D. (1927) Dokuchaiev's Ideas in the Development of Pedology and the Cognate Sciences. *Russian Pedological Investigations* No. 1, USSR Academy of Sciences, Leningrad.

Gobran, G. R. and Clegg, S. (1996) A conceptual model for nutrient availability in the mineral soil–root system. *Canadian Journal of Soil Science*, **76**, 124–131.

Goerres, J. F., Dichiaro, M. J., Lyons, J. B. and Amador, J. A. (1998) Spatial and temporal patterns of soil biological activity in a forest and an old field. *Soil Biology and Biochememistry*, **30**, 219–230.

Goforth, B. R., Graham, R. C., Hubbert, K. R., Zanner, C. W. and Minnick, R. A. (2005) Spatial distribution and properties of ash and thermally altered soils after high-severity forest fire, southern California. *International Journal of Wildland Fire*, **14**, 343–354.

Goldammer, J. G. (1993) Historical biogeography of fire: tropical and subtropical. In: P. J. Crutzen and J. G. Goldammer (eds) *Fire in the Environment: The ecological, atmospheric, and climatic importance of vegetation fires*. John Wiley & Sons, Ltd, Chichester, pp. 298–314.

Goldich, S. S. (1938) A study of rock weathering. *Journal of Geology*, **46**, 17–58.

Golding, D. L. and Stanton, C. R. (1972) Water storage in the forest floor of subalpine forests of Alberta. *Canadian Journal of Forest Research*, **2**, 1–6.

Gonçalves, J. L. M., Barros, N. F., Nambiar, E. K. S. and Novais, R. F. (1997) Soil and stand management for short-rotation plantations. In: E. K. S. Nambiar and A. G. Brown (eds) *Management of Soil, Nutrients, and Water in Tropical Plantation Forests*. ACIAR Monograph #43, Canberra, pp. 379–417.

González-Pérez, J. A., González-Vila, F. J., Almendros, G. and Knicker, H. (2004) The effect of fire on soil organic matter – a review. *Environment International*, **30**, 855–870.

Goovaerts, P. (1997) *Geostatistics for Natural Resources Evaluation*. Oxford University Press, New York.

Gould, A. B., Hendrix, J. W. and Ferriss, R. S. (1996) Relationship of mycorrhizal activity to time following reclamation of surface mine land in western Kentucky. I. Propagule and spore population densities. Canadian Journal of Botany, **74**, 247–261.

Gracen, E. L. and Sands, R. (1980) Compaction of forest soils: a review. *Australian Journal of Soil Research*, **18**, 163–189.

Grady, K. C. and Hart, S. C. (2006) Influences of thinning, prescribed burning, and wildfire on soil processes and properties in southwestern ponderosa pine forests: A retrospective study. *Forest Ecology and Management*, **234**, 123–135.

Graham, R. C. and Wood, H. B. (1991) Morphologic development and clay redistribution in lysimeter soils under chaparral and pine. *Soil Science Society of America Journal*, **55**, 1638–1646.

Gray, L. E. (1971) Physiology of vesicular-arbuscular mycorrhizae. In: E. Haeskaylo (ed.) *Mycorrhizae*. USDA Forest Service Miscellaneous Publication 1189, pp. 145–150.

Gray, T. R. G. and Williams, S. T. (1971) *Soil Micro-organisms*. Longman, London.

Grayson, S. J., Vaughan, D. and Jones, D. (1997) Rhizosphere carbon flow in trees, in comparison with annual plants: the importance of root exudation and its impact on microbial activity and nutrient availability. *Applied Soil Ecology*, **5**, 29–56.

Grebe, C. (1852) *Forstliche Gebirgskunde*. Bodenkunde und Klimalehre. Vienna.

Green, R. N., Trowbridge, R. L. and Klinka, K. (1993) *Towards a Taxonomic Classification of Humus Forms*. Forest Science Monograph 29. Society of American Foresters, Bethesda, MD.

Greenwood, N. and Earnshaw, A. (1984) *Chemistry of the Elements*. Pergamon, Oxford.

Grier, C. (1975) Wildfire effects on nutrient distribution and leaching in a coniferous ecosystem. *Canadian Journal of Forest Research*, **5**, 599–607.

Griffith, B. G., Hartwell, F. W. and Shaw, T. E. (1930) The evolution of soil as affected by old field white pine-mixed hardwood succession in central New England. *Harvard Forest Bulletin*, 15.

Griffiths, R. P., Madritch, M. D. and Swanson, A. K. (2009) The effects of topography on forest soil characteristics in the Oregon Cascade Mountains (USA): Implications for the effects of climate change on soil properties. *Forest Ecology and Management*, **257**, 1–7.

Grigal, D. F. (1984). Shortcomings of soil surveys for forest management. In: J. G. Bockheim (ed.) *Forest Land Classification: Experience, problems, perspectives*. North Central Forest Soils Committee, Madison, WI.

Grigal, D. and McColl, J. (1977) Litter decomposition following forest fire in northeastern Minnesota. *Journal of Applied Ecology*, **14**, 531–538.

Grubb, P. J. (1995) Mineral nutrition and soil fertility in tropical rain forests. In: A. E. Lugo and C. Lowe (eds) *Tropical Forests: Management and ecology*, Springer-Verlag, New York, pp. 308–330.

Guerrini, I. A., Gonçalves, J. L. M. And Bôas, R. L. V. (1999) Uses of industrial residues in Brazilian plantation forestry. In: C. Henry, R. Harrison and B. Bastian (eds) *The Forest Alternative – Principles and practice residual of uses*. College of Forest Resources, University of Washington, Seattle.

Hackl, E., Pfeffer, M., Donat, C., Bachmann, G. and Zechmeister-Boltenstern, S. (2004) Microbial nitrogen turnover in soils under different types of natural forest. *Forest Ecology and Management*, **118**, 101–112.

Hagedorn, F., Spinnler, D., Bundt, M., Blaser, P. and Siegwolf, R. (2003) The input and fate of new C in two forest soils under elevated CO_2. *Global Change Biology*, **9**, 862–872.

Hagen-Thorn, A., Callesen, I., Armolaitis, K. and Nihlgård, B. (2004) The impact of six European tree species on the chemistry of mineral topsoil in forest plantations on former agricultural land. *Forest Ecology and Management*, **195**, 373–384.

Haines, L. W. and Pritchett, W. L. (1964) The effects of site preparation on the growth of slash pine. *Soil Crop Science Society of Florida Proceedings*, **24**, 27–34.

Haines, L. W., Maki, T. F. and Sanderford, S. G. (1975) The effects of mechanical site preparation treatments on soil productivity and tree (*Pinus taeda* L. and *P. elliottii* Engelm. Var. *elliottii*) growth. In: B. Bernier and C. Winget (eds) *Forest Soils and Forest Land Management*, Lavel University Press, Quebec City, Quebec, Canada, pp. 379–395.

Hall, S. J. and Raffaelli, D. G. (1993) Food webs: Theory and reality. *Advances in Ecological Research*, **24**, 182–239.

Hallbäcken, L. and Bergholm, J. (1997) Nutrient dynamics. In: *Imbalanced forest nutrition – vitality measures*. Section of Ecology, Swedish University of Agricultural Sciences, SNS Project 1993–1996, Uppsala, pp. 112–136.

Hamburg, S. P. (1984) Effects of forest growth on soil nitrogen and organic matter pools following release from subsistence agriculture. In: E. L. Stone (ed.) *Forest Soils and Treatment Impacts*. Proceedings of the 6th North American Forest Soils Conference. University of Tennessee, Knoxville, pp. 145–158.

Hamilton, W. J., Jr. and Cook, D. B. (1940) Small mammals and the forest. *Journal of Forestry*, **38**, 468–473.

Handley, W. R. C. (1954) *Mull and Mor Formation in Relation to Forest Soils*. UK Forestry Commission, Bulletin 23. London. 115 pp.

Hansen, K., Vesterdal, L., Kappel Schmidt, I., Gundersen, P., Sevel, L., Bastrup-Birk, A., Pedersen, L. B. and Bille-Hansen, J. (2009) Litterfall and nutrient return in five tree species in a common garden experiment. *Forest Ecology and Management*, **257**, 2133–2144.

Hanson, E. A. and Dawson, J. O. (1982) Effect of *Alnus glutinosa* on hybrid *Populus* height growth in a short-rotation intensively cultured plantation. *Forest Science*, **28**, 49–59.

Hansson, K. (2011) *Impact of tree species on carbon in forest soils*. Doctoral thesis #2011:71, Faculty of Natural Resources and Agricultural Sciences, Swedish University of Agricultural Sciences, Uppsala.

Hansson, K., Olsson, B. A., Olsson, M., Johansson, U. and Berggren Kleja, D. (2011) Differences in soil properties in adjacent stands of Scots pine, Norway spruce, and silver birch in SW Sweden. *Forest Ecology and Management*, **262**, 522–530.

Harpole, W. S., Ngai, J. T., Cleland, E. E., Seabloom, E. W., Borer, E. T., Bracken, M. E. S., Elser, J. J., Gruner, D. S., Hillebrand, H., Shurin, J. B. and Smith, J. E. (2011) Nutrient co-limitation of primary producer communities. *Ecology Letters*, **14**, 852–862.

Harrington, C. B. and DeBell, D. S. (1995) Effects of irrigation, spacing and fertilization on flowering and growth in young *Alnus rubra*. *Tree Physiology*, **15**, 427–432.

Harrington, T. B. and Edwards, M. B. (1996) Structure of mixed pine and hardwood stands 12 years after various methods and intensities of site preparation in the Georgia Piedmont. *Canadian Journal of Forest Research*, **26**, 1490–1500.

Harrison, R. B. and Johnson, D. W. (1992) Inorganic sulfate dynamics. In: D. W. Johnson and S. E. Lindberg (eds) *Atmospheric Deposition and Nutrient Cycling*, Springer-Verlag, New York, pp. 104–118.

Hart, S. C. (1999) Nitrogen transformations in fallen tree boles and mineral soil of an old-growth forest. *Ecology*, **80**, 1385–1394.

Hart, S. C. and Firestone, M. K. (1991) Forest floor–mineral soil interactions in the internal nitrogen cycle of an old-growth forest. *Biogeochemistry*, **12**, 103–127.

Hart, S. C., Binkley, D. and Campbell, R. (1986) Predicting loblolly pine current growth and growth response to fertilization. *Soil Science Society of America Journal*, **50**, 230–233.

Hart, S. C., Binkley, D. and Perry, D. A. (1997) Influence of red alder on soil nitrogen transformations in two conifer forests of contrasting productivity. *Soil Biology and Biochemistry*, **29**, 1111–1123.

Hart, S. C., Nason, G. E., Myrold, D. D. and Perry, D. A. (1994) Dynamics of gross nitrogen transformations in an old-growth forest: the carbon connection. *Ecology*, **75**, 880–891.

Hart, S. C., DeLuca, T. H., Newman, G. S., MacKenzie, D. and Boyle, S. I. (2005) Post-fire vegetative dynamics as drivers of microbial community structure and function in forest soils. *Forest Ecology and Management*, **220**, 166–184.

Harwood, C. and Jackson, W. (1975) Atmospheric losses of four plant nutrients during a forest fire. *Australian Forestry*, **38**, 92–99.

Haskins, K. E. and Gehring, C. A. (2004) Long-term effects of burning slash on plant communities and arbuscular mycorrhizae in a semi-arid woodland. *Journal of Applied Ecology*, **41**, 379–388.

Hättenschwiler, S., Aeschlimann, B., Coûteaux, M.-M., Roy, J. and Bonal, D. (2008) High variation in foliage and leaf litter chemistry among 45 tree species of a neotropical rainforest community. *New Phytologist*, **179**, 165–175.

He, Y., Chen, C., Xu, Z., Williams, D. and Xu, J. (2009) Assessing management impacts on soil organic matter quality in subtropical Australian forests using physical and chemical fractionation as well as ^{13}C NMR spectroscopy. *Soil Biology and Biochemistry*, **41**, 640–650.

Hedley, M. J., Stewart, J. W. B. and Chauhan, B. S. (1982) Changes in inorganic and organic soiil phosphorus fractions by cultivation practices and by laboratory incubations. *Soil Science Society of America Journal*, **46**, 970–976.

Heiberg, S. O. and Chandler, R. F. (1941) A revised nomenclature of forest humus layers for the northeastern United States. *Soil Science*, **52**, 87–99.

Helvey, J. D. and Patric, J. H. (1988) Research on interception loss. In: W. T. Swank and D. A. Crossley, Jr (eds) *Forest Hydrology and Ecology at Coweeta*, Springer-Verlag, New York, NY, pp. 130–137.

Hendricks, J. L. J. and Boring, L. R. (1999) N$_2$-fixation by native herbaceous legumes in burned pine systems of the southeastern United States. *Forest Ecology and Management*, **113**, 167–177.

Hendrix, P. (1995) *Earthworm Ecology and Biogeography in North America*. CRC Press, Boca Raton, FL.

Henry, E. (1908) *Les Sol Forestiers*. Paris.

Herbauts, J. (1982) Chemical and mineralogical properties of sandy and loamy-sochreous brown earths in relation to incipient podzolization in a brown earth-podzol evolutive sequence. *Journal of Soil Science*, **33**, 743–762.

Herbert, M. A. and Schönau, A. P. G. (1989) Fertilising commercial forest species in southern Africa: research progress and problems. *Southern Africa Forestry Journal*, **151**, 58–70.

Hesselman, H. (1917) *Studier over salteterbildningen i naturliga jordmaner och des betydelse i vaxtekologiskt avseende*. Medd Skogsforskosanst. Stockholm #13-14.

Hesselman, H. (1926) Studier over batrskogens humustache, dess egenskaper och beroende av skogsvarden. *Statens Skogsforsoksant Meddel*, **22**, 169–552.

Hewlett, J. D. (1972) *An analysis of forest water problems in Georgia*. Georgia Forest Research Council Report 30.

Heyward, F. and Barnette, R. M. (1934) *Effect of frequent fires on chemical composition of forest soils in the longleaf region*. Florida Agricultural Experiment Station Bulletin 265.

Hicks, W. T., Harmon, M. E. and Myrold, D. D. (2003) Substrate controls on nitrogen fixation and respiration in woody debris from the Pacific Northwest, USA. *Forest Ecology and Management*, **176**, 25–35.

Hilgard, E. W. (1906) *Soils*. Macmillan, New York.

Hillel, D. (1982) *Introduction to Soil Physics*. Academic Press, San Diego, CA.

Hills, G. A. (1952) *The classification and evaluation of sites for forestry*. Ontario Department of Lands and Forests Research Report 24.

Hobbie, S. E., Ogdahl, M., Chorover, J. and Chadwick, O. A. (2007) Tree species effects on soil organic matter dynamics: the role of soil cation composition. *Ecosystems*, **10**, 999–1018.

Hobbie, S. E., Oleksyn, J., Eissenstat, D. M. and Reich, P. B. (2010) Fine root decomposition rates do not mirror those of leaf litter among temperate tree species. *Oecologia*, **162**, 505–513.

Hockman, J. N. and Allen, H. L. (1990) Nutritional diagnoses in loblolly pine stands using a DRIS approach. In: S. P. Gessel, D. S. Lacate, G. F. Weetman and R. F. Powers (eds) *Sustained Productivity of Forest Soils*, University of British Columbia Faculty of Forestry Publication, Vancouver, pp. 500–514.

Hoffland, E., Kuyper, T. W., Wallander, H., Plassard, C., Gorbushina, A. A., Haselwandter, K., Holmström, S., Landeweert, R., Lundström, U. S., Rosling, A., Sen, R., Smits, M. M., vanHees, P. A. W. and vanBreemen, N. (2004) The role of fungi in weathering. *Frontiers in Ecology and Environment*, **2**, 258–264.

Hoffman, R. J. and Ferreira, R. F. (1976) *A reconnaissance of effects of a forest fire on water quality in Kings Canyon National Park*. U.S.D.I. Geological Survey Open File Report 76-497. 17 pp.

Hoffmann-Wellenhof, B., Lichtenegger, H. and Collins, J. (1994) *GPS: Theory and Practice*, 3rd edition, Springer-Verlag, New York, NY.

Högberg, M. N. and Högberg, P. (2002) Extramatrical ectomycorrhizal mycelium contributes one-third of microbial biomass and produces, together with associated roots, half the dissolved organic carbon in a forest soil. *New Phytologist*, **154**, 791–795.

Högberg, M. N., Bääth, E., Nordgren, A., Arnebrant, K. and Högberg, P. (2003) Contrasting effects of nitrogen availability on plant carbon supply to mycorrhizal fungi and saprotrophs – a hypothesis based on field observations in boreal forest. *New Phytologist*, **160**, 225–238.

Högberg, M. N., Briones, M. J. I., Keel, S. G., Metcalfe, D. B., Campbell, C., Midwood, A. J., Thornton, B., Hurry, V., Linder, S., Näsholm, T. and Högberg, P. (2010) Quantification of effects of season and nitrogen supply on tree below-ground carbon transfer to ecto-mycorrhizal fungi and other soil organisms in a boreal pine forest. *New Phytologist*, **187**, 485–493.

Högberg, P. (1997) ^{15}N natural abundance in soil–plant systems. *New Phytologist*, **137**, 179–203.

Högberg, P. and Read, D. J. (2006) Towards a more plant physiological perspective on soil ecology. *Trends in Ecology and Evolution*, **21**, 548–554.

Högberg, P., Fan, H. B., Quist, M., Binkley, D. and Tamm, C. O. (2006) Tree growth and soil acidification in response to 30 years of experimental nitrogen loading on boreal forest. *Global Change Biology*, **12**, 489–499.

Högberg, P., Nordgren, A., Buchmann, N., Taylor, A. F. S., Ekblad, A., Högberg, M. N., Nyberg, G., Ottosson, M. and Read, D. J. (2001) Large-scale forest girdling shows that current photosynthesis drives soil respiration. *Nature*, **411**, 789–792.

Holdridge, L. R. (1967) *Life Zone Ecology*. The Tropical Science Center, San Jose, Costa Rica.

Hole, F. D. (1981) Effects of animals on soils. *Geoderma*, **25**, 75–112.

Holmen, H. (1969) Afforestation of peatlands. *Skogso Lantdbrakad Tidshrit*, **108**, 216–235.

Holmen, H. (1971) Forest fertilization in Sweden. *Skogso Lantdbrakad*, **110**, 156–162.

Hoopmans, P. D., Flinn, W., Gear, P. W. and Tomkins, I. B. (1993) Sustained growth response of *Pinus radiate* on podzolised sands to site management practices. *Australian Forestry*, **56**, 27–33.

Hoosbeek, M. R. (1998) Incorporating scale into spatio-temporal variability: applications to soil quality and yield data. *Geoderma*, **85**, 113–131.

Hoover, M. D. (1949) Hydrologic characteristics of South Carolina piedmont forest soils. *Soil Science Society of America Proceedings*, **16**, 368–370.

Hopmans, P., Bauhus, J., Khanna, P. and Weston, C. (2005) Carbon and nitrogen in forest soils: Potential indicators for sustainable management of eucalypt forests in south-eastern Australia. *Forest Ecology and Management*, **220**, 75–87.

Hovmand, M. F. (1999) Cumulated deposition of strong acid and sulphur compounds to a spruce forest. *Forest Ecology and Management*, **114**, 19–30.

Hue, N. V., Craddock, G. R. and Adams, F. (1986) Effect of organic acids on aluminum toxicity in subsoils. *Soil Science Society of America Journal*, **50**, 28–34.

Huikari, O. (1973) *Results of fertilization experiments on peatlands drained for forestry*. Finnish Forest Research Institute Report 1.

Humphreys, F. R. and Craig, F. G. (1981) Effects of fire on soil chemical, structural, and hydrological properties. In: A. M. Gill, R. Groves and I. Nobel (eds) *Fire and the Australian Biota*, Australian Academy of Sciences Press, Canberra, pp. 177–200.

Humphreys, F. R. and Lambert, M. J. (1965) An examination of a forest site which has exhibited the ashbed effect. *Australian Journal of Soil Research*, **3**, 81–94.

Humphreys, F. R. and Pritchett, W. L. (1971) Phosphorus adsorption and movement in some sandy forest soils. *Soil Science Society of America Proceedings*, **35**, 495–500.

Hungerford, R. (1980) Micro environmental response to harvesting and residue management. In: *Environmental consequences of timber harvesting in Rocky Mountain coniferous forests*. USDA Forest Service General Technical Report INT-90, Ogden, UT, pp. 37–73.

Hunter, I. R. and Graham, J. D. (1983) Three-year response of *Pinus radiata* to several types and rates of phosphorus fertiliser on soils of contrasting phosphorus retention. *New Zealand Journal of Forestry Science*, **13**, 229–238.

Huston, M. and Smith, T. (1987) Plant succession: life history and competition. *American Naturalist*, **130**, 168–198.

Hyvönen, R., Persson, T., Andersson, S., Olsson, B., Ågren, G. and Linder, S. (2008) Impact of long-term nitrogen addition on carbon stocks in trees and soils in northern Europe. *Biogeochemistry*, **89**, 121–137.

Hüttl, R. F. and Schaaf, W. (eds) (1997) *Magnesium Deficiency in Forest Ecosystems*. Kluwer, Dordrecht.

Imaizumi, F., Sidle, R. C. and Kamei, R. (2008) Effects of forest harvesting on the occurrence of landslides and debris flows in steep terrain of central Japan. *Earth Surface Processes and Landforms*, **33**, 827–840.

Ingestad, T. (1981) Growth, nutrition, and nitrogen fixation in grey alder at varied rates of nitrogen addition. *Physiologia Plantarum*, **50**, 353–364.

Ingestad, T. (1982) Relative addition rate and external concentration: driving variables used in plant nutrition research. *Plant, Cell and Environment*, **5**, 443–453.

Isaac, L. A. and Hopkins, H. G. (1937) The forest soil of the Douglas fir region and changes wrought upon it by logging and slashburning. *Ecology*, **18**, 264–279.

Jackson, J. K. (1977) Irrigated plantations. In: *Savanna Afforestation in Africa*. Food and Agriculture Organization of the United Nations, Rome, pp. 277–285.

Jenkinson, D. S., Bradbury, N. J. and Coleman, K. (1994) How the Rothamsted classical experiments have been used to develop and test models for the turnover of carbon and soil nitrogen. In: R. A. Leigh and A. E. Johnston (eds) *Long-term Experiments in Agricultural and Ecological Studies*. CAB International, Oxon, pp. 117–138.

Jenny, H. (1961a) *E. W. Hilgard and the birth of modern soil science*. Collana Della Revista "Agrochemica," Pisa.

Jenny, H. (1961b) Derivation of state factor equations of soils and ecosystems. *Soil Science Society of America Proceedings*, **25**, 385–388.

Jenny, H. (1980) *The Soil Resource. Origin and behaviour*. Ecology Studies 37. Springer-Verlag, New York.

Jenny, H. (1984) My friend, the soil. A conversation with Hans Jenny, interview by K. Stuart. *Journal of Soil and Water Conservation*, **39**, 158–161.

Jiménez Esquilín, A., Stromberger, M. E., Massman, W. J., Frank, J. M. and Shepperd, W. D. (2007) Microbial community structure and activity in a Colorado Rocky Mountain forest soil scarred by slash pile burning. *Soil Biology and Biochemistry*, **39**, 1111–1120.

Jobbágy, E. G. and Jackson, R. B. (2000) The vertical distribution of soil organic carbon and its relation to climate and vegetation. *Ecological Applications*, **10**, 423–446.

Johnson, C. E., Johnson, A. H. and Siccama, T. G. (1991) Whole-tree clear-cutting effects on exchangeable cations and soil acidity. *Soil Science Society of America Journal*, **55**, 502–508.

Johnson, D. W. (1992) Nitrogen retention in forest soils. *Journal of Environmental Quality*, **21**, 1–12.

Johnson, D. W. and Lindberg, S. E. (1992) *Atmospheric Deposition and Forest Nutrient Cycling*. Springer-Verlag, New York.

Johnson, D. W. and Todd, D. (1985) Nitrogen availability and conservation in young yellow poplar and loblolly pine plantations fertilized with urea. *Agonomy Abstracts*, **77**, 220.

Johnson, D. W., Susfalk, R. B., Dahlgren, R. A. and Klopatek, J. M. (1998) Fire is more important than water for nitrogen fluxes in semi-arid forests. *Environmental Science and Policy*, **1**, 79–86.

Johnson, E. A. and Miyanishi, K. (2008) Testing the assumptions of chronosequences in succession. *Ecology Letters*, **11**, 419–431.

Johnson, K., Scatena, F. N. and Pan, Y. (2009a) Short- and long-term responses of total soil organic carbon to harvesting in a northern hardwood forest. *Forest Ecology and Management*, **259**, 1262–1267.

Johnson, K., Scatena, F. N., Johnson, A. H. and Pan, Y. (2009b) Controls on soil organic matter content within a northern hardwood forest. *Geoderma*, **148**, 346–356.

Jokela, E., Martin, T. A. and Vogel, J. G. (2010) Twenty-five years of intensive forest management with southern pines: important lessons learned. *Journal of Forestry*, **108**, 338–347.

Jones, D. L., Hodge, A. and Kuzyakov, Y. (2004) Plant and mycorrhizal regulation of rhizodeposion. *New Phytologist*, **163**, 459–480.

Jones, H. E., Högberg, P. and Ohlsson, H. (1994) Nutritional assessment of a forest fertilization experiment in northern Sweden by root bioassays. *Forest Ecology and Management*, **64**, 59–69.

Jurgensen, M. F., Larsen, M. J., Spano, S. D., Harvey, A. F. and Gale, M. R. (1984) Nitrogen fixation associated with increased wood decay in Douglas fir residue. *Forest Science*, **30**, 1038–1044.

Kadeba, O. and Boyle, J. R. (1978) Evaluation of phosphorus in forest soils: comparison of phosphorus uptake, extraction method and soil properties. *Plant and Soil*, **49**, 285–297.

Kaiser, C., Fuchleuger, L., Koranda, M., Gorfer, M., Stange, C. F., Kitzler, B., Rasche, F., Strauss, J., Sessitsch, A., Zechmeister-Boltenstern, S. and Richter, A. (2011) Plants control the seasonal dynamics of microbial N cycling in a beech forest soil by belowground C allocation. *Ecology*, **92**, 1036–1051.

Kane, E. S., Valentine, D. W., Turetsky, M. R. and McGuire, A. D. (2007) Topographic influences on wildfire consumption of soil organic carbon in interior Alaska: Implications for black carbon accumulation. *Journal of Geophysical Research*, **112**, G03017, doi:10.1029/2007JG000458.

Kaplan, E. D. (ed.) (1996) *Understanding GPS: Principles and Applications*. Artech House Publishers, Boston, MA.

Katznelson, H., Rowatt, J. W. and Peterson, E. A. (1962) The rhizosphere effect of mycorrhizal and nonmycorrhizal roots of yellow birch seedlings. *Canadian Journal of Botany*, **40**, 378–382.

Kaufman, C. M., Prichett, W. L. and Choate, R. E. (1977) *Growth of slash pine (Pinus elliottii Engel. var. elliottii) on drained flatwoods*. Florida Agricultural Experiment Station Bulletin 792.

Kaupenjohann, M. (1997) Tree nutrition. In: R. F. Hüttl and W. Schaaf (eds) *Magnesium Deficiency in Forest Ecosystems*, Kluwer Academic Press, Dordrecht, pp. 275–308.

Kaur, A., Chaudhary, A., Kaur, A., Choudhary, R. and Kaushik, R. (2005) Phospholipid fatty acid – A bioindicator of environment monitoring and assessment in soil ecosystems. *Current Science*, **89**, 1103–1112.

Kawana, A. and Haibara, H. (1983) Fertilization programs in Japan. In: R. Ballard and S. Gessel (eds) *IUFRO Symposium on Forest Site and Continuous Productivity*. USDA Forest Service General Technical Report PNW-163, Portland, OR.

Kay, B. D. (1997) Soil structure and organic carbon: a review. In: R. Lal, J. M. Kimble, R. F. Follett and B. A. Stewart (eds) *Soil Processes and the Carbon Cycle*. CRC Press, Boca Raton, FL, pp. 169–197.

Kaye, J. and Hart, S. C. (1997) Competition for nitrogen between plants and soil microorganisms. *Tree*, **12**, 139–143.

Kaye, J. P., Binkley, D. and Rhoades, C. (2003) Stable soil nitrogen accumulation and flexible organic matter stoichiometry during primary floodplain succession. *Biogeochemistry*, **63**, 1–22.

Kaye, J. P., Binkley, D., Zou, X. and Parrotta, J. A. (2002) Non-labile soil ^{15}Nitrogen retention beneath three tree species in a tropical plantation. *Soil Science Society of America Journal*, **66**, 612–619.

Kaye, J., Resh, S., Kaye, M. and Chimner, R. (2000) Nutrient and carbon dynamics in a replacement series of Eucalyptus and Albizia trees. *Ecology*, **81**, 3267–3273.

Keith, H. (1997) Nutrient cycling in eucalypt ecosystems. In: J. Williams and J. Woinarski (eds) *Eucalypt Ecology*, Cambridge University Press, Cambridge, pp. 197–226.

Kellman, M. and Kading, M. (1992) Facilitation of tree seedling establishment in a sand dune succession. *Journal of Vegetation Science*, **3**, 679–688.

Kellogg, C. E. (1951) Soils and land classification. *Journal of Farm Economics*, **33**, 499–513.

Kessell, S. L. (1927) Soil organisms: the dependence of certain pine species on a biological soil factor. *Empire Forestry Journal*, **6**, 70–74.

Kevan, D. K. McE. (1962) *Soil Animals*. Witherby, London.

Khanna, P. K. and Raison, R. J. (1986) Effect of fire intensity on solution chemistry of surface soil under a *Eucalyptus pauciflora* forest. *Australian Journal of Soil Research*, **24**, 423–434.

Khlosi, M., Cornelis, W. M., Douaik, A., van Genuchten, M. T. and Gabriels, D. (2008) Performance Evaluation of Models That Describe the Soil Water Retention Curve between Saturation and Oven Dryness. *Vadose Zone Journal*, **7**, 87–96.

Kieft, T. L., Soroker, E. and Firestone, M. K. (1987) Microbial biomass response to a rapid increase in water potential when dry soil is wetted. *Soil Biology and Biochemistry*, **19**, 119–126.

Kiese, R, Hewett, B., Graham, A. and Butterbach-Bahl, K. (2003) Seasonal variability of N_2O-emissions and CH_4-uptake from/by a tropical rainforest soil of Queensland. *Australia. Global Biogeochemical Cycles*, **17**, 1043–1057.

Kirchner, J. W., Finkel, R., Riebe, C. S., Granger, D. E., Clayton, J. L., King, J. G. and Megahan, W. F. (2001) Mountain erosion over 10 yr, 10 k.y., and 10 m.y. time scales. *Geology*, **29**, 591–594.

Klaminder, J., Lucas, R. W., Futter, M. N., Bishop, K. H., Köhler, S. J., Egnell, G. and Laudon, H. (2011) Silicate mineral weathering rate estimates: Are they precise enough to be useful when predicting the recovery of nutrient pools after harvesting? *Forest Ecology and Management*, **261**, 1–9.

Klawitter, R. A. (1970a) *Does bedding promote pine survival and growth on ditched wet sands?* USDA Forest Service Research Note NE-109.

Klawitter, R. A. (1970b) Water regulation on forest land. *Journal of Forestry*, **68**, 338–342.

Klein, C. and Dutrow, B. (2008) *Manual of Mineral Science*. John Wiley & Sons, Inc., New York.

Klinge, H. (1976) Nahrstoffe, Wasser, und Durchwurzelung von Podsolen und Latosolen unter tropischem Regenwald bei Manaus/Amazonien. *Biogeographica*, **7**, 45–58.

Klinka, K. and Carter, R. E. (1990) Relationships between site index and synoptic environmental factors in immature coastal Douglas fir stands. *Forest Science*, **36**, 815–830.

Klinka, K., Green, R. N., Trowbridge, R. L. and Lowe, L. E. (1981) *Taxonomic classification of humus forms in ecosystems of British Columbia*. BC Ministry of Forests Land Management Report #8, Victoria.

Klotzbücher, T., Kaiser, K., Guggenberger, G., Gatzek, C. and Kalbitz, K. (2011) A new conceptual model for the fate of lignin in decomposing plant litter. *Ecology*, **92**, 1052–1062.

Knicker, H. (2006) How does fire affect the nature and stability of soil organic nitrogen and carbon? A review. Biogeochemistry, **85**, 91–118.

Knight, D. H., Fahey, T. J. and Running, S. W. (1985) Water and nutrient outflow from contrasting lodgepole pine forests in Wyoming. *Ecological Monographs*, **55**, 29–48.

Knight, H. (1966) Loss of nitrogen from the forest floor by burning. *Forestry Chronicle*, **42**, 149–152.

Knight, P. J. and Nicholas, I. D. (1996) Eucalypt nutrition: New Zealand experience. In: P. M. Attiwill and M. A. Adams (eds) *Nutrition of Eucalypts*. CSIRO, Collingwood, Australia, pp. 275–302.

Korb, J., Johnson, N. C. and Covington, W. W. (2004) Slash pile burning effects on soil biotic and chemical properties and plant establishment: recommendations for amelioration. *Restoration Ecology*, **12**, 52–62.

Kormanik, P. P., Bryan, W. C. and Schultz, R. C. (1977) The role of mycorrhizae in plant growth and development. In: H. M. Vines (ed.) *Physiology of Root-Microorganisms Associations*. Proceedings of the Symposium of the Southern Section of the American Society of Plant Physiologists, Atlanta, pp. 1–10.

Koven, C. D., Ringeval, B., Friedlingstein, P., Ciais, P., Cadule, P., Khvorostyanov, D., Krinner, G. and Tarnocai, C. (2011) Permafrost carbon-climate feedbacks accelerate global warming. PNAS, **36**, 14769–14774.

Kozlowski, T. T. (1968) Soil water and tree growth. In: N. E. Linnartz (ed.) *17th Annual Forest Symposium: The Ecology of Southern Forests*. Louisiana State University Press, Baton Rouge, pp. 30–57.

Kraemer, J. F. and Hermann, R. K. (1979) Broadcast burning: 25-year effects on forest soils in the western flanks of the Cascade Mountains. *Forest Science*, **25**, 427–439.

Krammes, J. S. and DeBano, L. F. (1965) Soil wettability: a neglected factor in watershed management. *Water Resources Research*, **1**, 283–286.

Krasilnikov, P., Carré, F. and Montanarella, L. (2008) *Soil Geography and Geostatistics*. European Commission, Luxembourg.

Kreutzer, K. (1995) Effects of forest liming on soil processes. *Plant and Soil*, **168–169**, 447–470.

Kryszowska, A. J., Blaylock, M. J., Vance, G. F. and David, M. B. (1996) Ion-chromatographic analysis of low molecular weight organic acids in spodosol forest floor solutions. *Soil Science Society of America Journal*, **60**, 1565–1571.

Krzyszowska-Waitkus, A., Vance, G. F. and Preston, C. M. (2006) Influence of coarse wood and fine litter on forest organic matter composition. *Canadian Journal of Soil Science*, **86**, 35–46.

Kuhry, P., Nicholson, B. J., Gignae, L. D., Vitt, D. H. and Bayley, S. E. (1993) Development of sphagnum-dominated peatlands in boreal continental Canada. *Canadian Journal of Botany*, **71**, 10–22.

Kurz-Besson, C., Coûteaux, M.-M., Thiéry, J. M., Berg, B. and Remacle, J. (2005) A comparison of litterbag and direct observation methods of Scots pine needle decomposition measurement. *Soil Biology and Biochemistry*, **37**, 2315–2318.

Lafon, C. W., Huston, M. A. and Horn, S. P. (2000) Effects of agricultural soil loss on forest succession rates and tree diversity in east Tennessee. *Oikos*, **90**, 431–441.

Laganière, J., Angers, D. A. and Paré, D. (2010) Carbon accumulation in agricultural soils after afforestation: a meta-analysis. *Global Change Biology*, **15**, 439–453.

Laiho, R. and Prescott, C. E. (1999) The contribution of coarse woody debris to carbon, nitrogen and phosphorus cycles in three Rocky Mountain coniferous forests. *Canadian Journal of Forestry Research*, **29**, 1592–1603.

Laine, J., Minkkinen, K., Sinisalo, J., Savolainen, I. and Martikainen, P. J. (1997) Greenhouse impact of a mire after drainage for forestry. In: C. C. Trettin, M. F. Jurgensen, D. F. Grigal, M. R. Gayle and J. J. Jeglem (eds) *Northern Forested Wetlands: Ecology and management*, Lewis Publishers, Boca Raton, FL, pp. 437–447.

Lal, R., Kimble, J. M., Follett, R. F. and Stewart, B. A. (1998a) *Soil Processes and the Carbon Cycle*. CRC Press, Boca Raton, FL.

Lal, R., Kimble, J. M., Follett, R. F. and Stewart, B. A. (1998b) *Management of Carbon Sequestration in Soil*. CRC Press, Boca Raton, FL.

Landsberg, J. D. and Cochran, P. H. (1980) Prescribed burning effects on foliar nitrogen content in ponderosa pine. In: *Proceedings of the Sixth Conference on Fire and Forest Meteorology*, April 22–24. Society of American Foresters, Washington, pp. 209–213.

Larsen, I. J., MacDonald, L. H., Brown, E., Rough, D., Welsh, M. J., Pietraszek, J. H., Libohova, Z., de Dios Benavides-Solorio, J. and Schaffrath, K. (2009) Causes of post-fire runoff and erosion: water repellency, cover, or soil sealing? *Soil Science Society of America Journal*, **73**, 1393–1407.

Larson, E. R., Kipfmueller, K. F., Hale, C. M., Frelich, L. E. and Reich, P. B. (2010) Tree rings detect earthworm invasions and their effects in northern Hardwood forests. *Biological Invasions*, **12**, 1053–1066.

Lavelle, P. (1997) Faunal activities and soil processes: adaptive strategies that determine ecosystem function. *Advances in Ecological Research*, **27**, 93–133.

Leaf, A. L., Leonard, R. E., Berglund, J. V., Eschner, A. R., Cochran, P. H., Hart, J. B., Marion, G. M. and Cunningham, R. A. (1970) Growth and development of *Pinus resinosa* plantations subjected to irrigation-fertilization treatments. In: C. T. Youngberg and C. B. Davey (eds) *Tree Growth and Forest Soils*, Oregon State University Press, Corvallis, OR.

LeBauer, D. S. and Treseder, K. K. (2008) Nitrogen limitation of net primary productivity in terrestrial ecosystems is globally distributed. *Ecology*, **89**, 371–379.

Lehotsky, K. (1972) Sand dune fixation in Michigan – thirty years later. *Journal of Forestry*, **70**, 155–160.

Leick, A. (1995) *GPS Satellite Surveying*, 2nd edition. John Wiley & Sons, Inc., New York.

Levy, G. (1972) Premiers resultats concernant duex experiences d'assainissment du sol sur plantations de resineux. *Annual Science of Forestry*, **29**, 427–450.

Levy-Varon, J. H., Schuster, W. S. F. and Griffin, K. L. (2011) The autotrophic contribution to soil respiration in a northern temperate deciduous forest and its response to stand disturbance. *Oecologia*, DOI 10.1007/s00442-011-2182-y.

Lewis, W. M., Jr. (1974) Effects of fire on nutrient movement in a South Carolina pine forest. *Ecology*, **55**, 1120–1127.

Likens, G., Bormann, F. H., Pierce, R. S., Eaton, J. S. and Johnson, N. (1977) *Biogeochemistry of a Forested Ecosystem*. Springer–Verlag, New York. 146 pp.

Lin, H., Wheeler, D., Bell, J. and Wilding, L. (2005) Assessment of soil spatial variability at multiple scales. *Ecological Modelling*, **182**, 271–290.

Lindahl, B., Ihrmark, K., Boberg, J., Trumbore, S. E., Högberg, P., Stenlid, J. and Finlay, R. D. (2007) Spatial separation of litter decomposition and mycorrhizal nitrogen uptake in a boreal forest. *New Phytologist*, **173**, 611–620.

Linder, S. (1995) Foliar analysis for detecting and correcting nutrient imbalances in Norway spruce. *Ecological Bulletins*, **44**, 178–190.

Lindsay, W. (1979) *Chemical Equilbria in Soils*. John Wiley & Sons, Inc., New York. 449 pp.

Lindsay, W. and Vlek, P. L. G. (1977) Phosphate minerals. In: J. B. Dixon and S. B. Weed (eds) *Minerals in Soil Environments*, Soil Science Society of America, Madison, WI, pp. 639–672.

Little, S. and Ohmann, J. (1988) Estimating nitrogen lost from forest floor during prescribed fires in Douglas fir/western hemlock clearcuts. *Forest Science*, **34**, 152–164.

Litton, C. M., Raich, J. W. and Ryan, M. G. (2007) Carbon allocation in forest ecosystems. *Global Change Biology*, **13**, 2089–2109.

Löfvenius, M. O. (1993) *Temperature and radiation regimes in pine shelterwood and clear-cut areas*. PhD thesis, Swedish University of Agricultural Sciences, Umeå.

Lorio, P. L., Howe, V. K. and Martin, C. N. (1972) Loblolly pine rooting varies with microrelief on wet sites. *Ecology*, **53**, 1134–1140.

Lovett, G. M., Weathers, K. C., Arthur, M. A. and Schultz, J. C. (2004) Nitrogen cycling in a northern hardwood forest: Do species matter? *Biogeochemistry*, **67**, 289–308.

Lugo, A. E., Sanchez, M. J. and Brown, S. (1986) Land use and organic carbon content of some subtropical soils. *Plant and Soil*, **96**, 185–196.

Lundgren, B. (1978) *Soil conditions and nutrient cycling under natural and plantation forests in Tanzanian highlands*. Swedish University of Agricultural Science, Uppsala, Sweden.

Lundström, U. S., and van Breemen, N. (2000) The Podzolization Process. A Review. *Geoderma*, **94**, 91–107.

Lundström, U. S., van Breemen, N., Bain, D. C., van Hees, P. A. W., Giesler, R., Gustafsson, J. P., Ilvesniemi, H., Karltung, E., Melkerud, P.-A., Olsson, M., Riise, G., Wahlberg, O., Bergelin, A., Bishop, K., Finlay, R., Jongmans, A. G., Magnusson, T., Mannerkoski, H., Nordgren, A., Nyberg, L., Starr, M. and Tau Strand, L. (2000) Advances in understanding the podsolization process resulting from a multidisciplinary study of three coniferous forest soils in the Nordic Countries. *Geoderma*, **94**, 335–353.

Lutz, H. J. and Chandler, R. F. (1946) *Forest Soils*. John Wiley & Sons, New York.

Lyford, W. H., Jr (1943) The palatability of fallen forest tree leaves to millipeds. *Ecology*, **24**, 252–261.

Lyford, W. H. (1952) Characteristics of some podzolic soils of the northeastern United States. *Soil Science Society of America Proceedings*, **16**, 231–234.

Lyford, W. H. (1963) *Importance of ants to brown podzolic soil genesis in New England*. Harvard Forestry Paper 7.

Lyford, W. H. and MacLean, D. W. (1966) *Mound and pit microrelief in relation to soil disturbance and tree distribution in New Brunswick, Canada*. Harvard Forestry Paper 15.

Lyford, W. H. and Wilson, B. F. (1966) *Controlled growth of forest tree roots: techniques and applications*. Harvard Forestry Paper 16.

Lynch, J. M. (1990) *The Rhizosphere*. Wiley (Interscience), Chichester, 458 pp.

MacCarthy, P., Clapp, C. E., Malcolm, R. L. and Bloom, P. R. (1990) *Humic Substances in Soil and Crop Sciences: Selected readings*. American Society of Agronomy, Madison, WI.

MacDonald, L. H. and Larsen, I. J. (2009) Effects of forest fires and post-fire rehabilitation: a Colorado, USA case study. In: A. Cerda and P. R. Robichaud (eds) *Fire Effects on Soils and Restoration Strategies*. Science Publishers, Enfield, New Hampshire, pp. 423–452.

MacFarland, J. W., Ruess, R. W., Kielland, K., Pregitzer, K., Hendrick, R. and Allen, M. (2010) Cross-ecosystem comparisons of in situ plant uptake of amino acid-N and NH_4^+. *Ecosystems*, **11**, 177–193.

MacFarlane, I. C. (1969) *Maskey Engineering Handbook*. University of Toronto Press, Toronto.

MacKenzie, M. D., McIntire, E. J. B., Quideau, S. A. and Graham, R. C. (2008) Charcoal distribution affects carbon and nitrogen contents in forest soils. *Soil Science Society of America Journal*, **72**, 1774–1785.

MacKinley, D. C., Ryan, M. G., Birdsey, R. A., Giardina, C. P., Harmon, M. E., Heath, L. S., Houghton, R. A., Jackson, R. B., Morrison, J. F., Murray, B. C., Pataki, D. E. and Skog, K. E. (2011) A synthesis of current knowledge on forests and carbon storage in the United States. *Ecological Applications*, **21**, 1902–1924.

MacLean, A. J. (1961) Potassium-supplying power of some Canadian soils. *Canadian Journal of Soil Science*, **41**, 196–206.

Mader, D. L. (1964) Soil variability – a serious problem in soil-site studies in the Northeast. *Soil Science Society of America Journal*, **27**, 707–709.

Magnusson, T. (1992) *Temporal and spatial variation of the soil atmosphere in forest soils of northern Sweden*. PhD thesis, Swedish University of Agricultural Sciences, Umeå.

Mailly, D., Ndiaye, P., Margolis, H. A. and Pineau, M. (1994) Dune stabilization and reforestation with filao (*Casuarina equisetifolia*) in the northern coastal zone of Senegal. *Forestry Chronicle*, **70**, 282–290.

Makkonen-Spiecker, K. and Spiecker, H. (1997) Influence of magnesium supply on tree growth. In: R. F. Hüttl and W. Schaaf (eds) *Magnesium Deficiency in Forest Ecosystems*, Kluwer Academic Press, Dordrecht, pp. 215–254.

Malm, D. and Moller, G. (1975) Skillnader i volymtillvaxtokning efter godsling med urea resp ammoniumnitrat. In: *Foreningen Skogstradsforadling, 1974 arsbok*. Institutet for Skogsfofbattring, pp. 46–63.

Mann, W. F. and McGilvray, J. M. (1974) *Response of slash pine to bedding and phosphorus application in southeastern flatwoods*. United States Department of Agriculture Forest Service Research Paper SO-99.

Mant, R., Jones, D., Reynolds, B., Ormerod, S. and Pullin, A. S. (2011) *What is the impact of "liming" of streams and rivers on the abundance and diversity of fish and invertebrate populations?* Center for Environmental Evidence Review #09-015, available online: http://www.environmentalevidence.org/SR76.html

Marchand, D. E. (1971) *Chemical weathering, soil development, and geochemical fractionation in a part of the White Mountains, Mono and Inyo Counties, California*. Geological Survey Professional Papers (US) #352J.

Margesin, R. (2009) *Permafrost Soils*. Springer-Verlag, Berlin Heidelberg.

Marichal, R., Mathieu, J., Couteaux, M.-M., Mora, P., Roy, J. and Lavelle, P. (2011) Earthworm and microbe response to litter and soils of tropical forest plantations with contrasting C:N:P stoichiometric ratios. *Soil Biology and Biochemistry*, **43**, 1528–1535.

Marschner, H. (1995) *Mineral Nutrition of Higher Plants*. Academic Press, London. 889 pp.

Marx, D. H. (1972) Ectomycorrhizae as biological deterrents to pathogenic root infections. *Annual Review of Phytopathology*, **10**, 429–454.

Marx, D. H. (1977) The role of mycorrhizae in forest production. In: *TAPPI conference papers*, Atlanta, GA, pp. 151–161.

Marx, D. H., Bryan, W. C. and Davey, C. B. (1970) Influence of temperature on aseptic synthesis of ectomycorrhizae by *Thelephora terrestris* and *Pisolithus tinctorius* on loblolly pine. *Forest Science*, **16**, 424–431.

Mast, A. (1989) *A laboratory and field study of chemical weathering with special reference to acid deposition*. PhD thesis, University of Wyoming, 174 pp.

Matthews, S. W. (1973) This changing earth. *National Geographic Magazine*, **143**, 1–37.

Matziris, D. and Zobel, B. (1976) Effects of fertilization on growth and quality characteristics of loblolly pine. *Forest Ecology and Management*, **1**, 21–30.

May, B., Smethurst, P., Carlyle, C., Mendham, D., Bruce, J. and Baillie, C. (2009) *Review of fertiliser use in Australian forestry*. Final report Project No: RC072-0708, Forest and Wood Products Australia, Melbourne.

May, J. T., Arkss, C. C. and Perkins, H. F. (1969) Establishment of grass and tree vegetation on spoil from kaolin strip mining. In: *Ecology and Revegetation of Devastated Land*, Vol. 2. Gordon and Breach, New York, NY, pp. 137–147.

McClaugherty, C. A., Aber, J. D. and Melillo, J. M. (1982) The role of fine roots in the organic matter and nitrogen budgets of two forested ecosystems. *Ecology*, **63**, 148l–l490.

McColl, J. G. and Gressel, N. (1995) Forest soil organic matter: characterization and modern methods of analysis. In: W. W. McFee and J. M. Kelly (eds) *Carbon Forms and Functions in Forest Soils*. Soil Science Society of America, Madison, WI, pp. 13–32.

McColl, J. G. and Powers, R. F. (1998) Decomposition of small diameter woody debris of red fir determined by nuclear magnetic resonance. *Communications in Soil Science and Plant Analysis*, **29**, 2691–2704.

McFarlane, K. J., Schoenholtz, S. H. and Powers, R. F. (2010) Soil organic matter stability in intensively managed ponderosa pine stands in California. *Soil Science Society of America Journal*, **74**, 979–992.

McFee, W. W. and Stone, E. L. (1965) Quantity, distribution, and variability of organic matter and nutrients in a forest podzol in New York. *Soil Science Society of America Proceedings*, **29**, 432–436.

McHarg, I. L. (1969) *Design with Nature*. Natural History Press, Garden City, New York, NY.

McNabb, D. H., Gaweda, F. and Froehlich, H. A. (1989) Infiltration, water repellency, and soil moisture content after broadcast burning a forest site in southwest Oregon. *Journal of Soil and Water Conservation*, **44**, 87–90.

McQuilkin, W. E. (1935) Root development of pitch pine, with some comparative observations on shortleaf pine. *Journal of Agricultural Research*, **51**, 983–1016.

Mead, D. J. and Pritchett, W. L. (1971) A comparison of tree responses to fertilizers in field and pot experiments. *Soil Science Society of America Proceedings*, **35**, 346–349.

Mehlich, A. (1953) *Determination of P, K, Ca, Mg, and NH_4*. Soil Test Division Mimeo, North Carolina Department of Agriculture, Raleigh, NC. www.ncagr.com/agronomi/pdffiles/mehlich53.pd

Mehlich, A. (1984) Mehlich 3 soil test extractant: A modification of Mehlich 2 extractant. *Communication in Soil Science and Plant Analysis*, **15**, 1409–1416.

Mehrotra, V. S. (1998) Arbuscular mycorrhizal associations of plants colonizing coal mine spoil in India. *Journal of Agricultural Science*, **130**, 125–133.

Melillo, J. M., Aber, J. D., Linkins, A. E., Ricca, A., Fry, B. and Nadelhoffer, K. J. (1989) Carbon and nitrogen dynamics along the decay continuum: plant litter to soil organic matter. In: M. Clarholm and L. Bergström (eds) *Ecology of Arable Land*, Kluwer, Amsterdam, pp. 53–62.

Melillo, J. M., Steudler, P. A., Aber, J. D., Newkirk, K., Lux, H., Bowles, F. P., Catricala, C., Magill, A., Ahrens, T. and Morrisseau, S. (2002) Soil warming and carbon-cycle feedbacks to the climate system. *Science*, **298**, 2173–2176.

Melin, J., Nommik, H., Lohm, U. and Flower-Ellis, J. (1983) Fertilizer nitrogen budget in a Scots pine ecosystem attained by using root-isolated plots and ^{15}N technique. *Plant and Soil*, **74**, 249–263.

Menyailo, O. V., Hungate, B. A. and Zech, W. (2002) Tree species mediated soil chemical changes in a Siberian artificial afforestation experiment. *Plant and Soil*, **242**, 171–182.

Merriam, C. H. (1898) *Life zones and crop zones of the United States*. USDA Biological Survey Bulletin 10, Washington, D.C.

Merriam, C. H. and Steineger, L. (1890) *Results of a biological survey of the San Francisco mountain region and the desert of the Little Colorado, Arizona*. North American Fauna Report 3. U.S. Department of Agriculture, Washington, D.C.

Messina, M. G. and Conner, W. H. (1998) *Southern Forested Wetlands: Ecology and management*. Lewis Publishers, Boca Raton, FL.

Meyer, G. A. and Pierce, J. L. (2003) Climatic controls on fire-induced sediment pulses in Yellowstone National Park and central Idaho: a long-term perspective. *Forest Ecology and Management*, **178**, 89–104.

Mika, P. G., Moore, J. A., Brockley, R. P. and Powers, R. F. (1992) Fertilization response by interior forests: when, where, and how much? In: H. N. Chappell, G. F. Weetman and R. E. Miller (eds) *Forest Fertilization: Sustaining and improving nutrition and growth of western forests*. College of Forest Resources, University of Washington, Seattle, pp. 127–142.

Mikola, P. (1973) Application of mycorrhizal symbiosis in forestry practice. In: G. C. Marks and T. T. Kozlowski (eds) *Ectomycorrhizae: Their ecology and physiology*. Academic Press, New York, pp. 383–411.

Miller, H. G. (1981) Forest fertilization: some guiding concepts. *Forestry*, **54**, 157–167.

Miller, R. E. and Bigley, R. E. (1990) Effects of burning Douglas fir logging slash on stand development and site productivity. In: S. P. Gessel, D. S. Lacate, G. F. Weetman and R. F. Powers (eds) *Sustained Productivity of Forest Soils*, University of British Columbia Faculty of Forestry Publication, Vancouver, pp. 362–376.

Minderman, G. (1968) Addition, decomposition and accumulation of organic matter in forests. *Journal of Ecology*, **56**, 355–362.

Minkkinen, K. and Laine, J. (1998) Long-term effect of forest drainage on the peat carbon stores of pine mires in Finland. *Canadian Journal of Forest Research*, **28**, 1267–1275.

Minkkinen, K., Laine, J., Nykanen, H. and Martikainen, P. J. (1997) Importance of drainage ditches in emissions of methane from mires drained for forestry. *Canadian Journal of Forest Research*, **27**, 949–952.

Montes, R. A. and Christensen, N. L. (1979) Nitrification and succession in the Piedmont of North Carolina. *Forest Science*, **25**, 287–297.

Moore, D. G. (1975) *Effects of forest fertilization with urea on stream quality – Quilcene Ranger District, Washington*. USDA Forest Service Research Note PNW-241. 9 pp.

Moore, T. R., Trofymow, J. A., Prescott, C. E. and Titus, B. D. (2010) Nature and nurture in the dynamics of C, N, and P during litter decomposition in Canadian forests. *Plant and Soil*, **339**, 163–175.

Morby, F. E. (1984) Nursery site selection, layout and development. In: M. L. Duryea and T. D. Landis (eds) *Forest Nursery Manual*, Martinus Nijhoff/Dr W. Junk Publishers, Boston, MA, pp. 9–16.

Morin, G. and Calas, G. (2006) Arsenic in soils, mine tailings, and former industrial sites. *Elements*, **2**, 97–101.

Moro, L. and Gonçalves, J. L. M. (1995) Efeito da "cinza" de biomassa florestal sobre a produtividade de povoamentos puros de *Eucalyptus grandis* e avalião financeira. IPEF, *Piracicaba*, **48/49**, 18–27.

Morris, L. A., Pritchett, W. L. and Swindell, B. F. (1983) Displacement of nutrients into windrows during site preparation of a flatwoods forest. *Soil Science Society of America Journal*, **47**, 591–594.

Motavalli, P. P., Palm, C. A., Elliott, E. T., Frey, S. D. and Smithson, P. C. (1995) Nitrogen mineralization in humid tropical forest soils: mineralogy, texture, and measured nitrogen fractions. *Soil Science Society of America Journal*, **59**, 1168–1175.

Mroz, G. D., Jurgensen, M. F., Harvey, A. E. and Larsen, M. J. (1980) Effects of fire on nitrogen in forest floor horizons. *Soil Science Society of America Journal*, **44**, 395–400.

Muller, P. E. (1879) Studier over skovjord som bidrag til skodyrkiningens theori. *Tidsskrift for Skovbrug*, **3**, 1–124.

Munsell (1990) *Munsell soil color charts*. Munsell Color, 2441 North Calvert Street, Baltimore, MD.

Nachimuthu, G., King, K., Kristiansen, P., Lockwood, P. and Guppy, C. (2007) Comparison of methods for measuring soil microbial activity using cotton strips and a respirometer. *Journal of Microbiological Methods*, **69**, 322–329.

Nadelhoffer, K. J., Aber, J. D. and Melillo, J. M. (1983) Leaf-litter production and soil organic matter dynamics along a nitrogen-availability gradient in southern Wisconsin (U.S.A.). *Canadian Journal of Forest Research*, **13**, 12–21.

Nadelhoffer, K. J., Emmett, B. A., Gundersen, P., Kjrnaas, O. J., Koopmans, C. J., Schleppi, P., Tietema, A. and Wright, R. F. (1999) Nitrogen deposition makes a minor contribution to carbon sequestration in temperate forests. *Nature*, **398**, 145–148.

Nadkarni, N. M., Schaefer, D., Matelson, T. J. and Solano, R. (2002) Comparison of arboreal and terrestrial soil characteristics in a lower montane forest, Monteverde, Costa Rica. *Pedobiologia*, **46**, 24–33.

Nambiar, E. K. S. (1985) Increasing forest productivity through genetic improvement of nutritional characteristics. In: R. Ballard (ed.) *Forest Potential Productivity and Value*, Weyerhaeuser Science Symposium 4, Weyerhaeuser Co., Tacoma, WA, pp. 191–215.

Nambiar, E. K. S. and Brown, A. G. (1997) *Management of Soil, Nutrients and Water in Tropical Plantation Forests*. Australian Centre for International Agricultural Research Monograph 43, Canberra.

Nambiar, E. K. S. and Fife, D. N. (1991) Nutrient retranslocation in temperate conifers. *Tree Physiology*, **9**, 185–207.

Nambiar, E. K. S. and Sands, R. (1992) Effects of compaction and simulated root channels in the subsoil on root development, water-uptake and growth of radiate pine. *Tree Physiology*, **10**, 297–306.

Nannipieri, P., Ascher, J., Ceccherini, M. T., Landi, L., Pietramellara, G. and Renella, G. (2003) Microbial diversity and soil functions. *European Journal of Soil Science*, **54**, 655–670.

Näsholm, T. (1994) Removal of nitrogen during needles senescence in Scots pine (*Pinus sylvestris* L.). *Oecologia*, **99**, 290–296.

National Soil Survey Handbook, Part 622. (2011) http://soils.usda.gov/technical/handbook/contents/part622.html

Nave, L. E., Vance, E. D., Swanston, C. W. and Curtis, P. S. (2010) Harvest impacts on soil carbon storage in temperate forests. *Forest Ecology and Management*, **259**, 857–866.

Neal, J., Wright, E. and Bollen, W. B. (1965) *Burning Douglas fir slash: physical, chemical, and microbial effects in the soil*. Oregon State University, Forest Research Laboratory, Corvallis.

Negi, J. D. S. (1984) *Biological productivity and cycling of nutrients in managed and manmade ecosystems*. PhD thesis, Garhwal University, Srinagar (U.P.), India. 161 pp.

Newman, H. C. and Schmidt, W. C. (1980) Silviculture and residue treatments affect water use by a larch/fir forest. In: *Environmental Consequences of Timber Harvesting in Rocky Mountain Coniferous Forests*. USDA Forest Service General Technical Report INT-90, Ogden, UT, pp. 75–110.

Nicholas, J. R. J. and Hinkel, K. M. (1996) Cocurrant permafrost aggradation and degradation induced by forest clearing in central Alaska, USA. *Arctic and Alpine Research*, **28**, 294–299.

Nihlgård, B. (1971) Pedological influences of spruce planted on former beech forest soils in Scania, south Sweden. *Oikos*, **22**, 302–314.

Nihlgård, B. and Popovic, B. (1984) *Effekter av olika kalkningsmedel I skogsmark – en litteratur'versikt*. Statens Naturvardsverk PM 1851, Solna.

Nocentini, C., Certini, G., Knicker, H., Francioso, O. and Rumpel, C. (2010) Nature and reactivity of charcoal produced and added to soil during wildfire are particle-size dependent. *Organic Geochemistry*, **41**, 682–689.

Nohrstedt, H.-Ö. (1990) Effects of repeated nitrogen fertilization with different doses on soil properties in a *Pinus sylvestris* stand. *Scandinavian Journal of Forest Research*, **5**, 3–15.

Nohrstedt, H.-Ö, Ring, E., Klemedtsson, L. and Nilsson, Ö. (1994) Nitrogen losses and soil water acidity after clearfelling of fertilized experimental plots in a *Pinus sylvestris* stand. *Forest Ecology and Management*, **66**, 69–86.

North Carolina State Forest Nutrition Cooperative (1997) *Ten-year growth and foliar responses of mid-rotation loblolly pine plantations to nitrogen and phosphorus fertilization*. North Carolina State Forest Nutrition Cooperative Report #39, Raleigh.

Nykvist, N. and Rosén, K. (1985) Effect of clear-felling and slash removal on the acidity of northern coniferous soils. *Forest Ecology and Management*, **11**, 157–169.

Oades, J. M., Gillman, G. P. and Uehara, G. (1989) Interactions of soil organic matter and variable-charge clays. In: D. Coleman, J. M. Oades and G. Uehara (eds) *Dynamics of Soil Organic Matter in Tropical Ecosystems*, University of Hawaii, Honolulu, pp. 69–95.

O'Connell, A. M. and Sankaran, K. V. (1997) Organic matter accretion, decomposition and mineralisation. In: E. K. S. Nambiar and A. G. Brown (eds) *Management of Soil, Nutrients and Water in Tropical Plantation Forests*. ACIAR Monograph #43, Canberra, pp. 443–480.

Oheimb, G. von, Hardtle, W., Naumann, P. S., Westphal, C., Assmann, T. and Meyer, H. (2008) Long-term effects of historical heathland farming on soil properties of forest ecosystems. *Forest Ecology and Management*, **255**, 1984–1993.

Ohlson, M., Dahlberg, B., Økland, T., Brown, K. J. and Halvorsen, R. (2009) The charcoal carbon pool in boreal forest soils. *Nature Geoscience*, **2**, 692–695.

O'Loughlin, C. L. and Watson, A. (1981) Note on root-wood strength deterioration in *Nothofagus fusca* and *N. truncata* after clearfelling. *New Zealand Journal of Forest Science*, **11**, 183–185.

O'Loughlin, C. L., Rowe, L. K. and Pearce, A. J. (1982) *Exceptional storm influences on slope erosion and sediment yield in small forest catchments, North Westland, New Zealand*. Institute of Engineers, Australia. National Symposium on Forest Hydrology, Melbourne.

Olsen, S. R. and Sommers, L. E. (1982) Phosphorus. In: A. L. Page (ed.) *Methods of Soil Analysis, part 2: Chemical and microbiological properties*. American Society of Agronomy, Madison, WI.

Olsen, S. R., Cole, C. V., Watanabe, F. S. and Dean, L. A. (1954) *Estimation of available phosphorus in soils by extraction with sodium bicarbonate*. USDA Circular 939.

Olson, J. (1981) Carbon balance in relation to fire regimes. In: *Fire Regimes and Ecosystem Properties*. USDA Forest Service GTR-WO-26, Washington, D.C., pp. 327–378.

Olsson, M. T., Erlandsson, M., Lundin, L., Nilsson, T., Nilsson, A. and Stendahl, J. (2009) Organic carbon stocks in Swedish podzol soils in relation to soil hydrology and other site characteristics. *Silva Fennica*, **43**, 209–222.

Oren, R., Schulze, E.-D., Werk, K. S. and Meyer, J. (1988) Performance of two *Picea abies* (L.) Karst. stands at different stages of decline. VII. Nutrient relations and growth. *Oecologia*, **77**, 163–173.

Outcalt, K. W. (1994) *Evaluations of a restoration system for sandhills longleaf pine communities.* United States Department of Agriculture Forest Service General Technical Report 247.

Paces, T. (1986) Rates of weathering and erosion derived from mass balance in small drainage basins. In: S. M. Colman and D. P. Dethier (eds) *Rates of Chemical Weathering of Rocks and Minerals*, Academic Press, Orlando, pp. 531–551.

Papen, H., Rosenkranz, P., Butterbach-Bahl, K., Gasche, R., Willibald, G. and Brüggemann, N. (2005) Effects of tree species on C- and N-cycling and biosphere-atmosphere exchange of trace gases in forests. In D. Binkley and O. Menyailo (eds) *Tree Species Effects on Soils: Implications for Global Change*. NATO Science Series, Kluwer Academic Publishers, Dordrecht, pp. 165–172.

Paré, D. and Bernier, B. (1989a) Origin of phosphorus deficiency observed in declining sugar maple stands in the Quebec Appalachians. *Canadian Journal of Forest Research*, **19**, 24–34.

Paré, D. and Bernier, B. (1989b) Phosphorus-fixing potential of Ah and H horizons subjected to acidification. *Canadian Journal of Forest Research*, **19**, 132–134.

Parkinson, B. W. and Spilker, J. J. (eds) (1996) *Global Positioning System: Theory and Practice*. Volumes I and II. American Institute of Aeronautics and Astronautics, Inc. Washington, D.C.

Parrotta, J. A., Baker, D. D. and Fried, M. (1996) Changes in dinitrogen fixation in maturing stands of *Casuarina equisetifolia* and *Leucaena leucocephala*. *Canadian Journal of Forest Research*, **26**, 1684–1691.

Paton, T. R., Humphreys, G. S. and Mitchell, P. B. (1995) *Soils, A New Global View*. Yale University Press, New Haven, CN.

Paul, E. A. (ed.) (2007) *Soil Microbiology and Biochemistry*, 3rd edition. Academic Press, Burlington, MA.

Paul, E. A. and Clark, F. E. (1996) *Soil Microbiology and Biochemistry*, 2nd edition. Academic Press, New York, NY, 340 pp.

Pavich, M. J. (1986) Processes and rates of saprolite production and erosion on a foliated granitic rock of the Virginia Piedmont. In: S. M. Colman and D. P. Dethier (eds) *Rates of Chemical Weathering of Rocks and Minerals*. Academic Press, Orlando, pp. 552–590.

Pearson, H. L. and Vitousek, P. M. (2001) Stand dynamics, nitrogen accumulation, and nitrogen fixation in regenerating stands of *Acacia koa*. *Ecological Applications*, **11**, 1381–1394.

Pennock, D. J. (2004) Designing field studies in soil science. *Canadian Journal of Soil Science*, **84**, 1–10.

Pennock, D., Yates, T. and Braidek, J. (2008) Soil Sampling Designs. In: M. R. Carter and E. G. Gregorich (eds) *Soil Sampling and Methods of Analysis*, 2nd edition. CRC Press, Boca Raton, FL, pp. 1–14.

Perkins, P. V. and Vann, A. R. (1997) The bulk density amelioration of minespoil with pulverized fuel ash. *Soil Technology*, **10**, 111–114.

Persson, T., Lundkvist, H., Wirön, A., Hyvönen, R. and Wessön, B. (1989) Effects of acidification and liming on carbon and nitrogen mineralization and soil organisms in mor humus. *Water, Air and Soil Pollution*, **45**, 77–96.

Pessotti, J. E. S., Rizzo, L. T. B. and Vaillant, P. F. M. (1983) Caracteristicas do meio fisico do distrito florestal norte. *Salvador, INTEC, Copener Florestal Itda*, **1**, 1–3.

Piccolo, A. (1996) *Humic Substances in Terrestrial Ecosystems*. Elsevier, Amsterdam.

Pichtel, J. R., Dick, W. A. and Sutton, P. (1994) Comparison of amendments and management practices for long-term reclamation of abandoned mine lands. *Journal of Environmental Quality*, **23**, 766–772.

Pilbeam, C. J., Mahapatra, B. S. and Wood, M. (1993) Soil matric potential effects on gross rates of nitrogen mineralization in an Orthic Ferralsol from Kenya. *Soil Biology and Biochemistry*, **25**, 1409–1413.

Pimentel, D. and Pimentel, M. (1979) *Food, Energy and Society*. Edward Arnold, London. 165 pp.

Pliny the Elder (AD 77–79) *Naturalis Historia*, Rome.

Pollierer, M. M., Langel, R., Körner, C., Maraun, M. and Scheu, S. (2007) The underestimated importance of belowground carbon input for forest soil animal food webs. *Ecology Letters*, **10**, 729–736.

Ponge, J.-F., Jabiol, B. and Gégout, J.-C. (2011) Geology and climate conditions affect more humus forms than forest canopies at large scale in temperate forests. *Geoderma*, **162**, 187–195.

Ponomareva, V. V. (1964) *Theory of Podzolization*, trans. A. Gourevitch. Israel Program for Science Translations, Jerusalem.

Popovic, B. and Andersson, F. (1984) *Markkalkning och skogsproduktion – litteraturversikt och revision av svenska kalkningsforsk*. Statens NaturvDrdsverk PM 1792, Solna.

Posada, J. M. and Schuur, E. A. G. (2011) Relationships among precipitation regime, nutrient availability, and carbon turnover in tropical rain forests. *Oecologia*, **165**, 783–795.

Post, M. W., Emanuel, W. and Stangenberger, P. (1982) Soil carbon pools and world life zones. *Nature*, **298**, 156–159.

Powers, R. F. and Ferrell, G. T. (1996) Moisture, nutrient, and insect constraints on plantation growth: the "Garden of Eden" study. *New Zealand Journal of Forestry Science*, **26**, 126–144.

Powers, R. F., Mead, D. J., Burger, J. A. and Ritchie, M. W. (1994) Designing long-term site productivity experiments. In: W. J. Dyck and D. W. Cole (eds) *Impacts of Forest Harvesting on Long-term Site Productivity*, Chapman & Hall, London, pp. 247–286.

Powers, R. F., Busse, M. D., McFarlane, K. J., Zhang, J. and Young, D. H. (2012) Long-term effects of silviculture on soil carbon storage: does vegetation control make a difference? *Forestry*, in review.

Powers, R. F., Scott, D. A., Sanchez, F. G., Voldseth, R. A., Page-Dumroese, D., Elioff, J. D. and Stone, D. M. (2005) The North American long-term soil productivity experiment: Findings from the first decade of research. *Forest Ecology and Management*, **220**, 31–50.

Prescott, C. E. (1996) Influence of forest floor type on rates of litter decomposition in microcosms. *Soil Biology and Biochemistry*, **10**, 1319–1325.

Prescott, C. E. (2010) Litter decomposition: what controls it and how can we alter it to sequester more carbon in forest soils? *Biogeochemistry*, **101**, 133–149.

Prescott, C. E., Maynard, D. G. and Laiho, R. (2000) Humus in boreal forests: friend or foe? *Forest Ecology and Management*, **133**, 23–36.

Prescott, C. E., Thomas, K. D. and Weetman, G. F. (1995) The influence of tree species on nitrogen mineralisation in the forest floor: lessons from three retrospective studies. In: D. J. Mead and I. S. Cornforth (eds) Proceedings of the Trees and Soils Workshop, Lincoln University, Agronomy Society of New Zealand Special Publication #10, Lincoln University Press, Canterbury, pp. 58–59.

Prescott, C. E., Coward, L. P., Weetman, G. F. and Gessel, S. P. (1993) Effects of repeated nitrogen fertilization on the ericaceous shrub, salal (*Gaultheria shallon*), in two coastal Douglas fir forests. *Forest Ecology and Management*, **61**, 45–60.

Pritchett, W. L. (1972) The effect of nitrogen and phosphorus on the growth and composition of loblolly and slash pine seedlings in pots. *Soil Crop Science Society of Florida Proceedings*, **32**, 161–165.

Pritchett, W. L. and Smith, W. H. (1974) *Management of wet savanna soils for pine production*. Florida Agricultural Experiment Station Technical Bulletin 762.

Prosser, J. I. (2002) Molecular and functional diversity in soil micro-organisms. *Plant and Soil*, **244**, 9–17.

Pyatt, D. G. (1970) *Soil groups of upland forests*. Forestry Commission Forest Record 71. London.

Pyatt, D. G. and Suarez, J. C. (1997) *An ecological site classification for forestry in Great Britain with special reference to Grampian, Scotland*. United Kingdom Forestry Commission Technical Paper 20.

Qualls, R. G., Haines, B. L. and Swank, W. T. (1991) Fluxes of dissolved organic nutrients and humic substances in a deciduous forest. *Ecology*, **72**, 254–266.

Qualset, C. O. and Collins, W. W. (1999) *Biodiversity in Agroecosystems*. CRC Press, Boca Raton, FL.

Raich, J. W. and Nadelhoffer, K. J. (1989) Belowground carbon allocation in forest ecosystems: global trends. *Ecology*, **70**, 1346–1354.

Raich, J. W., Riley, R. H. and Vitousek, P. M. (1994) Use of root-ingrowth cores to assess nutrient limitations in forest ecosystems. *Canadian Journal of Forest Research*, **24**, 2135–2138.

Raison, R. J., Khanna, P. and Woods, P. (1985) Mechanisms of element transfer to the atmosphere during vegetation fires. *Canadian Journal of Forest Research*, **15**, 132–140.

Raison, R. J., Khanna, P.K., Connell, M. J. and Falkiner, R. A. (1990) Effects of water availability and fertilization on nitrogen cycling in a stand of *Pinus radiata*. *Forest Ecology and Management*, **30**, 31–43.

Raman, E. (1893) *Forstliche Bodenkunde und Standortslehre*. J. Springer, Berlin.

Rawls, W. J., Pachepsky, Y. A., Ritchie, J. C., Sobecki, T. M. and Bloodworth, H. (2003) Effect of soil organic carbon on soil water retention. *Geoderma*, **116**, 61–76.

Reich, P., Griegal, D., Aber, J. and Gower, S. (1997) Nitrogen mineralization and productivity in 50 hardwood and conifer stands on diverse soils. *Ecology*, **78**, 335–347.

Reich, P. B., Oleksyn, J., Modrzynski, J., Mrozinski, P., Hobbie, S. E., Eissenstat, D. M., Chorover, J., Chadwick, O. A., Hale, C. M. and Tjoelker, M. G. (2005) Linking litter calcium, earthworms and soil properties: a common garden test with 14 tree species. *Ecology Letters*, **8**, 811–818.

Reiners, W. A., Bouwman, A. F., Parsons, W. F. J. and Keller, M. (1994) Tropical rain forest conversion to pasture: changes in vegetation and soil properties. *Ecological Applications*, **4**, 363–377.

Remezov, N. P. and Progrebnyak, P. S. (1965) Forest Soil Science (English translation). U.S. Department of Commerce, Clearinghouse for Federal Scientific and Technical Information, Springfield, Virginia. 261 pp.

Remezov, N. P. and Pogrebnyak, P. S. (1969) *Forest Soil Science*. National Science Foundation, Washington, D.C.

Resh, S. D., Binkley, D. and Parrotta, J. (2002) Greater soil carbon sequestration under nitrogen-fixing trees compared with Eucalyptus species. *Ecosystems*, **5**, 217–231.

Reuss, J. O. (1989) Soil-solution equilibria in lysimeter leachates under red alder. In: R. K. Olson and A. S. Lefohn (eds) *Effects of Air Pollution on Western Forests*. Air and Waste Management Association, Pittsburgh, pp. 547–559.

Reuss, J. O. and Johnson, D. W. (1986) *Acid Deposition and the Acidification of Soils and Waters*. Springer-Verlag, New York. 119 pp.

Rhoades, C. (1997) Single-tree influences on soil properties in agroforestry: lessons from natural forest and savanna ecosystems. *Agroforestry Systems*, **35**, 71–94.

Rhoades, C. and Binkley, D. (1995) Factors influencing decline in soil pH in Hawaiian Eucalyptus and Albizia plantations. *Forest Ecology and Management*, **80**, 47–56.

Rhoades, C., Eckert, G. E. and Coleman, D. (1998) Effects of pasture trees on soil nitrogen and organic matter: implications for tropical montane forest restoration. *Restoration Ecology*, **6**, 262–270.

Rhoades, C., Binkley, D., Oskarsson, H. and Stottlemyer, R. (2008) Soil nitrogen accretion along a floodplain terrace chronosequence in northwest Alaska: influence of the nitrogen-fixing shrub *Shepherdia canadensis*. *Ecoscience*, **15**, 223–230.

Richards, A. E., Forrester, D. I., Bauhus, J. and Scherer-Lorenzen, M. (2010) The influence of mixed tree plantations on the nutrition of individual species: a review. *Tree Physiology*, **30**, 1192–1208.

Richards, B. N. (1961) Soil pH and mycunhiza development in Pinus. *Nature* (London), **190**, 105.

Richter, D. D. and Babbar, L. I. (1991) Soil diversity in the tropics. *Advances in Ecological Research*, **21**, 315–389.

Richter, D. D. and Markewitz, D. (2001) *Understanding Soil Change: Soil sustainability over millennia, centuries, and decades*. Cambridge University Press, Cambridge.

Richter, D., Ralston, C. W. and Harms, W. (1982) Prescribed fire: effects on water quality and forest nutrient cycling. *Science*, **215**, 661–663.

Richter, D. D., Markewitz, D., Trumbore, S. E. and Wells, C. G. (1999) Rapid accumulation and turnover of soil carbon in a re-establishing forest. *Nature*, **400**, 56–58.

Richter, D. D., Allen, H. L., Li, J., Markewitz, D. and Raikes, J. (2006) Bioavailability of slowly cycling soil phosphorus: major restructuring of soil P fractions over four decades in an aggrading forest. *Oecologia*, **150**, 259–271.

Richter, D. D., Comer, P. J., King, K. S., Sawin, H. A. and Wright, D. S. (1988) Effects of low ionic strength solutions on pH of acid forested soils. *Soil Science Society of America Journal*, **52**, 261–264.

Robertson, W. K., Smith, W. H. and Post, D. M. (1975) Effect of nitrogen and placed phosphorus and dolomitic limestone in an Aeric Haploquod on slash pine growth and composition. *Soil Crop Society of Florida Proceedings*, **34**, 58–60.

Robichaud, P. R. (2000) Fire effects on infiltration rates after prescribed fire in northern Rocky Mountain forests, USA. *Journal of Hydrology*, **231–232**, 220–229.

Robichaud, P. R., Beyers, J. L. and Neary, D. (2000) *Evaluating the effectiveness of postfire rehabilitation treatments*. USDA Forest Service General Technical Report RMRS-GTR-63, Fort Collins.

Rochelle, J. A. (1979) The effects of forest fertilization on wildlife. In: *Proceedings of Forest Fertilization Conference*. Contribution #40, Institute of Forest Resources, University of Washington, Seattle, pp. 164–167.

Rockwood, D. L., Windsor, C. L. and Hodges, J. F. (1985) Response of slash pine progenies to fertilization. *Southern Journal of Applied Forestry*, **9**, 37–40.

Romell, L. G. (1922) Die Bodenventilation als okologiseher Factor. *Meddel Statens Skogsforsoksanstalt*, **19**, 281–359.

Romell, L. G. (1935) *Ecological problems of the humus layer in the forest*. Cornell University Agricultural Experiment Station Memoir 170.

Romell, L. G. and Heiberg, S. O. (1931) Types of humus layer in the forests of northeastern U.S. *Ecology*, **12**, 567–608.

Rose, R., Haase, D. L. and Boyer, D. (1995) *Organic matter management in forest tree nurseries: theory and practice*. Nursery Technology Cooperative, Oregon State University, Corvallis, OR.

Rosling, A. (2009) Trees, mycorrhiza and minerals – field relevance of *in vitro* experiments. *Geomicrobiology Journal*, **26**, 389–401.

Rosling, A., Lindahl, B. D. and Finlay, R. D. (2004) Carbon allocation to ecctomycorrhizal roots and mycelium colonizing different mineral substrates. *New Phytologist*, **162**, 795–802.

Rothe, A, Huber, C., Kreutzer, K. and Weis, W. (2002) Deposition and soil leaching in stands of Norway spruce and European beech: Results from the Höglwald in comparison with other European case studies. *Plant and Soil*, **240**, 33–45.

Roulet, N. T. and Moore, T. R. (1995) The effect of forestry drainage practices on the emission of methane from northern peatlands. *Canadian Journal of Forest Research*, **25**, 491–499.

Rożen, A., Sobczyk, L., Liszka, K. and Weiner, J. (2010) Soil faunal activity as measured by the bait-lamina test in monocultures of 14 tree species in the Siemianice common-garden experiment, Poland. *Applied Soil Ecology*, **45**, 160–167.

Ruehle, J. L. and Marx, D. H. (1979) Fiber, food, fuel and fungal symbionts. *Science*, **206**, 419–422.

Ruiz, N., Lavelle, P. and Jiménez, J. (2008) *Soil Macrofauna Field Manual*. Food and Agriculture Organization of the United Nations, Rome.

Russell, A. E., Raich, J. W., Arrieta, R. B., Valverde-Barantes, O. and González, E. (2010) Impacts of individual tree species on carbon dynamics in a moist tropical forest environment. *Ecological Applications*, **20**, 1087–1100.

Rustad, L. E., Campbell, J. L., Marion, G. M., Norby, R. J., Mitchell, M. J., Hartley, A. E., Cornelissen, J. H. C. and Gurevitch, J. (2001) A meta-analysis of the response of soil respiration, net nitrogen mineralization, and above-ground plant growth to experimental ecosystem warming. *Oecologia*, **126**, 543–546.

Rütting, T., Boeckx, P., Müller, C. and Klemedtsson, L. (2011) Assessment of the importance of dissimilatory nitrate reduction to ammonium for the terrestrial nitrogen cycle. *Biogeosciences Discussions*, **8**, 1169–1196.

Ryan, M. G., Binkley, D., Fownes, J. H., Giardina, C. P. and Senock, R. (2004) An experimental test of the causes of forest growth decline with stand age. *Ecological Monographs*, **74**, 393–414.

Ryan, M. G., Stape, J. L., Binkley, D., Fonseca, S., Loos, R., Takahashi, E. N., Silva, C. R., Silva, S., Hakamada, R., Ferreira, J. M., Lima, A. M., Gava, J. L., Leita, F. P., Silva, G., Andrade, H. and Alves, J. M. (2010) Factors controlling Eucalyptus productivity: How resource availability and stand structure alter production and carbon allocation. *Forest Ecology and Management*, **259**, 1695–1703.

Sabey, B. R., Pendleton, R. L. and Webb, B. L. (1990) Effect of municipal sewage sludge application on growth of two reclamation shrub species in copper mine spoils. *Journal of Environmental Quality*, **19**, 580–586.

Sajedi, T., Prescott, C. E., Seely, B. and Lavkulich, L. M. (2012) Relationships among soil moisture, aeration and plant communities in natural and harvested coniferous forests in coastal British Columbia, Canada. *Journal of Ecology*, DOI: 10.1111/j.1365-2745.2011.01942.x.

Sallenave, H. (1969) The cultivation of maritime pine in southwest France. *Phosphorus Agricultural*, **54**, 17–26.

Šamonil, P., Král, K. and Hort, L. (2010) The role of tree uprooting in soil formation: A critical literature review. *Geoderma*, **157**, 65–79.

Sanchez, P. A. (1976) *Properties and Management of Soils in the Tropics*. John Wiley & Sons, Inc., New York, NY.

Sanchez, P. A., Gichuru, P. and Katz, L. B. (1982) Organic matter in major soils of the tropical and temperate regions. *Transactions of the 12th International Congress of Soil Science*, **1**, 99–114.

Sanford, R. L., Saldarriaga, J., Clark, K. E., Uhl, C. and Herrera, R. (1985) Amazon rain forest fires. *Science*, **227**, 53–55.

Saric, M. R. and Loughman, B. C. (eds) (1983) *Genetic Aspects of Plant Nutrition*. Martinus Nijhoff/Junk, The Hague. 495 pp.

Sartz, R. S. (1973) *Snow and frost depths on north and south slopes*. USDA Forest Service Research Note NC-157.

Sato, S. and Comerford, N. B. (2006) Organic anions and phosphorus desorption and bioavailability in a humid Brazilian Ultisol. *Soil Science*, **171**, 695–705.

Sayn-Wittgeostein, L. (1969) The northern forest. *Canadian Pulp and Paper Industrial*, **22**, 77–78.

Schaefer, M. and Schauermann, J. (1990) The soil fauna of beech forests: comparison between a mull and a moder soil. *Pedobiologica*, **34**, 299–314.

Schaetzl, R. J. and Anderson, S. (2005) *Soils: Genesis and Geomorphology*. Cambridge University Press.

Schenk, H. J. and Jackson, R. B. (2002) The global biogeography of roots. *Ecological Monographs*, **72**, 311–328.

Schimel, J. P., Jackson, L. E. and Firestone, M. K. (1989) Spatial and temporal effects on plant-microbial competition for inorganic nitrogen in a California grassland. *Soil Biology and Biochemistry*, **21**, 1059–1066.

Schindlbacher, A., Zechmeister-Boltenstern, S. and Jandl, R. (2009) Carbon losses due to soil warming: do autotrophic and heterotrophic soil respiration respond equally? *Global Change Biology*, **15**, 901–913.

Schlenker, G. (1964) Entwicklung des Sudwestdeutschland angewandten Verfahrns der forstuchen Standortskunde. In: *Standort Wald und Waldwirtschaft in Obersebwahen*, Oberschwabishe Fiehtenreviere, Stuttgart, Germany, pp. 5–26.

Schlesinger, W. H. (1995) An overview of the carbon cycle. In: R. Lal, J. M. Kimble, E. Levine and B. A. Stewart (eds) *Soils and Global Change*. CRC Press, Boca Raton, FL.

Schlesinger, W. (1997) *Biogeochemistry: An analysis of global change*. Academic Press, San Diego. 588 pp.

Schmalz, H. J. and Coleman, M. D. (2011) Foliar Sulfate-Sulfur as a nutrient diagnostic tool for Interior Douglas Fir. *Western Journal of Applied Forestry*, **26**, 147–150.

Schmidt, M. W. I., Torn, M. S., Abiven, S., Dittmar, T., Guggenberger, G., Janssens, I. A., Kleber, M., Kögel-Knabner, I., Lehmann, J., Manning, D. A. C., Nannipieri, P., Rasse, D. P., Weiner, S. and Trumbore, S. E. (2011) Persistence of soil organic matter as an ecosystem property. *Nature*, **478**, 49–56.

Schnitzer, M. (1991) Soil organic matter – the next 75 years. *Soil Science*, **151**, 41–48.

Schoch, P. and Binkley, D. (1986) Prescribed burning increased nitrogen availability in a mature loblolly pine stand. *Forest Ecology and Management*, **14**, 13–22.

Schoeneberger, P. J., Wysocki, D. A., Benham, E. C. and Broderson, W. D. (eds) (2002) *Field Book for Describing and Sampling Soils*, Version 2.0. Natural Resources Conservation Service, National Soil Survey Center, Lincoln, NE. http://soils.usda.gov/technical/fieldbook/

Schoenholtz, S. H., Burger, J. A. and Kreh, R. E. (1992) Fertilizer and organic amendment effects on mine soil properties and revegetation success. *Soil Science Society of America Journal*, **56**, 1177–1184.

Schubert, K. R. (1982) *The energetics of biological nitrogen fixation*. American Society of Plant Physiologists, Rockville, Maryland. 31 pp.

Schultz, R. P. (1972) Root development of intensively cultivated slash pine. *Soil Science Society of America Proceedings*, **36**, 158–162.

Schumacher, F. X. and Coile, T. S. (1960) *Growth and Yields of Natural Stands of the Southern Pines*. T. S. Coile, Inc., Durham, NC.

Schuur, E. A. G., Chadwick, O. A. and Matson, P. A. (2001) Carbon cycling and soil carbon storage in mesic to wet Hawaiian montane forests. *Ecology*, **83**, 3183–3196.

Schuur, E. A., Bockheim, J., Canadell, J. G., Euskirchen, E., Field, C. B., Goryachkin, S. V., Hagemann, S., Kuhry, P., Lafleur, P. M., Lee, H., Mazhitova, G., Nelson, F. E.,

Rinke, A., Romanovsky, V. E., Shiklomanov, N., Tarnocai, C., Venevsky, S., Vogel, J. G. and Zimov, S. A. (2008) Vulnerability of permafrost carbon to climate change: implications for the global carbon cycle. *BioScience*, **58**, 701–714.

Scott, N. (1996) *Plant species effects on soil organic matter turnover and nutrient release in forests and grasslands*. PhD thesis, Colorado State University, Fort Collins. 133 pp.

Scott, N. A. (1998) Soil aggregation and organic matter mineralization in forests and grasslands: plant species effects. *Soil Science Society of America Journal*, **62**, 1081–1089.

Scott, N. and Binkley, D. (1997) Litter quality and annual net N mineralization: comparisons across sites and species. *Oecologia*, **111**, 151–159.

Seastedt, T. R. and Knapp, A. K. (1993) Consequences of nonequilibrium resource availability across multiple time scales: the transient maxima hypothesis. *American Naturalist*, **41**, 621–633.

Selker, J. S., Kent Keller, C. and McCord, J. T. (1999) *Vadose Zone Processes*. CRC Press, Boca Raton, FL.

Sharma, E. (1988) Altitudinal variation in nitrogenase activity of the Himalayan alder naturally regenerating on landslide-affected sites. *New Phytologist*, **108**, 411–416.

Sharma, E., Ambasht, R. S. and Singh, M. P. (1985) Chemical soil properties under five age series of *Alnus nepalensis* plantations in the Eastern Himalayas. *Plant and Soil*, **84**, 105–113.

Shaw, C. F. (1930) Potent factors in soil formation. *Ecology*, **11**, 239–245.

Sheikh, M. I. (1986) *Afforestation of arid and semi-arid areas in Pakistan*. Food and Agricultural Organization and Pakistan Forest Institute, Peshawar, Pakistan.

Shetty, K. G., Hetrick, A. D., Figge, D. A. H. and Schwab, A. P. (1994) Effects of mycorrhizae and other soil microbes on revegetation of heavy metal contaminated mine spoil. *Environmental Pollution*, **86**, 181–188.

Shoulders, E. and Tiarks, A. E. (1984) Response of pines and native forage to fertilizer. In: N. E. Linnartz and M. K. Johnson (eds) *Agroforestry in the Southern United States*, Louisiana Agricultural Experiment Station, Baton Rouge, pp. 105–126.

Sibanda, H. M. and Young, S. D. (1989) The effect of humus acids and soil heating on the availability of phosphate in oxide-rich tropical soils. In: J. Proctor (ed.) *Mineral Nutrients in Tropical Forests and Savannas*. Blackwell Scientific, Oxford, pp. 71–83.

Siddique, I, Guimarães Vieira, I. C., Schmidt, S., Lamb, D., Reis Carvalho, C. J., de O. Figueiredo, R., Blomberg, S. and Davidson, E. A. (2010) Nitrogen and phosphorus additions negatively affect tree species diversity in tropical forest regrowth trajectories. *Ecology*, **91**, 2121–2131.

Silberbush, M. and Barber, S. A. (1983) Sensitivity of simulated phosphorus uptake to parameters used by mechanistic-mathematical model. *Plant and Soil*, **74**, 93–100.

Silver, W. L. and Miya, R. K. (2001) Global patterns in root decomposition: comparisons of climate and litter quality effects. *Oecologia*, **129**, 407–419.

Simmons, G. L. and Pope, P. E. (1988) Influence of soil water potential and mycorrhizal colonization on root growth of yellow poplar and sweet gum seedlings grown in compacted soil. *Canadian Journal of Forest Research*, **18**, 1392–1396.

Six, J., Conant, R. T., Paul, E. A. and Paustian, K. (2002) Stabilization of soil organic matter: implications for C-saturation of soils. *Plant and Soil*, **241**, 155–176.

Six, J., Paustian, K., Elliott, E. T. and Combrink, C. (2000) Soil structure and organic matter: I. Distribution of aggregate-size classes and aggregate-associated carbon. *Soil Science Society of America Journal*, **64**, 681–689.

Skovsgaard, J. P. and Vanclay, J. K. (2008) Forest site productivity: a review of the evolution of dendrometric concepts for even-aged stands. *Forestry*, **81**, 13–31.

Skyllberg, U., Raulund-Rasmussen, K. and Borggaard, O. K. (2001) pH buffering in acidic soils developed under *Picea abies* and *Quercus robur* – effects of soil organic matter, adsorbed cations and soil solution ionic strength. *Biogeochemistry*, **56**, 51–74.

Smith, D. W. (1970) Concentrations of soil nutrients before and after fire. *Canadian Journal of Soil Science*, **50**, 17–29.

Smith, S. E. and Read, D. J. (2008) *Mycorrhizal Symbioses*. Academic Press/Elsevier, New York.

Snowdon, P. and Benson, M. L. (1992) Effects of combinations of irrigation and fertilisation on the growth and biomass production of *Pinus radiata*. *Forest Ecology and Management*, **52**, 87–116.

Sohlenius, B. (1982) Short-term influence of clear-cutting on abundance of soil microfauna (nematode, rotatoria, and tardigrada) in a Swedish pine forest soil. *Journal of Applied Ecology*, **19**, 349–359.

Soil Survey Staff (1993) *Soil Survey Manual*. USDA Handbook 18. Washington, D.C.

Soil Survey Staff (1999) *Soil Taxonomy – A basic system of soil classification for making and interpreting soil surveys*, 2nd edition. USDA Handbook 435. Washington, D.C.

Sollins, P. (1982) Input and decay of coarse woody debris in coniferous stands in western Oregon and Washington. *Canadian Journal of Forest Research*, **12**, 18–28.

Sollins, P., Cromack, K. and Li, C. Y. (1981) Role of low-molecular-weight organic acids in the inorganic nutrition of fungi and higher plants. In: D. T. Wicklow and G. C. Carroll (eds) *The Fungal Community*, Marcel Dekker, New York, pp. 607–619.

Sollins, P., Cline, S. P., Verhoeven, T., Sachs, D. and Spycher, G. (1987) Patterns of log decay in old-growth Douglas fir forests. *Canadian Journal of Forest Research*, **17**, 1585–1595.

Sollins, P., Grier, C. C., McCorison, F. M., Cromack, K. Jr., Fogel, R. and Fredriksen, R. L. (1980) The internal element cycles of an old-growth Douglas fir ecosystem in western Oregon. *Ecological Monographs*, **50**, 261–285.

Sommers, L. E., Gilmour, C. V., Wildung, R. E. and Beck, S. M. (1980) The effect of water potential on decomposition processes in soils. In: *Water Potential Relations in Soil Microbiology*. Soil Science Society of America, Madison, WI, pp. 97–117.

Son, Y. and Gower, S. T. (1991) Aboveground nitrogen and phosphorus use by five plantation-grown trees with different leaf longevities. *Biogeochemistry*, **14**, 167–191.

Sort, X. and Alcaniz, J. M. (1996) Contribution of sewage sludge to erosion control in the rehabilitation of limestone quarries. *Land Degradation and Development*, **7**, 69–76.

Sparks, D. L. (ed.) (1996) *Methods of Soil Analysis, Part 3. Chemical Methods*. Soil Science Society of America Book Series No. 5, Madison, WI.

Spears, J. D. H., Holub, S. M., Harmon, M. E. and Lajtha, K. (2003) The influence of decomposing logs on soil biology and nutrient cycling in an old-growth mixed coniferous forest in Oregon, U.S.A. *Canadian Journal of Forest Research*, **33**, 2193–2201.

Spiecker, H. (1995) Growth dynamics in a changing environment – long-term observations. *Plant Soil*, **168–169**, 555–561.

Sposito, G. (1989) *The Chemistry of Soils*. Oxford University Press, New York. 277 pp.

Spurr, S. H. (1952) *Forest Inventory*. Ronald Press, New York, NY.

Staaf, H. (1987) Foliage litter turnover and earthworm populations in three beech forests of contrasting soil and vegetation types. *Oecologia*, **72**, 58–64.

Standish, R. J., Cramer, V. A. and Hobbs, R. J. (2008) Land-use legacy and the persistence of invasive *Avena barbata* on abandoned farmland. *Journal of Applied Ecology*, **45**, 1576–1583.

Stape, J. L., Gomes, A. N. and Assis, T. F. (1997) Estimativa da produtividade de povomentos monoclonias de *Eucalyptus grandis* x urophylla no Nordeste do Estato de Bahia-Brazil em função das variabilidades pluviométrica e edáfica. In: *Proceedings of IUFRO Conference on Silviculture and Improvement of Eucalyptus*. EMBRAPA/CNPF, El Salvador, pp. 192–198.

Stape, J. L., Binkley, D., Jacob, W. S. and Takahashi, E. N. (2006) A twin-plot approach to determine nutrient limitation and potential productivity in eucalyptus plantations at landscape scales in Brazil. *Forest Ecology and Management*, **223**, 358–362.

Stark, J. and Hart, S. C. (1997) High rates of nitrification and nitrate turnover in undisturbed coniferous forests. *Nature*, **385**, 61–64.

Stark, L. R., Wenerick, W. R., Williams, F. M., Stevens, S. E., Jr. and Wuest, P. J. (1994) Restoring the capacity of spent mushroom compost to treat coal mine drainage by reducing the inflow rate: a microcosm experiment. *Water, Air and Soil Pollution*, **75**, 405–420.

Stark, N. M. (1977) Fire and nutrient cycling in a Douglas fir/larch forest. *Ecology*, **58**, 16–30.

Steinbrenner, E. C. (1975) Mapping forest soils on Weterhaeuser lands in the Pacific Northwest. In: B. Brernier and C. H. Winget (eds) *Forest Soils and Forest Land Management*. Laval University Press, Quebec City, Quebec, pp. 513–525.

Stephens, F. R. (1965) Relation of Douglas fir productivity to some zonal soils in the north-west Cascades of Oregon. In: C. T. Youngberg (ed.) *Forest–Soil Relationships in North America*. Oregon State University Press, Corvallis, OR, pp. 245–260.

Sterner, W. R. and Elser, J. J. (2002) *Ecological Stoichiometery: The biology of elements from molecules to the biosphere*. Princeton University Press, Princeton.

Stevenson, F.J. (1994) *Humus Chemistry*, 2nd edition. John Wiley & Sons, Inc., New York, NY.

Steward, F. C. (1964) *Plants at Work*. Addison-Wesley, Reading, MA.

Stobbe, P. C. and Wright, J. R. (1959) Modern concepts of the genesis of Podzols. *Soil Science Society of America Proceedings*, **23**, 161–164.

Stoeckeler, J. H. (1960) *Soil factors affecting the growth of quaking aspen forests in the Lake States*. Minnesota Agricultural Experiment Station Technical Bulletin 323.

Stone, E. L. (1968) Microelement nutrition of forest trees: a review. In: G. W. Bengtson (ed.) *Forest Fertilization: Theory and practice*. Tennessee Valley Authority, Muscle Shoals, Alabama, pp. 132–179.

Stone, E. L. (1975) Effects of species on nutrient cycles and soil change. *Philosophical Transactions of the Royal Society, London (B)*, **271**, 149–162.

Stone, E. L. and Kalisz, P. J. (1991) On the maximum extent of tree roots. *Forest Ecology and Management*, **46**, 59–102.

Stork, N. E. and Eggleton, P. (1992) Invertebrates as determinates of soil quality. *American Journal of Alternatives in Agriculture*, **7**, 38–47.

Stottlemyer, R., Toczydlowski, D. and Hermann, R. (1998) *Biogeochemistry of a mature boreal ecosystem: Isle Royale National Park, Michigan*. USDI National Park Service Scientific Monograph NPS/NRUSGS/NRSM-98/01, Fort Collins, CO.

Strader, R., Binkley, D. and Wells, C. (1989) Nitrogen mineralization in high elevation forests of the Appalachians. I. Regional patterns in spruce–fir forests. *Biogeochemistry*, **7**, 131–145.

Straus, S. E., Glasziou, P., Richardson, W. S. and Haynes, R. B. (2010) *Evidence-Based Medicine: How to Practice and Teach It*. Elsevier, Amsterdam.

Strickland, M. S. and Rousk, J. (2010) Considering fungal: bacterial dominance in soils – Methods, controls, and ecosystem implications. *Soil Biology and Biochemistry*, **42**, 1385–1395.

Sutton, R. F. (1991) *Soil properties and root development in forest trees: A review*. Forestry Canada. Information Report O-X-413.

Sutton, R. F. and Weldon, T. P. (1993) Jack pine establishment in Ontario: 5-year comparison of stock types with and without bracke scarification, mounding and chemical site preparation. *Forestry Chronicle*, **69**, 545–553.

Sutton, R. F. and Weldon, T. P. (1995) White spruce establishment in boreal Ontario mixed wood: 5-year results. *Forestry Chronicle*, **71**, 633–638.

Swanson, D. K. (1996) Susceptibility of permafrost soils to deep thaw after forest fires in Interior Alaska, U.S.A., and some ecologic implications. *Arctic and Alpine Research*, **28**, 217–227.

Swanson, F. J. (1981) Fire and geomorphic processes. In: H. A. Mooney, T. M. Bonnicksen, N. L. Christensen, J. E. Lotan and W. A. Reiners (eds) *Fire Regimes and Ecosystem Properties*. USDA Forest Service General Technical Report WO-26, pp. 421–444.

Swanston, D. N. and Dyrness, C. T. (1973) Stability of steep land. *Journal of Forestry*, **71**, 264–269.

Switzer, M. (1979) Energy relations in forest fertilization. In: *Proceedings of Forest Fertilization Conference*. Contribution #40, Institute of Forest Resources, University of Washington, Seattle, pp. 243–246.

Sykes, D. J. (1971) Effects of fire and fire control on soil and water relations in northern forests. In: C. W. Slaughter *et al.* (eds) *Fire in the Northern Environment*. USDA Forest Service Pacific Northwest Forest and Range Experiment Station, Portland, OR.

Szlavecz, K., McCormick, M., Xia, L., Saunders, J., Morcol, T., Whigham, D., Filley, T. and Csuzdi, C. (2011). Ecosystem effects of non-native earthworms in mid-Atlantic deciduous forests. *Biological Invasions*, **13**, 1165–1182.

Tan, K. H. (1993) *Principles of Soil Chemistry*, 2nd edition. Marcel Dekker, New York, NY.

Tanner, E. V. J., Vitousek, P. M. and Cuevas, E. (1998) Experimental investigation of nutrient limitation of forest growth on wet tropical mountains. *Ecology*, **79**, 10–22.

Taylor, C. M. A. (1991) *Forest fertilization in Britain*. Bulletin 95, Forestry Commission, Her Majesty's Stationery Office, London.

Teng, Y. and Timmer, V. R. (1995) Rhizosphere phosphorus depletion induced by heavy nitrogen fertilization in forest nursery soils. *Soil Science Society of America Journal*, **59**, 227–233.

Theodorou, C. and Bowen, G. D. (1969) The influence of pH and nitrate on mycorrhizal associations of *Pinus radiata* D. Don. *Australian Journal of Botany*, **17**, 59–67.

Tiedemann, A. R., Conrad, C. E., Dieterich, J. H., Hornbeck, J. W., Megahan, W. F., Viereck, L. A. and Wade, D. D. (1979) *Effects of fire on water: a state-of-knowledge review*. USDA Forest Service General Technical Report WO-10. 28 pp.

Tilki, F. and Fisher, R. F. (1998) Tropical leguminous species for acid soils: studies on plant form and growth in Costa Rica. *Forest Ecology and Management*, **108**, 175–192.

Timmer, V. R. and Morrow, L. D. (1984) Predicting fertilizer growth response and nutrient status of jack pine by foliar diagnosis. In: E. L. Stone (ed.) *Forest Soils and Treatment Impacts*. University of Tennessee, Knoxville, pp. 335–351.

Timmer, V. R. and Stone, E. L. (1978) Comparative foliar analysis of young balsam fir fertilized with nitrogen, phosphorus, potassium, and lime. *Soil Science Society of America Journal*, **42**, 125–130.

Tisdall, J. M. and Oades, J. M. (1982) Organic matter and water-stable aggregates in soils. *Journal of Soil Science*, **62**, 141–163.

Torbert, J. L. and Burger, J. A. (1984) Long-term availability of applied phosphorus to loblolly pine on a Piedmont soil. *Soil Science Society of America Journal*, **48**, 1174–1178.

Torn, M., Trumbore, S., Chadwick, O., Vitousek, P. and Hendricks, D. (1997) Mineral control of soil organic carbon storage and turnover. *Nature*, **389**, 170–173.

Toumey, J. W. (1916) *Foundations of silviculture upon an ecological basis*. Yale School of Forestry, New Haven, CT.

Trangmar, B. B., Yost, R. S. and Uehara, G. (1985) Application of geostatistics to spatial studies of soil properties. *Advances in Agronomy*, **38**, 45–94.

Trettin, C. C., Laiho, R., Minkkinen, K. and Laine, J. (2006) Influence of climate change factors on carbon dynamics in northern forested peatlands. *Canadian Journal of Soil Science*, **86**, 269–280.

Trettin, C. C., Jurgensen, M. F., Grigal, D. F., Gale, M. R. and Jeglum, K. (1997) *Northern Forested Wetlands: Ecology and management*. Lewis Publishers, Boca Raton, FL.

Trichet, P., Bakker, M. R., Augusto, L., Alazard, P., Merzeau, D. and Saur, E. (2009) Fifty years of fertilization experiments on *Pinus pinaster* in southwest France: the importance of phosphorus as a fertilizer. *Forest Science*, **55**, 390–402.

Troedsson, T. and Lyford, W. H. (1973) Biological disturbance and small-scale spatial variations in a forested soil near Garpenberg, Sweden. *Studia Forestalia Auccica*, **109**

Truax, B., Gagnon, D., Lambert, F. and Chevrier, N. (1994) Nitrate assimilation of raspberry and pin cherry in a recent clearcut. *Canadian Journal of Botany*, **72**, 1343–1348.

Turnbull, M. H., Goodall, R. and Stewart, G. R. (1995) The impact of mycorrhizal colonization upon nitrogen source utilization and metabolism in seedlings of *Eucalyptus grandis* Hill ex Maiden and *Eucalyptus maculata* Hook. *Plant, Cell and Environment*, **18**, 1386–1394.

Turner, J. and Lambert, M. J. (1996) Nutrient cycling and forest management. In: P. M. Attiwill and M. S. Adams (eds) *Nutrition of Eucalyptus*. CSIRO, Collingwood, Australia, pp. 229–248.

Turner, J., Lambert, M. J. and Gessel, S. P. (1979) Sulfur requirements of nitrogen fertilized Douglas fir. *Forest Science*, **25**, 461–467.

Ugolini, F. C., Dawson, H. and Zachara, J. (1977) Direct evidence of particle migration in the soil solution of a podzol. *Science*, **195**, 603–605.

Ulrey, A. L., Graham, R. C. and Bowen, L. H. (1996) Forest fire effects on soil phyllosilicates in California. *Soil Science Society of America Journal*, **60**, 309–315.

Ulrich, B. (1983) Interaction of forest canopies with atmospheric constituents. In: B. Ulrich and J. Pankrath (eds) *Effects of Accumulation of Air Pollutants in Forest Ecosystems*. D. Reidel, Boston, pp. 33–45.

UNFAO (1990) *Soil Map of the World*, revised legend. UNFAO, Rome.

UNFAO/ISRIC (2006) *World reference base for soil resources 2006: a framework for international classification, correlation and communication*. World Soil Resources Report 103, UNFAO/ISRIC, Rome.

Uroz, S., Calvaruso, C., Turpault, M.-P. and Frey-Klett, P. (2009) Mineral weathering by bacteria: ecology, actors and mechanisms. *Trends in Microbiology*, **17**, 378–387.

Ursic, S. J. (1970) *Hydrologic effects of prescribed burning and deadening upland hardwoods in northern Mississippi*. USDA Forest Service Research Paper SO-54. 15 pp.

Uselman, S. M., Qualls, R. G. and Thomas, R. B. (1999) A test of a potential short cut in the nitrogen cycle: The role of exudation of symbiotically fixed nitrogen from the roots of a N-fixing tree and the effects of increased atmospheric CO_2 and temperature. *Plant and Soil*, **210**, 21–32.

Valentine, D. and Allen, H. L. (1990) Foliar response to fertilization identifies nutrient limitation in loblolly pine. *Canadian Journal of Forest Research*, **20**, 144–151.

Valentine, D. and Binkley, D. (1992) Topography and soil acidity in an Arctic landscape. *Soil Science Society of America Journal*, **56**, 1553–1559.

Valentine, K. W. G. (1986) *Soil Resource Surveys for Forestry*. Clarendon Press, Oxford.

van Breemen, N. and Buurman, P. (2002) *Soil Formation*. Kluwer, The Netherlands.

van Breemen, N., Lundstrom, U. S. and Jongmans, A. G. (2000) Do plants drive podzolization via rock-eating mycorrhizal fungi? *Geoderma*, **94**, 163–171.

van Cleve, K. and Dyrness, C. T. (1985) The effect of the Rosie Creek Fire on soil fertility. In: G. Juday and C. T. Dyrness (eds) *Early Results of the Rosie Creek Fire Research Project 1984*. Miscellaneous Publication #85-2, Agricultural and Forestry Experiment Station, University of Alaska, Fairbanks, pp. 7–11.

van den Driessche, R. (1984) Soil fertility in forest nurseries. In: M. L. Duryea and T. D. Landis (eds) *Forest Nursery Manual*. Martinus Nijhoff/Dr W. Junk Publishers, Boston, MA, pp. 63–74.

van den Driessche, R. and Ponsford, D. (1995) Nitrogen induced potassium deficiency in white spruce (*Picea glauca*) and Engelmann spruce (*Picea engelmannii*) seedlings. *Canadian Journal of Forest Research*, **25**, 1445–1454.

van Genuchten, M. Th. (1980) A closed-form equation for predicting the hydrolic conductivity of unsaturated soils. *Soil Science Society of America Journal*, **44**, 892–898.

van Kessel, C., Farrell, R. E., Roskoski, J. P. and Keane, K. M. (1994) Recycling of the naturally-occurring ^{15}N in an established stand of *Leucaena leucocephala*. *Soil Biology and Biochemistry*, **26**, 757–762.

Van Lear, D. H. and Hosner, J. F. (1967) Correlation of site index and soil mapping units. *Journal of Forestry*, **65**, 22–24.

van Miegroet, H. and Cole, D. W. (1984) The impact of nitrification on soil acidification and cation leaching in a red alder ecosystem. *Journal of Environmental Quality*, **13**, 586–590.

van Miegroet, H., Cole, D. W., Binkley, D. and Sollins, P. (1989) The effect of nitrogen accumulation and nitrification on soil chemical properties in alder forests. In: R. K. Olson and A. S. LeFohn (eds) *Effects of Air Pollution on Western Forests*. Air and Waste Management Association, Pittsburgh, pp. 515–528.

van Rees, K. C. J. and Comerford, N. B. (1990) The role of woody roots of slash pine seedlings in water and potassium absorption. *Canadian Journal of Forest Research*, **20**, 1183–1191.

Verstraten, J. (1977) Chemical erosion in a forested watershed in the Oesling, Luxembourg. *Earth Surface Processes*, **2**, 175–184.

Vesterdal, L., Elberling, B., Riis Christiansen, J., Callesen, I. and Kappel Schmidt, I. (2012) Soil respiration and rates of soil carbon turnover differ among six common European tree species. *Forest Ecology and Management*, **264**, 185–196.

Vihnaneck, R. and Ballard, T. (1988) Slashburning effects on stocking, growth, and nutrition of young Douglas fir plantations in salal-dominated ecosystems of eastern Vancouver Island. *Canadian Journal of Forest Research*, **18**, 718–722.

Vimmerstedt, J. P. and Finney, J. H. (1973) Impact of earthworm introduction on litter burial and nutrient distribution in Ohio strip-mine spoil banks. *Soil Science Society of America Journal*, **37**, 388–391.

Vincent, A. B. (1965) *Black spruce: a review of its silvics ecology and silviculture*. Department of Forestry Canada, Publication 1100, Ottawa, Canada.

Vivanco, L. and Austin, A. T. (2008) Tree species identity alters forest litter decomposition through long-term plant and soil interactions in Patagonia, *Argentina. Journal of Ecology*, **96**, 727–736.

Voigt, G. K. (1965) Nitrogen recovery from decomposing tree leaf tissue and forest humus. *Soil Science Society of America Proceedings*, **29**, 756–759.

Vosso, J. A. (1971) Field inoculation with mycorrhizae fungi. In: E. Hacskaylo (ed.) *Mycorrhizae*. USDA Forest Service Miscellaneous Publication 1189, pp. 187–196.

Wackernagel, H. (2003) *Multivariate Geostatistics*. Springer Verlag, New York, NY.

Waisel, Y., Eshel, A. and Kafkafi, U. (1996) *Plant Roots: The hidden half*, 2nd edition. Marcel Decker, Inc., New York.

Waksman, S. A. (1952) *Soil Microbiology*. John Wiley & Sons, Inc., New York.

Walker, J., Raison, R. J. and Khanna, P. K. (1986) Fire. In: J. Russell and R. Isbell (eds) *Australian Soils: The human impact*. University of Queensland Press, pp. 185–216.

Wallwork, J. A. (1970). *Ecology of Soil Animals*. McGraw-Hill, New York.

Wang, W.-J., Qiu, L., Zu, Y.-G., Su, D.-X., An, J., Wang, H.-Y., Zheng, G.-Y., Sun, W. and Chen, X.-Q. (2011) Changes in soil organic carbon, nitrogen, pH and bulk density with the development of larch (*Larix gmelinii*) plantations in China. *Global Change Biology*, **17**, 2657–2676.

Wardle, D. A., Nilsson, M.-C. and Zackrisson, O. (2008) Fire-derived charcoal causes loss of forest humus. *Science*, **320**, 629.

Wardle, D. A., Hörnberg, G., Zackrisson, O., Kalela-Brundin, M. and Coomes, D. A. (2003) Long-term effects of wildfire on ecosystem properties across an island area gradient. *Science*, **300**, 972–975.

Waring, H. D. and Snowdon, P. (1977) Genotype–fertilizer interaction. In: *Annual report*, CSIRO, Division of Forest Research, Canberra, 1976–1977, pp. 19–22.

Warrick, A. W., Myers, D. E. and Nielsen, D. R. (1986) Geostatistical methods applied to soil science. In: *Methods of Soil Analysis, Part, 1., Physical and Mineralogical, Methods.*, Agronomy Monograph No. 9, 2nd edition. American Society of Agronomy, Madison, WI, pp. 53–82.

Way, D. A. and Oren, R. (2010) Differential responses to changes in growth temperature between trees from different functional groups and biomes: a review and synthesis of data. *Tree Physiology*, **30**, 669–688.

Weast, R. (1982) *CRC Handbook of Chemistry and Physics*. CRC Press, Boca Raton, FL.

Weaver, R. W., Angle, S. and Bottomley, P. (eds) (1994) *Methods of Soil Analysis, Part 2. Microbiological and Biochemical Properties*. Soil Science Society of America Book Series No. 5, Madison, WI.

Weber, M. G. (1987) Decomposition, litterfall, and forest floor nutrient dynamics in relation to fire in eastern Ontario jack pine ecosystems. *Canadian Journal of Forest Research*, **17**, 1496–1506.

Weetman, G. F. and Fournier, R. M. (1982) Graphical diagnoses of lodgepole pine responses to fertilization. *Soil Science Society of America Journal*, **46**, 381–398.

Wei, X. and Kimmins, J. P. (1998) Asymbiotic nitrogen fixation in harvested and wildfire-killed lodgepole pine forests in the central interior of British Columbia. *Forest Ecology and Management*, **109**, 343–353.

Wells, C. G. (1971) Effects of prescribed burning on soil chemical properties and nutrient availability. In: *Proceedings of the Prescribed Burning Symposium*. USDA Forest Service, Southeastern Forest Experiment Station, Asheville, NC, pp. 86–99.

Wells, C. G., Crutchfield, D. M., Berenyi, N. M. and Davey, C. B. (1973) *Soil and Foliar Guidelines for Phosphorus Fertilization of Loblolly Pine*. Research Paper SE-110. U.S. Department of Agriculture, Forest Service, Asheville, NC.

West, P. W. (2006) *Growing Plantation Forests*. Springer-Verlag, Berlin and Heidelberg.

White, E. H. and Pritchett, W. L. (1970) *Water-table control and fertilization for pine production in the flatwoods*. Florida Agricultural Experiment Station Technical Bulletin 743.

White, W. D. and Wells, S. G. (1981) Geomorphic effects of the La Mesa. In: *The La Mesa Fire Symposium*, Los Alamos National Laboratory LA-9236-NERP, pp. 73–90.

Whittaker, R. H. and Woodwell, G. M. (1972) Evolution of natural communities. In: J. A. Weins (ed.) *Ecosystem Structure and Function*, Proceedings of the 31st annual biological colloquium. Oregon State University Press, Corvallis, OR, pp. 137–159.

Wikum, D. A. and Shanholtzer, G. F. (1978) Application of the Braun-Blanquet cover-abundance scale for vegetation analysis in land development studies. *Environmental Management*, **2**, 323–329.

Wilde, S. A. (1946) *Forest Soils and Forest Growth*. Chronica Botanica, Waltham, MA.

Wilde, S. A. (1958) *Forest Soils*. Ronald Press, New York.

Wilde, S. A. and Voigt, G. K. (1967) The effects of different methods of tree planting on survival and growth of pine plantations on clay soils. *Journal of Forestry*, **65**, 99–101.

Will, G. M. (1972) Copper deficiency in radiata pine planted on sands at Mangawhai Forest. *New Zealand Journal of Forestry Science*, **2**, 217–221.

Will, G. M. (1985) *Nutrient deficiencies and fertiliser use in New Zealand exotic forests*. Forest Research Institute Bulletin #97, Rotorua. 53 pp.

Williams, R. and Frausto da Silva, J. (1996) *The Natural Selection of the Chemical Elements*. Oxford University Press, New York.

Wilson, S. McG., Pyatt, D. G., Malcolm, D. C. and Connolly, T. (2001) The use of ground vegetation and humus type as indicators of soil nutrient regime for an ecological site classification of British forests. *Forest Ecology and Management*, **140**, 101–116.

Wollum, A. G. (1973) Characterization of the forest floor in stands along a moisture gradient in southern New Mexico. *Soil Science Society of America Proceedings*, **37**, 637–640.

Wolt, J. D. (1994) *Soil Solution Chemistry: Applications to environmental science and agriculture*. John Wiley & Sons, Inc., New York. 345 pp.

Wood, H. B. (1977) Hydrologic differences between selected forested and agricultural soil in Hawaii. *Soil Science Society of America Journal*, **41**, 1332–1336.

Wood, P. J. A., Willens, F. and Willens, G. A. (1975) An irrigated plantation project in Abu Dhabi. *Commonwealth Forestry Review*, **54**, 139–146.

Wood, T. and Bormann, F. H. (1984) Phosphorus cycling in a Northern Hardwood forest: biological and chemical control. *Science*, **223**, 391–393.

Woodmansee, R. G. and Wallach, L. S. (1981) Effects of fire regimes on biogeochemical cycles. In: H. A. Mooney, T. M. Bonnicksen, N. L. Christensen, J. E. Lotan and W. A. Reiners (eds) *Fire Regimes and Ecosystem Properties*, USDA Forest Service General Technical Report WO-26, pp. 379–400.

Woods, R. V. (1976) *Early silviculture for upgrading productivity on marginal* Pinus radiata *sites in the southeast region of South Australia*. Wood Forestry Department Bulletin 24.

Woods, S. W., Birkas, A. and Ahl, R. (2007) Spatial variability of soil hydrophobicity after wildfires in Montana and Colorado. *Geomorphology*, **86**, 465–479.

Woodwell, G. M. and Whittaker, R. H. (1967) Primary production and the cation budget of the Brookhaven Forest. In: H. E. Young (ed.) *Symposium on Primary Productivity and Mineral Cycling in Natural Ecosystems*. University of Maine Press, Orono, pp. 151–166.

Wooldridge, D. D. (1970) Chemical and physical properties of forest litter layers in central Washington. In: C. T. Youngberg and C. B. Davey (eds) *Tree Growth and Forest Soils*. Oregon State University Press, Corvallis, OR, pp. 327–337.

Worst, R. H. (1964) A study of the effects of site preparation and spacing on planted slash pine in the coastal plains of southeast Georgia. *Journal of Forestry*, **62**, 556–557.

Wright, H. A. and Bailey, A. W. (1982) *Fire Ecology: United States and Southern Canada*. John Wiley & Sons, Inc., New York.

Wright, R. F. (1976) The impact of forest fire on the nutrient influxes to small lakes in northeastern Minnesota. *Ecology*, **57**, 549–663.

Yanai, R. D. (1992) Phosphorus budget of a 70-year-old northern hardwood forest. *Biogeochemistry*, **17**, 1–22.

Yanai, R. D., Currie, W. S. and Goodale, C. L. (2003a) Soil carbon dynamics after forest harvest: an ecosystem paradigm reconsidered. *Ecosystems*, **6**, 197–212.

Yanai, R. D., Stehman, S. V., Arthur, M. A., Prescott, C. E., Friedland, A. J., Siccama, T. G. and Binkley, D. (2003b) Detecting change in forest floor carbon. *Soil Science Society of America Journal*, **67**, 1583–1593.

Yavitt, J. B., Harms, K. E., Garcia, M. N., Wright, S. J., He, F. and Mirabello, M. J. (2009) Spatial heterogeneity of soil chemical properties in a lowland tropical moist forest, Panama. *Australian Journal of Soil Research*, **47**, 674–687.

Young, C. E. Jr. and Brendemuehl, R. H. (1973) *Response of slash pine to drainage and rainfall*. USDA Forest Service Research Note SE-186.

Youngberg, C. T. and Wollum, A. G. II (1976) Nitrogen accretion in developing *Ceanothus velutinus* stands. *Soil Science Society of America Journal*, **40**, 109–112.

Youngberg, C. T., Wollum, A. G. II and Scott, W. (1979) Ceanothus in Douglas fir clear-cuts: nitrogen accretion and impact on regeneration. In: J. C. Gordon, C. T. Wheeler and D. A. Perry (eds) *Symbiotic Nitrogen Fixation in the Management of Temperate Forests*. Forest Research Laboratory, Oregon State University, Corvallis, pp. 224–233.

Youssef, R. A. and Chino, M. (1987) Studies on the behavior of nutrients in the rhizosphere I: Establishment of a new rhizobox system to study nutrient status in the rhizosphere. *Journal of Plant Nutrition*, **10**, 1185–1196.

Zabowski, D., Skinner, M. F. and Payn, T. W. (2007) Nutrient release by weathering: implications for sustainable harvesting of *Pinus radiata* in New Zealand soils. *New Zealand Journal of Forestry Science*, **37**, 336–354.

Zanella, A., Jabiol, B., Ponge, J. F., Sartori, G., De Wall, R., Van Delft, B., Graefe, U., Cools, N., Katzensteiner, N., Hager, H. and Englisch, M. (2011) A European morpho-functional classification of humus forms. *Geoderma*, **164**, 138–145.

Zavitkovski, J. and Newton, M. (1968) Ecological importance of snowbrush *Ceanothus velutinus* in the Oregon Cascades. *Ecology*, **49**, 1134–1145.

Zechmeister-Boltenstern, S., Hackl, E., Bachmann, G., Pfeffer, M. and Englisch, M. (2005) Nutrient turnover, greenhouse gas exchange and biodiversity in natural forests of Central Europe. In: D. Binkley and O. Menyailo (eds) *Tree Species Effects on Soils: Implications for Global Change*. NATO Science Series, Kluwer Academic Publishers, Dordrecht, pp. 32–49.

Zeman, L. J. and Slaymaker, O. (1978) Mass balance model for calculation of ionic output loads in atmospheric fallout and discharge from a mountainous basin. *Hydrology Science Bulletin*, **23**, 103–117.

Zimmerman, M. H. and Brown, C. L. (1971) *Tree Structure and Function*. Springer-Verlag, New York.

Zobel, B. J. and Talbert, J. T. (1984) *Applied Forest Tree Improvement*. John Wiley & Sons, Inc., New York.

Zou, X. (1993) Species effects on earthworm density in tropical tree plantations in Hawaii. *Biology and Fertility of Soils*, **15**, 35–38.

Zou, X. and Bashkin, M. (1998) Soil carbon accretion and earthworm recovery following revegetation in abandoned sugarcane fields. *Soil Biology and Biochemistry*, **30**, 825–830.

Index

Ecology and Management of Forest Soils, Fourth Edition. Dan Binkley and Richard F. Fisher.
© 2013 John Wiley & Sons, Ltd. Published 2013 by John Wiley & Sons, Ltd.